BIOLOGICAL ROLE OF PLANT LIPIDS

BIOLOGICAL ROLE OF PLANT LIPIDS

Edited by

Péter A. Biacs
Central Food Research Institute
Budapest, Hungary

Katalin Gruiz
Technical University of Budapest
Budapest, Hungary

Tibor Kremmer
National Institute of Oncology
Budapest, Hungary

PLENUM PRESS · NEW YORK AND LONDON

Proceedings of the 8th International Symposium
on the Biological Role of Plant Lipids
held at
Budapest, Hungary, July 25—28, 1988

Library of Congress Catalog Card Number 88-43522

ISBN 0-306-43181-5

Joint edition by Plenum Press, New York and London
and Akadémiai Kiadó, Budapest, Hungary

Printed in Hungary

DEDICATION

The proceedings of the Eight International Symposium on the Biological Role of Plant Lipids are dedicated to

PROFESSOR HARTMUT K. LICHTENTHALER

who deserves acknowledgement for bringing together numbersome scientists from different fields of plant lipid reserch to exchange their results and ideas. In 1976 he organized one of the first international meetings in this field in Karlsruhe. Ever since, he encourages and helps organization, considering the successful performance as his own affair.
Hartmut K. Lichtenthaler studied pharmacy and plant physiology in Heidelberg and in Karlsruhe. He obtained his Ph.D in botany in 1961. Between 1962-64 he was research fellow on the University of California, Berkeley, with Professor Calvin. Since 1970 he is the head of the Botany Department of Karlsruhe University.

Engaged in international scientific societies e.g. within the Federation of European Societies of Plant Physiology (FESPP), where he served as a Council Member (1980-88) and as FESPP President (1984-86), his major efforts were devoted to establishing scientific links and cooperation between research groups of western and eastern countries. He provided guest scientists from both sides with opportunities to temporarily join his group.

Hartmut K. Lichtenthaler has made many fundamental and original contributions to plant lipid science, often performing pioneering work. His scientific activities in the last 30 years, in part performed with his students and associates or in cooperation with colleagues from foreign countries have been primarily concerned with the elucidation of the biochemistry and the physiological role of plant prenyllipids as well as their localization and compartmentation in the plant cell.

After having demonstrated the presence of several prenylquinones in the chloroplast thylakoids he elucidated in 1963 the complete prenyllipid and glycerolipid composition of the photosynthetic biomembrane. The thylakoid thus became the first biomembrane of which the lipid composition was known. Hartmut Lichtenthaler also isolated the osmiophilic globuli from chloroplasts and chromoplasts, coined them "plastoglobuli" and recognized their role as reservoirs of excess plastid lipids and secondary carotenoids. He studied the regulation by light (Pfr; blue) and phytohormones of the different classes of isoprenoid lipids, the "prenyllipids", a general name he proposed and which has been accepted by the scientific community ever since its introduction at the 1976 Karlsruhe Plant Lipid Symposium.

Further research activities have included the partial characterization of the plant HMG-CoA reductase and the study of the interrelationship between the cytoplasmic and organelle mevalonate pathways. He has also showed that mitochondrial biomembranes of most plants simultaneously contain two or more homologues of ubiquinones as electron carriers. More recently his research has been concerned with the de novo fatty acid and lipid biosynthesis as affected by xenobiotics. Here Hartmut Lichtenthaler has presented evidence that the plastidic acetyl-CoA carboxylase is the target for various classes of new herbicides.

His colleagues and friends in many countries in East and West wish him further success in his plant lipid research in the years to come.

PREFACE

The 8th International Symposium on the Biological Role of Plant Lipids took place at the Federation of Technical and Scientific Societies, Budapest, Hungary, July 25-28, 1988.

The host association was the Hungarian Scientific Society for Food Industry (MÉTE).

This was the first time that the Symposium was held in an Eastern European socialist country. Previous host cities were Norwich, Karlsruhe, Göteborg, Paris, Groningen, Neuchatel, Davis. The addition of Budapest to this distinguished list was the merit of the organizers, the Lipid Chemical Working Group of the Complex Committee of Food Science of the Hungarian Academy of Sciences and the Ministry of Agriculture and Food. The Symposium was organized by Professor Péter Biacs, Aladár Blaskovits, Tibor Farkas, Katalin Gruiz, Tibor Kremmer and Lászlò Vígh, with the assistance of Mrs. Márta Tardy-Lengyel.

The activity and the help of the staff of the Central Food Research Institute in Budapest is highly appreciated.

In organizing the Symposium, we started with the idea that a bridge between Eastern and Western scientists should be built in the field of plant lipid research as well. In fact, 180 participants were registered, 50 of them from Czechoslovakia, the German Democratic Republic, Poland, the USSR and Hungary.

The scientific program, divided into 14 sessions, included 11 general lectures of 30 minutes, 75 oral lectures of 15 minutes and 70 posters. Due to the large number of papers the Symposium was held in two parallel sections. Poster discussions at the site of the poster exhibition were a good solution giving a chance for free discussion and evaluation of scientific progress.

Lectures on traditional areas of plant lipid research, as lipid metabolism, structure and function of lipids were attended by a number of people. Special attention was paid to the session dealing with biocides and their interaction with plant lipids and the effect of environmental stress of plant lipids. Highlight of the Symposium were the new discoveries of molecular biology and the genetics of plant lipids. Given the growing importance of oleo-biotechnology, a special session was devoted to this subject for the first time in Budapest.

Subsequent to the four day meeting in Budapest, a round-table discussion on thermal acclimation and temperature stress was organized by T. Farkas, Ibolya Horváth and L. Vìgh, in the Biological Research Center of the Hungarian Academy of Sciences, in Szeged, on July 29-30. This program with about 40 participants provided a good opportunity to overview the present status of this field in a very informal way. Besides six lectures 10 posters were also presented and discussed.

This book presents the material discussed at the Symposium in Budapest and that of the roundtable discussion in Szeged. All papers presented in this volume were prepared camera-ready by the authors themselves and there was no proof-reading. The main reason for this was the wish of the publishers to ensure the speedy publication.

Participants attending the Symposium for the first time were delighted with the friendly atmosphere. The unique nature and character of plant lipid conferences was due to the group of distinguished scientists. We are indebted to the small but active scientific community, the members of which agreed to serve as chairmen of sessions or as moderators in poster discussions. In addition, we wish to thank Professor István Tòth-Zsiga, general secretary of MÉTE and his colleagues of the information office for their assistance in the organization of the conference and social events.

Finally, it is our pleasure and privilege to dedicate this volume to Professor Hartmut K. Lichtenthaler in honour of his outstandly scientific contributions and recognition of his efforts encouraging us to continue with such outstanding symposia.

Budapest, Hungary, August 1988

Peter A. Biacs

CONTENTS

CHAPTER 1
LIPID METABOLISM

CHAPTER 2
STRUCTURAL AND FUNCTIONAL ORGANIZATION OF LIPIDS

CHAPTER 3
BIOSINTHESIS AND FUNCTION OF PRENYLLIPIDS

CHAPTER 4
CARRIER PROTEINS, GENETICS OF PLANT LIPIDS

CHAPTER 5
BIOCIDES, INTERACTION WITH PLANT LIPIDS

CHAPTER 6
BIOTECHNOLOGY OF LIPIDS, NUTRITIONAL ASPECTS

CHAPTER 7
DEVELOPMENT, ENVIRONMENT, STRESS

CHAPTER 1

LIPID METABOLISM

BIOLOGICAL ROLE OF PLANT LIPIDS
P.A. BIACS, K. GRUIZ, T. KREMMER (eds)
Akadémiai Kiadó, Budapest and Plenum Publishing
Corporation, New York and London, 1989

FATTY ACID DESATURATION IN SOME PLANTS OF HORTICULTURAL INTEREST: PEA, SPINACH, POTATO, ETC...

Ch. Demandre, Hana Serghini-Caíd, Anne-Marie Justin and P. Mazliak

Université P. et M. Curie, UA 1180, Laboratoire de Physiologie Cellulaire
Tour 53/3, 4 Place Jussieu, 75252 Paris Cedex 05, France

ABSTRACT

Only molecular species of the eukaryotic type are present in the phosphatidylcholine (PC) from pea and spinach leaves or from potato tubers. In pea leaves or pea chloroplasts total MGDG is mostly represented by two eukaryotic molecular species : 18:3/18:3 (79 mol%) and 18:2/18:3 (9%). In spinach leaves the prokaryotic molecular species 18:3/16:3 forms 39 mol% of total MGDG ; however the eukaryotic species 18:3/18:3 (60 mol%) is dominant.
1 Oleate desaturation. We could follow *in vitro* the desaturation of oleate into linoleate with microsomes from aged potato tuber or those from developing pea leaves, starting from ^{14}C-oleoyl-CoA as a precursor. Six different molecular species of PC were involved in this desaturation, both in pea leaves and potato tubers. The main desaturation reactions could be written as follows : 18:3/^{14}C-18:1 PC ⟶ 18:3/^{14}C-18:2 PC ; 18:2/^{14}C-18:1 PC ⟶ 18:2/^{14}C-18:2 PC ; 16:0/^{14}C-18:1 PC ⟶ 16:0/^{14}C-18:2 PC.
In pea microsomes, 18:2/^{14}C-18:1 was the molecular species mostly desaturated whereas in potato microsomes it was 16:0/^{14}C-18:1 PC.
2 Linoleate desaturation. This desaturation could be observed *in vivo* with developing pea or spinach leaves. At the end of a 30 min pulse with ^{14}C-oleate, clear precursor-product relationships could be observed during the following 48 hr chase period, between ^{14}C-18:1 PC and ^{14}C-18:2 PC both in pea and spinach leaves ; active desaturation of ^{14}C-18:2 PC into ^{14}C-18:3 PC was observed in pea only whereas labelled ^{14}C-18:3-PC accumulated very slowly in spinach leaves.
In MGDG from pea leaves labelled linolenic acid accumulated steadily in 18:3/18:3 MGDG, after a lag time of 1 hr. In MGDG from spinach leaves labelled 18:3/18:2 MGDG was a precursor for labelled 18:3/18:3 MGDG.

It is concluded that linolenic acid is synthesized via PC molecular species in the eukaryotic pathway of pea leaves and via MGDG molecular species in the eukaryotic pathway of spinach leaves.

INTRODUCTION

Poly-unsaturated fatty acids (mainly linoleic (18:2), linolenic (18:3) and hexadecatrienoic (16:3) acids) are major constituents of plant membrane lipids. In green leaves, the C_{18} polyunsaturated fatty acids usually esterify the hydroxyl groups found either in position 1 or position 2 of the *sn* glycerol backbones of membranous phospholipids or galactolipids ; in contrast, hexadecatrienoic acid is always found esterified in position 2 of galactolipids.

It has been demonstrated that the polyunsaturated fatty acid content of plant membranes is a major parameter in the determination of their fluidity [1]. In turn, the ability of an organisms to maintain its membrane fluidity within an optimum range would determine its resistance to various environmental stresses (cold, frost, water-deficit, salinity, etc...) [2].

All these considerations show how essential would be a good understanding of the mechanisms of fatty acid desaturation in plants. This understanding is for instance absolutely required to study the regulation of membrane fluidity. Unfortunately the biosynthesis of polyunsaturated fatty acids in plants - particularly the biosynthesis of linolenic acid - still remains an unsolved problem, despite great advances performed these last years.

I A GENERAL FRAME FOR THE STUDY OF POLYUNSATURATED FATTY ACID BIOSYNTHESIS IN PLANTS : THE PROKARYOTIC AND EUKARYOTIC PATHWAYS [3 - 7].

The lipids which contain either saturated or unsaturated C_{16} fatty acids at the *sn*-2 position of their glycerol moieties are derived from phosphatidic acid synthesized within the chloroplast. The biosynthesis of lipids containing such a configuration is referred to as the prokaryotic pathway. The desaturation of palmitic acid to hexadecatrienoic acid is only observed in the prokaryotic pathway of the so-called $C_{16:3}$ plants.

On the other hand, the lipids which contain C_{18} fatty acids (mainly unsaturated) at the *sn*-2 position of glycerol are derived from diacylglycerol moieties synthesized in endoplasmic reticulum membranes. The synthesis of these lipids is referred to as the eukaryotic pathway and this metabolic route is predominant in plants containing no hexadecatrienoic acid, the so-called $C_{18:3}$ plants.

In this general frame of metabolic routes, it has been speculated that the biosynthesis of linolenic acid would occur in different cellular locations and _via_ different glycerolipid molecules used as substrates by plant ω^3 desaturases. For instance, in the prokaryotic pathway, it has been proposed that MGDG [8 - 12] or PG [13] molecules could be the substrates of the chloroplastic desaturases, acting on linoleic acid residues esterified in position 1 of _sn_ glycerol. On the other hand, in the eukaryotic pathway, it has been proposed that PC molecules containg linoleic acid in position 2 of _sn_-glycerol, could be desaturated into PC molecules containing linolenic acid [11, 14, 15] : this desaturation would occur in endoplasmic reticulum. Alternatively, MGDG molecules containing linoleic acid in position 2 and derived from cytoplasmic PC, could be desaturated within the chloroplast into MGDG molecules containing linolenic acid in position 2 [9, 10, 16, 17].

In our laboratory, we have taken advantage of the high-resolving power of radio-HPLC [18] to research eventual precursor-product relationships between the different molecular species of PC and MGDG found in the leaves of $C_{18:3}$ or $C_{16:3}$ plants. The glycerolipid molecular species we have studied had been labelled, either _in vivo_ (in entire leaves) or _in vitro_ (within microsomal preparations) from various radioactive precursors.

II MOLECULAR SPECIES OF PC AND MGDG FOUND IN THE LEAVES OF $C_{18:3}$ OR $C_{16:3}$ PLANTS

Combining HPLC and GLC, we were able to separate and identify from 5 to 9 different molecular species in the total PC from potato tuber of from pea, tobacco and spinach leaves (table I). Only molecular species of the eukaryotic type (18/18 or 16/18) are present in the PC of all plants. In potato tuber (a non-green tissue), molecular species containing linoleic acid (18:2/18:2, 16:0/18:2, 18:3/18:2 and 18:0/18:2) are largely predominant, forming all together 90% of total PC molecular species. In green leaves, either from the $C_{18:3}$ plant (pea) or from the two $C_{16:3}$ plant (tobacco and spinach), PC molecular species containing linolenic acid (18:3/18:3, 18:3/18:2, 16:0/18:3) are predominant.

MGDG molecular species distributions are different in $C_{18:3}$ (potato and pea) and in $C_{16:3}$ plants (tobacco and spinach). Only the latter contain typical prokaryotic molecular species (18:3/16:3 and 18:3/16:2) forming respectively 22 and 39% of total MGDG in tobacco and spinach leaves. In potato tuber, eukaryotic MGDG molecular species containing linoleic acid are predominant. In pea leaves, nearly 90% of total MGDG is represented by two eukaryotic molecular species : 18:3/18:3 (79 mol %) and 18:3/18:2 (9 mol %). In the leaves of the two $C_{16:3}$ plants (tobacco and spinach), the eukaryotic MGDG molecular species (mainly 18:3/18:3 and 18:3/18:2) remain largely dominant over prokaryotic molecular species. In all plant leaves analyzed in table I, two molecular species represent ca 90% of total MGDG : in pea these are two

TABLE I - MOLECULAR SPECIES COMPOSITION (IN MOLE %) OF PC AND MGDG from potato tuber or green leaves.

	PC				MGDG			
Molecular species	Potato tuber	Pea leaves	Tobacco leaves	Spinach leaves	Potato tuber	Pea leaves	Tobacco leaves	Spinach leaves
18:3/16:3	--	--	--	--	--	--	21.6	38.6
18:3/16:2	--	--	--	--	--	--	--	0.3
18:3/18:3	tr	11.1	14.5	28.3	16.8	79.3	71.9	59.3
18:3/18:2	13.7	24.4	17.9	13.8	36.4	8.9	3.5	0.6
18:2/18:2	32.5	18.9	8.9	11.6	46.7	6.8	--	0.2
16:0/18:3	10.4	26.7	41.4	15.5	--	5.0	3.1	0.4
18:3/18:1	--	--	--	9.0	--	--	--	0.6
16:0/18:2	36.3	7.0	17.3	4.7	tr	--	--	--
18:2/18:1	--	2.5	--	6.5	--	--	--	--
18:0/18:3	--	6.5	--	--	--	--	--	--
18:0/18:2	7.1	2.9	--	--	--	--	--	--
16:0/18:1	--	--	--	3.6	--	--	--	--
18:1/18:1	--	--	--	6.9	--	--	--	--

eukaryotic molecular species (18:3/18:3 and 18:3/18:2) and in tobacco and spinach, these two molecular species comprise one dominant eukaryotic MGDG (18:3/18:3) and one prokaryotic MGDG (18:3/16:3). These results point to the importance of the eukaryotic pathway for MGDG biosynthesis, even in the case of $C_{16:3}$ plants.

III OLEATE DESATURATION IN VITRO

Most evidence suggests that oleoyl phosphatidylcholine is the true substrate of the complex oleate-desaturase system present in the endoplasmic reticulum of higher plants [3-6].

To catch transient oleoyl-PC molecular species we have first incubated potato tuber or green leaf microsomes with ^{14}C-oleoyl CoA, in absence of NADH in the incubation medium. In those conditions, only the acylation of lysophosphatidylcholine can occur [18, 19] ; further desaturation is impaired by the omission of NADH. For incubation times comprised between 5 and 60 min, about 70% of total lipid radioactivity was recovered in PC. Table II shows the results obtained with the different microsomal preparations.

TABLE II - LABELLED MOLECULAR SPECIES OF PC appearing in potato tuber or green leaf microsomes after 10 min incubation with ^{14}C-oleoyl-CoA. No NADH was present in the incubation medium - Results are expresed as % of total ^{14}C-oleoyl PC.

Molecular species*	Potato	Pea	Sunflower	Safflower	Spinach
18:3/^{14}C-18:1	2.4	23.5	5.9	8.7	40.5
^{14}C-18:1/18:3	1.6	1.5	0.4	1.0	19.5
18:2/^{14}C-18:1	31.0	36.8	42.5	50.3	14.5
^{14}C-18:1/18:2	10.0	1.1	14.1	19.5	4.9
16:0/^{14}C-18:1	50.1	32.6	7.4	16.9	15.7
^{14}C-18:1/16:0	4.9	1.6	5.8	3.6	4.8
18:1/^{14}C-18:1	--	2.9	18.8	--	--
18:0/^{14}C-18:1	--	--	5.1	--	--

* Fatty acid residues indicated before the slash are esterified in position 1 of <u>sn</u>-glycerol ; fatty acid residues indicated after the slash are esterified in position 2.

It clearly appears that in each tissue analyzed several PC molecular species can be utilized as substrates by the microsomal oleate-desaturase system. The main species are the same in all tissues but the quantitative distribution of ^{14}C-oleoyl residues among these molecular species is very different from tissue to tissue. In potato and pea leaves, 16:0/^{14}C-18:1 PC was the main or a major substrate of the desaturase. In pea, sunflower and safflower leaf microsomes, the 18:2/^{14}C-18:1 PC was the most intensively labelled molecular species. In spinach, the 18:3/^{14}C-18:1 PC species appeared mostly labelled. Finally, if ^{14}C-oleoyl residues were esterified nearly exclusively (96%) in position 2 of sn glycerol backbones of PC molecules in pea, 16% esterification occurred in position 1 in potato, 21% in sunflower and 25% both in spinach and safflower.

From this table, it appears that the specificities of microsomal lysophosphatidylcholine acyltransferases are different from tissue to tissue.

When NADH was added to the incubation medium, oleate desaturation occurred in all molecular species containing ^{14}C-oleoyl residues. We could check with potato or pea microsomes that the desaturase proper did not diplay any selectivity towards any of the previously labelled PC molecular species. Time course of oleate-desaturation in pea or potato-microsomes indicated mainly the following reactions to occur :

$$16:0/^{14}C\text{-}18:1 \xrightarrow{-2H} 16:0/^{14}\text{-}18:2$$
$$18:2/^{14}C\text{-}18:1 \xrightarrow{-2H} 18:2/^{14}C\text{-}18:2$$
$$18:3/^{14}C\text{-}18:1 \xrightarrow{-2H} 18:3/^{14}C\text{-}18:2$$

In potato tuber and pea leaves, these newly synthesized molecular species comprise 82.5 mol % and 50.3 mol % of total PC molecular species (see table I). Remarkably, the remaining PC molecular species are nearly exclusively those containing linolenic acid.

IV LINOLEATE DESATURATION IN VIVO

The biosynthesis of linolenic acid, the dominant fatty acid in chloroplasts, has remained an unsolved problem despite continuous researches on this topic for more than forty years. Very little success has been obtained with attempts of in vitro linoleate desaturation [20, 21].

To overcome progressively the numerous difficulties linked to linoleate desaturation studies, we began with in vivo studies [22]. As a labelled precursor of linolenic acid, we have used ^{14}C-oleate, given externally to entire leaves, to follow the labelling of the metabolic intermediates of the sole eukaryotic pathway. To perform the labelling experiments, we have used two types of plants : on one hand pea, a typical $C_{18:3}$ plant where the eukaryotic pathway of galactolipid biosynthesis is the most active [5] ; on the other hand spinach, a $C_{16:3}$ plant, with a prokaryotic pathway as active as the eukaryotic one [5].

Following a 30 min pulse all the ^{14}C-oleate incorporated was found acylated to PC, in pea as well as in spinach leaves. In pea there was still a small increase in PC labelling during the first hour of chase, up to a maximum. Thereafter, PC radioactivity was constantly decreasing, while galactolipid labelling was regularly increasing, first in MGDG, then in DGDG. In spinach, the same evolutions were occurring more slowly. These results suggest strongly that the eukaryotic pathway is functionning well in both plants, the ^{14}C-oleoyl residues being first acylated to PC molecules, eventually desaturated, and then transferred to galactolipids.

- <u>Labelling of PC molecular species</u>. After a 30 min pulse with ^{14}C-oleate, pea leaves contained four heavily labelled PC molecular species : 18:3/^{14}C-18:1, 18:2/^{14}C-18:1, 18:1/^{14}C-18:1 and 16:0/^{14}C-18:1 - It is interesting to notice that these four molecular species are the same as those which were labelled *in vitro* from ^{14}C-oleoyl CoA with pea microsomes. Kinetics of labelling during the chase period clearly showed the progressive desaturation of each one of these PC molecular species, allowing to write the four following reactions :

18:3/18:1 PC $\longrightarrow$ 18:3/18:2 PC $\longrightarrow$ 18:3/18:3 PC
18:2/18:1 PC $\longrightarrow$ 18:2/18:2 PC $\longrightarrow$ 18:2/18:3 PC
18:1/18:1 PC $\longrightarrow$ 18:1/18:2 PC $\longrightarrow$ 18:1/18:3 PC
16:0/18:1 PC $\longrightarrow$ 16:0/18:2 PC $\longrightarrow$ 16:0/18:3 PC

In spinach leaves the same PC molecular species appeared labelled although at times different from those observed in pea leaves. After the 30 min pulse, only 16:0/^{14}C-18:1 and 18:3/^{14}C-18:1 were labelled ; two hrs later 18:2/^{14}C-18:1 appeared and 4 hrs later 18:1/^{14}C-18:1. Then the desaturation of ^{14}C-oleoyl residues produced ^{14}C-linoleoyl residues but ^{14}C-linolenoyl residues appeared (after 24 or 48 h) very slowly, as compared with desaturation in pea leaves.

- <u>Labelling of MGDG molecular species</u>

No labelled oleate was incorporated into the MGDG of pea leaves whereas ^{14}C-18:1-MGDG appeared in spinach leaves. The kinetics of labelling during the chase period of the two MGDG molecular species (18:3/18:3 and 18:3/18:2) which appeared labelled in pea after a lag time of 1 hr did not show any clear precursor-product relationships between these two species ; during all the chase period 18:3/18:3 MGDG accumulated steadily most of the ^{14}C-radioactivity. In MGDG from spinach leaves, four molecular species appeared labelled after 2 hr lag time. Labelling kinetics showed that 18:3/18:2 MGDG could be used as a precursor for 18:3/18:3 MGDG. The two other labelled molecular species (18:3/18:1 and 18:2/18:2) contained only 10% each of total MGDG radioactivity at the moment of their maximum labelling.

CONCLUSIONS

The schema of fig. 1 is proposed to summarize the results. In pea leaves only "eukaryotic pathways" would be present for MGDG biosynthesis and linolenic acid would be synthesized mostly in endoplasmic reticulum, while esterified to PC molecules. In spinach leaves, both "prokaryotic and eukaryotic pathways" would function simultaneously but linolenic acid would be formed mostly in chloroplasts, while esterified to MGDG molecules.

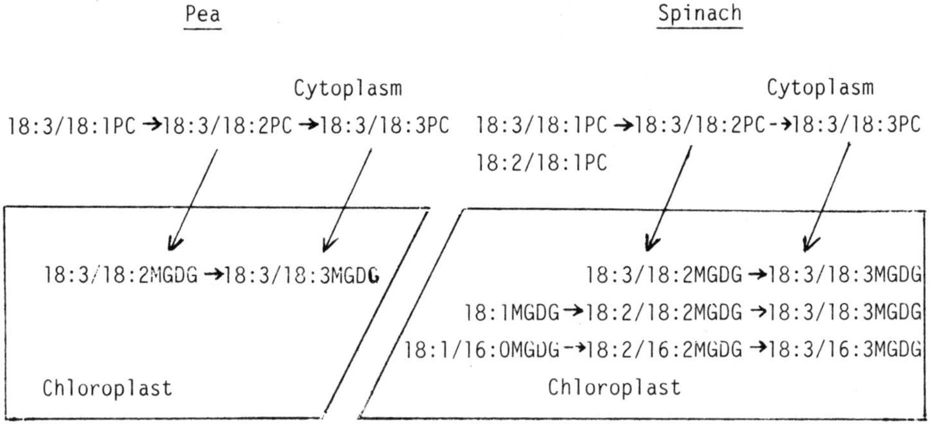

Fig. 1 - Main metabolic pathways proposed for the biosynthesis of PC and MGDG molecular species in pea and spinach leaves.

REFERENCES

[1] N.F. HADLEY - The adaptive role of lipids in biological systems, J. Wiley and Sons, New York, 1985, 319 pp.

[2] P. MAZLIAK - Environmental effects on fatty acid quality in Plant lipids: targets for manipulation, N.J. Pinfield, editor, Bristol, UK (in the press).

[3] P.G. ROUGHAN and C.R. SLACK - Cellular organization of glycerolipid metabolism. Ann. Rev. Plant Physiol., 33, 97-132 (1982)

[4] P.G. ROUGHAN and C.R. SLACK - Glycerolipid synthesis in leaves. Trends Biochem. Sc., 9, 383-386 (1984)

[5] M. FRENTZEN - Biosynthesis and desaturation of the different diacylglycerol moieties in higher plants. J. Plant Physiol., 124, 193-209 (1986).

[6] K. GOUNARIS, J. BARBER and J.L. HARWOOD - The thylakoid membranes of higher plant chloroplasts. Biochem. J., 237, 313-326 (1986).

[7] J.W.M. HEEMSKERK and J.F.G.M. WINTERMANS - Role of the chloroplast in the leaf acyl-lipid synthesis. Physiologia plantarum, 70, 558-568 (1987).

[8] B.W. NICHOLS, A.T. JAMES and J. BREUER - Interrelationships between fatty acid biosynthesis and acyl-lipid synthesis in Chlorella vulgaris. Biochem. J., 104, 486-496 (1967).

[9] N.W. LEM and J.P. WILLIAMS - Changes in the ^{14}C-labelling of molecular species of 3-monogalactosyl-1,2-diacylglycerol in leaves of Vicia faba treated with compound san 9785. Biochem. J., 209, 513-518 (1983).

[10] H.A. NORMAN and J.B. ST JOHN - Metabolism of unsaturated monogalactosyldiacylglycerol molecular species in Arabidopsis thaliana. Plant Physiol., 81, 731-736 (1986).

[11] H.A. NORMAN and J.B. ST JOHN - Differential effects of a substituted pyridazinone, BASF 13-338, on pathways of monogalactosyldiacylglycerol synthesis in Arabidopsis. Plant Physiol., 85, 684-688 (1987).

[12] J. ANDREWS and E. HEINTZ - Desaturation of newly synthesized monogalactosyldiacylglycerol in spinach chloroplasts. J. Plant Physiol., 131, 75-90 (1987).

[13] J.B. MUDD and R. DEZACKS - Synthesis of phosphatidylglycerol by chloroplasts from leaves of Spinacia oleracea L. Arch. Bioch. Biophys., 209, 584-591 (1981).

[14] J.P. WILLIAMS, G.R. WATSON and S.P.K. LEUNG - Galactolipid synthesis in Vicia faba leaves II - Formation and desaturation of long chain fatty acids in phosphatidylcholine, phosphatidylglycerol and galactolipids. Plant Physiol., 57, 179-184 (1976).

[15] N.W. LEM and J.P. WILLIAMS - Desaturation of fatty acids associated with monogalactosyldiacylglycerol : the effects of San 6706 and San 9785. Plant Physiol., 68, 944-949 (1981).

[16] A.V.M. JONES and J.L. HARWOOD - Desaturation of linoleic acid from exogenous lipids by isolated chloroplasts. Biochem. J., 190, 851-854 (1980).

[17] J.I. ONISHI and M. YAMADA - Glycerolipid sunthesis in Avena leaves during greening of etiolated seedlings. III Synthesis of α-linolenoyl-monogalactosyldiacylglycerol from liposomal linoleoylphosphatidylcholine by Avena plastids in the presence of phosphatidylcholine exchange protein. Plant Cell Physiol., 23, 767-773 (1982).

[18] C. DEMANDRE, A. TREMOLIERES, A.M. JUSTIN and P. MAZLIAK - Oleate desaturation in six phosphatidylcholine molecular species from potato tuber microsomes. Biochim. Biophys. Acta, 877, 380-386 (1986).

[19] H. SERGHINI-CAID, C. DEMANDRE, A.M. JUSTIN and P. MAZLIAK - Oleoyl-phosphatidylcholine molecular species desaturated in pea leaf microsomes. Possible substrates of oleate-desaturase in other green leaves. Plant Science, 54, 93-101 (1988).

[20] E.G. DAVIDOVA, L.D. BALASHOVA, A.P. BELOV - Desaturation of linoleic acid in psychrophilic yeast. Biokhimya, 51, 1655-1658 (1986).

[21] J. ANDREWS and E. HEINZ - Desaturation of newly synthesized monogalactosyldiacylglycerol in spinach chloroplasts. J. Plant Physiol., 131, 75-90 (1987).

[22] H. SERGHINI-CAID, C. DEMANDRE, A.M. JUSTIN and P. MAZLIAK - Linolenic acid biosynthesis via glycerolipid molecular species in pea and spinach leaves. Phytochemistry, 27 (1988) (in the press).

BIOLOGICAL ROLE OF PLANT LIPIDS
P.A. BIACS, K. GRUIZ, T. KREMMER (eds)
Akadémiai Kiadó, Budapest and Plenum Publishing
Corporation, New York and London, 1989

STUDIES ON THE BIOSYNTHESIS OF PHOSPHATIDYLSULFOCHOLINE AND DEOXYCERAMIDESULFONIC ACID IN THE DIATOM NITZSCHIA ALBA

M. Kates[1] and B.E. Volcani[2]

[1]Dept. of Biochemistry, Univ. of Ottawa, Ottawa K1N 9B4 Canada
[2]Scripps Institution of Oceanography, Univ. of California San Diego, La Jolla, CA 92093, USA

SUMMARY

Studies have been carried out on the biosynthesis of phosphatidylsulfocholine (PSC) and 1-deoxyceramidesulfonic acid (DCS) in the non-photosynthetic diatom Nitzschia alba and the photosynthetic diatom Nitzschia angularis. Sulfocholine was incorporated intact into PSC of N. alba and probably also of N. angularis, and choline was incorporated into phosphatidylcholine of both diatoms. Incorporation of choline occurs at a greater rate than sulfocholine in both diatoms, but sulfocholine inhibits the incorporation of choline in both diatoms. Carbon from serine was incorporated into PSC and not into PC in N. alba, but into both PSC and PC in N. angularis; ethanolamine was similarly incorporated but at much lower levels. In both diatoms, synthesis of PSC probably proceeds from serine to methionine to sulfocholine which is incorporated into PSC by the phosphocholine transferase; this enzyme can transfer either phosphocholine or phosphosulfocholine. The sulfur atom in DCS, but not carbons 1 and 2, appears to be supplied by cysteine (or cysteic acid?). However, ^{14}C from [^{14}C]serine was incorporated into DCS, but not in presence of cycloserine. Biosynthesis of DCS may thus occur by the serine-palmitoyl-CoA pathway with introduction of the sulfur atom by a cystathionine-type reaction with cysteine, followed by oxidation of the SH group to a sulfonic acid group.

INTRODUCTION

The non-photosynthetic diatom, Nitzschia alba, contains phosphatidylsulfocholine (PSC) as the major membrane phospholipid, which completely replaces phosphatidylcholine (PC) [1,2]. The sulfonium analogue (PSC) also occurs in photosynthetic diatoms, which however contain major proportions of PC [3]. All diatoms studied also contain another novel sulfolipid, 1-deoxyceramidesulfonic acid (DCS) [1,2,4,5].

Preliminary biosynthetic studies have shown that both cysteine and methionine are precursors of PSC, while cysteine but not methionine is also a precursor of DCS [6]. Methionine was also shown to supply both the sulfur and the S-methyl groups of PSC, suggesting that methionine is converted to sulfocholine which is incorporated into PSC by a pathway [6] analogous to the nucleotide (Kennedy) pathway for PC biosynthesis [7].

To test this hypothesis, the incorporation of [Me-^{3}H] and [^{35}S] sulfocholine into PSC has now been studied in *N. alba*. In addition, to help explain why *N. alba* contains no nitrogenous phospholipids while photosynthetic diatoms, like *Nitzschia angularis*, contain PC and phosphatidylethanolamine (PE) [4], the incorporation of labelled sulfocholine, choline, serine and ethanolamine into the phospholipids of these two diatoms was compared.

In regard to the biosynthesis of DCS, it was suggested previously [5,6] that either cysteine of cysteic acid might substitute for serine in the condensation step with palmitoyl-CoA in an analogous pathway to that for sphingosine [8]. Alternatively, condensation could occur with serine followed by introduction of the sulfur by a cystathionine-type reaction with cysteine and then oxidation to the sulfonic acid. These possibilities have now been tested by comparing the incorporation of [^{14}C]serine, [^{14}C] and [^{35}S]cysteine, and [^{14}C] and [^{35}S]cysteic acid into DCS of *N. alba* in presence or absence of cycloserine, an inhibitor of the serine-palmitoyl-CoA condensation [9].

RESULTS AND DISCUSSION

Biosynthesis of PSC and PC

When cells of *N. alba* were incubated for 44 hr with [^{3}H] and [^{35}S] sulfocholine (initial ratio ^{3}H/^{35}S, 8.4-12.2) both radioisotopes were incorporated into PSC (major) and lyso-PSC (minor) with ^{3}H/^{35}S ratio 8.3 and 9.5, respectively, indicating that sulfocholine is incorporated intact into PSC, presumably by a pathway analogous to the nucleotide pathway for PC biosynthesis in plants [10]: sulfocholine $\xrightarrow{\text{ATP, CTP}}$ CDP-sulfocholine + 1,2-diacylglycerol $\rightarrow$ phosphatidylsulfocholine (Eq. 1).

Table 1. INCORPORATION OF SULFOCHOLINE AND CHOLINE INTO PHOSPHATIDYL SULFOCHOLINE (PSC) AND PHOSPHATIDYLCHOLINE (PC) IN DIATOMS*

Precursor	Incorporation (nmol) into:		% of Control	
	PSC(+ lysoPSC)	PC(+ lysoPC)	PSC	PC †
	Nitzschia alba			
[Me-^{3}H]Sulfocholine	145		100	
[Me-^{14}C]Choline		219		100(150)
[Me-^{3}H]Sulfocholine	154		106	
+ [Me-^{14}C]Choline		62		28
	Nitzschia angularis			
[Me-^{3}H]Sulfocholine	90		100	
[Me-^{14}C]Choline		291		100(323)
[Me-^{3}H]Sulfocholine	144		160	
+[Me-^{14}C]		39		13

*Cultures of diatoms (50 ml; 1 x 10^{6} cells/ml) were grown in presence of 1.5 mM concentration of all labelled precursors, for 52 hr (*N. alba*) or 110 hr (*N. angularis*).
†Values in parentheses are % of PSC control.

To obtain further information on the pathway for PSC biosynthesis and to answer the question whether *N. alba* is capable of synthesizing PC, as does *N. angularis*, the incorporation of labelled sulfocholine, choline, serine and ethanolamine into the lipids of these two diatoms was examined. The results showed that *N. alba* could incorporate choline into PC about 1.5 times the rate of sulfocholine into PSC, but that sulfocholine inhibited the incorporation of choline to the extent of about 70% (Table 1). In *N. angularis*, however, incorporation of choline was about 3.2 times that of sulfocholine but sulfocholine inhibited the incorporation of choline to the extent of 87% (Table 1). Note that choline stimulates the incorporation of sulfocholine into PSC (+ lysoPSC) in *N. angularis* to a much lesser extent in *N. alba*.

These results are consistent with the incorporation of sulfocholine, as the CDP-sulfocholine derivative, into PSC by the phosphocholine transferase enzyme (Eq. 1). In both diatoms, the enzyme appears to have a higher affinity for choline than for sulfocholine and sulfocholine inhibits the incorporation of choline to about the same extent. This hypothesis could explain why N. angularis contains both PSC and PC [3] but not why N. alba contains only PSC [1,3].

However, in N. alba, ^{14}C from [^{14}C] serine was incorporated into PSC to the extent of 17% of the total incorporation into lipids but no ^{14}C could be detected in PC; with [^{14}C]ethanolamine much lower incorporation of ^{14}C into PSC and none into PC was observed (Table 2). A much higher incorporation of carbon from serine occurred in N. angularis, both into PSC and PC; ^{14}C from [^{14}C]ethanolamine was also incorporated into both PSc and PC but at lower levels (Table 2). These results suggest that in both diatoms the β-carbon of serine was incorporated into methionine via methylene tetrahydrofolate [11], the methionine being then converted to sulfocholine (via dimethyl-β-propiothetin [12]), which is incorporated into PSC as shown in Eq. 1.

Serine was also converted in N. angularis, but not in N. alba, to phosphatidylserine (PS) and phosphatidylethanolamine (PE), and thence, by methylation with SAM, into PC [10]. These results are thus consistent with the fact that N. alba contains PSC but not PC and PE [1,2], while N. angularis contains both PC and PE along with PSC [4].

Table 2. INCORPORATION OF SERINE AND ETHANOLAMINE INTO LIPIDS OF DIATOMS*

Diatom	Precursor	Total Incorporation nmol	% Incorporation into PSC	PC	PS	PE	DCS	FA†
N. alba	[^{14}C]Serine	333	17	tr	2.1	1.1	5.2	53
	[^{14}C]Ethanolamine	38	10		-	1.6	3.6	74
N. angularis	[^{14}C]Serine	1744	6.9	5.5	0.7	4.7	2.7	19
	[^{14}C]Ethanolamine	132	3.8	3.0	-	7.6	2.4	12

*Incubation conditions as in Table 1.

†Abbreviation: FA, fatty acids

Biosynthesis of DCS

Incorporation of [^{35}S]sulfate into DCS of *N. alba* was not affected by the presence of unlabelled cysteine but was inhibited to the extent of 57% by unlabelled cysteate (Table 3). Incorporation of ^{35}S from [^{35}S]cysteine was also inhibited (78%) in the presence of unlabelled cysteate, while that from [^{35}S]cysteate was only slightly inhibited in the presence of unlabelled cysteine (Table 3). However, little or no incorporation of ^{14}C from [^{14}C]cysteate or from [^{14}C]cystine into DCS was observed, most of the label in the lipid fraction appearing in free fatty acids (Table 3). With [^{14}C]serine as precursor, significant incorporation of ^{14}C into DCS did occur (Table 2), and may have resulted from direct incorporation by the serine-palmitoyl-CoA pathway [8]. This was further confirmed by the cycloserine inhibition [9] of

Table 3. INCORPORATION OF VARIOUS PRECURSORS INTO DEOXYCERAMIDE SULFONIC ACID (DCS) IN *NITZSCHIA ALBA**

Precursor	Inhibitor	Incorporation into DCS, pmol	% of Control†
[^{35}S]Sulfate (4 μM)	None	8.2	100 (19)
"	Cysteine (200 μM)	7.8	95 (20)
"	Cysteate (200 μM)	3.5	43 (3)
[^{35}S]Cysteine (40 μM)	None	12.7	100 (39)
"	Cysteate (200 μM)	2.8	22 (16)
[^{35}S]Cysteate (50 μM)	None	31.2	100 (85)
"	Cysteine (200 μM)	28.1	90 (77)
[^{14}C]Cysteate‡ (1.6 μM)	None	0.3	- (4)
[^{14}C]Cystine‡ (50 μM)	None	1.2	- (5)

*Incubation conditions as in Table 1.
†Values in parentheses are % of incorporation into total lipids
‡Incorporation occured mostly (80-90%) into free fatty acids.

Table 4. BIOSYNTHESIS OF DEOXYCERAMIDESULFONIC ACID FROM LABELLED SERINE, CYSTEINE OR CYSTEATE IN THE NON-PHOTOSYNTHETIC DIATOM, *N. ALBA**

Precursor	Inhibitor	Incorporation into DCS, nmol	% of Control
[^{14}C]Serine			
(0.6 mM)	None	2.56	100
"	cycloserine (1.66 mM)	0.13	5
[^{35}S]Cysteine			
(0.6 mM)	None	0.59	100
"	cycloserine (1.66 mM)	0.56	95
"	serine (0.8 mM)	0.72	122
"	serine + cycloserine	0.61	103
[^{35}S]Cysteate			
(0.6 mM)	None	0.47	100
"	serine (0.8 mM)	0.39	83
"	serine + cycloserine	0.13	27

*Cultures of *N. alba* (6 ml; 5 x 10^6 cells/ml) were incubated for 4 hr with stirring at 29°C in presence of indicated labelled precursors with or without the indicated inhibitors.

^{14}C incorporated from [^{14}C]serine but not ^{35}S incorporation from [^{35}S] cysteine and the stimulation of [^{35}S]incorporation from [^{35}S]cysteine, but not from [^{35}S] cysteate, by unlabelled serine (Table 4).

These results are consistent with the hypothesis that the sulfur atom of DCS is derived from cysteine and carbon atoms 1 and 2 are derived from serine. Biosynthesis of DCS would then follow the well-known [8]serine-palmityol-CoA pathway with introduction of the sulfur atom into sphingosine or ceramide by a cystathionine-type reaction with cysteine followed by oxidation of the SH group to a sulfonic acid group The role of cysteic acid is not clear, since it cannot participate in a cystathionine-type reaction, but it may be a more effective donor of sulfur than sulfate for synthesis of cysteine from serine.

In this connection it is of interest that biosynthesis of an analogue of DCS (capnine, 1-deoxy-15-methylhexadecasphinganine-1-

sulfonic acid) in the gliding bacterium Cytophaga johnsonae has been shown to involve cysteate preferentially to cystine or sulfate as precursor for the capnine sulfur atom and to cystine or serine for capnine carbons 1 and 2 [13]. The pathway for DCS biosynthesis in diatoms thus appears to be different from that of capnine in Cytophaga.

ACKNOWLEDGEMENT

This work was supported in part by grants from NSERC of Canada (A-5324) (M.K.) and by USA Public Health Service, NIH Grant GM-08229. (B.E.V.). The authors are grateful to Dr. Walter Godchaux for a generous gift of [^{14}C]cysteic acid and for helpful discussions.

REFERENCES

1. Anderson, R., Livermore, B.P., Kates, M. and Volcani, B.E. (1978) Biochim. Biophys. Acta 528, 77-88.
2. Anderson, R., Kates, M. and Volcani, B.E. (1978) Biochim. Biophys. Acta 528, 89-106.
3. Bisseret, P., Ito, S., Tremblay, P.-A., Volcani, B.E., Dessort, D. and Kates, M. (1984) Biochim. Biophys. Acta 796, 320-327.
4. Kates, M. and Volcani, B.E. (1966) Biochim. Biophys. Acta 116, 264-278.
5. Anderson, R., Livermore, B.P., Volcani, B.E. and Kates, M. (1975) Biochim. Biophys. Acta 409, 259-263.
6. Anderson R., Kates, M. and Volcani, B.E. (1979) Biochim. Biophys. Acta 573, 557-561.
7. Kennedy, E.P. (1962) The Harvey Lectures, Series 57, pp. 143-171, Academic Press, New York.
8. Stoffel, W. (1970) Chem. Phys. Lipids 5, 139-158.
9. Sundaram, K.S. and Lev, M. (1984) Biochem. Biophys. Res. Commun. 119, 814-819.
10. Kates, M. and Marshall, M.D. (1975) in Recent Advances in Chemistry and Biochemistry of Plant Lipids (Mercer, E.I. and Galliard, T., eds), pp. 115-159, Academic Press, London, New York.
11. Taylor, R.T. and Weissbach, H. (1973) in The Enzymes (Boyer, P.D. ed.), Vol. 9, pp. 121-165, Academic Press, New York.

12. Greene, R.C. (1962) J. Biol. Chem. 237, 2251-2254.

13. Abanat, D.R., Godchaux, W., Polychroniou, G. and Leadbetter, E.R. (1985) Biochem. Biophys. Res. Commun. 130, 873-878.

BIOLOGICAL ROLE OF PLANT LIPIDS
P.A. BIACS, K. GRUIZ, T. KREMMER (eds)
Akadémiai Kiadó, Budapest and Plenum Publishing
Corporation, New York and London, 1989

PROPIONYL-CoA GENERATION AND CATABOLISM IN HIGHER PLANT PEROXISOMES

H. Gerbling and B. Gerhardt

Botanisches Institut, Universität Münster, D-4400 Münster, FRG

Saturated, straight-chain even-numbered fatty acids are degraded by the peroxisomal ß-oxidation system in hihger plants (1). Unsaturated fatty acids appear also to be degraded by this system. The additional enzyme reactions required to link the catabolism of unsaturated fatty acids to the ß-oxidation sequence have recently been demonstrated in glyoxysomes (2). A third group of fatty acids of physiological importance comprises the branched-chain 2-oxo fatty acids (bFA). These fatty acids are intermediates of the catabolism of the branched-chain amino acids and can result as a source of substrate for fatty acid degradation in the course of the steady-state protein turnover. In *Lemna minor*, 50-60% of the leucine and isoleucine resulting from protein turnover is not recycled into protein synthesis (3). Of the amino acids in soluble leaf protein 15-20 % are branched-chain amino acids.

Pathways for the catabolism of the bFA have been elucidated in bacteria and in mitochondria of animal tissue. The pathways ultimately lead to acetyl-CoA and/or succinyl-CoA, compounds which are capable of entering the tricarboxylic cycle. The catabolism of the bFA in plants has received very little attention. We have studied it under the aspect that peroxisomes are the site of fatty acid degradation in higher plants and that bFA could be a substantial source of substrate in nonfatty plant tissues.

PROPIONYL-CoA - INTERMEDIATE OF THE BRANCHED-CHAIN 2-OXO FATTY ACID CATABOLISM IN PEROXISOMES

Peroxisomes from mung-bean (*Vigna radiata L.*) hypocotyls activate, by oxidative decarboxylation, 2-oxoisocaproate, 2-oxoisovalerate and 2-oxo-3-methylvalerate which are derived by transamination from leucine, valine and isoleucine, respectively (4). Activation of the bFA was not observed in mitochondria. The acyl-CoAs generated by the oxidative decarboxylation are oxidized by the peroxisomal acyl-CoA oxidase (4). In order to elucidate further steps of the peroxisomal catabolism of the bFA, assay mixtures were separated by reversed-phase HPLC, and the acyl-CoAs appearing on the HPLC elution profile were identified. In the course of these studies it became evident that the peroxisomal catabolism of the bFA leads to both propionyl-CoA and acetyl-CoA, independent of the nature of the bFA used as substrate for the organelles. Based on the results of kinetic studies, acetyl-CoA has to be considered as the ultimate end product of the pathways whereas propionyl-CoA appeared to be a general intermediate of the peroxisomal catabolism of the bFA. The peroxisomal degradation of propionyl-CoA was then investigated separately.

PEROXISOMAL CATABOLISM OF PROPIONYL-CoA

The propionate metabolism in higher plant tissues has already been studied by Stumpf and coworkers (5,6) and a pathway for propionate catabolism has been proposed (6,7). This pathway (modified ß-oxidation) has been supported by data published recently (8). However, of the intermediates and end products of the proposed pathway none of the acyl-CoAs has been demonstrated so far.

Peroxisomes were incubated with propionyl-CoA and the reaction mixture was analyzed for acyl-CoAs by HPLC as described previously (4). The reaction mixture contained 175 mM Tris-HCl, pH 8.5, 0.33 mM NAD, 0.3 mM CoASH, 0.1 mM propionyl-CoA, and 60-160 µg peroxisomal protein. Without incubation, the HPLC elution profile of the reaction mixture showed absorbance peaks below 10 min retention time, at 13.5 min (CoASH) and at 21.6 min (propionyl-CoA). Incubation of the reaction mixture ultimately resulted in three additional, distinct peaks on the HPLC elution profile. The intensity of these peaks changed during the incubation (Fig. 1). The retention time (18.4 min) of the additional peak appearing at last was identical with that of the reference standard acetyl-CoA. The retention times of the two other additional peaks were identical with those of acrylyl-CoA (33.2 min) and 3-hydroxypropionyl-CoA (20.6 min), intermediates of the propionyl-CoA catabolism by the proposed modified ß-oxidation pathway. Since both, acrylyl-CoA and 3-hydroxypropionyl-CoA were not available commercially reference standards have been prepared

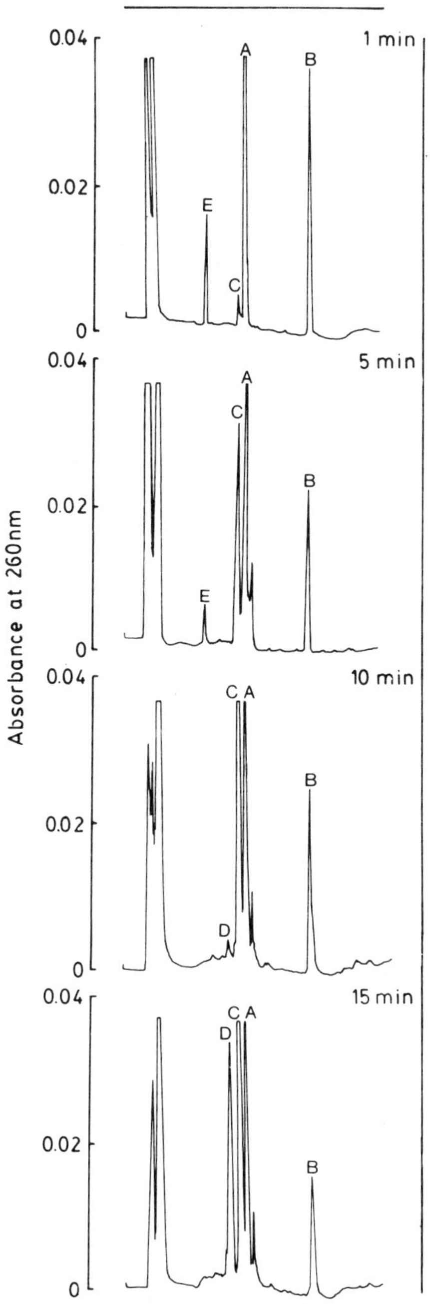

Fig. 1. HPLC elution profiles of a reaction mixture containing peroxisomes and propionyl-CoA. Aliquots of the reaction mixture were separated by HPLC after the indicated incubation times. Peak labeling: A - propionyl-CoA, B - acrylyl-CoA, C - 3-hydroxypropionyl-CoA, D - acetyl-CoA, E - coenzym A.

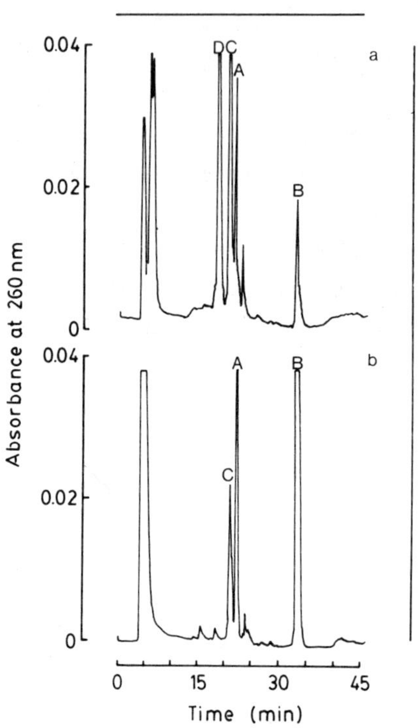

Fig. 2. HPLC elution profiles of reaction mixtures containing either propionyl-CoA and peroxisomes (a) or propionyl-CoA, acyl-CoA oxidase and crotonase (b). The incubation times were 30 min (a) and 15 min (b), respectively. Peak labeling as in Fig. 1.

from propionyl-CoA using commercial enzymes (Sigma, Munich FRG). Propionyl-CoA was treated either with acyl-CoA oxidase (1.7 nkat) or with acyl-CoA oxidase and crotonase (115 nkat). Aliquots of these reaction mixtures were then separated by HPLC to determine the retention times of the reaction product(s) (Fig. 2). - Confirmation of peak identity was made by prior mixing of aliquots of the reaction mixtures containing propionyl-CoA and peroxisomes on the one hand and propionyl-CoA and enzyme(s) on the other hand. and analysing this mixture by HPLC. The peaks to be identified and those of the reference standards showed complete overlap on the elution profile.

Malonyl-CoA as a possible intermediate of the proposed propionyl-CoA catabolism (6.7) was not detected unequivocally on the HPLC elution profiles.

The peroxisomes also possess the ability to activate propionate. With propionate as substrate the activity (7.8 nmol propionyl-CoA formed min^{-1}

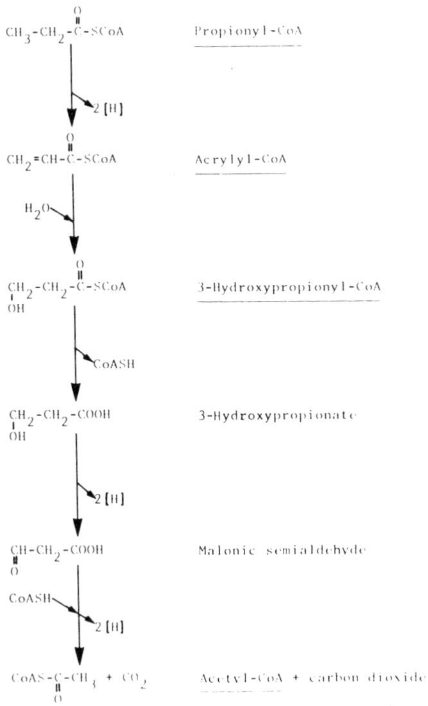

Fig. 3. Catabolism of propionyl-CoA by the proposed modified ß-oxidation (adapted from (7)). The underlined compounds are those shown to be formed from propionyl-CoA (and propionate) in the present investigation.

mg^{-1}) of their acyl-CoA synthetase amounts to 44% of the activity observed with palmitate. The propionyl-CoA is oxidized by the peroxisomal acyl-CoA oxidase with an activity of 54.4 nmol H_2O_2 formed min^{-1} mg^{-1}, corresponding to 69% of the activity observed with palmitoyl-CoA.

Mammalian mitochondria metabolize propionyl-CoA to succinyl-CoA. This pathway involves the carboxylation of propionyl-CoA in a biotin-dependent reaction to yield methylmalonyl-CoA. In a subsequent B_{12}-dependent mutase reaction, methylmalonyl-CoA is rearranged to succinyl-CoA. HPLC elution profiles of reaction mixtures containing propionyl-CoA, ATP and peroxisomes from mung-bean hypocotyls, did never show an absorbance peak with the retention time 16.4 min due to succinyl-CoA.

CONCLUSION

The data presented demonstrate that propionyl-CoA, intermediate of the bFA catabolism in higher plant peroxisomes, is metabolized to acetyl-CoA in these organelles. Intermediates of the pathway are acrylyl-CoA and 3-hydroxypropionyl-CoA. The results are consistent with and support the concept proposed by Stumpf and coworkers (6,7), that plant cells metabolize propionate to acetyl-CoA by a modified ß-oxidation. This modified ß-oxidation is located in the peroxisome.

REFERENCES

1. Gerhardt, B. (1987) Fatty acid ß-oxidation in higher plants. In: The Metabolism, Structure, and Function of Plant Lipids (P.K. Stumpf, J.B. Mudd and W.D. Nes, eds.). Plenum Press, New York, pp. 399-404.

2. Behrends, W., Thieringer, R., Engeland, K., Kunau, W.-H. and Kindl., H. (1988) The glyoxysomal ß-oxidation system in cucumber seedlings: identification of enzymes required for the degradation of unsaturated fatty acids. Arch. Biochem. Biophys. 263, 170-177.

3. Davis, D.D. and Humphrey, T.J. (1978) Amino acid recycling in relation to protein turnover. Plant Physiol. 61, 54-58.

4. Gerbling, H. and Gerhardt, B. (1988) Oxidative decarboxylation of branched-chain 2-oxo fatty acids by higher plant peroxisomes. Plant Physiol. (in press).

5. Giovanelli, J. and Stumpf, P.K. (1962) Fat metabolism in higher plants. X. Modified ß-oxidation of propionate by peanut mitochondria. J. Biol. Chem. 231, 411-426.

6. Hatch, M.D. and Stumpf, P.K. (1962) Fat metabolism in hiher plants. XVIII. Propionate metabolism by plant tissues. Arch. Biochem. Biophys. 96, 193-198.

7. Stumpf, P.K. (1976) Lipid metabolism. In: Plant Biochemistry (J. Bonner and J.E. Varner, eds.), 3. ed., Academic Press, New York, pp. 427-461.

8. Halarnkar, P.P., Wakayama, E.J. and Blomquist, G.J. (1987) Metabolism of propionate to 3-hydroxypropionate and acetate in the lima bean *Phaseolus limensis*. Phytochemistry 27, 997-999.

BIOLOGICAL ROLE OF PLANT LIPIDS
P.A. BIACS, K. GRUIZ, T. KREMMER (eds)
Akadémiai Kiadó, Budapest and Plenum Publishing
Corporation, New York and London, 1989

DE NOVO BIOSYNTHESIS OF GLYCEROLIPIDS IN PLANT MITOCHONDRIA

Margrit Frentzen[1], M. Neuburger[2] and R. Douce[2]

[1] Institut für Allgemeine Botanik, Universität Hamburg, 2000 Hamburg, FRG
[2] Laboratoire de Physiologie Cellulaire Végétale, UA CNRS 576, DRF/CENG, 38041 Grenoble, France

Glycerol 3-phosphate acyltransferase (G3P-AT) and 1-acylglycerol 3-phosphate acyltransferase (LPA-AT), which catalyse the first two steps in course of de novo biosynthesis of glycerolipids, are known enzymic activities not only of plastids and microsomes (Frentzen 1986) but also of mitochondria from higher plants (Douce 1971, Vick and Beevers 1977, Sparace and Moore 1979). In mitochondria of castor bean endosperm the two enzymic activities were localized in both the inner and outer membrane (Sparace and Moore 1979), but little is known about the enzymic properties.

We have started to characterize the mitochondrial G3P-AT and LPA-AT by using highly purified organelle fractions isolated from potato tubers and pea leaves according to the methods previously described (Neuburger et al. 1982, Day et al. 1985). Optimizing the assay conditions for the G3P-AT and the LPA-AT we observed different pH optima for the two enzymes. Similar to the microsomal AT system (Hares and Frentzen 1987), the LPA-AT of mitochondria showed maximal acylation rates at alkaline pH values between 8 and 11 whereas the G3P-AT had a pH optimum at 7.5. In whole mitochondrial fractions specific activities of about 50 and 200 pmole x min^{-1} x mg^{-1} protein were determined for the G3P-AT and LPA-AT, respectively.

In mitochondria from potato tubers both G3P-AT and LPA-AT are firmly bound to the membranes. They showed 20-30fold higher specific activities in outer than in inner membrane fractions. Hence, in potato mitochondria in contrast to the organelles from castor bean endosperm (Sparace and Moore 1979) G3P-AT and LPA-AT are mainly or even exclusively located in the outer membrane. According to first experiments with submitochondrial fractions from pea leaves, LPA-AT showed a similar distribution as in potato mitochondria whereas the G3P-AT behaved like a soluble protein. The G3P-AT is, however, not localized in the matrix, but presumably in the intermembrane space since the G3P-AT was released from the organelles even under conditions under which the mitoplast stayed intact. Consistent with the different submitochondrial localization we observed a different accessibility of the G3P-AT from pea and potato mitochondria to thermolysin.

The analysis of the reaction products formed by the mitochondrial G3P-AT and LPA-AT revealed that the G3P-AT exclusively directed the acyl groups to the C-1 position of G3P while the LPA-AT catalysed the acylation at position 2.

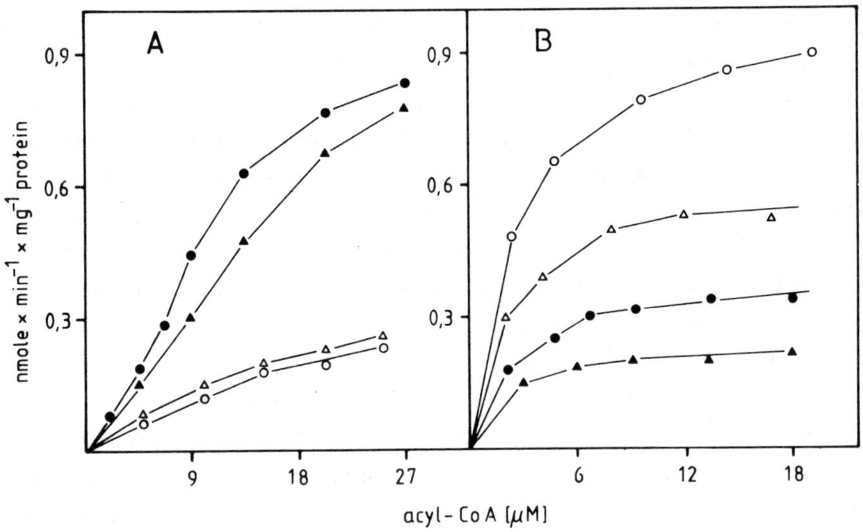

Fig. 1: Fatty acid specificity of the mitochondrial G3P-AT (A) and LPA-AT (B) from pea leaves. The acylation rates of G3P (A) und LPA (B) are given in dependence on the concentration of the different acyl-CoA thioesters (palmitoyl-CoA, stearoyl-CoA, oleoyl-CoA, linoleoyl-CoA).

Hence, the two mitochondrial AT displayed the same positional specificities as the corresponding enzymic activities of plastids and microsomes (Frentzen 1986).

For the acylation reactions the two mitochondrial enzymes from both potato tubers and pea leaves specifically utilized CoA-thioesters while ACP-thioesters, the preferred substrates of the plastidial enzymes (Frentzen et al. 1983), were not used.

Differences between the mitochondrial G3P-AT from potato tubers and pea leaves were not only observed with respect to their intraorganelle localization but also in regard of their fatty acid specificities. While the potato enzyme showed similar acylation rates with the different acyl-CoA thioesters, the pea enzyme exhibited distinctly higher activities with saturated than with unsaturated acyl-CoA thioesters (Fig. 1A). Thus, the mitochondrial G3P-AT from pea leaves, in contrast to that from potato tubers, possesses a pronounced fatty acid specificity for saturated acyl groups. Such a specificity has not been observed for G3P-ATs from other compartments.

In contrast to the G3P-AT, the mitochondrial LPA-ATs from the two different organs displayed very similar properties. Both enzymes did not show a fatty acid specificity with respect to the acyl groups in the acyl acceptor, but they showed one with respect to the acyl groups in the acyl donor.

Table: Fatty acid selectivity of the mitochondrial G3P-AT and LPA-AT from potato tubers. The fatty acid composition of 1-acylglycerol 3-phosphate and of position 2 of phosphatidic acid is given in % of the total incorporation formed from the equimolar acyl-CoA mixtures.

	acyl-CoA mixtures							
	16:0	18:1	16:0	18:1	18:2	18:0	18:1	18:2
					%			
LPA	68	32	45	22	33	43	22	35
C-2 of PA	9	91	9	48	43	8	48	45

The pea (Fig. 1B) as well as the potato LPA-AT preferentially used unsaturated acyl-CoA thioesters, especially oleoyl-CoA.

In order to determine the fatty acid selectivities of the mitochondrial G3P-AT and LPA-AT, acyl-CoA thioester mixtures were offered to submitochondrial fractions and the fatty acid composition of lysophosphatidic acid and of the C-2 position of phosphatidic acid were analysed.

Consistent with the observed fatty acid specificities, the G3P-AT from potato tubers utilized the acyl groups to almost the same proportion as they were offered whereas the LPA-AT from potato tubers strongly discriminated against saturated acyl groups (Table). The fatty acid selectivities of the two enzymes of pea mitochondria are under current investigations.

The financial support by the commissariat l'énergie atomique Grenoble and by the Deutsche Forschungsgemeinschaft is gratefully acknowledged.

References

Day, D.A., Neuburger, M., Douce, R. 1985: Biochemical characterization of chlorophyll-free mitochondria from pea leaves. Aust. J. Plant Physiol. 12, 219-228.

Douce, R. 1971: Incorporation de l'acide phosphatidique dans le cytidine diphosphate diglycéride des mitochondries isolées des inflorescences de chou-fleur. CR Acad. Sci. Ser. D. 272, 3146-3149.

Frentzen, M. 1986: Biosynthesis and desaturation of the different diacylglycerol moieties in higher plants. J. Plant Physiol. 124, 193-209.

Frentzen, M., Heinz, E., McKeon, T.A., Stumpf, P.K. 1983: Specificities and selectivities of glycerol-3-phosphate acyltransferase and monoacylglycerol-3-phosphate acyltransferase from pea and spinach chloroplasts. Eur. J. Biochem. 129, 629-636.

Hares, W., Frentzen, M. 1987: Properties of the microsomal acyl-CoA: sn-1-acyl-glycerol-3-phosphate acyltransferase from spinach (Spinacia oleracea L.) leaves. J. Plant Physiol. 131, 49-59.

Neuburger, M., Journet, E.-P., Bligny, R., Carde, J.-P., Douce, R. 1982: Purification of plant mitochondria by isopycnic centrifugation in density gradients of Percoll. Arch. Biochem. Biophys. 217, 312-323.

Sparace, S.A., Moore, T.S. Jr. 1979: Phospholipid metabolism in plant mitochondria, submitochondrial sites of synthesis. Plant Physiol. 63, 963-972.

BIOLOGICAL ROLE OF PLANT LIPIDS
P.A. BIACS, K. GRUIZ, T. KREMMER (eds)
Akadémiai Kiadó, Budapest and Plenum Publishing
Corporation, New York and London, 1989

ACCUMULATION OF LIPIDS IN DEVELOPING SEA BUCKTHORN (HIPPOPHAE RHAMNOIDES L.) FRUITS

A.G. Vereshchagin

Lipid Biochemistry Research Unit, Institute of Plant Physiology, Academy of Sciences, Moscow, USSR

INTRODUCTION

Plant reserve oils consisting for the most part of triacylglycerols (TAGs) are known to be localized mainly in the seeds. These oils play a major role during seed germination, being utilized to form seedling biomass /1/. However, there is a number of species (oil palm, avocado, olive, sea buckthorn etc.), which accumulate great amounts of oil not only in the seeds, but also in a water-rich mesocarp. These oils are referred to below as plants with oily mesocarp (POM). The biological role of mesocarp TAGs is unknown, and therefore studies in this field are of great interest. Sea buckthorn was chosen for these studies because it is widely distributed in our country, whereas other POM are not readily available in Russia; besides, its oil is extensively applied in medicine /2/.

MATERIALS AND METHODS

Mesocarp and seeds of mature fruits were extracted /3/ and TAGs were prepared by TLC. Total lipids and TAGs were converted into methyl esters (ME) of saturated (S) and unsaturated (U) fatty acids (FAs). FA composition of these lipids and their absolute content were determined by GLC combined with MS and internal standard technique.

RESULTS AND DISCUSSION

The lipid content in mature (105 days after flowering, DAF) mesocarp and seeds was 111-177 and 474-804 μmole/g fresh weight. The former values are close to those found earlier, but the latter greatly exceed them /2,4/. Up to 93% of total fruit oil is in its mesocarp. The FA composition of these lipids is shown in Table 1 (105 DAF, E and NE). It is seen that E lipids of both mesocarp and seeds include eight FA species each. However, these organs differ greatly in lipid FA composition. In mesocarp TAGs C_{16}-FAs, vz. hexadecenoic (H) and palmitic (P) predominate. These TAGs were for the first time shown to include hexadecadienoic acid (m/z of M^+ at GLC-MS = 266). Seed lipids contain mainly C_{18}-UFAs - oleic (O), linoleic (L) and linolenic (Le). All these results confirm those of other workers /2,5,6/. The NE mesocarp lipids are characterized by the presence of C_{20-22}-FAs comprising together ~30% of total FAs. All these differences are likely to be caused by various genotype of respective organs: cotyledons possess the embryo genotype including the genes of both parents, whereas mesocarp containes only maternal genes. Earlier the predominance of P and O has been regarded as being typical for mesocarp oils /1/; as shown here, this rule can now be supplemented by one more feature, vz. the high level of $C_{16:1}$-acid.

Thus, the FA composition of sea buckthorn fruit reserve lipids is rather unusual. Firstly, the mesocarp contains a highly saturated oil; such oils are present in few plant species /7,8/. Secondly, TAGs rich in C_{18}-UFAs accumulate in the seeds although the latter are devoid of chloroplasts which are generally believed to be responsible for UFA synthesis in oilseeds (linseed, soybean etc. /9/). Therefore, the study of fruit reserve lipid formation in sea buckthorn, as a representative of POM, is desirable. To begin this study, we determined the dynamics of biomass and lipid accumulation in 12-105 DAF fruits (Fig. 1). It is seen that its rate is maximal at 50-78 (seeds) and 78-88 DAF (mesocarp); the absolute rate in mesocarp is higher than in the seeds by an order of magnitude, but the

Table 1. DYNAMICS OF TOTAL LIPID FATTY ACID COMPOSITION IN FRUITS

W% of total FAMEs. Lipids also contained traces of $C_{14:0}$ and $C_{15:0}$ acids. E, NE - extractable and non-extractable lipids of mature fruits (105 DAF); the share of NE in both mesocarp and seeds does not exceed 0.3% of the sum N+NE. Mesocarp NE lipids at 105 DAF also contained $C_{20:0}$ 2.5%, $C_{22:0}$ 3.9%, $C_{22:2}$ 23.4%.

FAs	Mesocarp at 12-105 DAF					Seeds at 12-105 DAF				
	12	50	78	105		12	50	78	105	
	E	E	E	E	NE	E	E	E	E	NE
$C_{16:0}$	21.4	30.9	33.0	36.3	19.3	19.5	17.6	7.1	9.7	12.4
$C_{16:1}$	6.9	38.5	51.8	46.9	26.5	4.9	7.6	0.8	0.9	2.9
$C_{16:2}$	0.4	1.2	2.0	2.2	0.0	tr.	0.0	0.0	0.0	0.0
$C_{18:0}$	1.2	0.2	0.2	0.7	1.6	1.6	2.1	2.0	2.8	2.1
$C_{18:1}$	13.0	15.5	7.0	7.1	6.8	15.0	15.1	14.8	14.9	17.1
$C_{18:2}$	31.3	9.9	5.4	5.6	11.0	30.0	21.2	41.0	39.2	42.0
$C_{18:3}$	24.6	3.5	0.5	0.6	2.8	26.2	35.6	33.9	30.5	18.8

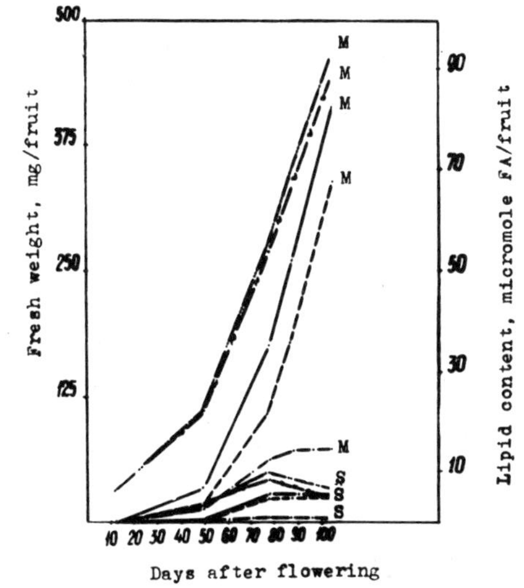

Fig. 1. Dynamics of the mesocarp (M) and seeds (S) fresh weight and the absolute content of their total lipids, TAGs and polar lipids during maturation —··—, —~— total and "defatted" fresh weight; ——, —·— total and polar lipids;--- TAGs

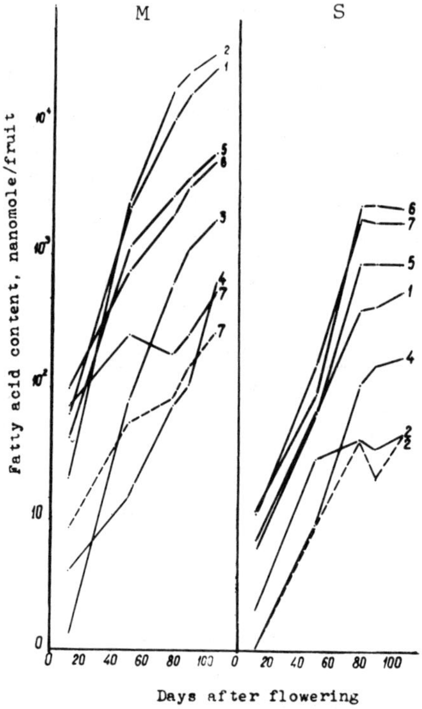

Fig. 2. Dynamics of the absolute content of individual esterified FAs in maturing fruits. FAs: 1-palmitic, 2-hexadecenoic, 3-hexadecadienoic, 4-stearic, 5-oleic, 6-linoleic, 7-linolenic; —— total lipid FAs, --- FAs of triacylglycerols.

relative rate (μmole FA/fruit/day/mg fresh weight) is three times less in mesocarp compared to seeds.

By 12th DAF both parts of the fruit are similar in their lipid FA composition (Table 1). Later however, the formation of mesocarp and seed lipids was accompanied with the rise of C_{16}-FAs and C_{18}-UFAs resp. After 78th DAF the characteristic composition of each organ remained nearly constant although mesocarp in this period was notable for rapid lipogenesis and the seeds - by total lack of oil accumulation (Fig. 1). Mesocarp polar lipids were basically similar in their FA composition to TAGs, being characterized by a high content of an "uncommon" $C_{16:1}$-acid; in other plants such FAs (erucic, petroselinic etc.) are usually absent from reserve organ polar lipids /10/.

In Fig. 2 are shown the results of calculating FA absolute molar amounts per single fruit, which give more accurate pattern of lipid formation dynamics than the data of Table 1. It is seen that with increasing maturity this value either rises or remains constant; its fall was observed very rarely.

Total lipids of a given organ of various maturity usually include 70-90% TAGs. Therefore we determined TAG composition using lipase hydrolysis /11/. It was shown that S_2U and SU_2 TAG types and PH_2 (30% of total TAGs), P_2H (20%) and H_3 (14%) TAG species predominate in mesocarp, and SU_2 and U_3 (mainly L_2Le, 14%, LLe_2, 17% and L_3, 7%)) were the major TAGs of seeds. All TAGs of the fruit, although of different composition, were similar in their structure to other plant TAGs /12/. Therefore, their composition could be accurately established without lipase hydrolysis, using only the formulae of Litchfield /12/. In very immature fruits (12th DAF) the composition thus calculated did not agree with the found one; later there was an almost complete coincidence between the both sets of data. Possibly, TAGs formed by 12th DAF are rapidly subjected to further enzymatic reactions, whereas TAGs produced during intense oil formation remain intact and do not participate in any metabolic processes until germination /9/.

REFERENCES

1. Hiditch, T.P. and Williams, P.N. (1964) The chemical constitution of natural fats, Wiley, New York, 745 pp.
2. Franke, W. und Müller, H. (1983) Ang. Botanik, 57, 77-83.
3. Zhukov, A.V. and Vereshchagin, A.G. (1974) Fisiologiya rastenii, 21, 659-663.
4. Filatov, I.I. (1976) Fruit and berry crops, Inst. Agric., Gorky, pp. 34-35.
5. Zham'yansan, Ya. (1978) Khimiya prir. soedin., No. 1, 133-134.
6. Zhmyrko, T.G. (1984) Khimiya prir. soedin., No. 3, 300-305.
7. Hitchcock, C. and Nichols, B.W. (1971) Plant lipid biochemistry, Academic Press, New York, 387 pp.

8. Mannan, A., Farooqi, J.A. and Asif, M. (1984) Chemistry and Industry, No. 23, p. 851-852.
9. Appelqvist, L.A. (1975) in Galliard, T. and Mercer, E.I. (Eds.) Recent advances in the chemistry and biochemistry of plant lipids, Academic Press, New York, pp. 247-286.
10. Zhukov, A.V. and Vereshchagin, A.G. (1980) Fisiologiya rastenii, 27, 390-398.
11. Ozerinina, O.V., Berezhnaya, G.A., Eliseev, I.P. and Vereschagin, A.G. (1987) Khimiya prir. soedin., No. 1, 52-57.
12. Litchfield, C. (1972) Analysis of triglycerides, Academic Press, New York, 355 pp.

BIOLOGICAL ROLE OF PLANT LIPIDS
P.A. BIACS, K. GRUIZ, T. KREMMER (eds)
Akadémiai Kiadó, Budapest and Plenum Publishing
Corporation, New York and London, 1989

MOBILISATION OF STORAGE LIPIDS IN GERMINATING OILSEEDS

D.J. Murphy and I. Cummins

Department of Biological Sciences, University of Durham, DH1 3LE, UK

INTRODUCTION

Storage triacylglycerols are broken down by triacylglycerol lipases following seed germination. Despite the expanding biotechnological uses of lipases and the importance of lipolysis for seedling growth, relatively little is known about the nature and regulation of plant lipases. Two important areas of controversy are (a) the possible regulation of lipases by proteinaceous inhibitors and (b) the subcellular distribution of oilseed lipases.

ROLE OF LIPASE INHIBITORS

Several authors have reported the occurrence of very active proteinaceous inhibitors of lipases in several oilseed species (Satouch & Matshushita, 1976; Wang & Huang, 1984; Chapman, 1987). This raises the possibility that lipases may be present, but inactive, in dry seeds and that regulation of lipase activity may therefore take place independently of its synthesis. We have recently measured both the activity of lipase (by enzymatic assay) and the amount of lipase (by ELISA) in seedlings of rapeseed at various times after germination (Murphy et al., 1988). At all times there was a very strong correlation between the two, indicating that lipase is only present in an active form in rapeseed. This has been investigated further by studying the effect of a soluble protein fraction from rapeseed seedlings of various ages on the activity of the microsomal lipase. The data in Table 1 show that even very large amounts of the soluble protein fraction had relatively little effect on lipase activity. It is concluded that lipases in germinating rapeseed are not regulated by soluble proteinaceous inhibitors.

Table 1 Effect of soluble fractions on microsomal lipase activity. Assays were performed in the presence of 15 µg microsomal protein ±

Microsomal fraction (h after germination)	Supernatant fraction (h after germination)	Relative lipase activity %
165	-	100
"	31	67
"	90	87
"	100	95
"	140	100
"	165	95
"	194	101

70 µg soluble protein from germinating seedlings of various ages. Lipase activity was determined using the copper soaps assay.

SUBCELLULAR LOCALISATION OF LIPASES

There appears to be considerable interspecific variation in the localisation of lipases in germinating tissue and they have been found to be associated with oil-body, endoplasmic reticulum and glyoxysomal membranes (Lin et al., 1983; Abigor et al., 1985; Hills & Beevers, 1985; Maeshima & Beevers, 1987; Murphy et al., 1988). These apparently conflicting data may have several causes including different assay methods, the use of seedlings of different ages or they may reflect genuine interspecific differences in localisation. The situation in rapeseed was investigated by assaying the two major subcellular sites of lipase activity at various times after germination. The results presented in Fig. 1 show that lipase activity is initially found in the oil-body fraction but is later associated mainly with the microsomal fraction. At even later times, there was evidence of lipase activity in the glyoxysomal fraction (data not shown). These results may explain some of the conflicting data concerning lipase localisation since the use of seedling of different ages will give rise to very different conclusions.

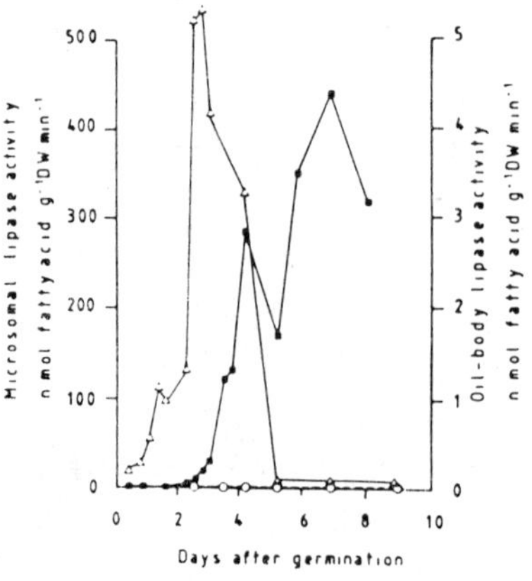

Figure 1

Changes in the lipase activity associated with oil-body (△———△) and microsomal (■———■) fractions following germination of rapeseed cotyledons. Microsomal fractions were obtained by ultracentrifugation at 100,000 xg for 2h. Oil-body fractions were prepared by repeated floatation following centrifugation at 20,000 xg for 15 min. The purity of the respective fractions was verified by marker enzyme assays and by SDS-PAGE as previously described (Murphy et al., 1988).

The data in Table 2 show that lipase activity was only present in germinating seeds and was absent from all other parts of the rapeseed plant. These data are consistent with our ELISA results which show that lipase protein is likewise confined to germinating seeds.

Table 2 Lipase activity in various organs of rapeseed
The various organs were homogenised rapidly and lipase activity was determined by the copper soaps method.

Organ	Relative lipase activity (g^{-1} dry wt. tissue)
Dry seed	0 tr
Germinating seed (4 days)	100
Germinating seed (12 days)	38
Developing embryo	0
Seed pod	0
Mature leaf	0
Root	0
Flower petal	0

It has recently been shown that an antibody raised against the glyoxysomal lipase of castor bean also cross-reacts with a single polypeptide of similar apparent molecular weight (60 kDa) in many different oilseed species (Hills & Beevers 1987; Murphy & Hills, 1988). This antibody inhibited lipase activity in rapeseed extracts and the amount of antigenic 60 kDa polypeptide closely followed the variations in lipase activity after germination. Analysis of sucrose density gradients

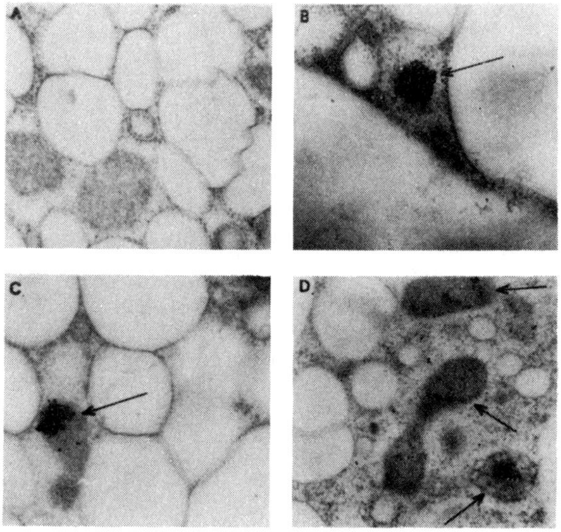

Figure 2

Immunogold-labelled sections of cells from germinating rapeseed cotyledons 5 days after germination A. Control section treated with pre-immune serum. No gold labels are visible. Many full or depleted oil-bodies and micro-bodies can be seen. B,C & D sections treated with anti-lipase IgG originally raised against purified castor bean glyoxysomal lipase. In all cases the gold labels are exclusively associated with small highly-stained microbodies (arrowed) which readily stained in the presence of diaminobenzidine, indicative of the presence of catalase. It is concluded that the gold labels are interacting with glyoxysomes.

revealed that the antigenic 60 kDa polypeptide was mainly present in oil-body and microsomal fractions at early stages of seedling growth but was increasingly associated with the glyoxysomal fraction at later stages. This was despite the presence of relatively little lipase activity in the glyoxysomal fraction. Analysis of sections of 5-day old seedlings by immunogold labelling using the anti-lipase reveals that gold particles were exclusively associated with microbodies identified as glyoxysomes, as shown in Fig. 2.

ACKNOWLEDGEMENTS

The support of the Agriculture & Food Research Council and NATO Science Council to DJM is gratefully acknowledged.

REFERENCES

Abigor, D.R., Opute, F.I., Opoku, A.R. & Osagie, A.U. (1985) Partial purification and some properties of the lipase present in oil palm (Elaeis guineesis) mesocarp, J. Sci. Food Agric. 36, 599-606.

Chapman, G.W. (1987) A proteinaceous competitive inhibitor of lipase isolated from Helianthus annus seeds, J. Sci. Food Agric.

Hills, M.J. & Beevers, H. (1987a) Ca^{2+} stimulated neutral lipase activity in castor bean lipid bodies, Plant Physiol. 84, 272-276.

Hills, M.J. & Beevers, H. (1987b) Binding of antilipase to 62 kDa protein in many species of oilseed, Plant Physiol.

Lin, Y., Wimer, L.T. & Huang, A.H.C. (1983) Lipase in the lipid bodies of corn scutella during seedling growth, Plant Physiol. 73, 460-463.

Maeshima, M. & Beevers, H. (1985) Purification properties of glyoxysomal lipase from castor bean, Plant PHysiol. 79, 489-493.

Murphy, D.J., Cummins, I. & Kang, A.S. (1988) Immunological investigation of lipases in germinating oilseed rape, Brassica napus, J. Sci. Food Agric. in press.

Murphy, D.J. & Hills, M.J. (1988) Immunological characterisation of lipases from a wide range of oilseed species. In: Biotechnology for the Fats & Oils Industry (Applewhite, T.E. ed.) Amer. Oil Chem Soc. in press.

Satouchi, K. & Matshushita, S. (1976) Proteinaceous inhibitors of lipase activity in germinating oilseeds, Agric. Biol. Chem. 40, 888-897.

Wang, S. & Huang, A.H.C. (1987) Inhibitors of lipase activities in soybean and other oilseeds, Plant Physiol. 76, 929-934.

BIOLOGICAL ROLE OF PLANT LIPIDS
P.A. BIACS, K. GRUIZ, T. KREMMER (eds)
Akadémiai Kiadó, Budapest and Plenum Publishing
Corporation, New York and London, 1989

LIPASE INHIBITION BY CoA AND OLEOYL-CoA

M. J. Hills[1] and D.J. Murphy[2]

[1]Federal Centre for Lipid Research, D-4400 Münster, FRG
[2]Botany Department, University of Durham, Durham DH1 3LE, UK

Abstract

The neutral lipase from the lipid bodies of the endosperm from young castor bean seedlings is inhibited substantially by the addition of both oleoyl-CoA and CoA. When present alone, CoA has little detectable effect on neutral lipase activity and oleoyl-CoA causes a much lesser degree of inhibition when present on its own.

Introduction

Early growth of young oilseed plants is supported by the conversion of storage oils to sugars which act as an energy source for the seedling. Lipase (triacylglycerolacylhydrolase) catalyses the first step in this pathway. At present there is little knowledge of the coordination of lipase activity with fatty acid oxidation although it is likely that effectors controlling activity of the lipase enzyme exist (Huang 1987).

Experimental

Neutral lipase in the lipid body membranes isolated from 4 d castor seedlings was used as the enzyme in these experiments (Hills and Beevers 1987). Aliquots of lipase were incubated at 30°C for 15 min in Bis-tripropane buffer (pH 7.5) with 2.5 mM emulsified triolein, 2 mM DTT and 2 mM $CaCl_2$. Products were extracted, separated by TLC procedures and counted for radioactivity.

Results and Discussion

Table 1 shows that the addition of 120 μM free oleoyl-CoA and 2 mM CoA cause nearly 90% inhibition of neutral lipase activity. When present on its own, 120 μM oleoyl-CoA caused only 35% inhibition and CoA had little detectable effect on lipase activity. Fig. 1 shows the effect of oleoyl-CoA concentration in the presence of CoA, on the kinetics of oleic acid, diolein and monoolein production from triolein. After an initial inhibition at lower concentrations, there is a range of oleoyl-CoA concentrations where oleic acid release is

Table 1. Effect of including oleoyl-CoA and Co-A in the standard assay, on the activity of the castor bean neutral lipase.

Experiment	nmol oleic acid assay^{-1}
Control	71 (100 %)
2 mM CoA	70 (99 %)
120 μM oleoyl-CoA	45 (65 %)
2 mM CoA + 120 μM oleoyl-CoA	8 (12 %)

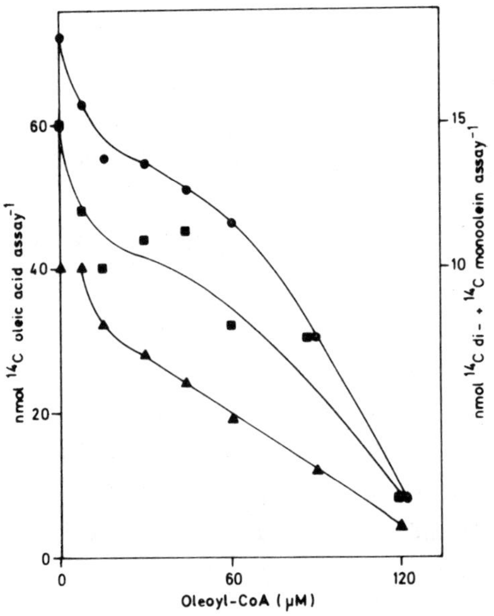

Fig. 1. Effect of oleoyl-CoA concentration on the release of oleic acid ( ), diolein and monoolein ( ) from triolein by castor bean lipid body neutral lipase in the presence of 2 mM Coenzyme A.

relatively unaffected. Then at higher concentrations, inhibition is very marked. The amounts of di-, and monoolein were similarly reduced with increasing oleoyl-CoA. It is proposed that the effect of oleoyl-CoA and CoA might be physiologically significant in controlling lipase activity in vivo.

Acknoledgements. We thank Dr H. Beevers, UC Santa Cruz USA for providing laboratory facilities and helpful advise during this work. Grant. NSF PCM 84-03542

REFERENCE

Huang AHC (1987) Lipases. In: Stumpf and Conn Biochemistry of plants 9 p 91
Hills MJ and Beevers H (1987) Ca^{2+} stimulated neutral lipase activity in castor bean lipid bodies. Plant Physiol. 84 272-276

BIOLOGICAL ROLE OF PLANT LIPIDS
P.A. BIACS, K. GRUIZ, T. KREMMER (eds)
Akadémiai Kiadó, Budapest and Plenum Publishing
Corporation, New York and London, 1989

PERSPECTIVES OF SEARCH FOR EICOSANOID ANALOGS IN PLANTS

I.A. Tarchevsky and A.N. Grechkin

Kazan Institute of Biology, USSR Academy of Sciences, P.O. Box 30, Kazan, 420084, USSR

Arachidonic acid ($20{:}4\omega^6$), the main component of membrane lipids in animal tissues, is the percursor of a number of highly active bioregulators, namely, prostaglandins, thromboxanes, leukotrienes and lipoxins. The success of research in this direction has led to attempts of searching for eicosanoids, particularly prostaglandins, in plant tissues |1-3|. One of the most thoroughly made works of such kind of Panosyan with co-workers has led to a negative result. Instead of prostaglandin $F_{2\alpha}$, screening of which was made by TLC, they have found in *Brionia alba* a number of trihydroxyacids, probably the products of linolenate hydroperoxides rearrangements. Successful screening of prostaglandins E and F in *Kalanchoe blossfeldiana* and *Larix sibirica* was described by Yanistyn and Levin |2, 3|. However, these works are characterized by deficiency of data confirming the identification of purified compounds. Prostaglandin A_1 was found in the *Allium cepa* bulbs |4|. However, it was recently shown that prostaglandin-like activity, present in bulbs, belongs to trihydroxyoctadecenoic acids |5|. Today the occurence of prostaglandins among plants has been clearly established only for red alga *Gracilaria lichenoides* |6|.

The little success of search for prostaglandins in plant tissues is probably not accidental. To evaluate the perspectives of search for eicosanoids or their analogs it is necessary to take into account the relative abundance of different polyenoic fatty acids in plants. Most plants, according to all available data, are free of metabolic precursors of eicosanoids,

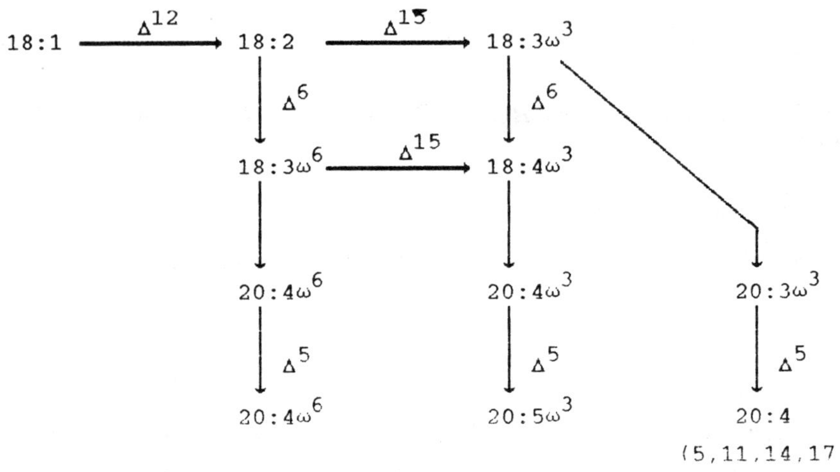

Scheme 1. The main pathways of polyenoic fatty acids formation in plant and animal systems.

namely $20:4\omega^6$ and other C_{20}-polyenoic fatty acids |7|. These acids are absent in Cyanophyta containing mainly $18:3\omega^3$ and $18:2\omega^6$. During the formation of eucaryotic plant cell, lipid metabolism has considerably got complicated in comparison with Cyanophyta. Euglenophyta, green and some other algae, mosses, ferns contain not only $18:2\omega^6$ and $18:3\omega^3$, but also $18:3\omega^6$, $20:3\omega^6$ and $20:4\omega^6$ |7|. These eucaryotes, being at low levels of evolutionary development, probably have two independent systems of lipid biosynthesis, namely plastidal and cytoplasmic ones. Thus, their cells contain glycolipids with high content of $18:3\omega^3$ typical for plastids and, on the other hand, phospholipids rich in $18:3\omega^6$ and $20:4\omega^6$ |8|.

During the eucaryotes evolution further changes took place. Conifers do not contain $20:4\omega^6$ |9|. That is probably caused by the loss of Δ^6-desaturase activity. Thus, in parallel with the absence of $20:4\omega^6$, conifers have 5,11,14-20:3 and 5,11,14,17-20:4 |9| (right side of Scheme 1). In members of *Boraginacede* and *Caryophyllacede* $20:4\omega^6$ was not found too, but $18:3\omega^6$ and $18:4\omega^3$ are present |10, 11|. Due to the absence of acids with Δ^5-double bond, it seems possible that $20:4\omega^6$ formation in members of these two families is limited by the lack of Δ^5-desaturase activity.

Up to date, there are no strongly evidential data suggesting the presence of $20{:}4\omega^6$ and $20{:}5\omega^3$ in flowering plants, where $18{:}3\omega^3$ is the main component. There are no data of biosynthetic experiments, confirming the presence of both Δ^6- and Δ^5-desaturases in some higher plant. Alien nature of $20{:}4\omega^6$ and $20{:}3\omega^6$ for plants is proved by the fact that these exogenous acids initiate the necrotization of the surface cell layer of potato tuber slices |12|.

At the same time, it cannot be neglected that the $20{:}4\omega^6$ formation as a "rudimentary" pathway may occure in the generative organs, or at the turning-points of ontogenesis. Thus, small amounts of $20{:}4\omega^6$ were found in commercial wheat germ oil |13| and in *Populus balsamifera* buds |14|. These data, however, are in need of thorough checking. In particular, germ oil used by Yanistyn in his work |13|, might contain the determined amount of $20{:}4\omega^6$ (12.5 pmole/g oil) as a strange admixture. Moreover, our data (not presented), obtained by HPLC, capillary GC and MS, show that *Populus* buds do not contain $20{:}4\omega^6$ in amount of more than 0.01% of total fatty acids.

The lack of convincing data on the presence of $20{:}4\omega^6$ and other C_{20}-polyenoic fatty acids allows to assume that the products of oxidative metabolism of $18{:}3\omega^3$ and $18{:}2\omega^6$, the main components of plant lipids, play a role of bioregulators in plants, like eicosanoids in animals. Some physiologically active metabolites of $18{:}2\omega^6$ and $18{:}3\omega^3$, formed by the lipooxygenase pathway, are already known. These are traumatic acid, traumatin |15| and jasmonic acid |16|. The first two compounds are the wound hormones, stimulating growth |15|. Jasmonate in the most cases acts like abscisic acid |17|. Trihydroxyacids described by Panosyan, similarly to prostaglandins, causes smooth muscle constriction |1|.

Studying the metabolization of exogenous $|1\text{-}^{14}C|18{:}2\omega^6$ in pea seedlings, we have found the formation of a great variety of metabolites, dozens of oxygenated derivatives are formed. 12-Hydroxydodec-9Z-enoic acid, one of the main products, was firstly isolated from natural sources |18|. A lot of oxiranyl compounds are also formed. Some of isolated substances were found to stimulate the growth of soybean callus. These results are the subject of another report at present Symposium. The observed broad variety of oxygenated $18{:}2\omega^6$ and $18{:}3\omega^3$ metabolites allows to consider the search of new bioregulators of this class to be very promising.

Oxygenated metabolites of $18{:}3\omega^3$ and $18{:}2\omega^6$ of plant origin can be considered as eicosanoid analogs in two respects: due to similarity of their biogenesis and of their regulatory activity. Taking this likeness into account, we propose a common trivial name "linolenoids" for this new class of plant bioregulators.

REFERENCES

1. Panosyan A.G., Avetisyan G.M., Mnatzakanyan V.A., Asatryan T.A., Vartasyan S.A., Boroyan R.G., Batrakov S.G. (1979) Bioorg Khim 5:242-253
2. Levin E.D., Alaudinov S.T., Cherepanova V.E. (1984) Khim Prir Soed 567-571
3. Janistyn B. (1982) Planta 154:485-487
4. Attrep K.A., Bellman W.P. (1980) Lipids 15:292-297
5. Claeys M., Ustunes L., Laekeman G., Herman A.G., Vlietinck A.J., Ozer A. (1986) Prog Lipid Res 25:53-58
6. Gregson R.P., Marwood J.F., Quinn R.J. (1979) Tetrahedron Lett 4505-4506
7. Smith C.R. (1971) Progr Chem Fats Other Lipids 11:137-177
8. Radunz A. (1968) Hoppe-Seyler's Z physiol Chem 349:303-309
9. Jamieson G.R., Reid E.H. (1972) Phytochemistry 11:269-275
10. Jamieson G.R., Reid E.H. (1969) Phytochemistry 8:1488-1494
11. Jamieson G.R., Reid E.H. (1971) Phytochemistry 10:1575-1577
12. Preisig C.L., Kuc J. (1985) Arch Biochem Biophys 236: 379-389
13. Janistyn B. (1982) Planta 155:342-344
14. Isayeva E.V., Levin E.D. (1987) Khim Prir Soed 513-516
15. Zimmerman D.C., Coudron C.A. (1979) Plant Physiol 63: 536-541

16. Vick B.A., Zimmerman D.C. (1983) Biochem and Biophys Res Communs 111:470-477
17. Ueda J., Kato J. (1980) Plant Physiol 66:246-249
18. Grechkin A.N., Korolev O.S., Kuramshim R.A., Yefremov Y.J., Musin R.M., Ilyasov A.V., Latypov S.K., Tarchevsky I.A. (1987) Dokl AN SSSR 297:1257-1260

BIOLOGICAL ROLE OF PLANT LIPIDS
P.A. BIACS, K. GRUIZ, T. KREMMER (eds)
Akadémiai Kiadó, Budapest and Plenum Publishing
Corporation, New York and London, 1989

EXPRESSION OF LIPOXYGENASE ISOZYMES IN SOYBEAN TISSUES

D.F. Hildebrand, K.M. Snyder, T.R. Hamilton-Kemp*, G. Bookjans, C.S. Legg and R.A. Andersen

Departments of Agronomy and *Horticulture, Univ. of Kentucky, Lexington, KY 40546, USA

Lipoxygenases (E.C. 1.13.11.12) catalyze the peroxidation of compounds that possess a *cis*, *cis*-1,4-pentadiene structure. Their principal substrates in higher plants are the di- and tri-unsaturated fatty acids, linoleic (C18:2) and linolenic (C18:3) acids. Lipoxygenases (LOXs) are also known to catalyze the cooxidation of chlorophylls and carotenoids (Axelrod et al, 1987; Vick and Zimmermann, 1987).

LOXs are apparently ubiquitous in higher plants and animals and are often (or always?) present as multiple forms or isozymes (Axelrod et al, 1981; Mack et al, 1987; Schewe et al, 1986). LOXs have been of interest because of their importance in flavor and aroma biogenesis and the importance of LOX metabolites in mammalian physiology (Hatanaka et al, 1987; Vick and Zimmerman, 1987; Parker, 1987).

In this study we report the expression and some properties of different LOX isozymes in various tissues of mutant and wild type soybean lines.

MATERIALS AND METHODS

The near-isogenic soybean lines mutant for combinations of the three principal seed LOXs (resulting in undetectable levels of these LOXs) were kindly provided by N. Nielsen (Davies and Nielsen, 1987).

Activity measurement and isoelectric focusing followed published techniques (Funk et al, 1985). Hexanal yield was de-

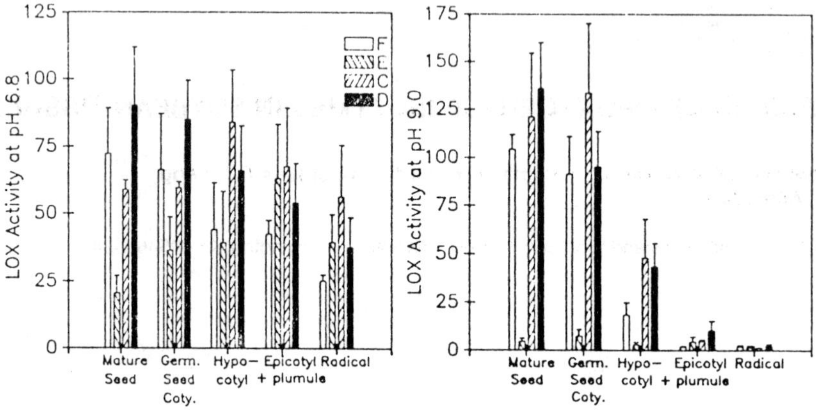

Figure 1. LOX activity in seed and seedling tissues of soybean mutants. Enzyme was extracted in ca. 4 volumes of 20 mM sodium phosphate buffer, pH 6.8. What we designate as line A is mutant for LOX 2 and 3; line B is mutant for LOX 1 and 3; line C is mutant for LOX 2; line D is mutant for LOX 3; line E mutant for LOX 1 and F is the wildtype recurrent parent, the soybean cultivar 'Century'.

termined by gas chromatography of trapped headspace volatiles (Andersen et al, 1986).

RESULTS AND DISCUSSION

LOX activity measured at pH 6.8 was readily detectable in all the soybean tissues tested (Fig. 1). At pH 9.0 very low LOX activity was found in the radical and epicotyl plus plumule with abundant activity in the other tissues. Furthermore, the LOX 1 mutation (line E) resulted in a large reduction in LOX activity at pH 9.0 and a small reduction at pH 6.8 in mature seeds and 7 day old seedling cotyledons and hypocotyls (Fig. 1). In contrast the LOX 2 and LOX 3 mutations (lines C and D) did not affect LOX activity at pH 6.8 and 9.0 in any of the tissues tested. So far, by Western immunobloting (data not shown) and in-gel activity staining of IEF gels (Fig. 2), LOXs known to be present in seeds could only be detected in 7 day old seedling cotyledons and perhaps hypocotyls. On the other hand, LOX

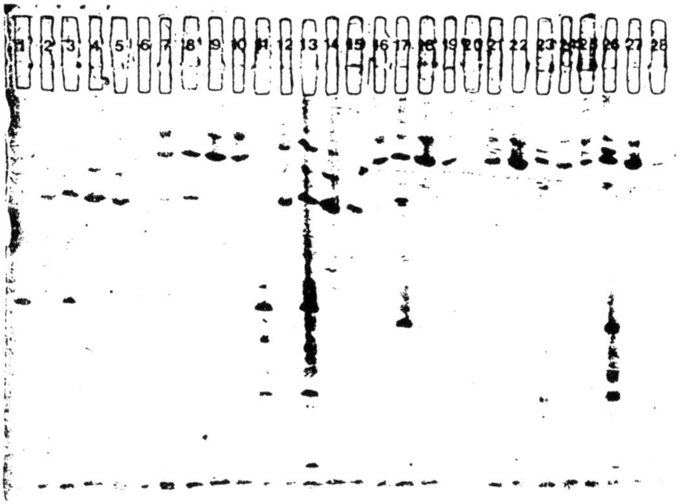

Figure 2. In-gel LOX activity staining of IEF gels of the tissues and lines described in Fig. 1 Lanes 1, 6, 11 and 20 have extracts from mature seeds; the other lanes contain extracts from 7-day-old seedlings: 2, 7, 12, 16, 21 and 25 = epicotyls + plumules; lanes 3, 8, 13, 17, 22 and 26 = cotyledons; 4, 9, 14, 18, 23 and 27 = hypocotyls; 5, 10, 15, 19, 24, 28 = radicals. Lanes 1-5 represent mutant backcross line A (see Fig. 1 legend); 6-10 B; 11-15 C; 16-19 D; 20-24 E and 25-18 F (or wild type).

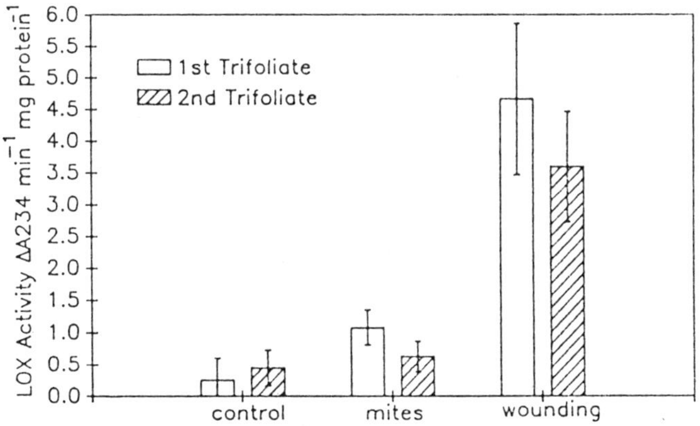

Figure 3. In the wounding experiment each leaflet of the first trifoliate of the soybean cultivar 'Williams' was crimped with a hemostat daily for 7 days. The mite treatment consisted of inoculating plants with 100 adult female spider mites (*Tetranychus urticae*) which were confined to the first trifoliate leaf with a ring of tanglefoot around the petiole. The mites were allowed to feed for 7 days.

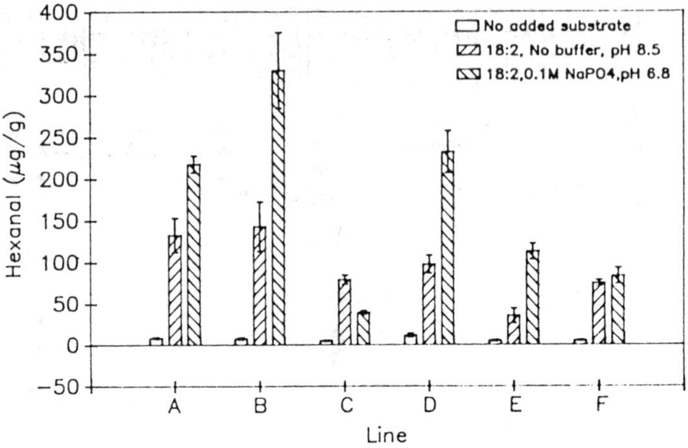

Figure 4. Hexanal formation of the soybean LOX null mutant backcross lines (see Fig. 1). The results shown represent means of at least four replications. Three sets of experiments were performed: 1. Finely ground powder from lyophilized seeds (0.5 g) was added to 20 ml water in a sealed reaction system and the volatiles produced during 60 min trapped and analyzed as described (Andersen et al, 1986). 2. As in 1. but 50 mg sodium linoleate was dissolved in the water prior to addition of seed powder and the reaction was only carried out for 20 min. 3. As in 2. but 0.1 M sodium phosphate, pH 6.8 was used instead of water.

isozymes with pIs more acidic than any of the 3 seed LOXs were seen in all tissues tested other than mature seeds (Fig. 2).

LOX activity was substantially increased in wounded and to a lesser extent in mite infested leaves (Fig. 3). Increased LOX was also found in the second, unwounded trifoliate of soybean plants, whose first trifoliates had been wounded (Fig. 3) indicating that LOX activity can be induced by translocatable factors. This induction of LOX activity was also seen in steady state levels of LOX protein as determined by immunoblots of SDS gels (data not shown). Preliminary evidence indicates that LOX poly A RNA levels are also increased. This wound induction of LOX may be related to possible roles in wound responses and/or pest resistance.

Measurement of hexanal production of seeds of the different LOX mutant backcross lines showed a significant effect of the presence or absence of LOX isozymes on hexanal formation (Fig. 4). The presence of either LOX 1 or LOX 2 stimulated hexanal production, whereas LOX 3 in some cases apparently reduced hexanal formation (Fig. 4) perhaps by catalyzing the conversion of the lipid hydroperoxides into nonvolatile oxidation products (Axelrod et al, 1981). However, the level of free fatty acid substrates (eg. C18:2) had the largest influence on hexanal yield (Fig. 4).

In conclusion it is apparent that LOX isozymes are qualitatively and quantitatively under stringent developmental control in soybeans. Additional isozymes are seen in vegetative tissues that appear different than any of the three well characterized seed isozymes. It is also evident that LOX isozymes in soybean leaves can be induced by wounding. The presence or absence of particular LOX isozymes in the seed significantly affect volatile lipid oxidation product formation. However, the properties of the 'new' LOX isozymes in vegetative tissues requires further investigation.

REFERENCES

Andersen, R.A., T.R. Hamilton-Kemp, P.D. Fleming and D.F. Hilderbrand. 1986. Volatile compounds from vegetative tobacco and wheat obtained by steam distillation and headspace trapping. In: *Biogenesis of aromas* (T.H. Parliment and R. Croteau, eds.). Amer. Chem. Soc. Symp. Series *317*: 99-111.

Axelrod, B., T.M. Cheesbrough and S. Laakso. 1981. Lipoxygenases from soybeans. Methods Enzymol. *71*: 441-451.

Davies, C.S. and N.C. Nielsen. 1987. Registration of soybean germplasm that lacks lipoxygenase isozymes. Crop Sci. *27*: 370-371.

Funk, M.O., M.A. Whitney, E.C. Hausknecht and E.M. O'brien. 1985. Resolution of the isozymes of soybean lipoxygenase using isoelectric focusing and chromatofocusing. Anal. Biochem. *146*: 246-251.

Hatanaka, A., T. Kajimara and J. Sekiya. 1987. Biosynthetic pathway for C_6-aldehyde formation from linolenic acid in green leaves. Chem. Phys. Lipids *44*: 341-361.

Mack, A.J., T.K. Peterman and J.N. Siedow. 1987. Lipoxygenase isozymes in higher plants: biochemical properties and physiological role. *Isozymes*: Current Topics Biol. Med Res. *13*: 127-154.

Parker, C.W. 1987. Lipid mediators produced through the lipoxygenase pathway. Ann. Rev. Immunol. *5*: 65-84.

Schewe, T., S.M. Rapoport and H. Kuhn. 1986. Enzymology and physiology of reticulocyte lipoxygenase: comparison with other lipoxygenases. *Advances in Enzymology and Related Areas of Molecular Biology 58*: 191-272.

Vick, B.A. and D.C. Zimmerman. 1987. Oxidative systems for modification of fatty acids: the lipoxygenase pathway. In: *The biochemistry of plants: a comprehensive treatise, vol. 9* (P.K. Stumpf, ed.). Academic Press, Orlando, FL, USA. pp. 53-90.

BIOLOGICAL ROLE OF PLANT LIPIDS
P.A. BIACS, K. GRUIZ, T. KREMMER (eds)
Akadémiai Kiadó, Budapest and Plenum Publishing
Corporation, New York and London, 1989

BIOSYNTHESIS OF HOMOSERINE LIPID

N. Sato

Department of Botany, Faculty of Science, University of Tokyo, Hongo, Tokyo, 113, Japan

INTRODUCTION

1,2-Diacylglyceryl-*O*-4'-(*N*,*N*,*N*-trimethyl)homoserine (DGTS) is a glycerolipid which is widely distributed in lower green plants (pteridophytes, bryophytes, and chlorophytes) and some chromophytes (Brown and Elovson 1974, Eichenberger 1982, Sato and Furuya 1984, 1985). Although the zwitterionic nature of this lipid suggests that it plays a role similar to the one played by PC, it is not clearly understood why it coexists in small amounts with PC in most species of pteridophytes and bryophytes. Although the intracellular localization of DGTS has not been systematically studied, DGTS has been found in the thylakoid membrane (Janero and Barrnett 1982) and the envelope membrane (Mendiola-Morgenthaler et al. 1985) of chloroplast in *Chlamydomonas*. It was also found as the major polar lipid in the plasma membrane of *Dunaliella* (Sheffer et al. 1985). Since the C-1 and C-2 positions of DGTS are occupied by palmitic acid and C_{18} or C_{20} unsaturated acids, respectively, the site of its synthesis is supposed to be in the cytoplasm but not in the plastid (Sato and Furuya 1983, Giroud et al. 1988).

Little is known about the biosynthesis of DGTS. Although tracer experiments have been performed to label DGTS by feeding algal cells with radioactive fatty acids (Schlapfer and Eichenberger 1983), nothing has been known about the biosynthesis of the head group of DGTS. As the first step toward the elucidation of the mechanism of the biosynthesis of DGTS, I have tried to identify precursor(s) to the head group of DGTS (Sato 1988, Sato and Kato 1988).

PRELIMINARY COMPETITION EXPERIMENT

As a preliminary survey, various C_4 amino acids were exogenously administered to intact cells of *Chlamydomonas* to see whether these compounds could compete with the photosynthetically assimilated ^{14}C (given as sodium bicarbonate) in the labelling of DGTS. The labelling of both polar head group (glyceryl-trimethylhomoserine) and acyl groups was determined. L-Homoserine (10 mM) lowered the labelling of the polar group to 85% and that of the acyl group to 57%. This result suggests that homoserine was degraded within the cell and used as a general carbon source. A similar result was obtained with L-threonine. L-Methionine (10 mM), on the other hand, lowered the labelling of the polar group of DGTS to 76% without affecting the labelling of the acyl groups (96% of the control). This result suggests that L-methionine was a precursor to the head group of DGTS (Fig. 1).

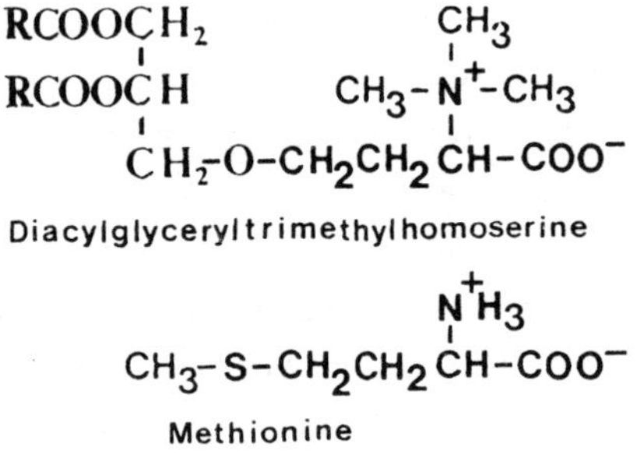

Fig. 1. Structures of DGTS and methionine

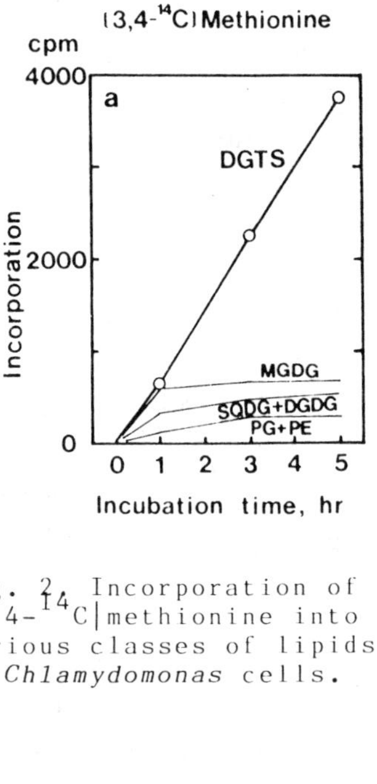

Fig. 2. Incorporation of |3,4-^{14}C|methionine into various classes of lipids in *Chlamydomonas* cells.

Table 1. Incorporation of |3,4-^{14}C|methionine (dpm)

DGTS	total	12333	(100%)
	acyl	692	(6%)
	polar*	11641	(94%)
MGDG	total	1116	(100%)
	acyl	999	(90%)
	polar**	117	(10%)

* glyceryltrimethylhomoserine
** glycerol + methyl galactoside

INCORPORATION OF METHIONINE IN *CHLAMYDOMONAS*

When the cells of *Chlamydomonas reinhardtii* strain C-238 were incubated with [3,4-^{14}C]methionine, a marked incorporation of the radioactivity into DGTS was observed (Fig. 2). The incorporation continued over the 5 h of experimental period. A low but significant amount of labelling was also detected in other classes of lipids. The location of the radioactivity within the DGTS molecule was studied by methanolysis and oxidation (Table 1). The radioactivity of DGTS was primarily (94%) localized in the polar group. When the polar group was oxidized with periodate, which is known to remove the glycerol moiety (Brown and Elovson 1974), most (88%) of the radioactivity remained. The radioactivity in MGDG, on the other hand, was localized in the acyl groups. The radioactivities in the acyl groups of DGTS and MGDG were at a similar level. This labelling of the acyl groups of lipids could be accounted for by the degradation of radioactive methionine to C_2 units, which was in turn utilized for *de novo* fatty acid synthesis. Essentially the same result was obtained with [1-^{14}C]methionine. These results indicate that the C_4 backbone of methionine is a precursor to the polar group of DGTS, probably its C_4 moiety. The incorporation of the amino group of methionine to DGTS has not been tested because of technical difficulty.

A similar experiment with [*methyl*-^{14}C]methionine showed that the S-methyl group of methionine was a precursor to the polar group of DGTS, presumably its N-methyl groups. It is interesting to note that the efficiency of incorporation of methionine into DGTS was remarkably high: 13% of the 3,4-labelled and 25% of the methyl-labelled methionine which were taken up by the cells were incorporated into DGTS during the 5 h period of incubation. This result attests to the major role of methionine in the synthesis of DGTS.

INCORPORATION OF METHIONINE IN LIVERWORT CELLS

Chlamydomonas is exceptional in that it contains a very high level of DGTS but no detectable PC (Sato and Furuya 1985, Giroud et al. 1988). In most of the organisms which contain DGTS, the content of DGTS is relatively low and a higher level of PC is present (Sato and Furuya 1985). It is therefore interesting to test if the same mechanism is operating in the synthesis of DGTS in these organisms. For this purpose, the cell culture of *Marchantia* is especially useful, because it is almost the only material available which is unicellular and free of contaminating bacteria or fungi.

When the liverwort cells were incubated with [3,4-^{14}C]methionine, the radioactivity was incorporated into DGTS. Although other classes of lipids such as MGDG and PC were also labelled to an appreciable extent, the specific radioactivity in DGTS was more than 10 times higher that that of PC and MGDG (DGTS was a minor component amounting to only 1.8% of total lipids). Degradation analysis showed that the radioactivity in DGTS was localized within the trimethylhomoserine moiety of the lipid. These results clearly suggest that methionine is a precursor to the polar group of DGTS in the liverwort cells, and confirm the generality of the role of methionine in the synthesis of DGTS.

REFERENCES

Brown, A. E., Elovson, J. (1974) Isolation and characterization of a novel lipid, 1(3),2-diacylglyceryl-(3)-O-4'(N,N,N-trimethyl)homoserine, from *Ochromonas danica*. Biochemistry 13: 3476-3482.

Eichenberger, W. (1982) Distribution of diacylglyceryl-O-4'-(N,N,N-trimethyl)homoserine in different algae. Plant Sci. Lett. 24: 91-95.

Giroud, C., Gerber, A., Eichenberger, W. (1988) Lipids of *Chlamydomonas reinhardtii*. Analysis of molecular species and intracellular site(s) of biosynthesis. Plant Cell Physiol. 29: 587-595.

Janero, D. R., Barrnett, R. (1982) Isolation and characterization of an ether-linked homoserine lipid from the thylakoid membrane of *Chlamydomonas reinhardtii* 137^{+}. J. Lipid Res. 23: 307-316.

Mendiola-Morgenthaler, L., Eichenberger, W., Boschetti, A. (1985) Isolation of chloroplast envelopes from *Chlamydomonas*. Lipid and polypeptide composition. Plant Sci. 41:97-104.

Schlapfer, P., Eichenberger, W. (1983) Evidence for the involvement of diacylglyceryl(N,N,N-trimethyl)homoserine in the desaturation of oleic and linoleic acids in *Chlamydomonas reinhardi* (Chlorophyceae). Plant Sci. Lett. 32: 243-252.

Sato, N.(1988) Dual role of methionine in the biosynthesis of diacylglyceryltrimethylhomoserine in *Chlamydomonas reinhardtii*. Plant Physiol. 86:931-934.

Sato, N., Furuya, M. (1983) Isolation and identification of diacylglyceryl-O-4'-(N,N,N-trimethyl)homoserine from the fern *Adiantum capillus-veneris* L. Plant Cell Physiol. 24:1113-1120.

Sato, N., Furuya, M. (1984) Distribution of diacylglyceryltrimethylhomoserine in selected species of vascular plants. Phytochemistry 23: 1625-1627.

Sato, N., Furuya, M. (1985) Distribution of diacylglyceryltrimethylhomoserine and phosphatidylcholine in non-vascular green plants. Plant Sci. 38: 81-85.

Sato, N., Kato, K. (1988) Analysis and biosynthesis of diacylglyceryl-N,N,N-trimethylhomoserine in the cells of *Marchantia* in suspension culture. Plant Sci. 55: 21-25.

Sheffer, M., Fried, A., Gottlieb, H. E., Tietz, A., Avron, M. (1986) Lipid composition of the plasma-membrane of the halotolerant alga, *Dunaliella salina*. Biochim. Biophys. Acta 857:165-172.

BIOLOGICAL ROLE OF PLANT LIPIDS
P.A. BIACS, K. GRUIZ, T. KREMMER (eds)
Akadémiai Kiadó, Budapest and Plenum Publishing
Corporation, New York and London, 1989

LIPIDS OF CHLAMYDOMONAS REINHARDTII: INCORPORATION OF ^{14}C-ACETATE, PALMITATE AND OLEATE INTO DIFFERENT LIPIDS AND EVIDENCE FOR LIPID-LINKED DESATURATION OF FATTY ACIDS

Ch. Giroud and W. Eichenberger

Institut für Biochemie der Universität Bern, Switzerland

INTRODUCTION

The lipids of *Chlamydomonas reinhardtii* mainly consist of MGDG, DGDG, SQDG, PG, PE, PI and the betaine lipid DGTS. PC, however, was shown to be strictly absent from this alga (Giroud et al. 1988). The chloroplast lipids are of the prokaryotic, DGTS and PE of the eukaryotic type. Each lipid class is characterized by an individual pattern of molecular species suggesting that fatty acids are desaturated in a lipid-linked process in which the different lipids individually act as substrates for the desaturase(s). In order to provide additional evidence for this process, cells were labelled with ^{14}C-acetate, ^{14}C-palmitate or ^{14}C-oleate and the radioactivity was measured after different times in each lipid class and in their fatty acids and molecular species.

MATERIALS AND METHODS

Chlamydomonas reinhardtii 137c arg-2 mt^{+} (The Culture Centre of Algae and Protozoa, Cambridge) was cultivated autotrophically under 10'000 lx of continuous fluorescent light for 2.5 days at 26-28°C. The nutrient was medium I (Sager and Granick 1954) with 0.47 mM arginine-HCl.

Cells were incubated (cell concentration 10^{8} cells/ml) for 1 hour with either 2-^{14}C-acetate (58 mCi/mmol, 20 µCi/10^{9} cells), 1-^{14}C-palmitate (54 mCi/mmol, 5 µCi/10^{9} cells) or 1-^{14}C-oleate (56 mCi/mmol, 5 µCi/10^{9} cells). After dilution with medium I, the suspension (cell concentration 10^6 cells/ml was kept under light conditions for another 23 hours.

Lipids were extracted with methanol and then separated by 2-dimensional TLC as described by Giroud et al. (1988).

Fatty acid phenacyl esters were prepared according to Borch (1975) and separated by HPLC (Giroud et al. 1988).

Molecular species of MGDG, DGDG and DGTS were separated by HPLC and the radioactivity of the effluent was measured in a Berthold LB 505 Monitor (Giroud et al. 1988).

RESULTS AND DISCUSSION

After 1 hour of incubation with ^{14}C-acetate, the label in MGDG was mainly found in 16:0 and in C_{16} and C_{18} monoene and diene fatty acids (Fig. 1). With time, the radioactivity shifted from the less to the more unsaturated fatty acids and molecular species accumulating finally in the α18:3/16:4 major species of MGDG. Since the 18:1/16:3 and 18:1/16:4 species were strongly and the 18:2/16:0 and α18:3/16:0 only weekly labelled, it is suggested that C_{16} fatty acids of MGDG were faster desaturated than the C_{18} fatty acids. In DGDG, the label was transferred from 18:1 to α18:3 to a much lesser extent, however, from 16:0 to α16:3 and 16:4 acids (Fig. 2). Correspondingly, the label shifted from the 18:1/16:0 and 18:2/16:0 species to the α18:3/16:0 major species of DGDG indicating that in this lipid, C_{18} fatty acids are desaturated predominantly. In DGTS, no radioactivity was transferred from saturated to unsaturated C_{16} acids (Fig. 3). In contrast, in the C_{18} series, the label rapidly shifted from 18:1 to 18:3 (5,9,12) and 18:4 fatty acids. Moreover, there was no distinct transfer of radioactivity from less to more unsaturated molecular species. Incubation with ^{14}C-palmitate caused an almost equal labelling of all kinds of DGTS molecular species (Fig. 4a). In contrast, with ^{14}C-oleate as a precursor, the radioactivity clearly shifted from the less to the more unsaturated species (Fig. 4b). Similar results were obtained also with PE (not shown).

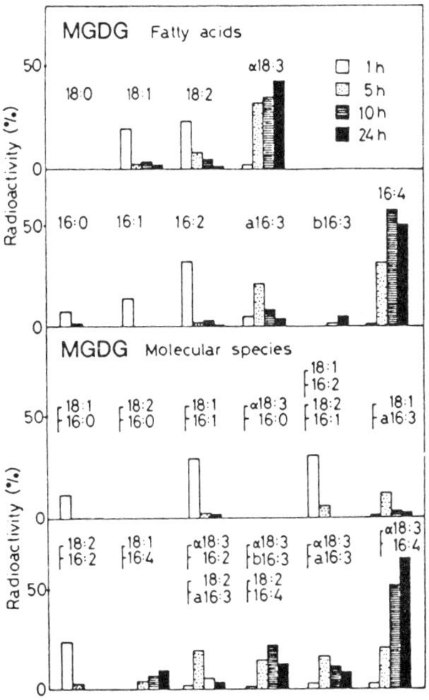

Fig. 1. Incorporation of 2-^{14}C-acetate into fatty acids and molecular species of MGDG.

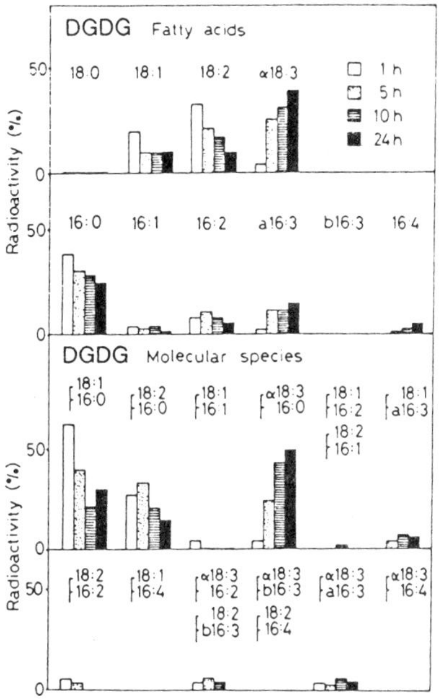

Fig. 2. Incorporation of 2-^{14}C-acetate into fatty acids and molecular species of DGDG.

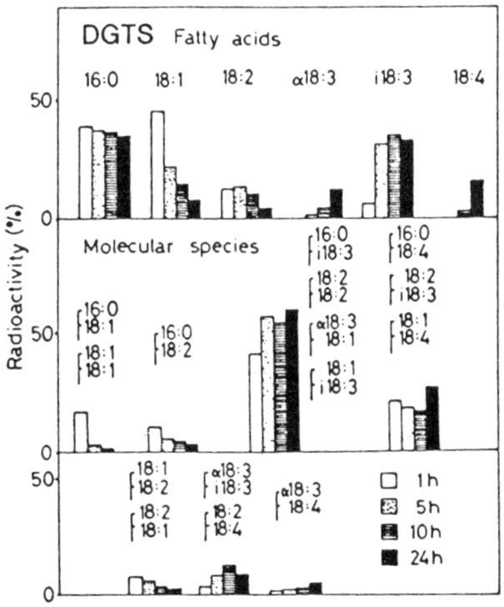

Fig. 3. Incorporation of 2-^{14}C-acetate into fatty acids and molecular species of DGTS.

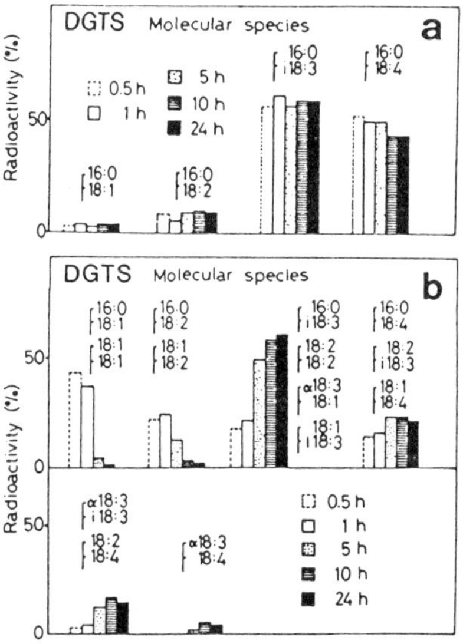

Fig. 4. Incorporation of a) 1-^{14}C-palmitate and b) 1-^{14}C-oleate into molecular species of DGTS.

a16:3 = 16:3(7,10,13); b16:3 = 16:3(4,7,10); i18:3 = 18:3(5,9,12)

Our results clearly demonstrate that in *Chlamydomonas*, fatty acid desaturation occurs in different ways in different cell compartments. In plastidal MGDG, both C_{16} (C-2 position) and C_{18} (C-1 position) fatty acids are desaturated leading to 16:4 and α18:3 acyl groups. In DGDG, only the C_{18} fatty acids (C-1 position) are desaturated, while the C_{16} acids in C-2 position remain unchanged. In DGTS and PE, both of which are of cytoplasmic origin, desaturation occurs exclusively on the C_{18} fatty acids which are located in the C-2 position, leading to 18:3(5,9,12) and 18:4(5,9,12,15) fatty acids. In the C-1 position which is mainly occupied by C_{16} fatty acids, a rapid exchange of acyl groups is strongly suggested.

The labelling kinetics of fatty acids and molecular species strongly supports the general view of a lipid-linked desaturation in which every lipid class is individually involved.

ACKNOWLEDGEMENTS

The work has been supported by the Swiss National Science Foundation.

REFERENCES

Borch, R.F. (1975) Separation of long chain fatty acids as phenacylesters by HPLC. Anal.Chem. 47: 2437-2439.

Giroud, C., Gerber, A. and Eichenberger, W. (1988) Lipids of *Chlamydomonas reinhardtii*. Analysis of molecular species and intracellular site(s) of biosynthesis. Plant and Cell Physiol. 29: in press.

Sager, R. and Granick, S. (1954) Nutritional control of sexuality in *Chlamydomonas reinhardtii*. J.Gen.Physiol. 37: 729-742

BIOLOGICAL ROLE OF PLANT LIPIDS
P.A. BIACS, K. GRUIZ, T. KREMMER (eds)
Akadémiai Kiadó, Budapest and Plenum Publishing
Corporation, New York and London, 1989

EPICUTICULAR WAX IN LUNARIA ANNUA : ASPECTS OF BIOSYNTHESIS AND SECRETION

K. Haas

Institut für Botanik, Univesität Hohenheim, Postafach 700562, D-7000 Stuttgart, FRG

INTRODUCTION

A comparative investigation was carried out on epicuticular and intracuticular waxes from silicles and leaves of Lunaria annua L. Young silicles (main axis ca 15 mm, developmental stage 1) and green premature silicles (main axis ca 40 mm, developmental stage 2) were examined using collodion films to separate epicuticular and intracuticular waxes for subsequent chemical analysis essentially as described earlier (Haas and Rentschler 1984). Waxes from the adaxial side of fully expanded leaves were also examined.

RESULTS AND DISCUSSION

The epicuticular wax of young silicles mainly consists of alkanes and free fatty acids (Table 1). Intracuticular wax composition indicates the increasing synthesis of alkanols and esters besides alkanes as dominant fraction. Aldehydes are not detectable in stage 1, they are formed in the further development of the silicles. The content of intracuticular wax remains low, a phase of enhanced wax synthesis between stage 1 and stage 2 preferably leads to increased epicuticular wax deposition. The appearance of the silicle surface concomitantly changes from glossy to glaucous.
Discrimination of wax components obviously takes place in the cuticle during the secretion process. This leads to a marked

enrichment of alkanes in the epicuticular wax, whilst higher relative amounts of alkanols and free fatty acids are retained in the intracuticular wax. This suggests their preferential partitioning in the cutin matrix (Haas and Rentschler 1984). In the leaves, no significant discrimination of wax components is visible and total wax yields are comparatively low.

Table 1. Composition (%) and yield of epicuticular and intracuticular waxes from silicles (developmental stages 1, 2) and leaves of *Lunaria annua*.

	Silicles (1)		Silicles (2)		Leaves	
	Epi	Intra	Epi	Intra	Epi	Intra
Free Fatty Acids	38	10	1	10	44	40
Free Alkanols	6	27	3	17	25	28
Aldehydes	-	-	6	5	18	19
Esters	4	15	8	7	2	4
Alkanes	52	48	82	61	11	9
Total Yield ($\mu g/cm^2$)	1.0	2.2	20.5	3.8	2.9	4.2

Table 2. Relative amounts of major esters in waxes from silicles and leaves of *Lunaria annua*.

	Homologues (%)								
	40	42	44	46	46:1	48	48:1	50	50:1
Silicles (stage 1)									
Epicuticular	3	22	41	15	-	9	-	4	-
Intracuticular	6	12	39	27	-	6	-	3	-
Silicles (stage 2)									
Epicuticular	1	2	13	16	10	7	32	2	9
Intracuticular	1	2	18	24	6	10	24	1	7
Leaves									
Epicuticular	14	10	12	25	-	17	-	8	-
Intracuticular	12	9	13	26	-	18	-	7	-

The dominant homologue of alkanes is C_{29} in waxes from silicles and leaves of all developmental stages examined. Only saturated esters in the range of C_{40}- C_{50} are found in the leaves and in young silicles (Table 2). However, high relative amounts of esters of monounsaturated fatty acids 22:1 and 24:1 are contained in epicuticular and intracuticular waxes from older silicles of *Lunaria annua* (Table 3). These unsaturated fatty acids are constituents of the seed triacylglycerols in unusual high percentages (Mukherjee and Kiewitt 1986). In the silicles, the unsaturated acyl moieties of wax esters are probably derived from the same elongating system which produces the corresponding acyl moieties of the seed triacylglycerols. Synthesis of wax esters may therefore take place using acyl-CoAs (Avato 1984) which are suggested also to be involved in tracylglycerol formation (Mukherjee and Kiewitt 1984).

Table 3. Relative amounts of major fatty acids, alkanols and aldehydes in epicuticular waxes from silicles (developmental stages 1, 2) of *Lunaria annua*.

	Homologues (%)									
	16	18	20	22	22:1	24	24:1	26	28	30
Stage 1										
Free Fatty Acids	19	18	14	8	-	4	-	12	9	13
Ester Fatty Acids	6	3	21	44	-	11	-	10	-	-
Free Alkanols			4	70	-	12	-	6	3	-
Ester Alkanols			1	54	-	23	-	14	4	-
Stage 2										
Free Fatty Acids	6	3	2	2	2	3	4	8	18	44
Ester Fatty Acids	3	2	12	25	4	4	43	1	2	2
Free Alkanols				5	-	52	-	25	7	3
Ester Alkanols				18	-	49		21	3	1
Aldehydes						2	-	7	38	46

Furthermore, chain length distributions of the saturated wax constituents are indicative of at least 2 elongating systems: one providing carbon chains in the range of C_{20}- C_{24}, and a system which produces the very long chains in the range of C_{26}- C_{30} and higher. However, further differentiation of elongating systems is likely to occur (Avato et al. 1985, Lessire et al. 1985). A developmental tendency to higher chain lengths often found within wax constituent classes is also visible in the silicle waxes. This may result from different elongases which subsequently act during ontogenetic development.

REFERENCES

Avato, P. (1984) Synthesis of wax esters by a cell-free system from barley (*Hordeum vulgare* L.). Planta 162, 487-494

Avato, P., Bianchi, G., Salamini, F. (1985) Absence of long chain aldehydes in the wax of the glossy 11 mutant of maize. Phytochemistry 24, 1995-1997

Haas, K., Rentschler, I. (1984) Discrimination between epicuticular and intracuticular wax in blackberry leaves: ultrastructural and chemical evidence. Plant Sci. Lett. 36, 143-147

Lessire, R., Juguelin, H., Moreau, P., Cassagne, C. (1985) Elongation of acyl-CoAs by microsomes from etiolated leek seedlings. Phytochemistry 24, 1187-1192

Mukherjee, K.D., Kiewitt, I. (1984) Changes in fatty acid composition of lipid classes in developing mustard seed. Phytochemistry 23, 349-352

Mukherjee, K.D., Kiewitt, I. (1986) Lipids containing very long chain monounsaturated acyl moieties in seeds of *Lunaria annua*. Phytochemistry 25, 401-404

BIOLOGICAL ROLE OF PLANT LIPIDS
P.A. BIACS, K. GRUIZ, T. KREMMER (eds)
Akadémiai Kiadó, Budapest and Plenum Publishing
Corporation, New York and London, 1989

PLASMA MEMBRANE BIOGENESIS IN HIGHER PLANTS: THE LIPID ROUTES

C. Cassagne, R. Lessire, P. Moreau, J.-J. Bessoule, P. Bertho and L. Maneta Peyret

IBCN-CNRS, Université de Bordeaux II. 1 Rue Camille Saint-Saëns, 33077 Bordeaux Cédex, France

INTRODUCTION

The study of the biogenesis of the plasma membrane in higher plants has lagged behind that of animal cells, though the endomembrane concept, as well as the indications concerning the membrane traffic, resulted from biochemical and morphological studies devoted to both plant and animal cells (see for example 1). In most cases, the bulk of the data available in the literature results from studies of the biosynthesis, processing, sorting and intracellular transfer of plasmalemma (or viral) glycoproteins. Compared to this situation, little is known on the *in vivo* biosynthesis and intracellular transfer of the lipids en route to the plasma membrane (2, 3).

Reasonable hypotheses for the lipid routes have been proposed and a debate has been opened as to whether these hypotheses were able to account for the variety of situations likely to be met in the various types of cells. From the debate, three main mechanisms concerning the way by which the plasma membrane may acquire its lipid components have emerged (1-3).

1 - The vesicular transport, now very well exemplified in the case of some glycoproteins. In this case it is also required that membrane lipids be transferred to the plasma membrane during the membrane flow. Which lipids, and how, has not been matter of concern, and the question remains almost entirely unanswered ;

2 - However, the vesicular transfer is not the only potential mechanism of lipid acquisition by the plasma membrane. The cytosolic transport of monomers, either spontaneous, or by means of lipid transfer proteins (LTP) has been hypothesized from *in vitro* experiments (4-6). The involvement of LTP is a so plausible -and consequently so popular- hypothesis that it is cited in any review as one of the favorite modes of transfer of intracellular lipids (2,3). But, however satisfying it may be as a working hypothesis, this mechanism is still awaiting the experimental data which could support the role of the LTPs in the process of *in vivo* plasma membrane biogenesis. On the other hand, it is assumed that the spontaneous, non protein-dependent, lipid transfer is extremely slow, if present at all, and is thus assigned no role in intracellular lipid transfer. Conversely, as excellently pointed out by SLEIGHT (2), much of the indirect evidence suggesting that LTPs are active in intracellular lipid transport consists of the finding of the rapid lipid translocation.

3 - The lateral diffusion could also be involved in plasma membrane biogenesis : the phospholipids could move in the plane of a membrane, and then, through a transient interconnection, reach a new membrane. As in the preceding mechanisms there is still a lack of *in vivo* evidence (2,3).

It is remarkable, as far as lipids are concerned, that there are so many well-designed hypotheses supported by so few conclusive data and this discrepancy is probably one of the challenges to be faced in the next decade in order to improve our knowledge on plasma membrane biogenesis in higher plants. In addition, these mechanisms, even if they can be taken for established, are not likely to coexist at the same rate throughout the cell life, or to involve the different lipid classes at the same extent in the various domains of the plasmalemma. It follows that there is a danger of over interpretation of the results in favour of one mechanism or an other. It has been estimated that there are more than 1,000 chemically distinct phospholipid species in eukaryotic cells. More than 200 are expected in the plasma membrane. Added to the fact that there are very few lipid markers of the plasmalemma, and that obviously these lipids will be preferentially studied, the chance is that the lipid "markers", considered as representative of a very complex phenomenon will only shed some light on their own fate, which may differ largely from that of the bulk of the other membrane lipids. A strategy to analyze the necessary lipid involvement must therefore consider as many as possible of the above mechanisms and try to evaluate their respective involvement.

DEFINITION OF THE APPROACHES

Basically, the strategy for studying the lipid routes of plasma membrane biogenesis implies that two complementary approaches be developed at the same time :
a) A molecular approach, which aims to study the *in vitro* biosynthesis of the plasma membrane lipids. This includes the study of the mechanisms of lipid formation, as well as the determination of the site(s) of their biosynthesis. The study, which in a first phase may be restricted to lipid "markers", has to be extended to all the lipids of the plasma membrane including the most ubiquitous.
A second step of this approach is an extensive study of the enzymes responsible for the lipid biosynthesis. As most of the synthesizing machinery is located in membranes, this step requires a solubilization and purification of the enzymes followed by a theoretical and experimental study of the functioning of the reconstituted enzymes and the modulation of their activity by the surrounding lipids.
b) A cellular approach will investigate the *in vivo* biosynthesis and the intracellular transfer of the lipids to the plasmalemma. The method to study these events *in vivo* has been described (7). Briefly, the method implies pulse-chase experiments, membrane subfractionation, purification of the various membrane compartments (and particularly the plasmalemma), and a subsequent analysis of the radiolabeled lipids. This procedure should allow to show the occurrence of intermembrane transfer of the lipids to the plasma membrane and, combined with the use of perturbants of membrane traffic, should help elucidate the eventual involvement of the Golgi complex in these transfers.

THE LEEK SYSTEM

It was shown years ago that C20-C24 fatty acids are chiefly located in the plasma membrane of leek cells, but are synthesized in (by) the endomembrane system. Two acyl-CoA elongases were detected, one elongating C18-CoA chiefly in the ER, the second elongating C20-CoA chiefly in the GA. The elongases were solubilized and partially purified. The kinetic parameters have been studied and the role of the surrounding membrane lipids on the C20-C24 fatty acid biosynthesis is being investigated (see also LESSIRE et al., this volume ; BESSOULE et al., this volume).
In vivo , these fatty acids are quantitatively transferred from a "light" membrane fraction to a "heavy" one. This transfer, in terms of kinetics of pulse-chase, starts from a ER-containing fraction and, through lipid-rich vesicular membranes, allows the C20-C24 fatty acids to reach the plasma membrane after a lag of about 30 min.
The use of monensin, a carboxylic ionophore known to disturb the vesicular traffic of proteins en route to the plasma membrane at the level of the GA budding gave the first evidence that C20-C24 fatty acids may partly transit through the GA (see also BERTHO et al., this volume).

PERSPECTIVES

Taking advantage of the improvements of the methods and concepts, based on results such as those summarized above, important breakthroughs are to be expected in the next decade in three main directions.

1 - The enzymes of the plasma membrane lipid biosynthesis.
From a biochemical point of view, it is important that the functioning of the enzymes synthesizing the lipids of the plasma membrane be largely reinvestigated in terms of lipid and protein interactions. The study is complicated by the fact that most of these enzymes are membrane-bound and often multienzyme complexes. The acyl-CoA elongases are a typical example of the difficulties encountered in this research area : the elongation process is far from clear, its regulation is unknown, the modulation of the synthesis by the membrane lipids is just beginning to be understood, but, above, all there is an absolute lack of data on the synthesis of these enzymes and its regulation and on how these events are integrated in the general pattern of the plasma membrane biogenesis.
2 - *In vivo* analysis of plasma membrane biogenesis
The transfer of C20-C24 fatty acids (and probably of shorter fatty acyl chains) involves, at least in part, a transit through the Golgi apparatus. To what extent, and associated with which polar heads remains to be elucidated. However, one has to keep in mind that, even if the forthcoming work definitely establishes this point, this will not imply that all lipids must use this pathway and that a cytosolic transport via LTPs is not also operative. It is thus extremely important that an *in vivo* approach able to study this pathway be developed, and that a visualization of the intracellular journey of the lipids destined to the plasmalemma be obtained *in situ* .
A possible answer to the question is the use of fluorescent lipids (2). Though there are limitations to this technique, which is probably more adapted to the study of endocytosis, it is likely that it will bring new informations in a near future. Another way of *in situ* visualization of the lipids is, in theory, the immunocytolocalization with antibodies raised against lipids. Antibodies against fatty acids have been prepared and, under well-designed experimental conditions, may recognize fatty acyl chains *in situ* (8). Specific anti-PS antibodies, recently prepared in the laboratory (9), could be of particular interest, as it is known that in higher plants C20-C24 fatty acids are chiefly found in PS (10). These new tools, added to the immunolocalization of LTPs *in situ* , could help to explore the role of LTPs in plant cells.
3 - *In vitro* cell-free analysis of membrane transfers
The complexity of the processes involved in plasma membrane biogenesis probably exceeds the possibilities of the mere *in vivo* analysis. In theory, cell-free systems should facilitate the study of the individual steps of the interaction between the various cell comparments. ROTHMAN et al. (11) provided the first evidence of the feasability of the approach and, recently, NOWACK et al. (12) described a cell-free system able to give informations on the budding from ER and the vesicle fusion with the GA. This system is suitable for the study of protein transfer, but could also be adapted to the case of the lipids.
In conclusion, all of these approaches aiming to investigate the overall plasma membrane biogenesis from the biosynthesis of the lipid enzymes, to the insertion of their products into the plasmalemma, may help to shed some light on the lipid routes, for which so much has been supposed and so little is known.

Abbreviations :
LTP : lipid transfer protein ; ER : endoplasmic reticulum ; GA : Golgi apparatus ; PS : phosphatidylserine

REFERENCES

1- Morré, D.J., Kartenbeck, J., Franke, W.W. (1979). Biochim. Biophys. Acta, 559:71-152

2 - Sleight, R.G. (1987). Ann. Rev. Physiol., 49:193-208

3 - Dawidowicz, E.A. (1987). Ann. Rev. Biochem., 56:43-61

4 - Rickers, J., Tober, I., Spener, F. (1984). Biochim. Biophys. Acta, 794:313-319

5 - Kader, J.C., Douady, D., Grosbois, M., Guerbette, F., Vergnolle, C. (1984). In "Structure, Function and Metabolism of Plant Lipids" (P.A. Siegenthaler, W. Eichenberger, eds), Elsevier, Amsterdam, pp. 283-290

6 - Watanabe, S., Yamada, M. (1986). Biochim. Biophys. Acta, 876:116-123

7 - Moreau, P., Juguelin, H., Lessire, R., Cassagne, C. (1988). In "Cell free analysis of membrane traffic" (D.J. Morré, K.E. Howell, G.M.W. Cook, W.H. Evans, eds), Alan Liss Inc., New-York, p. 303

8 - Maneta-Peyret, L., Onteniente, B., Geffard, M., Cassagne, C. (1986). Neurosci. Lett., 69:121-125

9 - Maneta-Peyret, L., Bessoule, J.J., Geffard, M., Cassagne, C. (1988). J. Immun. Methods, 108:123-127

10 - Murata, N., Sato, N., Takahashi, N. (1984). Biochim. Biophys. Acta, 795:147-150

11 - Rothman, J.E. (1988). In "Cell free analysis of membrane traffic" (D.J. Morré, K.E. Howell, G.M.W. Cook, W.H. Evans, eds). Alan Liss Inc, New-York, pp. 311-316

12 - Nowack, D.D., Morré, D.M., Paulik, M., Keenan, T.W., Morré, D.J. (1987). Proc. Natl. Acad. Sci. USA, 84:6098-6102

BIOLOGICAL ROLE OF PLANT LIPIDS
P.A. BIACS, K. GRUIZ, T. KREMMER (eds)
Akadémiai Kiadó, Budapest and Plenum Publishing
Corporation, New York and London, 1989

MONENSIN: A TOOL TO STUDY INTRACELLULAR TRANSPORT OF LIPIDS IN HIGHER PLANTS

P. Bertho, P. Moreau, Hélène Juguelin and C. Cassagne

IBCN-CNRS, Université de Bordeaux II, 1 Rue Camille Saint-Saëns, 33077 Bordeaux Cédex, France

INTRODUCTION

The biogenesis of the plasma membrane has been extensively studied in both animal and plant cells by following the biosynthesis, maturation and intracellular transport of glycoproteins. However, little is yet available concerning the intracellular transfer of lipids to the plasma membrane, and particularly in the plant cells.

A methodology was devised to investigate intermembrane transfer of lipids and very long chain fatty acids (VLCFA) in etiolated leek seedlings in vivo (1).

Pulse-chase experiments, followed by membrane subfractionation of microsomal pellets on sucrose gradients and the isolation of the plasma membrane by phase partition, have demonstrated the transfer of some lipids and of the VLCFA to the plasma membrane (2). The results suggested that an endoplasmic reticulum - Golgi - plasma membrane pathway could be one of the routes taken by the lipids to be transferred to the plasma membrane (3). The carboxylic ionophore monensin could be a particularly good tool to test this hypothesis, since it is known to disturb the traffic of glycoproteins at the level of the Golgi complex (4). The morphological consequences reported following the use of monensin are dilatation and vacuolisation of the Golgi cisternae.

For such a study, it was necessary to define the experimental conditions allowing the drug to penetrate into the seedlings and to determine the concentrations of monensin for which only slight effects on lipid metabolism levels were observed.

METHODOLOGY

Penetration of monensin into leek seedlings.

Extraction, isolation and quantitation techniques were devised to study the penetration of monensin into leek seedlings. As a preliminary step, the complete solubility of monensin in chloroform / methanol (2 :1, v/v) was demonstrated. The extraction of monensin from the seedlings was carried out with this mixture,utilized classically for extracting lipids from biological material. The best procedure for isolating monensin was obtained by tlc, with the solvent system composed of hexane/diethyloxide/acetic acid (20:80:2, v/v). In this system, the drug migrates with an hRf of 27.8 ± 0.7 (mean value ± sd, n=35). The resolution of monensin (Fig.1) allowed us to quantitate the drug by a densitometric technique capable of detecting as little as 50 pmoles of monensin.

Fig.1 : Tlc separation of monensin

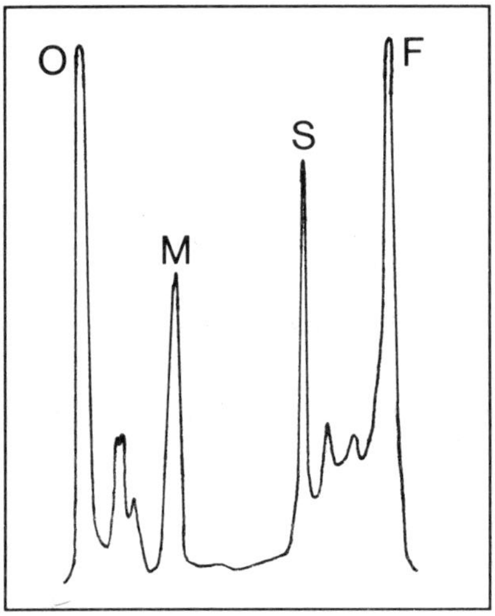

The resolution of monensin was carried out on silica gel plates. Lipids and monensin were visualized by copper acetate / phosphoric acid charring and scanned at 366 nm.
O : origin ; M : monensin
S : sterols ; F : front

The penetration of monensin into the seedlings, at various drug concentrations and incubation times, was studied. The quantity of monensin taken up by the seedlings increased with the incubation period up to 30 min, after which time it reached a plateau. The ratios of the amount of internalized monensin to the total quantity of monensin ranged from 0.04 to 0.08, and appeared to be independant of the drug concentration (between 10 and 350 μM). Moreover, in vivo and in vitro experiments showed that (i) there is only a small binding of the drug to microsomal membranes, leading to a monensin to lipid molar ratio of 10^{-4} in these membranes, and (ii) that the drug was not degraded within the cells up to 120 min.

Lipid metabolism and monensin.

The lipid metabolism was studied in the presence of monensin in order to determine the quantities of drug that induce a decrease of the incorporation of labelled precursor by the seedlings. After incubation of the seedlings with (1-^{14}C)acetate and at various concentrations of monensin (0.1 to 10 μM) for

30 and 120 min, the lipids were extracted and analysed. Monensin up to 10 µM had no effect on the global incorporation of radioactivity into the lipids for a 30 min incubation, whereas, for a 120 min incubation, inhibitions of 5 and 25% were recorded for monensin concentrations of 0.5 and 5 µM respectively.The quantitative analysis by tlc of the different lipid classes (phosphatidylcholine, phosphatidylethanolamine, phosphatidylserine, phosphatidic acid, diglycerides, sterols ...) did not exhibit any marked changes in the label distribution of the various lipids. Monensin, at concentrations having little effect on acetate incorporation into the lipids, was further used to investigate the involvement of the Golgi apparatus in the intracellular transfer of lipids.

Intracellular lipid distribution and monensin.

Two batches of seedlings were incubated with (1-^{14}C)acetate for 30 min, with or without 2µM monensin respectively. After homogenization and density gradient subfractionation of the microsomal pellets, the total lipid and fatty acid labels were determined.

Table 1 : Effect of monensin on the distribution of the lipid label in the different subcellular membrane fractions.

Membrane fractions g.cm -3	Characterization according to ref. 2 & 3	lipid label		fatty acid label	
		- mon	+ mon	- mon	+ mon
1.08 / 1.09	Intermediary membranes E.R. and Golgi markers	275	236	177	169
1.12	Main site of lipid synthesis E.R. markers	146	150	79	126
1.13 / 1.14	Enriched Golgi fraction Golgi marker	65	111	36	83
1.16 / 1.18	Contaminated P.M. P.M., E.R. and Golgi markers	66	77	43	53
Whole Gradient	Microsomes	98	114	69	91

Two batches of seedlings were incubated with (1-14C) acetate for 30 min, with or without 2 µM monensin . After subcellular fractionation , the total lipid and fatty acid labels and protein content of the different fractions were determined. The results are expressed as cpm.µg-1.

The specific radioactivities of lipids and fatty acids from each previously characterized fraction (2, 3) were calculated. Monensin leads to an accumulation of neosynthesized lipids in a Golgi-enriched fraction. Moreover, in the presence of monensin, the fatty acid label, analyzed by radio - glc , exhibits a much more important accumulation in this fraction (see table 1).

PROSPECTS

Using pulse-chase experiments, some authors hypothesized an intracellular lipid transport by membrane flow (3, 5, 6), but did not answer the question of the implication of the Golgi apparatus. The preliminary results, obtained by the methodology described in this paper, indicate that some lipids could be routed through the Golgi complex prior to their subsequent distribution throughout the cell.
This methodology appears to be an appropriate tool for investigating, both quantitatively and qualitatively, the role of the Golgi complex in intracellular lipid transport in higher plant cells.

ABBREVIATIONS

glc : gas-liquid chromatography
tlc : thin-layer chromatography

REFERENCES

(1) Moreau, P., Juguelin, H., Lessire, R. & Cassagne, C. (1986). Intermembrane transfer of long chain fatty acids synthesized by etiolated leek seedlings. Phytochemistry 25, 387-391.
(2) Moreau, P., Juguelin, H., Lessire, R. & Cassagne, C. (1988). Plasma membrane biogenesis in higher plants : in vivo transfer of lipids to the plasma membrane. Phytochemistry 27, 1631-1638.
(3) Moreau, P., Bertho, P., Juguelin, H. & Lessire, R. (1988). Intracellular transport of very long chain fatty acids in etiolated leek seedlings. Plant Physiol. Biochem. 26, 173-178.
(4) Tartakoff, A.M. (1983). Perturbation of the structure and function of the Golgi complex by monovalent carboxylic ionophores. Methods in Enzymol. 98, 47-59.
(5) Mills, J.T., Furlong, S.T. & Dawidowicz, E.A. (1984). Plasma membrane biogenesis in eukaryotic cells : translocation of newly synthesized lipids. Proc. Natl. Acad. Sci. USA 81, 1385-1388.
(6) De Silva, N.S. & Siu, C.H. (1981). Vesicle-mediated transfer of phospholipids to plasma membrane during cell aggregation of *Dictyostelium discoideum*. J. Biol. Chem. 256, 5845-5850.

BIOLOGICAL ROLE OF PLANT LIPIDS
P.A. BIACS, K. GRUIZ, T. KREMMER (eds)
Akadémiai Kiadó, Budapest and Plenum Publishing
Corporation, New York and London, 1989

TWO MECHANISMS COULD BE INVOLVED IN LIPID BIOSYNTHESIS IN SUNFLOWER SEEDS

M. Mancha, J.M. García and R. Garcés

Instituto de la Grasa, Apartado 1078, E-41080, Sevilla, Spain

INTRODUCTION

Developing sunflower seeds of normal genotype and "high oleic acid" mutant (90% oleic acid, 1% linoleic acid) (1) have been used to study the incorporation of $[2\text{-}^{14}C]$ acetate into neutral and polar lipids as a function of the incubation temperature.

MATERIALS AND METHODS

"High oleic acid" sunflower seeds were suplied by J.M. Fernández-Martínez (CIDA, Córdoba, Spain). This material and the normal cultivar HA-89 were grown at 34/20ºC (day/night), light intensity 300 $\mu Em^{-2}s^{-1}$ with 14 h photoperiod. Developing sliced seeds (0.3 g) were incubated with $[2\text{-}^{14}C]$ acetate (4 μCi; 57 mCi/mmol) for 4 h at the indicated temperature. Lipids were extracted into hexane-isopropanol and fractionated by TLC on silica gel into triacylglycerols (TAG), diacylglycerols (DAG) and polar lipids (PL). Fatty acid methyl esters were separated by TLC on silica gel-$AgNO_3$ into saturated, monoenes and dienes. The incorporation into lipid classes and different types of fatty acids was determined by scintillation counting and TLC-linear analyzer.

Results were expressed as nmols of acetate incorporated / g fresh tissue.

RESULTS AND DISCUSSION

The incorporation of $[2\text{-}^{14}C]$ acetate into total lipids of both normal genotype and "high oleic" sunflower seeds of 25 days after flowering (DAF) showed two peaks at 15ºC and 30-35ºC respectively and a minimun at 25ºC (results not shown). Fig. 1 shows the incorporation of acetate into TAG, DAG and PL. The effect of the incubation tempe-

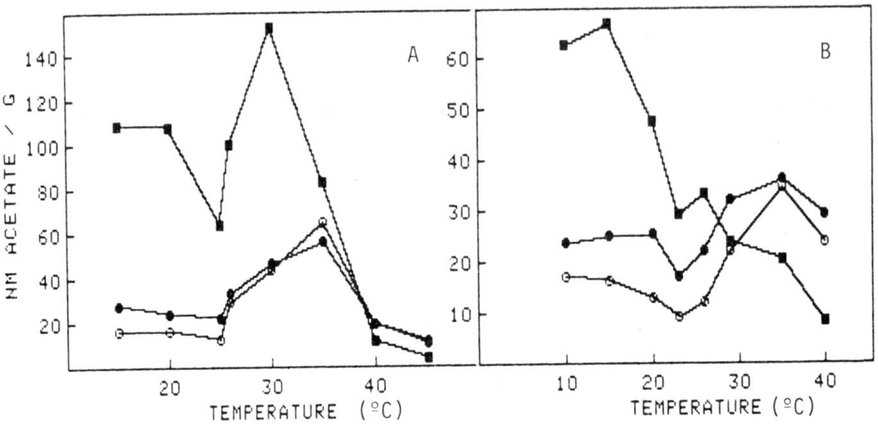

Fig. 1.- Incorporation of [2-^{14}C] acetate into TAG (■), PL (●), and DAG (○) from high oleic (A) and normal (B) sunflower seeds.

rature on DAG and PL synthesis was similar in both genotypes, the maximun of incorporation being at 35°C. However, the incorporation into TAG depend on the genotype. In the normal seeds the synthesis of TAG took place preferently at low temperature (15°C). On the contrary, in the mutant seeds the same peak was observed at 15-20°C although the main activity was detected at 30°C. The incorporation into fatty acids also showed differences (results not shown). As expected in normal seeds oleic acid was synthesized preferently at low temperature while the saturated were synthesized at high temperature. However, in the high oleic mutant seeds the synthesis of oleic showed a peak at 15-20°C although the main activity was again at 30°C while the synthesis of saturated acids was little affected by temperature. In the mutant seeds oleic acid was not desaturated to linoleic and apparently was synthesized and incorporated into TAG acording to two different mechanisms. The first one was common to the normal seeds and the second was characteristic of the "high oleic" genotype. These results suggest that two mechanisms could be involved in fatty acids and lipid synthesis in sunflower seeds.

REFERENCE

1. Fernández-Martínez, J., Jiménez-Ramírez, A., Domínguez-Giménez, J. Alcántara, M. 1986 Temperature effect on the oleic and linoleic acids of three genotypes in sunflower. Grasas y Aceites 37,326-331.

BIOLOGICAL ROLE OF PLANT LIPIDS
P.A. BIACS, K. GRUIZ, T. KREMMER (eds)
Akadémiai Kiadó, Budapest and Plenum Publishing
Corporation, New York and London, 1989

BIOCHEMICAL CHARACTERIZATION OF A HIGH OLEIC ACID MUTANT FROM SUNFLOWER

R. Garcés, J.M. García and M. Mancha

Instituto de la Grasa, Apartado 1078, E-41080, Sevilla, Spain

INTRODUCTION

The lipids of normal and a high oleic acid mutant (1) sunflower seeds have been biochemically characterized. "In vivo" and "in vitro" incubations with $[1-^{14}C]$ oleate showed a lack of oleate desaturase activity in developing mutant seeds.

MATERIALS AND METHODS

High oleic acid sunflower seeds were supplied by J.M. Fernández-Martínez (CIDA, Cordoba, Spain). This material and the normal cultivar RHA-274 were grown at 34/22ºC or 21/14ºC (day/night), light intensity 300 $\mu E\ m^{-2}\ s^{-1}$, with 14 h photoperiod. Seed lipids were extracted into hexane-isopropanol and fractionated by TLC into triacylglycerols (TAG), diacylglycerols (DAG) and polar lipids (PL). Fatty acid methyl esters were analysed by GLC or separated on silica gel-$AgNO_3$ plates. Seeds were incubated with $[1-^{14}C]$ oleate. Microsomes were obtained by differential centrifugation and incubated with $[1-^{14}C]$ oleoyl-CoA.

RESULTS AND DISCUSSION

The expression of the high oleic character took place exclusively in the seed tissues of the mutant during the active synthesis of reserve lipids. No other tissues like leaf, stem, root, pollen and even hull or seed membrane had high levels of oleic acid. Young seeds of both normal and high oleic mutant had similar fatty acids composition in the different lipid classes (table 1). However, the fatty acid composition of adult seeds depended on the

Table 1.- Fatty acid composition of lipid classes from high oleic acid and normal sunflower developing seeds.

			High oleic mutant mol %				Normal genotype mol %			
DAF*	Growth temperature	Lipid	16:0	18:0	18:1	18:2	16:0	18:0	18:1	18:2
10	34/22ºC	TAG	22.9	10.5	34.8	31.7	25.8	14.4	32.3	27.4
		PL	19.8	6.4	34.0	39.8	15.3	7.7	26.8	50.0
40	34/22ºC	TAG	3.4	6.0	89.3	1.2	5.4	3.8	59.9	30.8
		PL	7.5	5.3	83.9	3.3	11.4	7.1	44.4	37.0
40	21/14ºC	TAG	3.1	4.9	86.8	5.2	5.5	5.7	26.1	62.6
		PL	5.9	3.1	82.6	8.2	13.2	8.2	18.3	60.2

* Days after flowering.

genotype and growth temperature. At high temperature (34/22ºC) the main component was oleic acid in both genotypes. At low temperature however, linoleic acid was the main component in the normal seeds being oleic acid the main one in the mutant. In these seeds TAG contained always less linoleic acid than PL.

"In vivo" incubations of young seeds (less than 15 DAF) with [1-^{14}C] oleate showed preferential incorporation into PL and similar levels of desaturation in both genotypes. Seeds of more than 15 DAF incorporated fatty acids mainly into TAG. Normal seeds showed a maximun of oleate desaturation at 25 DAF while mutant seeds lacked this capacity.

"In vitro" incubations of microsomes with [1-^{14}C] oleoyl-CoA were in agreement with "in vivo" experiments. Microsomes from mutant seeds lacked oleate desaturase activity after 15 DAF.

REFERENCE

1. Fernández-Martínez, J., Jiménez-Ramírez, A., Domínguez-Giménez, J. and Alcántara, M. 1986 Temperature effect on the oleic and linoleic acids of three genotypes in sunflower. Grasas y Aceites 37, 326-331.

BIOLOGICAL ROLE OF PLANT LIPIDS
P.A. BIACS, K. GRUIZ, T. KREMMER (eds)
Akadémiai Kiadó, Budapest and Plenum Publishing
Corporation, New York and London, 1989

STUDIES ON THE IN VIVO GLYCEROLIPID AND FATTY ACID METABOLISM IN PEA ROOTS

Kathryn F. Kleppinger-Sparace and S.A. Sparace

Plant Science Department, Macdonald College of McGill University, Ste Anne de Bellevue, Quebec H9X 1C0, Canada

INTRODUCTION

Relatively little information is available concerning patterns of lipid and fatty acid metabolism in roots in comparison to leaves. We report here our findings on the *in vivo* lipid metabolism from 14C-acetate and 14C-glycerol in pea roots as part of a comprehensive approach towards under-standing root lipid metabolism.

METHODS

One-cm excised root tips from 3 day old germinating pea seeds were incu-bated with 0.3mM 14C-acetate or 0.4mM 14C-glycerol for up to 5 h. Lipids were extracted and analyzed by TLC. 14C-Labelled fatty acids were analyzed by radio-GLC. Details of methodology are as described elsewhere (3).

RESULTS

Excised pea root tips actively incorporated both exogenously supplied acetate and glycerol into lipid at linear rates for the first 1.5 to 2 h which greatly diminished for longer periods (Table I). Radioactivity of both precursors was incorporated into glycerolipids (70-85%) and sterols (15-20%) with only small changes over the 5h period. Glycerolipids labelled with acetate or glycerol were predominantly PC with lesser amounts of PE, PG, PI, PA, and TAG. With acetate as the precursor, the proportion of radioactivity in PC gradually diminished from 46 to 35% and the remaining glycerolipids increased slightly. With glycerol as the precursor, PC (plus approx. 10% unresolved PI) contained a larger proportion of radioactivity than when acetate was the precursor and showed no change with time.

The radioactivity of glycerolipids from 1h incubations was recovered

TABLE I. Summary of timecourse analysis for the incorporation of 14C-acetate and 14C-glycerol into pea root glycerolipids and fatty acids.

	INCUBATION TIME, HOURS													
	.3	.7	1	1.5	2	3	5	.3	.7	1	1.5	2	3	5
	NANOMOLES/GRAMM FRESH WEIGHT													
	ACETATE							GLYCEROL						
	3	8	8	12	15	11	22	10	13	21	31	30	29	40
	GLYCEROLIPIDS, %													
PC	46	45	47	46	44	38	35	(PC+PI)						
PI	14	14	14	13	12	11	11	61	59	57	58	60	59	59
PE	14	14	13	15	16	16	17	5	5	8	9	11	11	11
PG	11	11	11	10	11	14	15	15	15	12	13	10	11	13
TG	9	9	9	9	9	11	14	4	5	7	5	6	7	8
PA	6	6	6	7	8	9	7	15	16	16	14	13	12	9
	FATTY ACIDS, %													
14:0	tr	0	1	0	tr	tr	1							
16:0	40	40	39	36	44	41	35							
18:0	24	24	20	20	19	23	25							
18:1	31	28	29	34	24	19	18							
18:2	0	0	2	3	3	2	3							
20:0	2	4	5	2	4	6	5							
22:0	3	4	4	4	5	9	14							

in their fatty acids (56%) or head group moieties (99%) when acetate or glycerol, respectively, was the precursor. Radioactive fatty acids were predominantly 16:0, 18:0 and 18:1 which showed relatively small changes. Small, but increasing proportions of 20:0 and 22:0 were also synthesized.

DISCUSSION

Rapidly growing excised pea root tips are fully competent in glycerolipid and fatty acid biosynthesis from exogenously supplied 14C-acetate or glycerol. PC is the dominantly labelled lipid with both precursors. The proportion of PC remains constant with glycerol, but diminishes by 10% with acetate. Similar observations have been made with corn roots (4). These data suggest a central role for PC in glycerolipid and fatty acid metabolism in roots and that the concepts of "18:3 plants" and "prokaryotic vs. eukaryotic" lipid metabolism described for leaf tissue (1,2) may partially apply to roots. However, peas have considerably less 18:3 and galactolipid in their root tissues. Thus, the relevance of these concepts in roots remains to be determined.

REFERENCES

1. Gardiner, S.E., P.G. Roughan, C.R. Slack. 1982. Plant Physiol. 70: 1316-1320.
2. Roughan, G., R. Slack. 1984. TIBS. 9: 383-386.
3. Sparace, S.A., R. Menassa, K.F.Kleppinger-Sparace. 1988. Plant Physiol. 87: 134-137.
4. Willemot, C., J. Labrecque. 1982. Plant Physiol. 70: 1526-1529.

BIOLOGICAL ROLE OF PLANT LIPIDS
P.A. BIACS, K. GRUIZ, T. KREMMER (eds)
Akadémiai Kiadó, Budapest and Plenum Publishing
Corporation, New York and London, 1989

THE MONOOXYGENASE PATHWAY OF LINOLEIC ACID OXIDATION IN PEA SEEDLINGS

A.N. Grechkin, T.E. Gafarova, O.S.Korolev, R.A. Kuramshin and I.A. Tarchevsky

Kazan Institute of Biology, USSR Academy of Sciences, P.O. Box 30, Kazan 420084, USSR

Lipoxygenases (LOX's) represent the most complitely investgated enzymes of lipid metabolism in plants. Animals have two more types of oxygenases, along with LOX's, involved in metabolism of polyenoic fatty acids - cyclooxygenases and monooxygenases (MOX's). In this respect the question whether LOX oxidation is the only pathway of oxidative metabolism of unsaturated fatty acids in plants is of great interest.

The aim of present study was the search of NADPH-dependent MOX pathway of linoleate oxidation in pea seedlings.

Homogenate or isolated microsomes from leaves of 7 days-old seedlings were incubated with [1-^{14}C]linoleate in the medium containing 0.33 M sorbitol, 1mM $MgCl_2$ and 50 mM tris/HCl, pH 7.5. The reaction was performed within 30 min, starting with the addition of labelled acid solution in ethanol (0.1% of total volume). Reaction products were extracted by 2-butanone and analyzed by radio-HPLC on reversed phase column. Separate metabolites after semi-preparative collection were finally purified on the normal phase column. DIP-mode CI and EI mass spectra and ^{1}H-NMR spectra were recorded.

In Fig. 1 the typical radiochromatogram of 18:2 oxidation products is presented (column Partisil 5ODS-3, linear gradient from 60:40:0.1 to 96:4:0.1 (v/v) in the mixture of MeOH-H_2O-AcOH within 55 min, then hold isocratically, flow rate 0.4 ml/min.). Oxidation rate was found to increase both in homogenate and in microsomes in presence of 1 mM NADPH. This observation, as well as inhibitory effect of metyrapone, shows that oxidation may be, at least partly, due to MOX activity. Oxidation occured effectively in isolated microsomes. The reaction in isolated microsomes was strongly inhibited by CO.

Two products, identified as 12-hydroxydodec-12Z-enoic acid (RT = 31 min, Fig. 1) [3], unknown natural substance, and 9,10-epoxy-12Z-18:1 (RT = 57 min, Fig. 1), were found to stimulate the growth of soybean callus. A group of peaks at RT 53 min contains hydroxy-, hydroperoxy- and oxo-derivatives of 18:2. Epoxy- and hydroxy-derivatives have been shown earlier

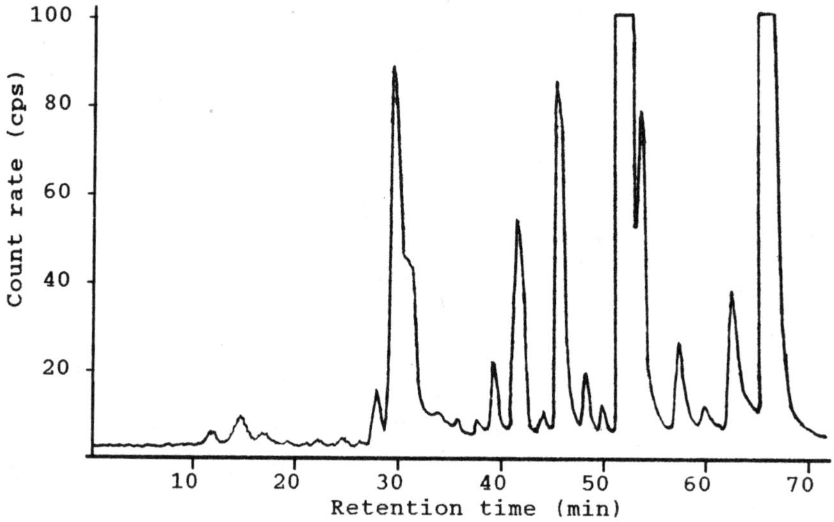

Fig. 1. Radiochromatogram of metabolites formed after 18:2 incubation with leaf homogenate in presence of 1 mM NADPH.

to be basic products of 20:4 oxidation in animal liver microsomes [1,2]. Studying the metabolization of 9,10- and 12,13--epoxides of 18:2, we have found formation of vicinal diols, double epoxide and epoxydiols.

The presented results show that the MOX oxidation of 18:2 coexists in pea seedlings with the well-known LOX pathway. Quite probably that some of 18:2 metabolites are formed as a result of cooperative action of LOX and MOX.

REFERENCES

1. Capdevila J, Marnett L et al (1982) Proc Nat Acad Sci USA 79:767-770
2. Chacos N, Falck JR et al (1982) Biochem and Biophys Res Communs 104:916-922
3. Grechkin AN, Korolev OS et al (1987) Dokl AN SSSR 297: 1257-1260

BIOLOGICAL ROLE OF PLANT LIPIDS
P.A. BIACS, K. GRUIZ, T. KREMMER (eds)
Akadémiai Kiadó, Budapest and Plenum Publishing
Corporation, New York and London, 1989

ENZYMATIC OXIDATION OF UNSATURATED FATTY ACIDS AND CAROTENOIDS IN TOMATO FRUIT

P.A. Biacs, H.G. Daood and Beatrix Czinkotai

Central Food Research Institute, Herman Ottó út 15, 1022 Budapest, Hungary

Introduction

Enzyme-catalyzed decolourization of naturally occurring pigments through coupled oxidation with unsaturated fatty acids was observed more than 40 years ago. The enzyme lipoxygenase /EC.1.13.11.12/ which catalyzes the oxidation of polyunsaturated fatty acids has been found to cause a loss in the colour intensity of many foods of plant origin during storage and processing[1].

Moreover, lipixygenase can effect by alternative ways the synthesis and stability of plant pigments[2].

Materials and Methods

Lipids of tomato fruit /Lycopersicon esculentum Ventura cv./ extracted with chloroform-methanol /2:1/. Methyl esters of fatty acids were prepared and analysed by gas-liquid chromatographical /GLC/, as reported earlier[3].

Tomato pigments were extracted and investigated by high-performance liquid chromatography /HPLC/ for their identification and also to measure relative changes during ripening[4]. Carotenoid bleaching activity of LOX was investigated by a newly elaborated HPLC method[5]. Extraction and enzyme assay of LOX from tomato fruit was carried out by a previously described method[6].

Results and Discussion

Fatty acid composition of tomato fruit and the changes occuring during ripening can be seen on fig. 1. Unsaturated fatty acids such as linoleic were abundant. At the first stage of ripening, when chlorophyll decomposed, fatty acid content of the fruit decreased to a high extent. The metabolic pathways of ripening led to an increase in fatty acids especially in linoleic acid concentration. Tomato plants when treated with Titavit /ascorbic acid complex of Titanium[7]/ the ripe fruits contained much less linoleic acid than the fruits of untreated ones. The decrease in linoleic acid content of Titavit treated fruits paralleled

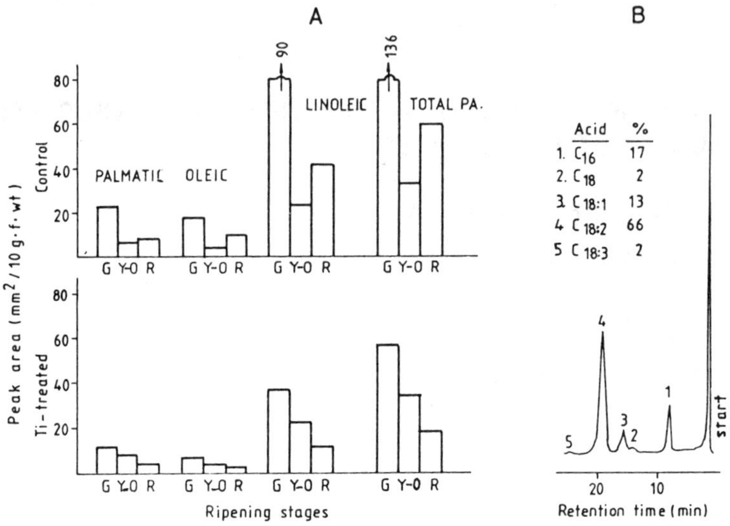

Fig. 1. A: Changes in fatty acid content of tomato fruit during ripening. B: Fatty acid composition of tomato fruit /GLC/

the high increase in LOX activity /Fig. 2/ and the retardation in the further development of lycopene and lycoxanthin, the characteristic red pigments in tomato[7]. These results suggest that LOX when oxidizing unsaturated fatty acids it cooxidizes cratoneoid pigments or inhibits their development by alternative pathways.

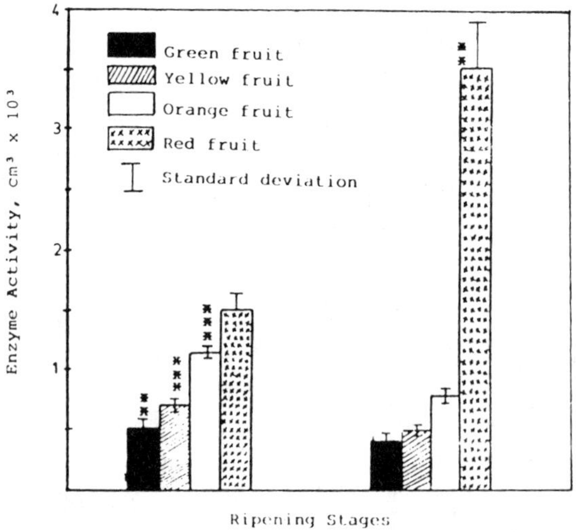

Fig.2.Change in LOX activity during ripening of tomato fruit

Different pigments showed different susceptibility to LOX-catalyzed cooxidation when investigated by HPLC method /Fig.3/. Most of pigments extracted from ripe red fruit were sensitive to the oxidation reaction. In the presence of chlorophylls /green tomato extract/ β-carotene and some pheophytins were stable towards LOX catalyzed linoleic acid oxidation. In general, it could be concluded that polar pigments such as chlorophylls and oxygen-containing xanthophylls are more susceptibel to the oxidation process than those of lower polarity.

1. Oxidation product
2. β-carotene
3. Pyrophephytin A
4. Cis-neurosporene
5. Chlorophyll B
6. Lutein
7. Chlorophyll A

1. Lutein
2. Neoxanthin
3. Lycopen
4. β-carotene
5. Lycoxanthin

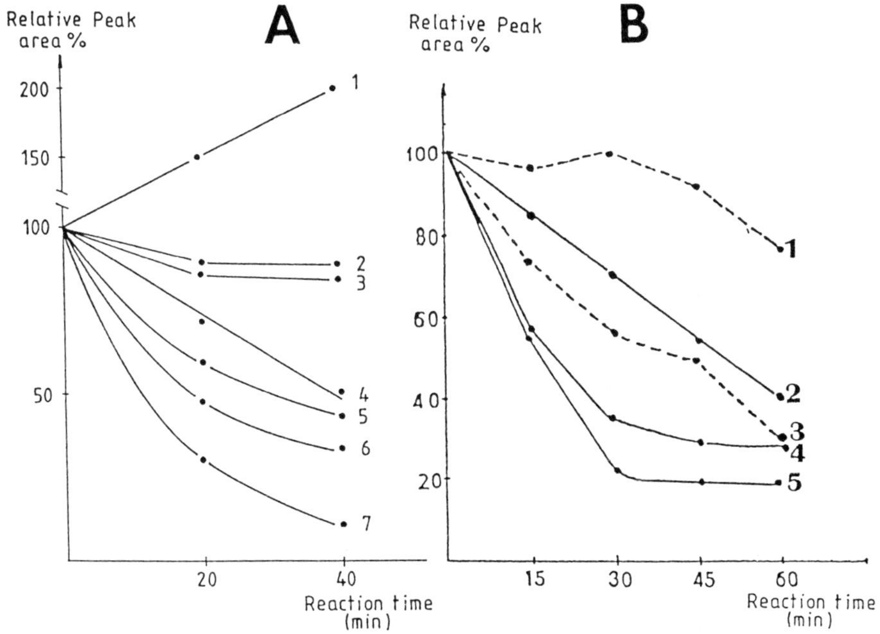

Fig.3.Degradation of (A) green tomato pigments and (B) red tomato pigments through LOX-catalyzed linoleic acid oxidation (HPLC analysis)

REFERENCES

1. Eskin,N.A., Grossman,S & Pinsky,A.(1977) CRC.Crit.Rev.Fd.Sci.Nutr.9,1
2. Vick,B.&Zimmerman,D.(1987): In Stumpf P.K., Mudd J.B. and Nes W.D.eds) Plenum Press New York and London p.383.
3. Daood,H.&Al-Ani,H.(1986). Acta Aliment. 15,319.
4. Daood,H. Biacs,P,Hoschke,A,Hajdu,F.&Vinkler,M.(1987)Acta Aliment 16,339.
5. Daood,H.&Biacs,P.(1988) Acta Aliment. 17, in press
6. Daood,H.&Biacs,P. (1988) Acta Aliment 17, p.55.
7. Biacs,P., Daood,H., Czinkotai,B.&Hajdu,F.(1987) Acta Hort. 220,443.

BIOLOGICAL ROLE OF PLANT LIPIDS
P.A. BIACS, K. GRUIZ, T. KREMMER (eds)
Akadémiai Kiadó, Budapest and Plenum Publishing
Corporation, New York and London, 1989

SODIUM CHLORIDE EFFECT ON GLYOXYSOMAL ENZYME ACTIVITIES IN MEDICAGO SEEDLINGS

Đ. Ben Miled and A.Cherif

Centre de Biologie et de Ressources Génétiques, INRST, B.P. 95 Hammam-Lif 2050, Tunisia

INTRODUCTION

The present work continued a previous one concerning the effect of salt on reserve lipid degradation of Medicago cotyledons during germination (1) considering in one hand the importance of fodder crops in Tunisia and the irrigation water quality on the other hand, we want to study modification induced by salt stress on lipid catabolism enzymes of Medicago seeds during germination

MATERIEL AND METHODS

Seeds of Medicago orbicularis L. used came from a station located in the North of Tunisia. Seed germination conditions were discribed elsewhere in a previous work (1). Determination of catalase (CAT EC 1.11.1.6) and isocitrate lyase (ICL EL 4.1.3.1) and malate synthetase (MS EC 4.1.3.2) activities have been made according to the method of TCHANG and al.(2). Enzymes activities were expressed in µmoles. min^{-1}. 10^{-1} pair of cotyledons.

RESULTS

activities studied from the first days of seeds imbibition. At the 2nd day CAT reached its maximum of activity whereas MS and ICL reached their ones at the 3rd day (Fig.1). For the following days, all the three activities declined rapidely and after 10 days CAT lost 80% of its maximal activity , MS an ICL lost about 90%.

When NaCl concentration in the culture medium was increased, changes in the enzymic activities of Medicago cotyledons occured. The CAT maximum activity was delayed of one day every time when NaCl concentration in creased of 2g/l. The maximum value of CAT activity was not affected at the concentration of 2g/l NaCl; but it decreased from 4g/l. Then the important decrease of activity observed in the control after reaching the peack was sharply attenuated. In treated cotyledons by 8g/l of NaCl, the maximal activity was reached the 6th day and remain unchanged until the 10th day (Fig.1A).

NaCl produced the same modification on MS and ICL activities evolution, so all salt concentrations used modified the enzymatic activities in comparaison to control (Fig.1B and C). The MS and ICL maximal amplitude activity was reached at the 4th day for 2 and 4g/l of NaCl, it was obtained later at the 5th and 6th day for 6 and 8g/l respectively. As for CAT, there was an attenuation of decrease activity after the peack activity in treated cotyledons, so at the 10th day and ICL activities represent each one only 40% of control maximal activity.

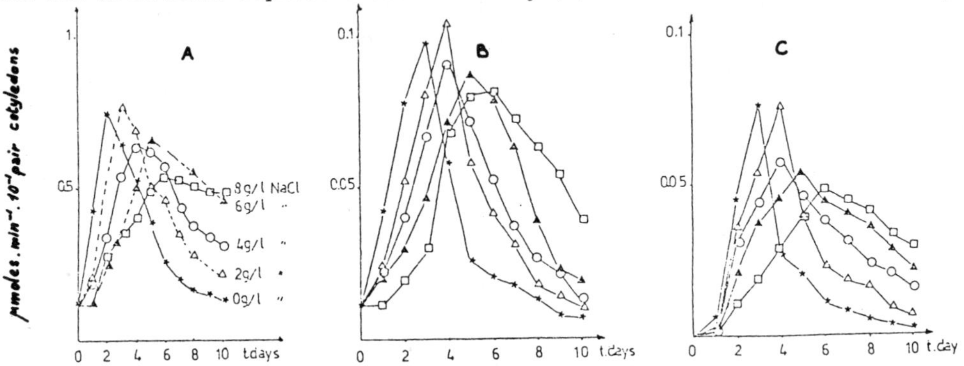

Figure 1 : NaCl effect on evolution of Catalase (A), malate synthetase (B) and isocitrate lyase (C) activities in the cotyledons of Medicago orbicularis. (each point represent the average of three essays).

CONCLUSION

The increase in glyoxysomal enzymatic activities of control cotyledons was paralleled with lipid degradation (1,2). These activities were maximal the first five days of seeds imbibition, afterwards they decreased considerably simultaneously to using up of lipid reserves. Salt delayed optimal activities of studied enzymes, this was paralleled with the decrease in lipidic reserve utilisation (1); NaCl concentration more than 2g/l inhibited paratially enzymes since their maximal activities were reduced in comparaison to control ones. Two hypothesis may be proposed : a) Salt acted directly on enzymes activities of glyoxylic cycle therefore it inhibited enzymes. b) Enzymes were synthetisized progressively to demand and if lipid degradation was decreased then synthesis of glyoxysomal enzymes was also delayed, in these conditions salt inhibited lipasic activity. More investigations were needed to check these hypthesis

BIBLIOGRAPHIE : (1) Ben Miled D. et A. Cherif-1986- Effet de NaCl sur la degradation des lipides de réserves au cours de la germination des graines de Medicag Ben Hamida F., ed, CNRS, Paris, 451-453. (2) Tchang F., Robert O. et Mazliak P.-1980-Utilisation des réserves lipidiques et formation des glyoxysomes et d'éti plastes dans les cotylédons de Tournesol. Physiol. Vég., 18, 117-130.

BIOLOGICAL ROLE OF PLANT LIPIDS
P.A. BIACS, K. GRUIZ, T. KREMMER (eds)
Akadémiai Kiadó, Budapest and Plenum Publishing
Corporation, New York and London, 1989

GLYCEROLIPID SYNTHESIS IN MICROSOMES AND OIL BODIES OF OIL PALM MESOCARP

K.C. Oo[1], Y.H. Chew[1] and A.S.H. Ong[2]

[1]Department of Biochemistry, University of Malaya, 59100 Kuala Lumpur, Malaysia
[2]Palm Oil Research Institute of Malaysia (PORIM), Bangi, Selangor, Malaysia

INTRODUCTION

Storage oils in seeds are synthesized from *sn*-glycerol-3-phosphate (G3P) by acyltransferase enzymes which are associated with microsomes but have also been reported in oil body preparations. We have studied acyltransferase activities in oil palm mesocarp.

MATERIALS AND METHODS

Washed oil bodies and microsomes were prepared from 18 week fruit mesocarp as in (1). Reaction mixture (0.1M CAPS-NaOH, pH 10.5, 2 mM $MgCl_2$, 5 mM DTT, 0.2% fat-free BSA, 0.05% gelatine, 0.39 nmol [^{14}C]palmitoyl-CoA (58 Ci/mol) and 0.2 mg enzyme protein in 1 ml) was incubated at 30°C for 1 hr. Lipids were analysed as in (2) and (3).

RESULTS AND DISCUSSION

Microsomes and oil bodies from palm mesocarp incorporated 65% and 83% of [^{14}C]palmitoyl-CoA into lipid products. In both cases the products were about 44% polar lipids, 47% fatty acids and only 4% TG. Total acylating activity was 1.62 nmol/mg protein for oil bodies and 0.63 nmol/mg protein for microsomes. The oil bodies were free of contamination by microsomal particles as shown by antimycin A-insensitive cytochrome C reductase activity, a marker enzyme for endoplasmic reticulum (25.2 nmol/min/mg protein for microsomes and 0.6 nmol/min/mg protein for oil bodies).

Diacylglycerol acyltransferase (DGAT) was active at two pH ranges. A peak at pH 8.5 was higher for microsomes than for oil bodies. A smaller peak occurred at pH 10.5. DGAT in both preparations was inhibited by deoxycholate and had maximal activity in the presence of gelatine, Mg^{2+},

DTT and BSA. Although Mg^{2+} can affect acyl-CoA solubility (4), under conditions used (METHODS) over 93% of the palmitoyl-CoA remained soluble. N-Ethylmaleimide and phenylmethylsulfonylfluoride slightly reduced palmitoyl-CoA hydrolysis in the oil bodies but both inhibited DGAT activity. Reactivity of microsomal DGAT towards acyl-CoA decreased in the order 18:1 > 16:0 > 14:0 > 18:0. Reaction rate increased with added dipalmitin with a maximum at 4 mM concentration.

Table 1 Effect of exogenous acceptors on acyltransferases

Acceptor added	nmol [^{14}C]palmitoyl-CoA incorpd/hr/mg protein					
	Oil Bodies			Microsomes		
	LPA	PA	PC	LPA	PA	PC
None	0.20	0.07	0.14	0.08	0.03	0.03
0.5 mM G3P	0.32	0.08	0.07	0.08	0.06	0.02
0.5 mM LPA	0.07	0.26	0.09	0.03	0.10	0.02
0.5 mM LPC	0.09	0.03	0.29	0.04	0.02	0.06

Three other acyltransferases (G3P-AT, LPA-AT, LPC-AT) were also present in oil bodies and microsomes (Table 1). Addition of an appropriate exogenous acyl acceptor showed incorporation into the corresponding lipid product.

REFERENCES

1. Slack, C.R., W.S. Bertaud, B.D. Shaw, R. Holland, J. Browse & H. Wright (1980) Biochem J. 190, 551-561.
2. Murphy, D.J. & K.D. Mukherjee (1987) Lipids 22, 293-298.
3. Roughan, P.G., C.R. Slack & R. Holland (1978) Lipids 13, 497-503.
4. Constantinides, P.P. & J.M. Steim (1986) Arch Biochem Biophys 250, 267-270.

BIOLOGICAL ROLE OF PLANT LIPIDS
P.A. BIACS, K. GRUIZ, T. KREMMER (eds)
Akadémiai Kiadó, Budapest and Plenum Publishing
Corporation, New York and London, 1989

LIPID SYNTHESIS IN OIL PALM KERNEL

K.C. Oo[1] and A.S.H. Ong[2]

[1]Department of Biochemistry, University of Malaya, Kuala Lumpur, Malaysia
[2]Palm Oil Research Institute of Malaysia (PORIM), Bangi, Selangor, Malaysia

INTRODUCTION

The oil palm *(Elaeis guineensis)* fruit produces two oils. Kernel (endosperm) oil is synthesized in the nut from 13-16 weeks after flowering (WAF) followed by deposition of mesocarp oil from 16-20 WAF. This paper reports some experiments on synthesis of kernel oil, a major source of medium chain fatty acids.

MATERIALS AND METHODS

Freshly harvested palm fruits were used. Incubation of kernel slices with labelled substrates and analysis of fatty acids and lipid products were by standard procedures (1).

RESULTS AND DISCUSSION

Oil deposition in the kernel started about 13 WAF and ended by 17 WAF with about 0.5 g oil/g dry wt (Table 1). Lipid synthesis, most active at 15 WAF, was clearly of storage oil due to the high level (60-70%) of C12/C14 acids and TG (30-50%). Polar lipids (10%) consisted of PA (32%), PC (12%), PE (11%) and their lyso compounds.

Acyl-ACP and acyl-CoA fractions were isolated (2) after incubation with [^{3}H]acetate (4 Ci/mmol). The saturated acids were isolated by $AgNO_3$-TLC followed by reverse phase TLC. The acyl-CoA pool contained 81.7% C16, 8.9% C18, 5.5% C14 and 0.9% C12 while the acyl-ACP pool contained 46.5% C14, 25.0% C12, 19.5% C16 and 5% C18. This resembled the coconut endosperm system (3) in contrast to *Cuphea*, where both pools contained much medium chain acids (4).

Table 1 Kernel slice incubation (6 hr) with [^{14}C]acetate

WAF	g oil/g tissue		nmol acetate incorpd.	% [^{14}C] in lipid product				
	wet wt	dry wt		TG	DG	MG	FA	PL
13	0.03	0.025	0.93	50.87	20.02	3.98	18.92	6.19
15	0.29	0.52	4.34	33.14	35.43	1.60	19.31	10.53
17	0.31	0.48	1.26	26.94	42.62	4.47	14.25	11.72

WAF	% ^{14}C in total fatty acids							
	8:0	10:0	12:0	14:0	16:0	18:0	18:1	18:2
13	2.3	6.3	45.3	26.2	12.0	5.3	1.3	1.3
15	5.9	5.5	34.3	22.4	5.5	9.0	3.4	14.1
17	not analysed							

[^{14}C] Fatty acids (12:0, C16:0, C18:0 and C18:1) were also incorporated by endosperm slices. Over 93% stearic acid remained unesterified. Lauric acid was esterified into TG (66.9%), PL (13%) and DC (5.6%). The corresponding figures for palmitic acid were 14.8%, 19.2% and 19.7% and for oleic acid 15.8%, 19.3% and 27.2%. The greater esterification of lauric acid probably reflected some specificity by endosperm acyltransferases in medium chain TG synthesis. RadioGC showed each substrate retained its original chain length without any modification.

Cell-free endosperm extract actively incorporated [^{14}C] malonylCoA (3) into lipids (45.5%), acyl-ACP (36.6%) and acyl-CoA (17.6%). About 98% of the lipids were free acids. Total product fatty acids were stearic (53.7%), palmitic (23.3%) and myristic (9.8%).

REFERENCES

1. Oo, K.C., S.K. Teh, H.T. Khor & A.S.H. Ong (1985) Lipids 20, 205-210.
2. Mancha, M., G.B. Stokes & P.K. Stumpf (1975) Anal Biochem 68, 600-608.
3. Oo, K.C. & P.K. Stumpf (1979) Lipids 14, 132-143.
4. Singh, S., T.Y. Nee & Pollard (1986) Lipids 21, 143-149.

BIOLOGICAL ROLE OF PLANT LIPIDS
P.A. BIACS, K. GRUIZ, T. KREMMER (eds)
Akadémiai Kiadó, Budapest and Plenum Publishing
Corporation, New York and London, 1989

LIPOXYGENASE ACTIVITY AND DEGRADATION OF ESSENTIAL FATTY ACIDS IN POPPYSEED ON STORAGE

J. Pokorný, T. Meshehdani, J. Pánek, J. Davídek and H. Pařízková*

Prague Institute of Chemical Technology, Prague
*Research Institute of Food Indusrtry, Prague, Czechslovakia

Poppyseed used as food filling and flavouring has high content of linoleic acid rich oil. Both lipoxygenase-catalyzed oxidation and autoxidation proceed at storage conditions, particularly in damaged seed, deteriorating the sensory quality.

Material: Full ripe poppyseed samples, grown in Bohemia and Moravia in 1984 - 1987, either manually, or machine harvested were stored at room temperature in the dark and at 30 - 50% relative moisture of air.

Analytical methods: Total lipids were determined by chloroform - methanol extraction after Folch,free lipids by rapid washing of whole seeds with hexane.The extract was analyzed for hydroperoxides (iodometrically), 2-thiobarbituric active substances (in 1-butanol), conjugated dienes (at 234 nm), and total oxidation products by reverse phase HPLC. The lipoxygenase activity was determined in defatted seeds with use of linoleate as substrate, and measuring changes of conjugated dienes. The sensory value was determined by grading and profiling (odour profile of whole seeds, and flavour profile of ground seeds) with use of unstructured graphical scales.

The content of damaged seeds was substantially higher in machine harvested than in manually harvested seeds. The degree of lipid oxidation in hexane extracts (higher in damaged seed) was significantly higher than in total lipids.

The lipoxygenase I activity was higher in poppyseed than the lipoxygenase II activity. After addition of buffer, the oxidation of linoleic acid was very rapid but the decomposition of hydroperoxides was also very fast so that the peroxide value attained the maximum in 5 - 15 min, and started to decrease again, the reaction rates depending on the content of unreacted linoleic acid.During storage of seeds containing less than 10% water,the content of conjugated dienes was doubled in 2 - 3 months,but decreased again by secondary reactions on prolonged storage.

The hydroperoxide decomposition may be attributed to the action of hydroperoxide lyases, and still more, to interaction of hydroperoxides or their carbonylic decomposition products with nonlipidic components,such as phenolic substances, amino acids or proteins.The interaction with proteins (which constituted nearly 50% of defatted seeds) is very rapid, taking place in a few minutes. About 80% of linoleic acid hydroperoxides reacted with aqueous solutions of poppyseed or model proteins within 20 min at 20°C. Lipids were bound mainly by hydrogen bonds but 20 - 30% were bound by covalent bonds, Schiff bases being primary reaction products. Some interaction products remain liposoluble as evident by weak fluorescence of lipid extracts (excitation maximum between 381 - 385 nm, emission maximum between 435 - 450 nm).

Under storage conditions, the flavour deterioration was only slow in undamaged, manually harvested seeds but 2 - 3 times faster in machine harvested seeds which were more or less damaged.The typical poppyseed and sweet flavour notes of fresh seeds weakened on storage while intensities of rancid, acidic and foreign flavour notes increased. Bitter and stale flavour notes were due to interaction products with protein. The intensity of off-flavours correlated with the degree of oxidation of poppyseed lipids except after long storage where secondary reactions prevailed. The bitter taste produced on storage was associated with insoluble reaction products and not with oxidized lipids even when free oxidized fatty acids taste bitter.

BIOLOGICAL ROLE OF PLANT LIPIDS
P.A. BIACS, K. GRUIZ, T. KREMMER (eds)
Akadémiai Kiadó, Budapest and Plenum Publishing
Corporation, New York and London, 1989

LIPID BIOSYNTHESIS AND COMPOSITION IN MATURE AND GERMINATED OLIVE POLLEN

Maria P. Rodriguez-Rosales, Marta Roldán, A. Belver and J.P. Donaire

Department of Biochemistry, Estación Experimental del Zaidín, C.S.I.C., Granada, Spain

It has been proposed that a decrease in membrane integrity might play a significant role in the deterioration of pollen and in the germination capacity, which can be related to losses of viability |1|. In the pollen cell a strong biosynthesis of reserve and membrane lipids take place during germination and tube growth |2, 3|. The work described here attempts to follow the viability of olive pollen by studying their lipid content and biosynthesis at maturity and during germination process.

MATERIAL AND METHODS

The procedures of harvest, storage and *in vitro* germination of pollen from *Olea europaea* L. cv. Marteño have been described in a previous work |4|. Mature or 24 h germinated pollen (2 g) were incubated at 27 ^{o}C in 40 ml of a medium containing 10% sucrose, 0.2 mM $CaSO_4$, 2 mM K_2SO_4, 0.01% H_3BO_3, 100 μg/ml tetracyclin and $|1-^{14}C|$-acetate (30 nmol, specific radioactivity 58 mCi, $mmol^{-1}$). Air was bubbling continuously through the medium and the pH was maintained at 5.50 $\pm$ 0.02 by automatic addition of 2.5 mM KOH or H_2SO_4 (Metrohm system pH-stat). After incubation with $|1-^{14}C|$-acetate, mature and germinated pollen are fixed in boiling MeOH and the lipids are extracted |5|, resolved by TLC and fatty acids and free sterols analyzed by GC. Each separated radioactive lipid was determined by liquid scitillation counting (Packard Tri-Carb).

RESULTS AND DISCUSSION

There was a significant increase in the amount of PI, PE, MG and DG, while the levels of PC and, above all, FFA decreased during germination of olive pollen. PC was identified as the major lipid synthetized in mature and germinated pollen, Germination provokes a decrease of ^{14}C incorporation into PC and an increase in PI, MG and DG. In this respect, Helsper *et al.* |6| indicated that the biosynthesis of PI is especially important during the germination of *Lilium* pollen. It is also possible to suggest that the small values of germination capacity observed in olive pollen |4| could be related to losses of PC from the membranes. However, further experimental data are needed in order to throw light on the relative importance of some lipids in these processes.

Table 1. *Content and distribution of radioactivity in different lipids formed after 3 h of incubation of* |*1-*^{14}C|*-acetate with mature (M) or 24 h germinated (G) olive pollen.* Values are the average of three independent experiments, not differing by more than 10% from the mean.

Lipid class	Lipid content ($\mu g \cdot g\ F\ w^{-1}$)		Total ^{14}C incorporation ($pmol \cdot g\ F\ w^{-1}$)		Radioactivity (% of total lipids)	
	M	G	M	G	M	G
PI	307.3	400.2	73.9	193.1	4.2	7.3
PC	4037.2	2275.2	577.3	663.9	32.8	25.1
PE	712.4	737.3	174.2	296.2	9.9	11.2
MG	684.4	758.4	86.2	190.4	4.9	7.2
DG	2053.4	2591.1	161.9	343.8	9.2	13.0
Free Sterols	1188.2	752.2	49.3	87.3	2.8	3.3
FFA	3222.9	1643.2	362.6	478.7	20.6	18.1
TG	2374.7	1432.5	198.9	309.5	11.3	11.7

REFERENCES

1. Shivanna, K.R. and Heslop-Harrison, J. (1981). Membrane state and pollen viability. *Ann. Bot.* *47*, 759-770.
2. Whipple, A.P. and Mascarenhas, J.P. (1977). Lipid synthesis in germinating *Tradescantia* pollen, *Phytochemistry*, *17*, 1273-1274.
3. Helsper, J.P.F.G. and Pierson, E.S. (1986). The effect of lectins on germinating pollen of *Lilium longiflorum* II. Effect of concanavalin A on phospholipid turnover and on biosynthesis of pectic polysacarides, *Acta Bot, Neer.*, *35*, 257-263.
4. Rodríguez-Rosales, M.P. and Donaire, J.P. (1988). Germination-induced changes in acyl lipids and free sterols of olive pollen, *New Phytol.*, *108*, 509-514.
5. Bligh E.G. and Dyer W.S. (1959). A rapid method of total lipid extraction and purification, *Can. J. Biochem. Biophys.* *37*, 911-917.
6. Helsper, P.F.G., Groot, F.M., Linskens, H.F. and Jackson, J.F. (1986). Phosphatidylinositol monophosphate in *Lilium* pollen and turnover of phospholipid during pollen tube extension, *Phytochemistry*, *25*, 2193-2199.

ACKNOWLEDGEMENTS

Support was obtained from the C.S.I.C. and C.A.Y.C.I.T. Project 1-179-2ID 181 (Spain).

BIOLOGICAL ROLE OF PLANT LIPIDS
P.A. BIACS, K. GRUIZ, T. KREMMER (eds)
Akadémiai Kiadó, Budapest and Plenum Publishing
Corporation, New York and London, 1989

CONTROL OF ACYL LIPID DESATURATION IN RHODOTORULA GRACILIS

Carole E. Rolph* and J.L. Harwood

Department of Biochemistry, University College, Cardiff, UK
*Present adress: Dept. of Biochemistry, University of Liverpool, Liverpool, UK

When grown under conditions of nitrogen-limitation, the carotenoid pigmented yeast *Rhodotorula gracilis* has been shown to accumulate large quantities of triacylglycerols. However, to date, the commercial utilization of this storage lipid has been limited by its unsaturated nature.

The major unsaturated fatty acids synthesized by N-limited *R.gracilis* are oleic, linoleic and α-linolenic acids, indicating the presence of Δ9, Δ12 and Δ15 desaturase enzymes [1]. In this report, we discuss experiments in which the activities of these desaturase systems have been altered by temperature and sterculate treatment.

MATERIALS & METHODS

R.gracilis was grown at either 15° or 30°C in N-limiting media (+/- methyl sterculate) in baffled shake flasks in an orbital incubator operating at 150 r.p.m. [2]. Stationary-phase cultures were harvested by centrifugation, freeze-dried and lipids extracted [3]. Individual lipid classes were separated by t.l.c. and analysed by g.l.c. [2].

In temperature-shift studies, cultures were grown at 15°C until mid log-phase and then shifted to 30°C. Changes in lipid desaturation were monitored by direct saponification of total lipids followed by g.l.c. analysis [4].

RESULTS & DISCUSSION

A shift in growth temperature from 15° to 30°C resulted in dramatic changes in the relative proportions of C_{18} polyenoic fatty acids present in total lipid extracts. The observed reduction in the linoleate and α-linolenate content of extracts from the shifted cultures demonstrated that both Δ12 and Δ15 desaturases could be controlled by growth temperature (see refs. [5] and [6]).

Further studies showed that the acyl compositions of the major phospholipids could be significantly altered by changes in growth temperature, with phospholipids extracted from 15°C grown yeast containing higher proportions of C_{18} polyenoic fatty acids than those extracted from 30°C grown cells. However, growth temperature was found to have little effect on the acyl composition of the accumulated triacylglycerols.

In contrast to the above results, treatment of either 15° or 30°C grown cells with the Δ9 desaturase inhibitor, sterculate, resulted in significant changes in the acyl compositions of all of the lipid classes studied. Typical changes included an increase in stearate / oleate ratios plus an increase in α-linolenate content. Furthermore, the former trend was most pronounced in the triacylglycerol pool.

In addition, with respect to phospholipid metabolism, changes in both growth temperature and sterculate treatment resulted in altered phospholipid proportions. The most significant difference was observed between cultures grown at different temperatures, with cells grown at 15°C exhibiting higher phosphatidylcholine / phosphatidylethanolamine ratios than those grown at 30°C.

CONCLUSIONS

Throughout these studies, mechanisms involved in the control of membrane fluidity have been shown to play a major role in the regulation of the degree of unsaturation exhibited by the major acyl lipid classes (see [7]). As both 15° and 30°C cultures possessed very active Δ9 desaturase systems, treatment of cultures with sterculate proved to be the only method by which the synthesis of unsaturated triacylglycerols could be significantly controlled.

ACKNOWLEDGEMENT - We gratefully acknowledge the financial support of Cadbury-Schweppes plc and the S.E.R.C.

REFERENCES

[1] Rolph, C.E., Moreton, R.S., Small, I.S. & Harwood, J.L. (1986) Biochem. Soc. Trans. **14**, 712

[2] Rolph, C.E. (1988) Ph.D. Thesis. University of Wales, U.K.

[3] Moreton, R.S. (1985) Appl. Microbiol. Biotechnol. **22**, 41-45

[4] Bolton, P. & Harwood, J.L. (1977) Biochem. J. **168**, 261-269

[5] Kates, M., Pugh, E.L. & Ferrante, G. (1984) in "Biomembranes 12. Membrane Fluidity" Kates, M. & Manson, L.A. Eds. Plenum Press, New York & London

[6] Thompson Jnr.,G.A. & Nozawa, Y. (1984) in "Biomembranes 12. Membrane Fluidity" Kates, M. & Manson, L.A. Eds. Plenum Press, New York & London

[7] Cook, H.W. (1985) in "Biochemistry of Lipids and Membranes" Vance, D.E. and Vance, J.E., Eds. Benjamin / Cummings Publishing Co., Inc., Menlo Park, California

BIOLOGICAL ROLE OF PLANT LIPIDS
P.A. BIACS, K. GRUIZ, T. KREMMER (eds)
Akadémiai Kiadó, Budapest and Plenum Publishing
Corporation, New York and London, 1989

GALACTOLIPID BIOSYNTHESIS IN CHLOROPLASTS OF UNHARDENED AND FROSTHARDENED SEEDLINGS OF SCOTS PINE

Eva Selstam

Department of Plant Physiology, Umeå University, 901 87 Umeå, Sweden

INTRODUCTION

In needles of Scots pine the galactolipid content is similar in summer and in winter needles, while the molar ratio between monogalactosyldiacylglycerol (MGDG) and digalactosyldiacylglycerol (DGDG) is approx. 2.0 in summer and 0.8 in winter (Selstam, unpublished). The low ratio of MGDG to DGDG found in winter needles of Scots pine and other extremely frost resistant plants may be an important characteristic of thylakoid membrane acclimated to freezing temperatures since MGDG and DGDG have very different physicochemical properties.

To investigate whether the MGDG to DGDG ratio is actively regulated, the galactolipid biosynthesis was investigated in unhardened and frosthardened Scots pine.

MATERIALS AND METHODS

Unhardened and frosthardened seedlings of Scots pine were cultivated hydroponically (1). Chloroplasts were isolated in a medium containing 0.4 M sucrose, 0.05 M Hepes pH 7.6, 0.01 M KCl, 0.001 M EDTA and 20% PEG-4000. The chloroplast pellet was resuspended in reaction buffer containing 0.4 M sorbitol, 0.004 M $MgCl_2$, 0.025 M Hepes pH 7.4 and 0.0025 M DTT. Assay for galactolipid biosynthesis was performed in 250 µl of reaction buffer containing 0.6 mg Chl/ml and 250 µM UDP-(^{14}C)-galactose (370 kBq/umol). Assay for galactolipid: galactolipid galactosyltransferase activity was performed in 200 µl reaction buffer containing 0.5 mg Chl/ml and 670 µM (^{14}C)-MGDG and 1.5 mM sodium deoxycholate (2).

RESULTS AND DISCUSSION

Measurements of temperature dependent galactolipid biosynthesis showed that frosthardened chloroplasts were about 3 times more active than unhardened chloroplasts (Table 1). The higher activity was found at all temperatures and was thus not due to a change in temperature optimum of the enzymes. The higher activity in frosthardened chloroplasts could be due to an augmentation of envelope membranes or to increased specific activity of enzymes. The galactolipid biosynthesis of frosthardened chloroplasts at 7C was only slightly lower than that of unhardened chloroplasts at 20C. The frosthardened Scots pine seedlings thus have a capability for galactolipid biosynthesis at low temperature, important for maintenance of the chloroplast during winter.

Table 1.

Temperature dependent incorporation (5 min.) of UDP-(^{14}C)-galactose into total galactolipids (pmol galactolipids · µg Chlorophyll^{-1}) in chloroplasts isolated from unhardened and frosthardened Scots pine seedlings.

	Temperature °C				
	7	15	20	25	30
Unhardened	0.5	1.7	3.5	4.4	6.3
Frosthardened	2.3	6.0	11.2	14.3	19.1

The incorporation of galactose in MGDG and DGDG at 20 and 10C showed that frosthardened chloroplasts have a relatively larger incorporation in DGDG than had unhardened chloroplasts. The ratios between synthesized MGDG and DGDG were at 20 and 10C 3.1 and 2.2, respectively in frosthardened chloroplasts and 5.6 and 5.8 in unhardened chloroplasts. The ratio was calculated at the end of the linear part of MGDG formation. The lower ratio in frosthardened chloroplasts may be an acclimation of the galactolipid biosynthesis to synthesize proportionally more DGDG in winter needles. This acclimation is dependent on light intensity since the ratio was also low, 3.3 in chloroplasts isolated from shaded unhardened

Table 2.

The activity of galactolipid: galactolipid galactosyltransferase (pmol lipid · µg Chlorophyll^{-1}) after 5 minutes at 20°C in chloroplasts isolated from unhardened and frosthardened Scots pine seedlings (H H = higher homologs of galactolipids).

	DGDG	H H	Total
Unhardened	47	58	105
Frosthardened	83	51	134

seedlings. The pathway for the biosynthesis of DGDG is still an open question (3, 4). Measurements of DGDG synthesis from (^{14}C)-MGDG showed that in frosthardened chloroplast 1.8 times more DGDG was formed than in unhardened chloroplasts (Table 2). Also the total activity of the galactolipid: galactolipid galactosyltransferase was 12 and 32 times higher than the UDP-galactose dependent galactosyltransferase activity in frosthardened and unhardened chloroplasts respectively. These results support the hypothesis that DGDG is synthesized by the galactolipid: galactolipid galactosyltransferase, and that this enzyme is more active in frosthardened seedlings.

REFERENCES

1. Selstam, E. and Öquist, G. 1985. Plant Science 42: 41-48.

2. Heemskerk, J.W.M., Wintermans, J.F.G. M., Joyard, J., Block, M.A., Dorn, A.-J. and Douce, R. 1986. Biochim. Biophys. Acta 877; 281-289.

3. Heemskerk, J.W. M. and Wintermans, J. F. G. M. 1987. Physiol. Plant. 70: 558-568.

4. Joyard, J. and Douce, R. 1987. Galactolipid synthesis. In The Biochemistry of Plants. Vol 9, ed 2 (P.K. Stumpf and E.E. Conn eds.) pp. 215-274 Academic Press Inc. Orlando.

BIOLOGICAL ROLE OF PLANT LIPIDS
P.A. BIACS, K. GRUIZ, T. KREMMER (eds)
Akadémiai Kiadó, Budapest and Plenum Publishing
Corporation, New York and London, 1989

COMPARISON OF ACETATE AND/OR PYRUVATE DEPENDENT FATTY ACID SYNTHESIS BY SPINACH CHLOROPLASTS

Jutta Springer and K.-P. Heise

Institut für Biochemie der Pflanze, University of Göttingen, D-3400 Göttingen, FRG

INTRODUCTION

With the acetyl-CoA synthetase (ACS; EC 6.2.1.1) and the pyruvate dehydrogenase complex (PDC) chloroplasts possess two pathways for the synthesis of acetyl coenzyme A. In the meantime, the physiological levels of acetate and pyruvate in spinach chloroplasts have been estimated by nonaqueous fractionation (Treede et al. 1986) and were found to be 5-fold higher in acetate (0.2-0.6 mM). However, the stromal pool size of acetyl coenzyme A, calculated by the same procedure, proved to be too low (about 20 µM) for comparing determinations of the plastidial acetyl coenzyme A level intermediarily formed from both substrates. Thus, in order to check the physiological relevance of acetate and pyruvate for fatty acid synthesis in chloroplasts, which at present is still controversial (Yamada et al. 1975; Roughan et al. 1978), their incorporation into fatty acids by spinach chloroplast suspensions (80-90% intact; O_2-evolution rate: 70-130 $\mu mol \cdot mg^{-1}\ Chl \cdot h^{-1}$) at approximately physiological concentrations has been reinvestigated. For comparison, the incorporation experiments have been performed with spinach leaf sections, in order to get an idea of the flux of acetate and pyruvate into different lipid compounds.

RESULTS AND DISCUSSION

Incorporation of ^{14}C into long-chain fatty acids of spinach chloroplasts

As shown in Fig. 1, at physiological concentrations (below 1 mM) the K_M for the incorporation into fatty acids was about 0.1 mM for both metabolites and thus agreed with the values for the ACS and the PDC (Treede et al. 1986). However, acetate was incorporated with a 3-fold higher V_{max}. Apart from the absolute incorporation rates, these results are in

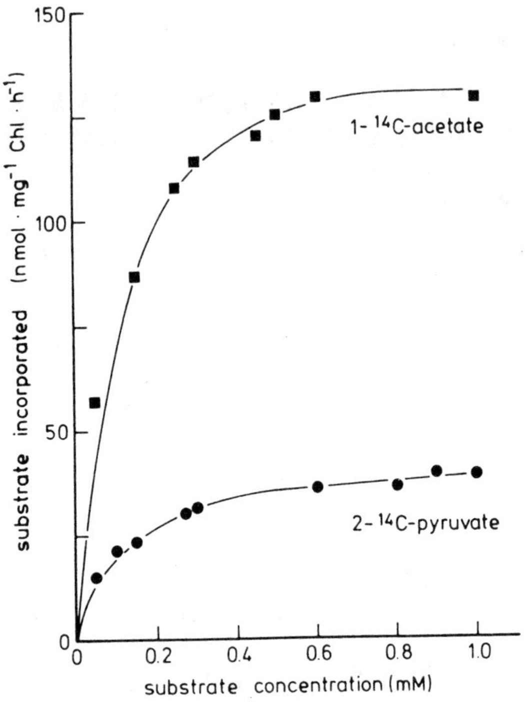

Fig. 1: Incorporation of ^{14}C into long-chain fatty acids of intact spinach chloroplasts incubated with 1-^{14}C-acetate (0.44-1.96 GBq·mmol^{-1}) or 2-^{14}C-pyruvate (0.11-0.42 GBq·mmol^{-1}). Preilluminated chloroplasts (5-10 min, 1 mg Chl·ml^{-1}) were incubated under saturating white light (15 min, 25 °C) in 250 µl assays (with 20 µg Chl) according to Sauer and Heise (1983).

full accordance with Roughan et al. 1978. In contrast, saturation for pyruvate incorporation into the fatty acid fraction was achieved only at physiological pyruvate concentrations (<1.0 mM). The observed diffusion kinetics at higher concentrations (data not shown) may be due to contaminations with derivates of the labeled substrate (Von Korff 1969). The latter observation corresponds well with the finding, that under comparable conditions pyruvate uptake by pea chloroplasts showed biphasic kinetics following a saturation at lower (<1 mM) but a diffusion mechanism at higher (>1 mM) concentrations (Proudlove and Thurman, 1981). The portion of label in coenzyme A-bound fatty acids made up about 16% after incubation with 1-^{14}C-acetate and about 30% after 2-^{14}C-pyruvate incorporation.

Competition experiments, in which spinach chloropasts were supplied with one precursor in the ^{14}C-labeled- and with the other in the cold form (Roughan et al. 1978) have been supported by double labeling with ^{3}H-acetate and 2-^{14}C-pyruvate and subsequent differentiation between the ^{3}H- and ^{14}C-fixed portion of freshly synthesized fatty acids (Fig. 2). As shown in Fig. 2 in the presence of the same acetate and pyruvate concentrations, 75% of the fatty acids were synthesized from ^{3}H-acetate. These double labeling experiments clearly demonstrate the predominant role of acetate as substrate in long-chain fatty acid synthesis at least in spinach chloroplasts. As recently discussed, this observation cannot be generalized, however, to other types of chloroplasts (Treede et al. 1986).

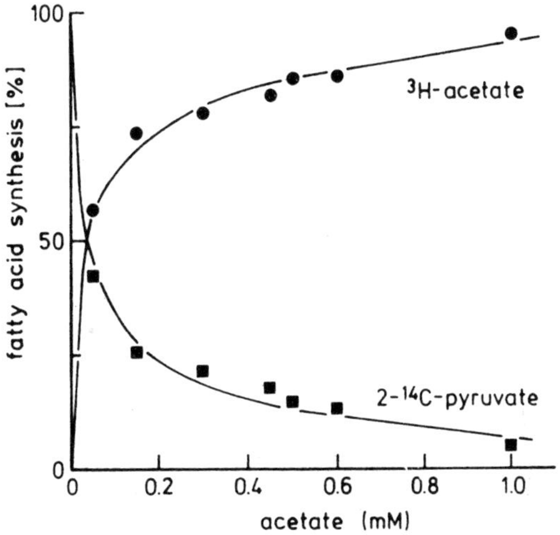

Fig. 2: Relative involvement of ^{3}H-acetate (1.26 GBq·mmol^{-1}) and 2-^{14}C-pyruvate (0.42 GBq·mmol^{-1}) in the fatty acid synthesis of intact spinach chloropasts in double labeling experiments at increasing acetate- and constant pyruvate concentrations (0.15 mM) (conditions see Fig. 1).

The incorporation of 1-^{14}C-acetate and 2-^{14}C-pyruvate by circular spinach leaf sections (ø 0.5 cm), which had been freed of the lower epidermis and infiltrated in buffer medium is shown in Table 1. Similar to chloroplast suspensions the efficiency of lipid incorporation was significantly higher from acetate. Furthermore, radioactivity from 1-^{14}C-acetate appeared to accumulate in glycerolipids while that from

Table 1. Incorporation of ^{14}C from 1-^{14}C-acetate* and 2-^{14}C-pyruvate** into different lipid fractions of spinach leaf sections after 30 min (conditions see Fig. 1).

lipid species	Relative distribution (%) of incorporation from 1-^{14}C-acetate	2-^{14}C-pyruvate
phospholipids	47	18
steryl glucosides	5	22
monogalactosyl diglycerides	21	10
neutral lipids***	19	32
acyl-CoA-thioesters	3	1
others	5	17
total incorporation ($nmol \cdot mg^{-1}$ $Chl \cdot h^{-1}$)	140	21

*(0.3 mM; 0.49 $GBq \cdot mmol^{-1}$) **(0.3 mM; 0.28 $GBq \cdot mmol^{-1}$)

***pigments and acylated steryl glucosides are included

2-^{14}C-pyruvate appeared to be shifted in favour of lipid fractions originating from isoprenoid metabolism.

Acknowledgement

This work was supported by the Deutsche Forschungsgemeinschaft

REFERENCES

Proudlove, M.O. and Thurman, D.A. (1981), New Phytol. 88, 255-264

Roughan, P.G., Holland, R., Slack, C.R. and Mudd, J.B. (1978), Biochem. J. 184, 565-569

Sauer, A. and Heise, K.-P. (1983), Plant Physiol. 73, 11-15

Treede, H.-J., Riens, B. and Heise, K.-P. (1986), Z. Naturforsch 41c, 733-740

Von Korff (1969), Methods Enzymol. 13, 519-523

Yamada, M. and Nakamura, Y. (1975), Plant Cell Physiol. 16, 151-162

BIOLOGICAL ROLE OF PLANT LIPIDS
P.A. BIACS, K. GRUIZ, T. KREMMER (eds)
Akadémiai Kiadó, Budapest and Plenum Publishing
Corporation, New York and London, 1989

BIOSYNTHETIC PATHWAY OF α-LINOLENIC ACID IN OLIVE PLANT LEAVES

M. Zarrouk, B. Marzouk and A. Cherif

Centre de Biologie et de Ressources Génétiques, I.N.R.S.T Borj-Cedria B.P 95, Hamman-Lif, Tunisie

INTRODUCTION

It is now well established that the synthesis of α-linolenic acid ($C_{18:3}$) in higher plant leaves occurs by sequential desaturation of oleic acid (1,2). Althought some uncertainties remain about the subcellular sites and the substrates involved in the two desaturation steps of oleic acid. According to the hypothesis of cooperation between cellular compartments for the synthesis of linolenic acid (eucaryotic pathway) (3,4,5), PC located in the ER is the predominant substrate for desaturation of $C_{18:1}$ to $C_{18:2}$.
For the desaturation of $C_{18:2}$ to $C_{18:3}$ two molecules are thought to be the substrates DGG and DAG.
In the present study, we try to investigate the $C_{18:3}$ synthesis in olive plant leaves ($C_{18:3}$ - plant) by labelling kinetics with sodium (1-^{14}C) acetate as precursor.

MATERIAL AND METHODS

Labelling experiments were carried out on young olive plant one year aged. Microdroplets of the radioactive solution (specific radioactivity 1,94GBq/mmole) were deposited on the surface of the first leaf from the top. Leaves were harvested after different times of incubation (2,4,6,12,24 and 48 h) and rinsed with distilled water. Lipids were extracted in chloroform - methanol (1:2, v:v) then separated by thin layer chromatography. The obtained fatty acids were analysed by radio gas chromatography. Details of the methods have been given elsewhere (6).

Kinetics of (1-^{14}C) acetate incorporation into total fatty acids of olive plant leaves.

The distribution of the radioactivity incorporated from (1-^{14}C) acetate into leaf total fatty acids (data not shwon) **showed** that oleic acid is highly labelled at 2 h but its relative radioactivity decreased regularly with time of incorporation while that of linoleic and linolenic acids (little or none labelled at 2 h) increased throughout the experiment. This confirms the progressive desaturation pathway for the bioynthesis of linolenic acid (7).

Kinetics of (1-^{14}C) acetate incorporation into main polar lipids of olive plant leaves.

During the first 6 hours of incorparation, phosphadidylocholine (PC) was the most highly labelled lipid, it contained more than 50% of the incorporated radioactivity of polar lipids. Afterwards its percentage of radioactivity declined until 48 h, while that of diacylgalactosylglycerol (DGG) anddiacyl-digalactosylglycerol (DDG) increased (Fig 1 A).

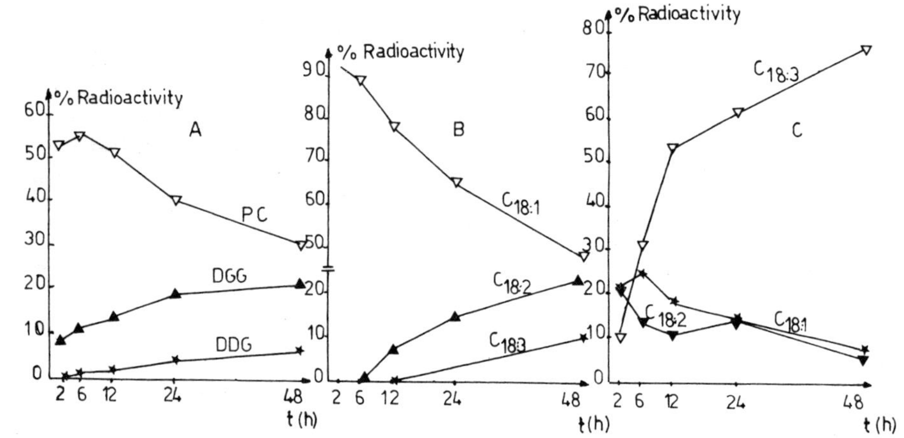

Fig.1 _ Incorporation of the radioactivity
(A) in PC, DGG and DDG - (B) in the fatty acids of PC -
(C) in the fatty acids of DGG.

This suggested a transfer of radioactivity from PC to galactolipids. Based on a dual labelling experiments with (1-^{14}C) acetate and (2-^{3}H) glycerol developping maïze and *Avena* leaves, SLACK et al. (8) and OHNISHI and YAMADA (3) showed that there is a transfer of DAG from PC to DGG following the scheme :

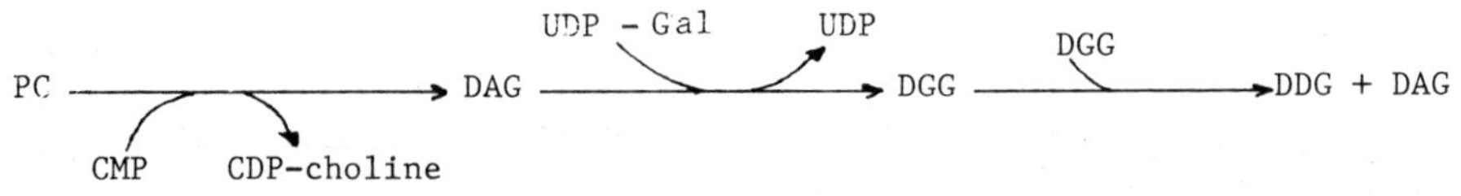

The very slow rate of labelling of DDG (Fig. 1A) suggested that it was produced by galactosylation of DGG (9).

Kinetics of $(1-^{14}C)$ acetate incorparation into fatty acids of PC and DGG of olive plant leaves.

In PC, oleic acid was initially the most heavily labelled fatty acid, it contained the major part of the radioactivity incorporated into PC (92,5%); its label declined throughout the experiment. Whereas both the linoleic and linolenic acids of PC became labelled after a lag of 6 and 12h respectively, and gradually gained label until 48h (Fig. 1B). This result suggested a slow sequential desaturation of oleic acid to linolenic acid via linoleic one within the PC molecule.

For DGG, the distribution of the radioactivity among fatty acids was characterized by a very rapid labelling of linolenic acid which reached the radioactivity level of 77% after 48h of incubation (Fig. 1C).

CONCLUSION

The comparison of labelling patterns of fatty acids acids of PC and DGG (Fig. 2A and 3A) showed that for the first hours of incorparation (until 12h) the linoleic acid of PC accumulated little or none radioactivity while both the linoleic and linolenic acids of DGG were largely labelled.This finding Suggested that oleic acid could be the major fatty acid donated by PC to DGG (probably as $C_{18:1}$-DAG), therefore, in olive plant leaf, PC didn't seem to serve as substrate for the desaturation of oleic acid to linoleic one as it was generally accepted by authors in other plant species.

One could believe that $C_{18:1}$ was desaturated on DAG molecule before galactosylation and formation of DGG as it was proposed by KESRI-BEN HASSAINE(5) on Carthamus cotyledons, but the distribution of the radioactive precursor into DAG fatty acids in olive plant leaves (data not shown) indicated that only palmitic and oleic acids were labelled at any time of incubation. This lead us to think that the two desaturation steps of oleic acid could take place on DGG molecule (see the following scheme). However in the absence of a direct evidence this must remain merely as a hypothesis.

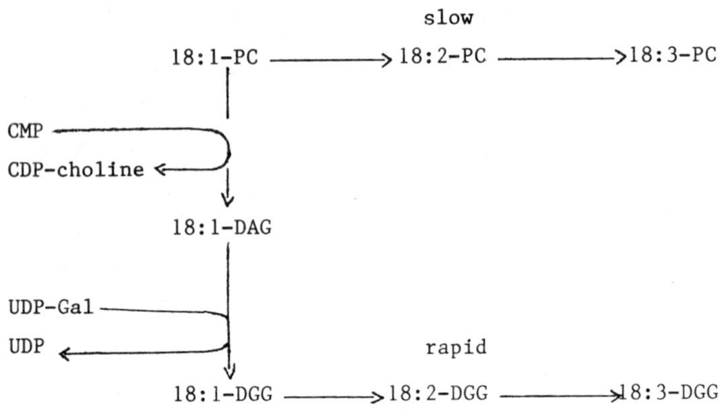

REFERENCES

1. Harris R.V. and James A.T. (1965) Linoleic and linolenic acid biosynthesis in plant leaves and a green algae.
Biochim. Biophys. Acta, 106,456-464

2. Chérif A., Dubacq J.P, Mache A., Oursel A. et Trémolières A. (1975) Biosynthesis of α-linolenic acid by desaturation of oleic and linoleic acids several organs of higher and lower plants.
Phytochem., 14, 703-706.

3. Ohnishi J.I. and yamada M. (1980) Glycerolipid synthesis in Avena leaves during greening of etiolated seedlings III.
Synthesis of α-linolenoyl-monogalactosyl diaclyglycerol from liposomal linoleoyl-phosphatidylcholine by Avena plastids in the presence of phosphatidylcholine exhange protein.
Plant and Cell physiol., 23,767-773.

4. Roughan P.G. and Slack C.R. (1982) Cellular organization of glycerolipid metabolism. Ann. Rev. Plant Physiol., 33, 97-132.

5. Kesri-Benhassaine G. (1985) Etude du metabolisme des lipides au cours du developpement du carthame (Carthamus tinctorius L.)
Thèse de Doctorat d'Etat, Université des Sciences de la Technologie d'Alger, 152 p.

6. Zarrouk M. (1988) Action du chlorure de sodium sur quelques aspects du métabolisme lipidique de l'oliver (Olea europea L.)
Thèse de Doctorat de Spécialité, Faculté des Sciences de Tunis, 119 p.

7. Trémolières A. et Mazliak P. (1974) Biosynthetic pathway of α-linolenic acid in developping pea leaves . In vivo and in vitro study.
Plant Science letters, 2, 193-201.

8. Slack C.R. , Roughan P.G. and Balasingham N. (1977) Labelling sutudies
in vivo on the metabolism of the acyl and glycerol moities of the glycerolipids in the developping maïze leaf.
Biochem. J. , 162,289-296.
9. Heemskerk J.W.M. , Jacobs F.H.M. , Bogemann G. and Wintermans J.F.G.M.
(1986) Galactosyltransferase acitivities in intact spinach chloroplasts and envelope membranes.
In *the Metabolism, Structure and Function of Plant Lipids*, P.K. Stumpf, J.B. Mudd and W.D. Nes Eds., Plenum Press, California, 305-307.

BIOLOGICAL ROLE OF PLANT LIPIDS
P.A. BIACS, K. GRUIZ, T. KREMMER (eds)
Akadémiai Kiadó, Budapest and Plenum Publishing
Corporation, New York and London, 1989

METABOLISM OF EXOGENOUS FATTY ACIDS BY LEAVES: POSITIONAL SPECIFICATIONS

P.G. Roughan[1] and G.A. Thompson Jr.[2]

[1]Division of Horticulture and Processing, DSIR, Mt. Albert Research Centre, Private Bag, Auckland, NewZealand
[2]Department of Botany, University of Texas, Austin, TX 78108, USA

There is good reason to believe that unesterified fatty acids (ufa) of chain length greater than C14 cannot be ligated to ACP or CoA within the chloroplast stroma. Therefore, cellular ufa may not be used for the synthesis of procaryotic glycerolipids since the acyltransferases responsible for initiating the synthesis of such glycerolipids are localized within the stroma. If a convenient way can be found for introducing labelled long chain fatty acids into leaf cells it should be possible to follow their eucaryotic metabolism in the absence of concomitant procaryotic metabolism. This is of particular interest in the case of palmitic acid metabolism by 16:3 plants.

We have already compared the incorporations of exogenous acetate, palmitic (16:3), stearic (18:0), and oleic (18:1) acids into the leaf glycerolipids of spinach (Roughan et al, 1987). Following application of the precursors in liquid paraffin to the upper surfaces of very young leaves on very young plants, and placing particular emphasis on the positional localization of labelled fatty acids, the important findings were:

(1) Saturated fatty acids were initially incorporated only into the sn-1 positions of glycerolipids, principally phosphatidylcholine (PC) and triacylglycerol (TAG).

(2) There was a slow transfer of 16:0 from PC to chloroplast glycolipids with the positional specificity remaining intact.

(3) Diacylgalactosylglycerol (DGG) and diacyldigalactosylglycerol (DDG) were labelled slowly and about equally by exogenous 16:0.

(4) Both 18:0 and 16:0 were desaturated to their respective trienes at the sn-1 position of DGG.

(5) Exogenous 16:0 and 18:1 were incorporated predominantly (>90%) into

eucaryotic species of phosphatidylglycerol (PG) and diacylsulphoquinovosylglycerol (DSG).

(6) Exogenous 18:1 was incorporated EQUALLY into both positions of PC, and was there desaturated to 18:2.

(7) The 18:2 was rapidly transferred from PC to both positions equally of DGG, and then desaturated to 18:3.

(8) DDG was labelled much more slowly than DGG by exogenous 18:1.

(9) Fatty acids synthesised in situ from exogenous acetate were incorporated equally into both positions of glycerolipids, predominantly PC, DGG and PG

(10) Fatty acid analyses by radioactivity and by mass were equivalent for DGG and PG, but not for PC, within 24 h of labelling with acetate.

The spinach data are compatible with the biosynthetic schemes shown in Figure 1.

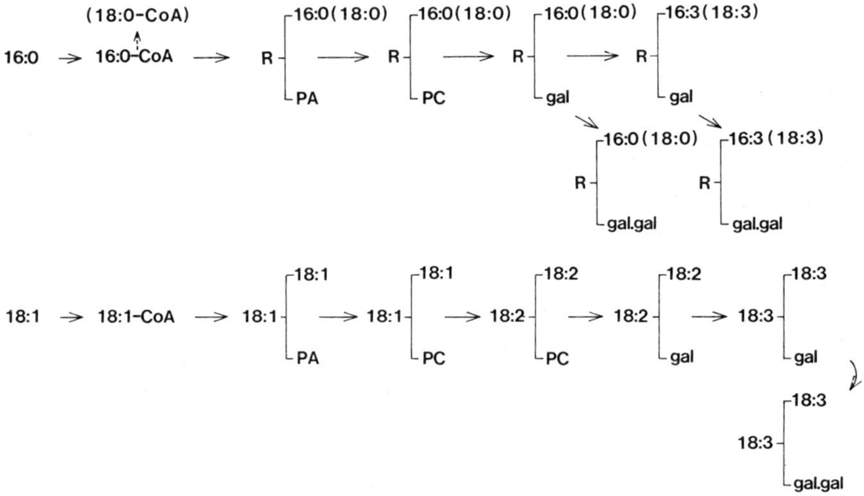

Figure 1. Major pathways for incorporation of exogenous palmitic (16:0), stearic (18:0) and oleic (18:1) acids into leaf glycerolipids.

Studies on the incorporation of exogenous 16:0 into the glycerolipids of expanding leaves have now been extended to pepino (<u>Solanum muricatum</u>), which like spinach contains about 50% procaryotic DGG, and to celery (<u>Apium graveoleus</u>), which contains about 75% procaryotic DGG. Uptake of the tracer from liquid paraffin was much slower from these leaves compared with spinach, even though all plants were grown in the same

TABLE 1. RADIOCARBON WITHIN LIPIDS FOLLOWING APPLICATION OF PALMITIC ACID TO PEPINO AND CELERY LEAVES

	Days	PI	PC	SQD	DDG	PG	PE	DGG	UFA	NP	% ^{14}C recovery
					% of total radiocarbon						
Pepino	1	6	17	2	3	2	7	3	49	10	80
	2	7	23	5	10	3	11	6	19	14	70
	4	5	15	6	14	3	7	6	19	19	48
	8	4	11	10	24	2	5	8	13	21	45
Celery	1	5	16	*	*	3	6	*	52	17	90
	2	5	22	*	2	3	10	2	27	28	76
	4	4	23	1	3	3	11	2	15	37	75

Lipids were extracted from leaves at the times indicated following application of the tracer and were separated by two dimensional tlc for scintillation counting. * = <1% NP = non-polar lipids.

conditions, and 20-30% of the fatty acid was unmetabolized after two days (Table 1). Also, a greater proportion of the 16:0 was elongated to 18:0 prior to incorporation into glycerolipids (Table 2). Incorporation into galactolipids was proportionally even lower, particularly in celery, than in spinach. DDG of pepino was labelled 2-3 times more rapidly than was DGG. With the exception of DGG in celery, all glycerolipids contained > 90% of their label at sn-1 for up to 4 days after application of the tracer. A high turnover of eucaryotic DGG for DDG synthesis probably

TABLE 2. RADIOCARBON WITHIN PHOSPHOLIPID FATTY ACIDS FOLLOWING APPLICATION OF PALMITIC ACID TO PEPINO AND CELERY LEAVES.

Fatty acid	PC		PE		PG	
	Pepino	Celery	Pepino	Celery	Pepino	Celery
			% of total fatty acid radiocarbon			
16:0	83	90	85	89	91	93
18:0	16	9	14	10	8	6
18:1	*	*	*	*	*	*
18:2	*	*	*	*	*	*
% ^{14}C at sn-1	96	93	95	94	97	92

Phospholipids were isolated by preparative tlc and were treated with phosphlipase A2. Released fatty acids and fatty acids from the lyso-lipids were resolved by argentation tlc. Values for pepino are averages for 4 and 8 days and those for celery are averages for 2 and 4 days. * = <1%.

TABLE 3. RADIOCARBON WITHIN FATTY ACIDS OF GALACTOLIPIDS FOLLOWING APPLICATION OF PALMITIC ACID TO PEPINO AND CELERY LEAVES

	Pepino				Celery			
	3 days		6 days		2 days		4 days	
Fatty acid	DGG	DDG	DGG	DDG	DGG	DDG	DGG	DDG
	% of total fatty acid radiocarbon							
16:0	49	83	44	76	41	82	34	77
16:1	2	1	*	*	4	1	5	2
16:2	2	*	*	*	13	1	12	2
16:3	7	*	5	*	21	2	24	4
18:0	10	11	11	18	4	6	4	9
18:1	1	*	*	*	*	2	*	*
18:2	3	1	*	*	3	3	3	2
18:3	26	4	35	4	13	3	18	4
% ^{14}C at _sn_-1	91	95	93	96	80	93	71	92

Galactolipids were isolated by preparative tlc and were hydrolysed using Rhizopus lipase. Released fatty acids and fatty acids from the lyso-lipids were separated by a combination of argentation and reverse-phase tlc.
* = < 1%

accounted for the relatively high proportion of radioactivity at the _sn_-2 position of celery DGG. The 18:0 that was incorporated into sn-1 of DGG of both species was efficiently desaturated to 18:3 whereas the 16:0 was desaturated relatively slowly, particularly in pepino (Table 3). Different labelled fatty acids in DGG and DDG, even though the latter is synthesised by the addition of galactose to the former, confirm that certain molecular species of DGG are preferred or selected for DDG synthesis. No hexadeca-_trans_-3-enoate was synthesised within PG of either plant from the supplied palmitic acid. These results nicely complement those obtained using spinach.

Reference

Roughan PG, Thompson GA Jr, Cho SH 1987. Arch Biochem Biophys _258_ 481-496

BIOLOGICAL ROLE OF PLANT LIPIDS
P.A. BIACS, K. GRUIZ, T. KREMMER (eds)
Akadémiai Kiadó, Budapest and Plenum Publishing
Corporation, New York and London, 1989

CONCENTRATION OF LONG-CHAIN ACYL-(ACYL-CARRIER PROTEIN) IN MESOPHYL CHLOROPLASTS FROM THE CHILLING-SENSITIVE PLANT AMARANTHUS LIVIDUS L.

I. Nishida[1] and P.G. Roughan[2]

[1]National Institute for Basic Biology, Myodaiji, Okazaki 444, Japan
[2]DSIR, Private Bag, Auckland, New Zealand

Correlation has been suggested between phosphatidylglycerol (PG) and the chilling sensitivity of higher plants (Murata 1983, Roughan 1985). Although acyl-(acyl-carrier protein) (acyl-ACP) is a physiological acyl-donor for the biosynthesis of PG in chloroplasts, its concentration in chloroplasts of chilling-sensitive plants is not thoroughly investigated (Roughan 1986).

Illuminated Amaranthus chloroplasts synthesized fatty acids from [1-^{14}C]acetate (17.5 mmol/Ci) at constant rates (800-2,200 nmol acetate h^{-1}/mg chlorophyll) for at least 5 min, and the label was incorporated into free fatty acids, PA, DG, lyso-PA and acyl-ACP. The labeled acyl-ACPs were recovered from the chloroplasts by a precipitation method modified from that of Mancha et al. (1975): The incubation was stopped by adding 1 volume of 5% acetic acid/2-propanol, and acyl-ACPs were precipitated by successive additions of 25 µl of saturated $(NH_4)_2SO_4$ and 5 ml of $CHCl_3/CH_3OH$ (2:1, v/v) containing 1% (v/v) of acetic acid. Radioactivity of the acyl-ACPs, which was counted after their acyl groups were converted to methyl esters by treatment with $NaOCH_3$, reached the saturation level within 1 min, and the level remained almost constant until the rate of fatty acid synthesis began to fall off after 5-min incubation or more. When the rate of fatty acid synthesis was at 2,200 nmol acetate h^{-1}/mg chlorophyll after 5-min incubation, [18:1-ACP], [18:0-ACP] and [16:0-ACP] were calculated to be 44, 99 and 71 pmol/mg chlorophyll, respectively.

The labeled acyl-ACPs were also recovered from the chloroplasts by silicic acid column chromatography modified from the silica gel thin-layer chromatography of Jackowski and Rock (1981): The incubation was stopped by adding 5 volumes of 1-butanol/acetic acid (4:1, v/v), and the resultant mixture was applied to a silicic acid column (0.9ϕ cm x 2.5 cm) equilibrated with 1-butanol/acetic acid/water (4:1:1, by volumes). The column was washed with 10 ml of the equilibrium solvent, and then with 10 ml of 1-butanol/acetic acid/water (5:2:4, by volumes). Fractions which were eluted with the latter solvent were combined for acyl-ACP, since it was confirmed that [^{3}H]alanine-labeled E. coli ACP was eluted in these fractions. Essentially equivalent results were obtained for [acyl-ACPs]. Aliquots of the lyophilized acyl-ACP were treated with $NaBH_4$ in 30% tetrahydrofuran/20 mM K-phosphate (pH 7.2) and with hydroxylamine (pH 8.0), respectively, to yield fatty alcohols and hydroxamates in 90-95% and 65-75% on the basis of total radioactivity of the acyl group. These results demonstrated that esterified fatty acids in the acyl-ACP fraction from the silicic acid column were solely acyl-thioesters.

When illuminated chloroplasts were transferred to darkness, amounts of radioactivity of acyl-ACPs decreased to less than 10% within 1 min. The radioactivity of acyl-ACPs as well as fatty acid synthesis resumed to almost their original levels within 1 min when the chloroplasts incubated in darkness were illuminated again. The quick response of changes in radioactivity of acyl-ACPs to fatty acid synthesis is consistent with the metabolic property which has been demonstrated for acyl-ACPs in spinach chloroplasts (Soll and Roughan 1982).

REFERENCES

Jackowski, S. and Rock, C. O. (1981). Regulation of coenzyme A biosynthesis. J. Bacteriol. 148, 926-932.

Mancha, M., Stokes, G. B. and Stumpf, P. K. (1975) Fat metabolism in higher plants. The determination of acyl-acyl carrier protein and acyl coenzyme A in a complex lipid mixture. Anal. Biochem. 68, 600-608.

Murata, N. (1983). Molecular species composition of phosphatidylglycerols from chilling-sensitive and chilling-resistant plants. Plant Cell Physiol. 24, 81-86.

Roughan, P. G. (1985). Phosphatidylglycerol and chilling sensitivity in plants. Plant Physiol. 77, 740-746.

Roughan, P. G. (1986). Acyl lipid synthesis by chloroplasts isolated from the chilling-sensitive plant Amaranthus lividus L. Biochim. Biophys. Acta 878, 371-379.

Soll, J. and Roughan, P. G. (1982). Acyl-acyl carrier protein pool sizes during steady-state fatty acid synthesis by isolated spinach chloroplasts. FEBS Lett. 146, 189-192.

BIOLOGICAL ROLE OF PLANT LIPIDS
P.A. BIACS, K. GRUIZ, T. KREMMER (eds)
Akadémiai Kiadó, Budapest and Plenum Publishing
Corporation, New York and London, 1989

TECHNIQUES FOR THE PURIFICATION OF ACYL-CoA ELONGASE FROM ALLIUM PORRUM LEAVES

J.-J. Bessoule, R. Lessire and C. Cassagne

Institut de Biochimie Cellulaire et Neurochimie-C.N.R.S. 1, Rue Camille Saint-Saëns, 33077 Bordeaux Cédex, France

INTRODUCTION

In Allium Porrum cells, the very long chain fatty acids (VLCFA), containing from 20 to 30 carbon atoms, are synthesized by the elongation of exogenous acyl-CoAs by malonyl-CoA (1). The reaction products are released as acyl-CoAs, incorporated into complex lipids and then transported to the plasma membrane (2,3). It has been previously demonstrated that the synthesis of VLCFA is catalysed by two enzymes that are distinguished from one another by the nature of their acyl-CoA primers (4) : a C18-CoA elongase and a C20-CoA elongase, and by their subcellular localization: the endoplasmic reticulum and the Golgi apparatus respectively (3). Nevertheless, despite several studies using plant and animal microsomal fractions (5,6), the solubilization of these enzymes has not been successful. Consequently, no purification of elongases has ever been carried out and their quaternary structure (one polypeptide or several subunits) is still unknown. In this paper we describe the techniques which can be used for the solubilization and the purification of the C18-CoA elongase from the microsomal fraction of Allium Porrum epidermal cells.

RESULTS AND DISCUSSION

Optimal conditions for the solubilization of the elongase.

In order to solubilize the C18-CoA elongating activity localized in the microsomal fraction from Allium porrum epidermal cells, three different detergents (Triton X-100; n Octyl ß-glucopyranoside; deoxycholate) were separately incubated at different concentrations with microsomes (obtained as previously described (7)). After 1 hour at 4°C, the mixtures were sedimented at 150 000g and the supernatants were used as solubilized enzyme sources. The optimal conditions for the solubilization have been defined as the conditions giving the highest elongating activity in the solubilized fraction. These optimal conditions were obtained when Triton X-100 was used. As indicated in Table 1, they do

TABLE 1: Triton concentration for optimal solubilisation of the acyl-CoA elongase

MICROSOMAL PROTEINS (μg)	[TRITON X-100] for optimal solubilization (mM)	TRITON X-100 microsomal protein (mg/mg)
160	1.20	0.98
210	1.65	0.98
650	5.00	1.00

not directly depend on the detergent concentration, but rather on the microsomal protein/detergent ratio (w/w) and the optimum solubilization required a Triton X-100/ protein (w/w) ratio of 1.

Purification of the C18-CoA elongase

In order to purify the C18-CoA elongase,the solubilized proteins were loaded onto a 5 ml DEAE cellulose column previously equilibrated with 0.08M Hepes buffer (pH 7) containing 0.02%Triton, 10mM ß-mercaptoethanol and 2% glycerol. 30% of the loaded proteins and the great majority of solubilized lipids were not retained. The column was then eluted with a 0 to 1 M gradient of NaCl. The fractions containing the C18-CoA elongating activity (generally three fractions) eluated for 0.3M NaCl, were pooled and then concentrated.
For the concentration step, several techniques have been used: for example, ammonium sulfate fractionation. The results reported in Table 2, show that the 0-35% ammonium sulfate fraction contained the largest amount of proteins, as well as the C18-CoA elongating activity. 73.2% of the total activity was recovered, but, unfortunately, the problem of eliminating the ammonium sulfate has not been solved. An overnight, or 1 hour , dialysis led to severe activity losses. The recovery of the total activity was only 12% and the specific activity was reduced more than 3-fold. Finally, the method giving the best results was to perform the concentration of the proteins using dialysis against dry G-200 Sephadex, which absorbs the moisture from the dialysis tube. As indicated in Table 2, under these conditions, practically no protein losses occurred. Moreover, the observed decrease of the acyl-CoA elongase activity was only due to the increase of the detergent/protein ratio resulting from the fact that the Triton X-100 was not dialysable and inhibited the activity when its concentration increased (4). Since this decrease is compensated during the next purification step, when the Triton concentration is readjusted to 0.02% (w/v), this method of concentration has been retained.
The concentrated proteins were then loaded onto a gel filtration column (Ultrogel AcA 34) previously equilibrated with 0.08M Hepes buffer (pH 7) con-

TABLE 2: Concentration studies of acyl-CoA elongase.

FRACTION	PURIFICATION FACTOR	YIELD (%)	PROTEIN RECOVERY(%)
Pooled fractions	1.O	100.0	100.0
NH2SO4			
0-35%	1.1	73.2	67.6
35-65%	_	0.9	15.1
0-35 % Dialysat	0.31	12.4	40.6
0-35% and Sephadex G25	0.75	32.9	43.9
Sephadex G200 Dialysat	0.81	80	98.5

taining 0.02%Triton, 10mM ß-mercaptoethanol and 2% glycerol.The elution of the Ultrogel column was carried out with the same buffer and fractions of 500µl were collected. The C18-CoA elongating activity eluted near the void volume, indicating an apparent molecular weight of 300kDa for this enzyme, in good agreement with our previous results (4).

After the filtration step on the Ultrogel column, the isolated proteins represented 0.7% of the total microsomal proteins used to carry out the purification (Table 3). These purified proteins were greatly delipidated : it could be calculated that, at the end of the purification, the relative lipid content (mg lipid/ mg protein) was only 8 % of the value observed in the original microsomal fraction.

Despite the high degree of delipidation, the techniques used have allowed to observe an increase in the specific activity. Based on this sole increase of the specific activity, the enrichment could be estimated at 4-5-fold, with a 1.44% yield.

TABLE 3 : Purification of the C18-CoA elongating activity.

STEP	PROTEINS (mg)	RELATIVE Phospholipid/	SPECIFIC ACTIVITY	YIELD (%)
MICROSOMES	10.0	1	0.33	100
SOLUBILISAT	5.0	1	0.54	82
DEAE	0.47	0.1	1.21	17
ULTROGEL	0.07	0.08	1.44	3

Electrophoresis analysis

In order to determine the structure of the elongases, the purified 300kDa complex was analysed by electrophoresis in the presence of sodium dodecyl sulfate. It appeared that the purified fraction was constituted by three major proteins of 56, 61 and 65 kDa respectively. Taking into account that during the purification, the proteins were incorporated into Triton X-100 micelles of 90 kDa (8), the molecular weights observed by electrophoresis were compatible with the 300 kDa estimated by gel filtration. Moreover, protein quantification by SDS electrophoresis/densitometry allowed to estimate that the 3 major protein bands represent about 0.8% of the microsomal proteins. This analysis gives a higher value for the enrichment factor (125-fold) for the acyl-CoA elongase, which is certainly a multi-molecular complex, than that obtained when the specific activity is used as the reference parameter. This difference is probably due to the high degree of delipidation of the enzymatic complex during the purification steps. Reconstitution studies have validated this hypothesis (9).

REFERENCES:

1: CASSAGNE C. and LESSIRE R. (1978) , Arch. Biochem. Biophy, 191, 146-152
2: LESSIRE R., JUGUELIN H., MOREAU P. and CASSAGNE C. (1985), Arch. Biochem.Biophy, 239, 260-269
3: MOREAU P., BERTHO P.,JUGUELIN H.and LESSIRE R. (1988), Plant Physiol Biochem., 26, 173-178
4: LESSIRE R., BESSOULE J -J. and CASSAGNE C. (1985), FEBS Lett., 187, 314-320
5: BOLTON P. and HARWOOD J.L. (1977), Biochem. J., 168, 261-269
6: BERNERT J.T. and SPRECHER H. (1979), J. Biol. Chem., 254, 11584-11590
7: AGRAWAL V.P., LESSIRE R. and STUMPF P.K. (1984), Arch. Biochem. Biophy, 239, 260 - 269
8: HJEMELAND L.M. and CHRAMBACH A. (1984), Methods in Enzymol., 104, 305-318
9: LESSIRE R. , BESSOULE J-J and CASSAGNE C. In this volume.

BIOLOGICAL ROLE OF PLANT LIPIDS
P.A. BIACS, K. GRUIZ, T. KREMMER (eds)
Akadémiai Kiadó, Budapest and Plenum Publishing
Corporation, New York and London, 1989

PROPERTIES OF PARTIALLY PURIFIED ACYL-CoA ELONGASE FROM ALLIUM PORRUM LEAVES

R. Lessire, J.-J. Bessoule and C. Cassagne

Institut de Biochimie Cellulaire et Neurochimie C.N.R.S. 1, Rue Camille Saint-Saëns, 33077 Bordeaux Cédex, France

In higher plants, the VLCFA biosynthesis, taking place in the microsomes, has been demonstrated in numerous materials (1-5). The existence of different acyl-CoA elongases has been postulated (3,4) and the separation of the C18-CoA elongase and C20-CoA elongase activities from leek epidermal cell microsomes have actually been carried out (7). Moreover, using Allium porrum epidermal cell microsomes, we solubilized and partially purified the acyl-CoA elongase (8). The electrophoresis analysis of the purified elongase, presenting an apparent molecular weight of 300 kD by gel filtration chromatography, demonstrated the presence of 3 major protein bands . The measurements of partial activities strongly suggested that the acyl-CoA elongase was composed of several separate enzymes (8). In this paper, the properties, the lipid requirement and the nature of the reaction product of the partially purified acyl-CoA elongase have been further investigated.

RESULTS:

1-Acyl-CoA specificity:

Saturated and unsaturated acyl-CoAs were tested as substrates for the partially purified enzyme (Figure 1). In the absence of acyl-CoA, no synthesis occurred, and, whatever the acyl-CoA chain length or saturation, the malonyl-CoA incorporation (expressed as nmoles synthesized /mg of proteins /h) was of the same order of magnitude. C16, C18 and C20-CoA were the more effective substrates, as noticed earlier using microsomes as enzyme source, but the differences of activities were not high. C18:1-CoA, which is not elongated at all by the microsomes, is accepted by the purified elongase.

2-Cofactor requirements:

The cofactor requirements for the acyl-CoA elongase were investigated (Figure 2). DTT and MgCl2 were required and their omission de creased the synthesis measured in the control by 34 and 48% respectively. NADPH seemed to be required, but NADH was not. In the absence of both reductants, the fatty acid synthesis remained unchanged. These results were not as clear as those obtained using the microsomes as enzyme source (4), and the fact that activity was detected in the absence of both reductants could be due, for instance, to the fact that the reductases involved in the elongation process were missing, or inactivated, as a consequence of the delipidation during the purification.

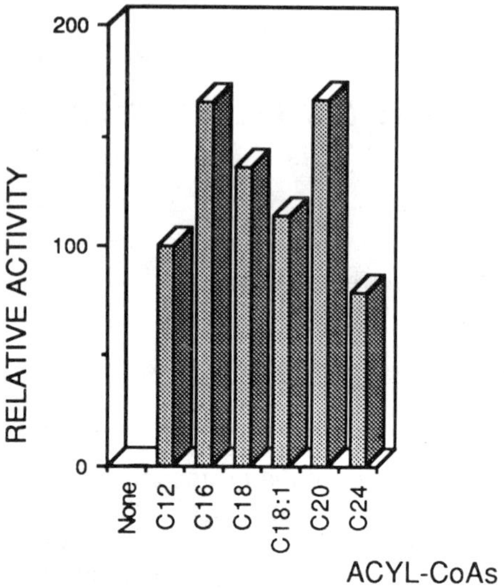

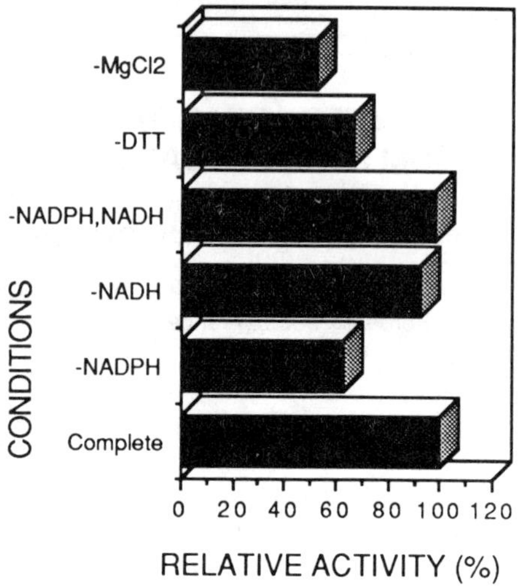

Fig 1: ACYL-CoAs SPECIFICITY OF C18-CoA ELONGASE.
About 1µg of partially purified elongase was incubated with each acyl-CoA (9µM) for 4h,at 30°C.

Fig 2: COFATOR REQUIREMENTS OF C18-CoA ELONGASE.
About 1µg of partially purified elongase was incubated for 4h,at 30°C.

3 Double-reciprocal plots for malonyl-CoA and stearoyl-CoA:

The activity of the acyl-CoA elongase was studied at malonyl-CoA concentrations in the range of 1 to 200 µM and at a stearoyl-CoA concentration of 6.9 µM. The double-reciprocal plots of Figure 3A shows the relationships between malonyl-CoA and/or stearoyl-CoA concentration and malonyl-CoA condensation to stearoyl-CoA.The apparent Km was 160 µM for the malonyl-CoA.The activity of the (at least partially restored) acyl-CoA elongase was studied at different concentrations of C18-CoA (Figure 3B), in the presence of 13.2 µg of soybean PC vesicules and 4.6

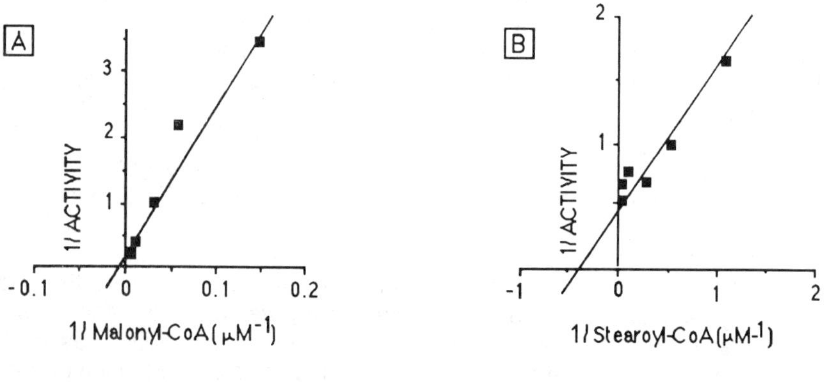

Fig 3: DOUBLE-RECIPROCAL PLOTS FOR MALONYL-CoA and STEAROYL-CoA.

μg of "leek lipids" . Under these conditions the apparent Km was 1.7μM.

4. Evidence of a long chain acyl-CoA formation:

After incubation of the proteins in the presence of (2-14C) malonyl-CoA and C18-CoA, the reaction was stopped by adding 6 volumes CHCl3/CH3OH 1:1 (v/v) . The resulting homogeneous phase was analyzed by TLC using BuOH/CH3COOH/H2O (5/2/3) as the developing solvent (9). The presence of labelled long chain acyl-CoAs was detected (Table 1), in good agreement with the fact that the acyl-CoAs have been demonstrated as being the products of the elongase (10). The formation of labelled long chain acyl-CoAs was strongly stimulated (4-fold) by the addition of lipids, indicating a lipid dependence of the acyl-CoA elongase. In this case, the activity jumped to about 2.7 nmoles/mg proteins/h, to reach the same range of activity as in the native microsomes (2.3 nmoles/mg/h) , suggesting a correct insertion of the enzyme(s) into the lipid vesicles.

TABLE 1: REACTION PRODUCTS OF THE PARTIALLY PURIFIED ACYL-CoA ELONGASE.

	RADIOACTIVITY (%)			
	Acetyl-CoA	Malonic acid	Long chain Acyl-CoAs	Unreacted Malonyl-CoA
Control	7.2	20.0	nd	72.8
Elongase	7.5	12.4	0.2	79.9
Elongase+ Lipids	7.2	12.4	0.9	79.5

4.5μg of proteins were incubated 150 min ,at 30°C, in the presence of 34μM of (2-14C) malonyl-CoA, NADPH (0.5mM),NADH (0.5mM),MgCL2 (1mM),and DTT (1mM) in the conditions described in Ref (4) .40 μl of" leek lipids" were added, and the control was made using (2-14C)malonyl-CoA in the presence of heat denaturated proteins. The products were separated by TLC and the results are expressed as percentages of initial radioactivity.

5-Lipid dependence of the acyl-CoA elongase:

As preliminary experiments, the influence of the addition of different amounts of lipids excluded from the DEAE-column ("leek lipids") on the C18-CoA elongase activity was studied (Table 2). The phospholipid composition of the lipid extract was determined by TLC densitometry as 3.31, 0.86 and 0.35 μg for PC, PS and PE respectively. The role of PC was chiefly studied and the influence of the addition of soybean PC vesicles on C18-CoA elongation was checked. In the presence of 19.8 μg of PC, the activity increased from 0.4 to 1.2 nmoles/mg/h. The highest activities were observed in the presence of 4.6 μg "leek lipids" with 13.2 μg of PC. Under these conditions the C18-CoA elongation was increased 6.5-fold. These results indicated that the addition of PC alone, which was the predominant lipid in the microsomes, is not sufficient to completely restore the acyl-CoA elongase activity.The ad-

TABLE 2: INFLUENCE OF LIPIDS ON THE ACYL-CoA ELONGASE ACTIVITY.

LIPID ADDITION	ACTIVITY (nmoles/mg/h)
None	0.4
6.9 µg "Leek Lipids"	1.6
19.8 µg PC	1.2
6.9 µg "Leek Lipids"+6.6 µg PC	2.3
4.6µg "Leek Lipids" + 13.2 µg PC	2.6

Acyl-CoA elongase activity was measured , using 1.5 µg of partially purified enzyme in the presence of different quantities of non-retained DEAE leek lipids and soy-bean PC.

dition of PC led to a stimulation of the activity lower than that observed in the presence of PS and PE. Furthermore, PS and PE were the most abundant lipids in the purified enzyme fraction, suggesting they could play an important role in the modulation of the elongase activity.

REFERENCES:

1. KOLATTUKUDY P.E. (1980) in The Biochemistry of Plants (Stumpf P.K. and Conn E.E.,eds), Vol. 4., p571,Academic Press,London.
2. HARWOOD J.L. (1988) Ann. Rev. Plant Physiol. Plant Mol. Biol.,39,101-137.
3. CASSAGNE C. and LESSIRE R. (1978) Arch. Biochem. Biophys. 191,146-152.
4. AGRAWAL V.P., LESSIRE R. and STUMPF P.K. (1984) Arch. Biochem. Biophys.,230,580-589.
5. LESSIRE R., JUGUELIN H.,MOREAU P. and CASSAGNE C. (1985) Phytochem.,24, 1187-1192.
6. AGRAWAL V.P. and STUMPF P.K. (1985) Arch. Biochem. Biophys.,240,154-165.
7. LESSIRE R.,BESSOULE J.J. and CASSAGNE C. (1985) FEBS Lett. 187,314-320.
8. BESSOULE J.J.,LESSIRE R. and CASSAGNE C. in this volume.
9. JUGUELIN H. and CASSAGNE C. (1984) Anal. Biochem.,142,329-335.
10. LESSIRE R., JUGUELIN H.,MOREAU P. and CASSAGNE C. (1985) Arch. Biochem. Biophys. ,239,260-269.

BIOLOGICAL ROLE OF PLANT LIPIDS
P.A. BIACS, K. GRUIZ, T. KREMMER (eds)
Akadémiai Kiadó, Budapest and Plenum Publishing
Corporation, New York and London, 1989

STORAGE OF MEDIUM AND LONG-CHAIN FATTY ACIDS IN TWO SPECIES OF DEVELOPING CUPHEA SEEDS

H.J. Treede, S. Deerberg and K.-P. Heise

Institut für Biochemie der Pflanze, University of Göttingen, D-3400 Göttingen, FRG

INTRODUCTION

Medium-chain fatty acids (MCFA) are formed in seeds of a number of plant species. Developing seeds of Cuphea wrightii, which mainly synthesize medium-chain fatty acids (C_8-C_{14}) show higher amounts of fatty acids than seeds of Cuphea racemosa, which primarily synthesize long-chain (C_{16}-C_{18})molecules. This difference is correlated with significant higher rates in Cuphea wrightii

1. of acetyl-CoA carboxylase in seed extracts
2. of lipid deposition within the embryo
3. of fatty acid synthesis, measured as acetate incorporation into fatty acids of the embryo.

In embryos of the medium-chain type, mainly fatty acids with chainlengths of 10 and 12 carbon atoms are synthesized and incorporated (up to 60%) into triglycerides. C_{16}- and C_{18}-fatty acids in the long-chain type, on the other hand, were bound up to 50% to polar lipids. The kinetic parameters of the acetyl-CoA-and malonyl-CoA-synthesizing enzymes in seed extracts showed similarities with spinach chloroplasts. Relative to the protein content this observation holds also for the level of some important substrates (ATP, ADP and pyruvate) for these key enzymes. An exception is the relatively high acetate concentration in both seed species. Non-aqueous fractionation has been applied to estimate the level of these substrates in the plastid.

Methods for metabolite measurements in Cuphea seed extracts have been described elsewhere (Treede et al. 1986). For acetate incorporation halved Cuphea seeds were incubated with 5 µCi 1-^{14}C acetate (53 Ci/mol, specific activity), in 0.2 ml of 50 mM MES, pH 6.0 at 37°C for 2 hours. After incubation, the embryos were separated from their coat and lipids were transmethylated according to the procedure of Thies (1974). Mass and radioactivity of lipids and fatty acids were measured by GLC-and TLC-techniques.

RESULTS AND DISCUSSION

The rate of lipid deposition during maximum fatty acid synthesis (between 20 th and 30 th day after flowering) was 8-fold higher for the short-chain- than for the long-chain type (Fig. 1). Similar high capacities in MCFA-formation are detected only in other Cuphea species (Singh et al. 1986). The higher rates of fatty acid deposition within short time periods in developing Cuphea wrightii embryos, could be interpreted as an important criterion for chain termination.

The de novo synthesized fatty acids in Cuphea wrightii were predominantly of medium chain length and were mainly deposited in triglycerides of the embryo (up to 60 %). In Cuphea racemosa, in contrast, longer chain fatty acids were the major component and were found (up to 50 %) in polar lipids within both embryo and seed coat (Table 1). In order to get an idea for the different lipid synthesizing capacities in the seeds of both Cuphea species important substrate- (acetate and pyruvate) and cofactor concentrations (ATP and ADP) for key enzymes of fatty acid synthesis have been measured in seed extracts and compared with corresponding ones in leaf tissues (Treede et al.1986). Except for the higher acetate levels in both Cuphea seeds, the other endogenous metabolite levels showed no significant differences between seed and leaf tissues.

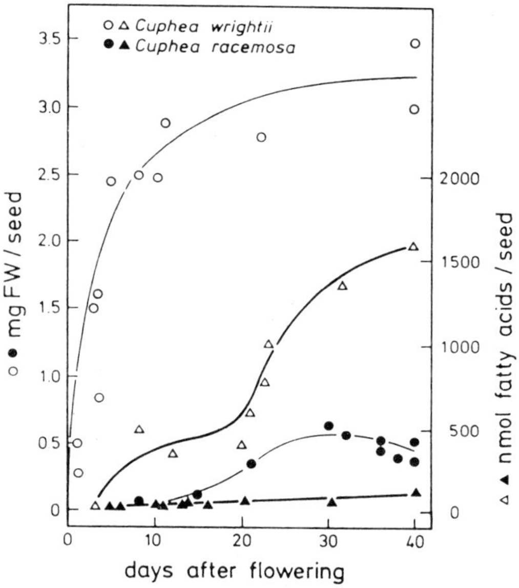

Fig. 1

Fatty acid deposition in Cuphea wrightii and Cuphea racemosa as a function of seed development

Table 1 Comparison of relative incorporation of ^{14}C from (1-^{14}C acetate) into fatty acids of different chain length with their mass (values in brackets) in halved seeds (30 days after flowering) of two *Cuphea* species

species	activity $\frac{\text{pmol(1-}^{14}\text{C)acetate}}{\text{mg fresh weight.h}}$	Percentage incorporation (mass) in fatty acids of the chain length 10:0	12:0	16:0	18:1	18:2
Cuphea wrightii						
embryo	33	33(37)	26(48)	13	18	(6)
coat	3	17(32)	26(16)	10(8)	37(18)	(20)
Cuphea racemosa						
embryo	7			43(20)	27(22)	22(55)
coat	15			56(19)	22(29)	14(52)

A rough estimation of the cellular distribution of these substrates, derived from non-aqueous fractionation studies according to Gerhardt and Heldt (1984), give no indication for differences in the control of the key enzymes of fatty acids by above metabolites under physiological conditions (Table 2).

Table 2 Determination of acetate, pyruvate and adenylate concentrations (nmol.mg^{-1}prot) from Cuphea seed- and spinach leaf extracts. An estimation of plastidial metabolite levels are given in brackets.

species	ATP	acetate	pyruvate	ATP/ADP-ratio
seeds:				
Cuphea wr.	1-3(0.3-0.7)	7-30(2-7)	0.7(0.3)	1-5 (2)
Cuphea rac.	1 (0.2-0.3)	15-23(3-5)	1.6(0.4)	1-4 (1)
leaves:				
spinach	5 (2)	2- 4(0.2-0.6)	0.5-1.5(0.04-0.12)	1-4 (2)

In *in vitro* experiments the chain length of newly synthesized fatty acids could be manipulated with different acetyl-CoA- and malonyl-CoA-ratios. Thus, the chain elongation could be reduced, if the acetyl-CoA (starter unit) concentration was increased (Singh et al. 1984). Under physiological conditions above metabolite ratio may be controlled by the activities of the acetyl-CoA synthetase (ACS) and acetyl-CoA carboxylase (ACC). Therefore, both enzyme activities have been compared in extracts (Table 3). Both seed materials showed similar ACS-activities as spinach chloroplasts. It appears of interest, however, that- like the total fatty acid synthesizing capacity- the ACC- activities of Cuphea wrightii seeds were 10-fold higher than in Cuphea racemosa and spinach chloroplasts.

Table 3 Kinetic constants of the acetyl-CoA synthetase and the acetyl-CoA carboxylase in extracts from Cuphea seeds and from spinach chloroplasts

enzyme activities*	spinach	Cuphea rac.	Cuphea wrightii
ACS	190	100	230
ACC	18	5-15	200
ACS/ACC	11	13	1

* nmol/mg protein.h

From our point of view at the present state of information for the deposition of medium-chain fatty acids in storage lipids of seeds two parameters appear of main interest.

(1) A very active fatty acid synthetase complex, an expression for which are the investigated parameters listed above.

(2) Typical transfer mechanisms (thioesterases; acyltransferases) which are specialized for the incorporation of these medium-chain fatty acids into storage lipids.

REFERENCES

1. Gerhardt, R. and Heldt, H. W. (1984) Plant Physiol. 75, 542-547
2. Sauer, A. and Heise, K.-P. (1983) Z. Naturforsch. 39, 268-275
3. Shimakata, T. and Stumpf, P.K.(1982) Proc.Natl.Acad.Sci.USA 79,5808-5812
4. Singh, S. S., Nee, T. and Pollard, M.R. (1984) in Structure, Function and Metabolism of Plant Lipids (Siegenthaler, P.-A. and Eichenberger, W., eds.) pp. 161-165, Elsevier Science Publishers BV, Amsterdam
5. Singh, S. S., Nee, T. and Pollard, M.R. (1986) Lipids 21, 143-149
6. Thies, W. (1974) Proc. 4 th Int. Rapeseed Conf. p. 275-283 Giessen, FRG
7. Treede, H.-J. and Heise, K.-P. (1985) Z. Naturforsch. 40, 496-502
8. Treede, H.-J., Riens, B. and Heise, K.-P. (1986), Z. Naturf. 41c 733-740

Acknowledgement This work was supported by BMFT 0318974 A7

BIOLOGICAL ROLE OF PLANT LIPIDS
P.A. BIACS, K. GRUIZ, T. KREMMER (eds)
Akadémiai Kiadó, Budapest and Plenum Publishing
Corporation, New York and London, 1989

THE ONTOGENY OF OIL STORAGE BODIES DURING SEED DEVELOPMENT IN CRUCIFERS

D.J. Murphy, I. Cummins and Jane N. O'Sullivan

Department of Biological Sciences, University of Durham, DH1 3LE, UK

INTRODUCTION

The seeds of oleiferous plants, such as rapeseed, soybean and maize, can contain up to 50% lipid by weight, in the form of triacylglycerols. In mature seeds, the triacylglycerols are contained in oil storage bodies of diameter 0.5-1.5 μm. The mechanism of oil-body formation has been investigated in several oilseed species including rapeseed (Murphy and Cummins, 1988), soybean (Herman, 1987), safflower (Ichihara, 1982) and maize (Huang *et al.*, 1987) but as yet no consensus view has emerged. The two main hypotheses are (i) oil-bodies are formed as mature entities by direct budding off from endoplasmic reticulum (ER) (Frey-Wyssling *et al.*, 1965; Wanner *et al.*,1981) and (ii) oil-bodies accumulate in the cytosol as naked droplets which subsequently acquire an osmiophilic membrane (Rest and Vaughan, 1972; Bergfeld *et al.*, 1978; Murphy and Cummins, 1988). In the present study, the ontogeny of oil-bodies has been studied using a combined biochemical, immunological and ultrastructural approach, in a wide range of oilseeds of the *Cruciferae*.

OLEINS - THE OIL-BODY MEMBRANE PROTEINS

We have reported previously that, in developing rapeseed embryos, the oil-bodies appear initially to lack a strongly electron-dense boundary layer (Murphy and Cummins, 1988). These immature oil-bodies contain little or no protein, but do contain some phospholipids. Close to the end of embryogenesis the oil-bodies acquire an osmiophilic layer, and simultaneously are found to contain about 10% w/w protein, most of which is made up of a single hydrophobic polypeptide of 19 kDa.

This protein has now been purified, and monospecific polyclonal antibodies have been raised against it. A similar antigenic polypeptide has been found in the seeds of all 15 species of crucifer tested to date (see Fig. 1). In contrast, there was no immunological cross-reactivity between the rapeseed anti-19kDa IgG and extracts from a total of 25 non-cruciferous oilseeds, including maize, soybean, sunflower, jojoba and linseed. The antigenic polypeptide in each crucifer species was localized in the oil-body fraction of the seed extracts. This was further confirmed by immunogold labelling (Fig. 2). Antibodies to the rapeseed 19 kDa oil-body protein specifically labelled oil-body membranes in ultrathin sections of mature rapeseed embryos, and also in mature embryos from radish, mustard and other cruciferous oilseeds. It is concluded that there is a family of structurally and functionally related proteins common to all oleiferous crucifers. Since these proteins are associated with oil storage, we propose the term OLEINS to describe this family of hydrophobic seed proteins. Preliminary proteolytic mapping studies have shown that the oleins from different crucifers give rise to similar patterns of cleavage fragments, which probably indicates a high degree of structural homology.

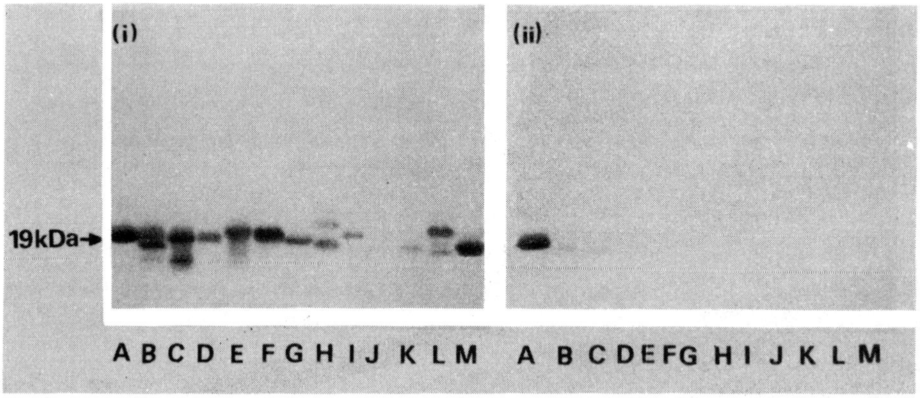

Fig.1 Immunoblot of the total oilseed proteins probed with IgG raised against the rapeseed 19kDa "olein". The antibodies, which recognise a single 19kDa polypeptide in rapeseed (lane A), cross-react with similar polypeptides in 15 different species of the *Cruciferae*, ie. (i) lanes A-M, (ii) lanes A-C. No cross-reactivity was observed with *Reseda* spp. (lanes D-F), *Pisum* spp. (G,H), maize (I), wheat (J) and *Linum* spp. (K-M).

The oleins are major seed proteins in crucifers, comprising up to 20% w/w of total seed protein, compared with about 20% and 45% respectively in the soluble seed storage proteins, napin and cruciferin. Hydrophobic proteins of 15-25 kDa are also associated with the oil-bodies of many other non-cruciferous species. While these proteins do not share common antigenic determinants with the oleins from crucifers, they do have similar functions. The interrelationship of oleins from a wide range of oilseed species in now under investigation in our laboratories.

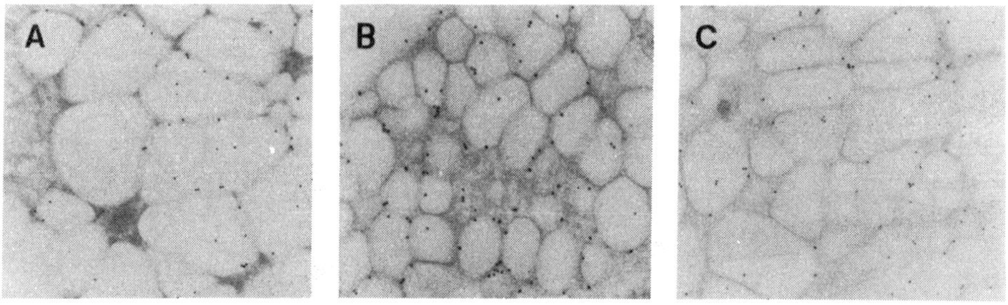

Fig.2 Immunogold labelled sections of cells from maturing cotyledons of (A) rapeseed; (B) mustard; (C) radish. Sections were treated with IgG raised against the rapeseed 19kDa "olein". Note specific labelling of the oil-body membranes of all three oilseed species. No gold labelling was observed in control sections treated with pre-immune serum.

SYNTHESIS OF OLEIN

Olein synthesis in rapeseed was followed by SDS-PAGE, ELISA and by [^{3}H]-leucine incorporation experiments. Data from SDS-PAGE showed that olein only accumulated at the latter stages of embryogenesis (Murphy and Cummins, 1988). These results were confirmed by an ELISA using antiserum raised against purified rape seed olein. Finally, olein synthesis was studied directly following *in vivo* translations in the presence of [^{3}H]-leucine, as shown in Fig. 3. Little or no trace of olein (19 kDa) was observed in embryos at the early or middle stages of development, whereas abundant olein synthesis was evident at the later stages.

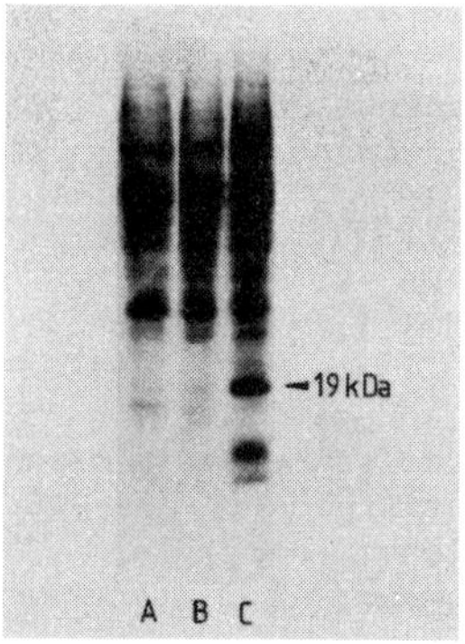

Fig.3 Translation products following incubation of embryos at (A) early, (B) middle and (C) late stages of maturation with [^{3}H]-leucine, and subsequent separation by SDS-PAGE. Note that olein (19 kDa) is only synthesised close to the end of embryogenesis.

MECHANISM OF OIL-BODY FORMATION

We were unable to observe budding-off of oil-bodies from ER, as proposed by Frey-Wyssling *et al.* (1963) and Wanner *et al.* (1981). Stobart *et al.* (1986) have reported that isolated safflower microsomes sectrete naked oil droplets into the medium. We have found that, in cruciferous seeds, oil-bodies were initially formed as large droplets of 1-3 μm diameter.

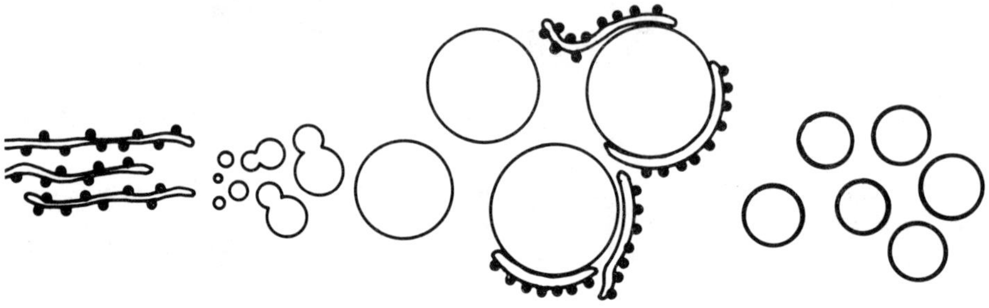

Fig.4 Model for oil-body formation. (A) Assembly of triacylglycerols on ER, (B) Sectretion and coalescence of small oil droplets, (C) Assembly of olein on oil-body boundary, (D) Appearance of mature oil-bodies with osmiophilic coat containing olein.

These then became surrounded by rough ER and acquired an electron-dense boundary layer, mostly made up of the 19 kDa hydrophobic protein, olein. Based upon these and other observations, we propose the scheme depicted in Fig.4 for oil-body formation in crucifers, and perhaps in other oilseeds.

ACKNOWLEDGEMENTS

This work was supported by grants from the Agriculture and Science Research Council and NATO Science Council. We thank Deborah Au for her assistance with parts of this work.

REFERENCES

Bergfeld, R., Hong, Y.N., Kuhnl, T. and Schopfer, P. (1978): Formation of oleosomes (storage oil bodies) during embryogenesis and their breakdown during seedling development in cotyledons of *Sinapsis alba* L. Planta **143**: 297-303.

Frey-Wyssling, A., Greishaber, E. and Mühlethaler, K. (1963): Origin of spherosomes in plant cells. J. Ultrastruct. Res. **8**: 506-516.

Herman, E.M. (1987): Immunogold localization and synthesis of an oil-body membrane protein in developing soybean seeds. Planta **172**: 336-345.

Huang, A.H.C., Qu, R., Wang, S., Vance, V.B., Cao, Y. and Lin, Y. (1987): Synthesis and degradation of lipid bodies in the scutella of maize. In: The Metabolism, Structure and Function of Plant Lipids pp 239-246. Stumpf, P.K., Mudd, J.B. and Nes, W.D. eds. Plenum, N.Y.

Ichihara, K. (1982): Formation of oleosomes in maturing safflower seeds. Agric. Biol. Chem. **46**: 1767-1773.

Murphy, D.J. and Cummins, I. (1988): Oilbody formation in developing rapeseed. In: Biotechnology for the Fats and Oils Industry. American Oil Chemists' Society, Champaign, USA, in press.

Rest, J.A. and Vaughan, J.G. (1972): The development of protein and oil bodies in the seed of *Sinapsis alba* L. Planta **105**: 245-262.

Stobart, A.K., Stymne, S. and Höglund, S. (1986): Safflower microsomes catalyse oil accumulation *in vitro*: a model system. Planta **169**: 33-37.

Wanner, G., Formanek, H. and Theimer, R.R. (1981): The ontogeny of lipid bodies (spherosomes) in plant cells. Ultrastructural evidence. Planta **151**: 109-123.

BIOLOGICAL ROLE OF PLANT LIPIDS
P.A. BIACS, K. GRUIZ, T. KREMMER (eds)
Akadémiai Kiadó, Budapest and Plenum Publishing
Corporation, New York and London, 1989

BIOSYNTHESIS OF TRIACYLGLYCEROLS CONTAINING VERY LONG CHAIN ACYL MOIETIES IN SEEDS

D.J. Murphy[1] and K.D. Mukherjee[2]

[1]University of Durham, Departement of Biological Sciences, Durham DH1 3LE, UK
[2]Federal Center for Lipid Research, Piusallee 68/76, D-4400 Münster, FRG

INTRODUCTION

Very long chain (n-9)-(*Z*)-monounsaturated fatty acids, such as gadoleic (20:1), erucic (22:1), and nervonic (24:1) acids, are major constituents of the seed triacylglycerols of several plants [1-5] that are of interest as renewable resources for the oleochemical industry [6]. We have shown recently that very long chain monounsaturated fatty acids are synthesized by the 15 000 x g particulate fractions from developing seeds of mustard (*Sinapis alba*), honesty (*Lunaria annua*), and nasturtium (*Tropaeolum majus*) by successive condensations of oleoyl-CoA with malonyl-CoA [7] in a similar manner as observed in several seeds [1,3,4,8,9]. We report here some properties of the elongase system catalyzing the synthesis of very long chain monounsaturated fatty acids and the pathways of their conversion to triacylglycerols.

EXPERIMENTAL

Methods used for isolation of the 15 000 x g particulate fraction from homogenates of the developing seeds, incubation of these fractions with [1-^{14}C]oleoyl-CoA or [2-^{14}C]malonyl-CoA in the presence of Mg^{2+}, CoASH, NADH and/or NADPH, isolation of the very long chain monounsaturated fatty acids, and the analysis of their methyl esters by radio gas chromatography were essentially as described before [4,7]. In several experiments, the total lipids were extracted from the products of incubation, fractionated into lipid classes, and analyzed by radio gas chromatography according to procedures described elsewhere [10].

RESULTS AND DISCUSSION

Time course of formation of very long chain monounsaturated fatty acids from [1-^{14}C]oleoyl-CoA or [2-^{14}C]malonyl-CoA by the 15 000 x g particulate fractions from *S. alba, L. annua, and T. majus* [11] corroborates the pathways involving successive condensations [1,3,4,8,9] in a similar manner as observed in the elongation of stearoyl-CoA [12].

Both NADH and NADPH were used as reductants in the condensation reactions, however, NADH was found to stimulate the formation of 20:1, whereas the synthesis of 22:1 was stimulated by NADPH [11]. Similar observations have been reported for a 15 000 x g particulate fraction from *B. juncea* seed [9]. These findings strongly suggest the existence of separate elongases.

Detergents, such as Triton X-100 and octyl thioglucoside, stimulate the elongases at low concentrations, but inhibit them at high concentrations [11]. Specifically, the formation of 22:1 is stimulated at low concentrations of Triton X-100 and inhibited at a higher concentration, whereas the formation of 20:1 is not affected at the higher detergent concentration. These findings support the concept of existence of separate elongases.

Treatment of the 15 000 x g particulate fraction from *S. alba* or *L. annua* with either detergent yielded solubilized fractions having substantial elongase activity compared to the controls [11]. Although the elongation to both 20:1 and 22:1 was observed in the controls, only 20:1 was synthesized by the solubilized fractions, which suggests that the entire elongase system catalyzing the formation of 20:1 has been selectively solubilized [11].

The very long chain acyl-CoA derivatives formed from radioactive oleoyl-CoA or malonyl-CoA did not accumulate in the acyl-CoA pool, instead, they were rapidly utilized for the formation of triacylglycerols and other glycerolipids. This is reminiscent of "metabolite channelling", as observed earlier in the desaturation of oleoylphosphatidylcholine [13].

The very long chain monounsaturated fatty acids synthesized from radioactive oleoyl-CoA or malonyl-CoA were readily incorporated into the major intermediates of the Kennedy-pathway, i.e. phosphatidic acids and diacylglycerols. Concomitantly the synthesis of triacylglycerols containing very long chain acyl moieties occurred, which is consistent with the operation of the Kennedy pathway.

With exception of minor proportions of 20:1, the very long chain monounsaturated fatty acids were not incorporated into either phosphatidylcholines or lysophosphatidic acids.

The mode of transfer of acyl moieties from very long chain acyl-CoA derivatives to glycerol-3-phosphate or lysophosphatidic acids deserves further investigation.

ACKNOWLEDGMENT

This work was supported by a research grant provided by Bundesministerium für Forschung und Technologie (BMFT), Bonn, Federal Republic of Germany.

REFERENCES

[1] Downey, R. K., Craig, B. M. (1964): Genetic control of fatty acid biosynthesis in rapeseed (*Brassica napus* L.). J. Am. Oil Chem. Soc. 41, 475-478.

[2] Gurr, M. I., Blades, J., Appleby, R. S.,Smith, C. G., Robinson, M. P., Nichols, B. W. (1974): Studies on seed oil triglycerides. Triglyceride biosynthesis and storage in whole seeds and oil bodies of *Crambe abyssinica*. Eur. J. Biochem. 43, 281-290.

[3] Pollard, M. R., Stumpf, P. K. (1980): Long chain (C_{20} and C_{22})fatty acid biosynthesis in developing seeds of *Tropaeolum majus*. An *in vivo* study. Plant Physiol. 66, 641-648.

[4] Mukherjee, K. D. (1986): Elongation of (n-9) and (n-7) *cis*-monounsaturated and saturated fatty acids in seeds of *Sinapis alba*. Lipids 21, 347-352.

[5] Mukherjee, K. D., Kiewitt, I. (1986): Lipids containing very long chain monounsaturated acyl moieties in seeds of *Lunaria annua*. Phytochemistry 25, 401-404.

[6] Princen, L. H., Rothfus, J. A. (1984): Development of new crops for industrial raw materials. J. Am. Oil Chem. Soc. 51, 282-289.

[7] Murphy, D. J., Mukherjee, K. D. (1988): Biosynthesis of very long chain monounsaturated fatty acids by subcellular fractions of developing seeds. FEBS Lett. 230, 101-104.
[8] Appleby, R. S., Gurr, M. I., Nichols, B. W. (1974): Studies on seed-oil triglycerides. Factors controlling the biosynthesis of fatty acids and acyl lipids in subcellular organelles of maturing *Crambe abyssinica* seeds. Eur. J. Biochem. 48, 209-216.
[9] Agrawal, V. P., Stumpf, P. K. (1985): Elongation systems involved in the biosynthesis of erucic acid from oleic acid in developing *Brassica juncea* seeds. Lipids 20, 361-366.
[10] Mukherjee, K. D. (1983): Lipid biosynthesis in developing mustard seed. Formation of triacylglycerols from endogenous and exogenous fatty acids. Plant Physiol. 73, 929-934.
[11] Murphy, D. J., Mukherjee, K. D. (1988): Properties of elongases synthesizing very long chain monounsaturated fatty acids in developing seeds. Lipids, submitted.
[12] Agrawal, V. P., Lessire, R., Stumpf, P. K. (1984): Biosynthesis of very long chain fatty acids in microsomes from epidermal cells of *Allium porrum* L. Arch. Biochem. Biophys. 230, 580-589.
[13] Murphy, D. J., Mukherjee, K. D., Woodrow, I. E. (1984): Functional association of a monoacylglycerophosphocholine acyltransferase and the oleoylglycerophosphocholine desaturase in microsomes from developing leaves. Eur. J. Biochem. 139, 373-379.

BIOLOGICAL ROLE OF PLANT LIPIDS
P.A. BIACS, K. GRUIZ, T. KREMMER (eds)
Akadémiai Kiadó, Budapest and Plenum Publishing
Corporation, New York and London, 1989

LIPID SYNTHESIS IN DEVELOPING COTYLEDONS OF LINOLENIC ACID DEFICIENT MUTANTS OF LINSEED

S. Stymne[1], A. Green[2] and M.L. Tonnet[2]

[1]Department of Plant Physiology, Swedish University of Agricurtural Sciences, P.O. Box 7047, 75007 Uppsala, Sweden
[2]CSIRO, Division of Plant Industry, G.P.O. Box 1600, Canberra City, ACT 2601, Australia

The biosynthetic pathways leading to alpha-linolenic acid in plants are poorely understood. This is partly due to the lability of the Δ15 desaturase(s) (catalysing the conversion of 18:2 to 18:3) which has prohibited any detailed *in-vitro* study of the enzyme(s). The present concept is that linolenic acid in plants is synthesised from linoleate while this acid is esterified in a complex lipid. In the leaf system the major lipid undergoing Δ15 desaturation is believed to be monogalactosyldiacylglycerol [4] and in oil seeds, phosphatidylcholine [5,6].

Alpha-linolenic acid (18:3) content is particularly high in linseed oil (45-65%). However, following EMS (ethyl methanesulphonate) mutagenesis of linseed (cv. Glenelg) two mutant lines, M1589 and M1722 were isolated which had a 30 - 40 % reduction in linolenate content in their seed oils [1]. By recombination of the two mutant lines a genotype was obtained with a near zero (<2%) linolenic acid level in the seed oil [2] referred to as "Zero". Only the ratio of linoleate (18:2) to linolenate was affected in the mutant lines which thus had normal content of the other fatty acids in their seeds. It was further shown that the two mutations exhibited codominant gene action and were on separate chromosomes [3].

In order to elucidate the mechanisms of alpha-linolenic acid synthesis in oil-seeds we have studied the biosynthesis and occurence of this acid in developing cotyledons of the high linolenic linseed (cv. Glenelg) and the linolenic acid deficient mutants, M1589, M1722 and "Zero" derived from this variety.

MATERIALS AND METHODS

Linseed plants were grown in a glasshouse in controlled enviroment of 22^{0}C/14^{0}C (12 hr day/12 hr night). Developing linseed cotyledons were carefully removed from their seedcoats 12 or 17 days after pollination (DAP) and subsequently used for the experiments or lipid analysis. *In-vivo* labelling experiments with [^{14}C] oleate and [^{14}C] linoleate were carried out essentially according to [7] with 10-15 cotyledon pairs and 1.8 µCi (ca 30 nmol) of ^{14}C fatty acid in each incubation. *In-vitro* assays of phosphatidylcholine desaturases were performed with homogenates of cotyledons (equivalent to 17 cotyledon pairs per incubation) prepared and incubated as described earlier for borage [8]. Extraction, separation and analytical procedures for the different lipids were carried out essentially according to [7] except that argentation t.l.c. [5] was used to separate the radioactive fatty acids according to unsaturation.

RESULTS

Dissected developing linseed cotyledons (12 and 17 DAP) of the parent variety (Glenelg) and the mutants lines (M1589, M1722 and Zero) were supplied with ammonium salt of

[1-^{14}C]linoleate for 90 min. Providing that the incubations were performed in bright light, the radioactive linoleate was readily taken up by the cotyledons, incorporated into various lipids and also partly destaurated to [^{14}C]linolenic acid. No significant difference in incorporation patterns and amounts were seen between the different lines (data not shown). About 80% of the total label that was incorporated into complex lipids were found in the three lipids involved in triacylglycerol biosynthesis, phosphatidylcholine, diacylglycerols and triacylglycerols. Distribution of the label between the sn-positions of sn-phosphatidylcholine (the presumed substrate for the Δ15 desaturase) did not differ between the lines (18-21% of the total label in phosphatidylcholine was found at position sn-1). The same analysis were performed in similar *in-vivo* labelling experiments for 40 min and also with [^{14}C]oleate as substrate. Again, no significant differences in incorporation pattern and amounts were seen between the lines (data not shown). Thus it appears that the mutations did not affect the enzymes channelling the precursors for linolenate synthesis, the oleate and the linoleate, into the different lipids.

Analysis of the radioactive distribution of label between [^{14}C]linoleate and [^{14}C]linolenate in various lipids after [^{14}C]linoleate feeding to developing cotyledons (17 DAP and 90 min of incubation) revealed significant lower levels of [^{14}C] linolenate in phospholipids and neutral lipids in the mutant lines compared to the wild type. The distribution of radioactivity between linoleate and linolenate at the individual sn-positions of sn-phosphatidylcholine also differed significantly between all the three mutant lines. At position sn-2, "Zero" had a 83% reduction in linolenate content compared to the wild type, while for both M1589 and M1722 corresponding reduction was 36% (Table 1). The "Zero" line had negligible radioactive linolenate at position sn-1, the M1589 had a 73% reduction compared to wild type and the M1722 did not differ significantly from the wild type in [^{14}C]linolenate content (Table 1). Similar, but less pronounced differences in [^{14}C]linolenate content in position sn-1 of sn-phosphatidylcholine between the M1589 and M1722 lines were also seen after the shorter period of incubation (data not shown).

Table 1. Distribution of [^{14}C] acyl groups in the sn-positions of phosphatidylcholine after [^{14}C]linoleate feeding *in-vivo* to developing cotyledons, 17 DAP. The results are means of two incubations.

	Relative radioactive distribution of fatty acids within each position (%)			
	Position sn-1		Position sn-2	
Line:	[^{14}C]18:2	[^{14}C]18:3	[^{14}C]18:2	[^{14}C]18:3
Glenelg	73.9	26.1	63.4	36.6
Zero	98.0	2.0	93.7	6.3
M1589	92.9	7.1	76.4	23.6
M1722	77.4	22.6	76.8	23.2

Table 2. Positional distribution of endogenous fatty acids in phosphatidylcholine from developing cotyledons, 17 DAP

	Relative distribution of fatty acids within each position (mol%)					
	Position sn-1			Position sn-2		
Line:	18:1	18:2	18:3	18:1	18:2	18:3
Glenelg	21.3	14.2	26.4	27.5	25.6	41.7
Zero	18.0	41.1	1.4	23.2	64.2	3.5
M 1589	20.6	29.3	11.6	26.3	38.3	26.5
M1722	22.4	20.3	20.0	27.4	43.8	20.8

Also, repeated analysis of the endogenous fatty acid composition at the different sn-positions of sn-phosphatidylcholine isolated from 17 DAP old cotyledons consitently revealed a difference between the M1589 and M1722 lines. Linolenate content of M1589 was somewhat higher at position sn-2 and substantially lower at position sn-1 then in the M1722 (Table 2).

Since phosphatidylcholine is considered to be a likely substrate for the Δ15 desaturase in oil-seeds[5,6] *in-vitro* assays of the phosphatidylcholine desaturases in homogenates from the different lines were carried out. When diluted homogenates (prepared from developing cotyledons according to Materials and Methods) were incubated in presence of air and NADH, endogenous oleate and linoleate of phosphatidylcholine were desaturated. This was evident by a NADH and air dependent decrease in the oleate content and an increase in the linolenate whereas the saturated fatty acids in the lipid remained unchanged. In incubations in the absence of NADH and under nitrogen the fatty acid distribution in phosphatidylcholine was indentical with that in the homogenates at the onset of the incubation. In homogenates of Glenelg, the desaturation of oleate occured with nearly the same rates at both sn-positions of sn-phosphatidylcholine and was virtually finished after 20 min of incubation (data not shown). Linoleate desaturation, on the other hand continued throughout a 60 min incubation period and the rates where about 3 times higher at position sn-2 than sn-1 (data not shown).

Desaturation rates of oleate and linoleate at the individual sn-positions of sn-phosphatidylcholine were compared after 40 min. of incubations with homogenates prepared from the different lines. Oleate desaturation occured to a similar extent in all lines and with a somewhat higher rate at position sn-2 (table 3). Linoleate desaturation at position sn-1 were variable but low in all lines. Because of the low amounts of phosphatidylcholine in each incubation, the low actual desaturation and a varying degrec of contamination with saturated fatty acids in the g.l.c. analysis of position sn-1, it is doubtful if the variation seen between the lines in linoleate desaturation at that position was significant (table 3). Only the Glenelg and M1589 lines showed significant desaturation of linoleate at position sn-2 (Table 3).

Table 3 *In-vitro* desaturation of endogenous oleate and linoleate at the different sn-positions of sn-phoshatidylcholine by homogenates of developing cotyledons.

Oleate desaturation rates were calculated by substracting the mol% of oleate found in the individual sn-positions in homogenates incubated under nitrogen from the mol% of oleate found in homogenates incubated with air and NADH. Linoleate desaturation rates are the increase in mol% of linolenate at the different sn-positions in incubations with air and NADH compared with incubations under nitrogen. Incubations were carried out at 25°C for 40 min. The results are the means of two incubations.

	nmol desaturated/(40 min x 100 nmol of phosphatidylcholine)			
	Position sn-1		Position 2	
	Oleate	Linoleate	Oleate	Linoleate
Line:				
Glenelg	8.3	5.1	9.9	16.0
Zero	6.0	0.5	10.1	0.0
M1589	8.5	1.4	12.7	9.7
M1722	8.3	1.2	11.5	1.7

DISCUSSION

Thus from *in-vivo* radioactive labelling experiment as well as from analysis of endogenous fatty acids in phosphatidylcholine it can be concluded that the two mutations M1589 and M1722 affect two different enzymes with somewhat different substrate specificity. The M1589 mutation inactivates an enzyme which is responsible for linolenic acid occurence at position sn-1 and sn-2 of phosphatidylcholine while the M1722 mutation primarly affects linolenate content at position sn-2. *In-vitro* assays of the linoleoyl-phophatidylcholine desaturase showed significant desaturation only in the M1589 and Glenelg lines. The majority (more that 75%) of that desaturation occured at position sn-2. This can be compared with the $\Delta 6$ linoleoyl-phosphatidylcholine desaturase (converting linoleate to gamma-linolenate) which is even more specific for linoleate at position sn-2 [9]. The present data suggest therefore that there are two different enzymes responsible for the synthesis of linolenate for the triacylglycerols in linseed. One enzyme is a linoleoyl-phosphatidylcholine desaturase which has a preference for linoleate at position sn-2. The activity of this enzyme is affected by the mutation in the M1722 line. The other enzyme, which is important for linolenate occurence in both position sn-1 and sn-2 of phosphatidylcholine might be a positional non-specific phosphatidylcholine desaturase which activity is lost in the *in-vitro* assays, or it might be a linoleoyl-desaturase with other subtrate than phosphatidylcholine. The M1589 line contains a mutation affecting the activity of this enzymes. Both enzymes have approximately the same importance in the synthesis of linolenate for the triacylglycerols in linseed.

ACKNOWLEDGEMENT

S. Stymne is grateful for financial support from CSIRO, Australia, The Swedish Natural Science Research Council and The Karlshamn Foundation (Sweden) to work as a visiting scientist at the Division of Plant Industry, CSIRO, Canberra.

REFERENCES

1. Green, A.G. and Marshall, D.R. (1984). Isolation of induced mutants in linseed (*Linum usitatissimum*) having reduced linolenic acid content. Euphytica **33**, 321-328.
2. Green, A.G. (1986). A mutant genotype of flax *(Linum usitatissimum. L.)* containing very low levels of linolenic acid in its seed oil. Can. J. Plant Sci. **66**, 499-503.
3. Green, A.G. (1986) Genetic control of polyunsaturated fatty acid biosynthesis in flax (*Linum usitatissimum*) seed oil. Theor. Appl. Genet. **72**, 654-661.
4. Roughan, P.G., Mudd, J.B., McManus, T.T. and Slack, C.R. (1979). Linoleic and alpha-linolenate synthesis by isolated spinach (*Spinacia oleracea*) chloroplasts. Biochem. J. **184**, 571-574.
5. Stymne, S. and Appelqvist L-Å. (1980). The biosynthesis of linoleate and alpha-linolenate in homogenates from developing soya bean cotyledons. Plant Sci. Lett. **17**, 287-294.
6. Browse, J.A. and Slack, C.R. (1981). Catalase stimulates linoleate desaturation in microsome preparations from developing linseed cotyledons. FEBS Lett. **131**, 111-114.
7. Stymne, S. and Stobart, A.K. (1986). Biosynthesis of gamma-linolenic acid in cotyledons and microsomal preparations of the developing seeds of common borage (*Borago officinalis*). Biochem. J. **240**, 385-393
8. Griffiths, A.K. Stobart, A.K. and Stymne, S. (1988) Delta-6 and delta-12 desaturase activities and phosphatidic acid formation in microsomal preparations from the developing cotyledons of common borage (*Borago officinalis*). Biochem. J. **252**, 641-647.

BIOLOGICAL ROLE OF PLANT LIPIDS
P.A. BIACS, K. GRUIZ, T. KREMMER (eds)
Akadémiai Kiadó, Budapest and Plenum Publishing
Corporation, New York and London, 1989

SYNTHESIS OF OCTADECATETRAENOIC ACID (OTA) IN BORAGE (BORAGO OFFICINALIS)

G. Griffiths[1], E.Y. Brechany[2], W.W. Christie[2], S. Stymne[1] and K. Stobart[3]

[1]Department of Plant Physiology, Swedes University of Agricultural Sciences, Uppsala, Sweden
[2]The Hannah Research Institute, Ayr, Scotland
[3]Department of Botany, Bristol University, England

Members of the *Boraginaceae* often have uncommon C18-polyunsaturated fatty acids. The seed TAGs are enriched in gamma-linolenic acid (GLA; C18:3,n-6)[1] and leaves may contain an octadecatetraenoic acid (OTA;C18:4,)[2]. Humans are unable to form linoleic acid (C18:2,n-6) and alpha-linolenic acid (ALA;C18:3,n-3). It is considered that the more direct prostaglandin precursors, GLA and, perhaps, OTA, may alleviate some clinical and atopic disorders. There is, therefore, interest in producing new plant varieties, perhaps from those of an already established agricultural nature , which synthesise oils with GLA and OTA, and so yield a more valuable commodity. Before such genetic transformations are feasible, however, it is essential to establish how such acids are synthesised and how this is integrated with complex lipid assembly. Previous work has dealt with GLA in developing seeds of borage[3-5] and we have speculated on the mode of OTA synthesis in leaves[5]. We report here further experiments on OTA in plants.

RESULTS AND DISCUSSION

Previously, the OTA in borage leaf lipids was identified by glc, Ag-tlc and hydrogenation. The chain length and double-bond distribution has now been examined by GC-MS of it's picolinyl ester derivative. Significant ions were present at m/z= 367, 352 and 338 and confirmed the empirical formula, the terminal methyl and the following methylene group, respectively. Prominant ions at m/z= 299, 338 and 312 confirmed the first double bond at position 15. Further gaps of 14 and 26 a.m.u. in sequence showed that the remaining double bonds were in positions 12, 9 and 6, respectively. The OTA was, therefore, C18:4, n-3. The OTA in the total leaf lipid of borage was relatively high (25%) and far greater than found in related spp. Borage leaf lipids, MGDG and DGDG, contained 35% and 13% OTA, respectvely, whereas it was absent in PG. OTA was also present in PC and PE (9% and 6%, respectively). GLA was present in all lipids with only traces in PG (MGDG, 6%; DGDG, 5%; PC, 7%; PE, 7%). ALA was found in all lipids examined. The OTA and GLA was only in position sn-2 of MGDG (79 and 10%, respectively) and absent at the sn-1. ALA accounted for nearly 90% of the fatty acids at the sn-1 of MGDG, with only 10% at the sn-2 position. The distribution of the fatty acids in PC indicated that GLA and OTA were almost totally restricted to the sn-2 position. ALA and GLA were also found in the seeds of some related spp. whose cotyledons also lacked chloroplasts. Under these circumstances the oil also contained OTA (eg. *Cynoglossum*: ALA, 8%; GLA, 8%; OTA, 3%. *Anchusa*: ALA, 17%; GLA, 17%; OTA, 5%). The results suggest that tissue with ALA and GLA will also contain OTA.

OTA Biosynthesis. Microsomes from the developing seeds of borage, which have a Δ6-desaturase [3,4], were incubated with ^{14}C-acylCoA substrates. The results (table 1) show that 18:2 was efficiently converted to GLA and no OTA was formed. Labelled ALA readily gave rise to OTA. No conversion of supplied GLA occurred. The microsomes lack a Δ15 desaturase but can utilise exogenously supplied ALA and convert it, through the action of the Δ desaturase, to OTA. Similar results were obtained with *Anchusa* microsomes

Table 1. OTA synthesis in borage microsomes

SUBSTRATE	RADIOACTIVITY RECOVERED (%)			
	18:2	GLA	ALA	OTA
[^{14}C]18:2CoA	65	35	-	0
[^{14}C]ALACoA	-	-	76	24
[^{14}C]GLACoA	-	100	-	0

It was necessary to establish whether tissues which can synthesise ALA and have no Δ6-desaturase, can form OTA from supplied GLA. Unfortunately, *in-vitro* membranes from seed and leaf sources did not retain Δ15 activity and leaf tissues rapidly oxidised C18-trienoic acid substrates. Experiments, therefore, used cotyledons from the immature seeds of linseed which form oil rich in ALA. The cotyledons had good Δ12 and Δ15 desaturase activity and readily utilised the ammonium salts of labelled C18:1 and C18:2 and formed ALA (table 2). [^{14}C]GLA was incorporated into PC, however, no radioactive OTA was detectable. The Δ15-desaturase appears, therefore, unable to utilise GLA as a substrate and introduce a 4th double bond. The OTA is formed, therefore, in a precise fashion and only via ALA and not GLA (scheme 1).
While ALA is possibly synthesised to some extent in the ER of the leaf, it is probable that most of the OTA in MGDG arises through the action of Δ6-, and Δ15-desaturases that are resident in the chloroplast with the Δ6 exhibiting specificity for substrate at position sn-2 [see 4]. This would account for the exclusion of OTA from position sn-1. The question still arises, however, of how the desaturases act in the specific sequence rather than randomly with

the formation of a mixed population of GLA and OTA at position sn-2 of MGDG. The observations also raise questions regarding substrate recognition for desaturases. The Δ12-desaturase may align with the double bond at position 9 and the methyl end[6]. Here, however, the Δ15 desaturase may be rather non-specific and attach to the 1st double bond from the carboxyl end and hence no OTA will arise from GLA. The Δ6-desaturase, on the other hand, may recognise the acyl chain from the double bond at position 9 and the carboxyl end, resulting in OTA from ALA. Substrate recognition for desaturases which utilise complex lipids requires further study.

Table 2. GLA utilisation in the developing cotyledons of linseed

A. In total-lipid fatty acids

SUBSTRATE	RADIOACTIVITY RECOVERED(%)				
	18:1	18:2	GLA	ALA	OTA
[14C]18:1	64	27	-	9	
[14C]18:2	-	66	-	34	-
[14C]GLA	-	-	100	-	0
B. In phosphatidylcholine fatty acids*					
[^{14}C]18:1	37(0.31)	50(0.42)	0	13(0.11)	0
[^{14}C]18:2	0	53(0.21)	0	47(0.19)	0
[^{14}C]GLA	0	0	100(0.42)	0	0

*, Figures in parentheses are nmol radioactive substrate incorporated.

Scheme1. Biosynthesis of OTA

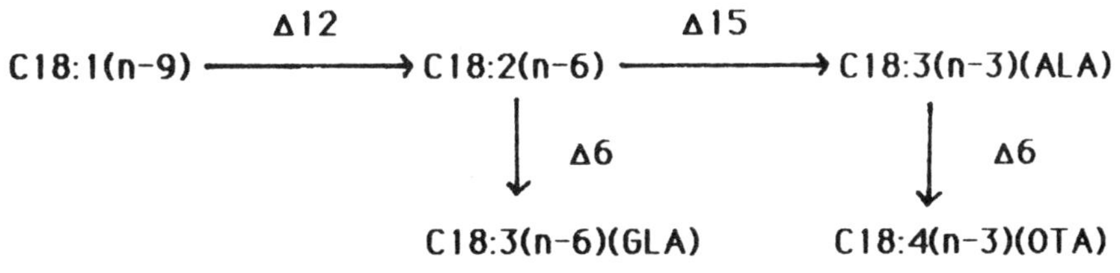

References

[1]Kleiman,R., Earle,F.R. & Wolff, I.A. (1964) J. Am. Oil Chem. Soc. **41**, 459-460.
[2]Jamieson,G.R. & Reid,E.H. (1969) Phytochem. **8**, 1489-1494.
[3]Stymne,S. & Stobart, A.K. (1986) Biochem. J. **240**, 385-393.
[4]Griffiths, G., Stobart, A. K. & Stymne,S. (1988) Biochem. J. **252**, 641-647.
[5]Stymne,S., Griffiths, G. & Stobart, A.K. (1987) in **The Metabolism, Structure & Function of Plant Lipids**, eds., P.K. Stumpf, J.B. Mudd & W.D. Nes, Plenum Press, N.Y., pp. 405-412.
[6]Howling,D., Morris,L.J., Gurr,M.I. & James,A.T. (1972) Biochim. biophys. Acta **260**, 10-19.

BIOLOGICAL ROLE OF PLA. T LIPIDS
P.A. BIACS, K. GRUIZ, T. KRE! 'MER (eds)
Akadémiai Kiadó, Budapest an. Plenum Publishing
Corporation, New York and Lond. ı, 1989

ANALYSIS OF SPECIFIC, UNSATURATED PLANT FATTY ACIDS

Emma Dabi-Lengyel, I. Zámbó, P. Tétényi and Eva Héthelyi

Research Institute for Medicinal Plants, Budakalász, Hungary

INTRODUCTION

For the selective epoxidation of specific fatty acids we expected to find such seed oils, containing cis-5-icosenoic (Eicosenoic) acid as the main component. On the base our examination we had chosen a plant with short life cycle, with good genetic flexibility, *Limnanthes alba* Benth (LIMNANTHACEAE) (Meadowfoam), which is native in North-America. It is well adapted to the our climate, and has good oil-yield (25.9%), with high concentration of icosenoic acid (46.8%) in total fatty acids.

Because the classical column chromatographic separation is difficult, we decided to work up a simplier enrichment and production of cis-5-icosenoic acid (1).

RESULTS, CONCLUSION

For the enrichment of the specific fatty acid we succeeded in producing fatty acid-mixtures, containing dominantly (73.3%) 5-cis-icosenoic acid ($C_{20:1}/5$) from *Limnanthes alba* seed oil.

Procedure was worked up to small and large scale enrichment of 5-icosenoic acid by low-temperature crystallization in acetone at laboratory condition as follows:

Enrichment-procedure of cis-5-icosenoic acid:

Limnanthes alba seeds
extr. by petrol (40 °C) 8 hrs.

↓

saponification,
soln. in acetone

↓

cooling -72 °C; 2.5 hrs. ←

↓

vacuum-filtration
(cooled) — Twice

↓

Precipitate
20 °C; 16 hrs.

↓

vacuum evaporation →
soln. in acetone

The analytical control of products was carried out by HPLC of the corresponding bromophenacyl esters (2). Concentrated product, containing 80% cis-5-icosenoic acid was produced by micropreparative HPLC method.

REFERENCES

1. Chang, Shu-Pei, Rothfus, J.A.: Enrichment of Eicosenoic and Docosadienoic Acids from Limnanthes Oil, "Journal of the American Oil Chemists' Society" Vol. *54*. 549-552, 1977.
2. Tweeten, T.N., Wetzel, D.L.: High Performance Liquid Chromatographic Analysis of Fatty Acid Derivates from Grain and Feed Extracts. "Cereal Chem.," *56*, 398. 1979.

BIOLOGICAL ROLE OF PLANT LIPIDS
P.A. BIACS, K. GRUIZ, T. KREMMER (eds)
Akadémiai Kiadó, Budapest and Plenum Publishing
Corporation, New York and London, 1989

GC/MS INVESTIGATION ON DIFFERENT FATTY ACIDS FROM SEEDS OF SOME MEDICINAL PLANTS

Eva Héthelyi, P. Tétényi, I. Zámbó, P. Kaposi, B. Dános and Emma Dabi-Lengyel

Research Institute for Medicinal Plants, Budakalász, Hungary

INTRODUCTION

Seed oils of various species belonging mainly to the ASTERACEAE, BORAGINACEAE, CUCURBITACEAE, ONAGRACEAE and PAPAVERACEAE families were investigated. The characteristic composition of fatty acids in plant seeds were determined by mass-spectrometry, using GC/MS method in our experiments (JEOL JMS-D300, EI ion source). The investigation was extended on the recognition of vegetable oils used as aliments too.

RESULTS AND CONCLUSION

Analysed seed oils could be arranged into four groups:

- oleic,
- linoleic-,
- γ -linolenic, and
- special structure acid types.

- Oleic acid rate varies between 46-64 percentage in oil of *Olea europaea*, *Asarum europaeum*, *Asclepias verticillata*, *Angelica archangelica*, *Aralia spinosa*.

- Linoleic acid proved to be a main component in seed oils of *Papaver somniferum*, *P. bracteatum*, *Carthamus tinctorius*, *Glycine hispida*, *Juglans regia*, *Cucurbita pepo*, *Cichorium intibus*, *Zea mays*. The germ oils of wheat and maize contain linoleic acid as main component as well.

- γ-linolenic acid was identified from *Borago officinalis* (1), *Oenothera lamarckiana*, *O. biennis*, *Lappula squarrosa* and from several species belonging to the families BORAGINACEAE. *Borago officinalis* contains about 22% γ-linolenic acid.

- Linoleic and linolenic acids with unsaturated bonds of special situation - differing from the formers - were identified by their spectra from seed oils of *Aquilegia vulgaris*, *Consolida regalis* (2) *Calendula officinalis* and *Hippophae rhamnoides*.

It was established unambigously that these seed oils contain always fatty acids with paired number of carbon. The unsaturated fatty acids dominate (min. 75%) in the composition of seed oils.

REFERENCES

1. Héthelyi, É., Pobozsny, K., Tétényi, P., Kaposi, P.: Analysis of the Essential Fatty acid content in Boraginaceae seeds. "Acta Alimentaria" *8*. 103-104. 1979.
2. Dabi, E., Héthelyi, É., Zámbó, I., Tétényi, P., Simonides, V.: Unsaturated Fatty Acid from seed oil of Consolida regalis. "Phytochemistry" Vol. *25* (5), 1221-1222. 1986.

BIOLOGICAL ROLE OF PLANT LIPIDS
P.A. BIACS, K. GRUIZ, T. KREMMER (eds)
Akadémiai Kiadó, Budapest and Plenum Publishing
Corporation, New York and London, 1989

SOME ASPECTS OF THE ROLE OF PHOTOSYNTHETIC ELECTRON TRANSPORT COMPONENTS IN CHLOROPLAST FATTY ACID DESATURATION

Tatiana E. Gafarova

Kazan Institute of Biology, USSR Academy of Sciences, P.O. Box 30, Kazan 420084, USSR

According to the current concept, plant and animal desaturase systems include flavoprotein, Fe-containing protein and desaturase. Thus, oleate desaturase of potato microsomes is assumed to consist sequentially of NADH-ferricyanide reductase, cytochrome b_5 and desaturase |1|. At present the involvement of photosynthetic electron transport components in the linoleate desaturation remains unknown. To study this problem the effect of 3-(3,4-dichlorophenyl)-1,1-dimethylurea (DCMU), the classical inhibitor of photosynthetic electron transport, on oleate and linoleate desaturation was investigated.

The shoots of 10-day old pea seedlings were cut and transferred to DCMU solution ($2*10^{-4}$ M) for 6 hours. Then, microdroplets of $|1-^{14}C|$ oleate solution (47 mCi/mmol) in 2-methoxyethanol (25 μCi/ml) were deposited on the pea upper leaf surfaces for 6 and 24 hours. Chloroplast were isolated, disrupted by hypoosmotic shock and incubated with 0.2 mM $|1-^{14}C|$ acetate (42 mCi/mmol) as described earlier |2| and superoxide dismutase (0.25 mg/ml, 2000u/mg). Methods used for lipid extraction and fatty acid methyl ester analysis are given in |2|.

DCMU inhibitory effect on linoleate desaturation appeared in 24 hours after the beginning of pea shoot incubation with $|1-^{14}C|$ oleate: the percentage of ^{14}C-linolenate decreased twice in the presence of DCMU (Tabl. 1). On the one hand, the linolenate percentage decrease may be caused by the decrease of the reduced pyridine nucleotide level in the cells in the presence of DCMU. Besides, the possibility of electron transfer to linoleate desaturase both from NADPH and from H_2O may be

Table 1. Effect of DCMU on |1-^{14}C|oleate desaturation in pea leaves

Incubation time (h)	Control (without DCMU)			DCMU ($2*10^{-4}$ M)		
	Fatty acid radioactivity (% of total)					
	18:1	18:2	18:3	18:1	18:2	18:3
6	71.5	24.8	3.7	66.8	29.0	4.2
24	33.6	51.2	15.2	43.2	48.8	8.0

Table 2. Effect of superoxide dismutase on ^{14}C-oleate biosynthesis in pea chloroplasts

Superoxide dismutase	Fatty acid radioactivity (cpm * 10^4)		
	16:0	18:1	18:1/16:0
-	3.5 ± 0.1	9.0 ± 0.2	2.31
+	3.3 ± 0.1	7.4 ± 0.3	2.26

assumed because DCMU is known to inhibit photosynthetic electron transport at the reducing side of photosystem 2. Our earlier data on |1-^{14}C|linoleoyl-CoA desaturation in pea thylakoids in the absence of exogenous electron donors also indicate the possibility of electron transfer to the chloroplast linoleate desaturase from H_2O |2|. So, chloroplast desaturase can apparently accept electrons both from NADPH and from H_2O through photosystems 2 and 1. In this connection the existence of chloroplast desaturase complexes is doubtful.

It is well known, that electrons from ferredeoxin reduce oxygen in Mehler reaction. The resulting superoxide-anion participation in desaturation reactions can be proposed, as already suggested by Sreekrishna and Joshi for animal tissue |3|. However, superoxide dismutase addition to chloroplast incubation media with |1-^{14}C|acetate (Tabl. 2) has not led to any alteration of stearate desaturase activity.

REFERENCES

1. Bonnerot C., Galle A.-M., Jolliot A., Kader J.-C. (1984) In: Siegenthaler P.-A., Eichenberger W. (eds) Structure, Function and Metabolism of Plant Lipids. Elsevier, Amsterdam, pp. 55-58
2. Grechkin A.N., Gafarova T.E., Tarchevsky I.A. (1984) Biokhimiya *49*: 1887-1889
3. Sreekrishna K., Joshi V.C. (1980) Biochim Biophys Acta *619*: 267-273

BIOLOGICAL ROLE OF PLANT LIPIDS
P.A. BIACS, K. GRUIZ, T. KREMMER (eds)
Akadémiai Kiadó, Budapest and Plenum Publishing
Corporation, New York and London, 1989

DEVELOPMENT OF FATTY ACID SYNTHESIS IN GREENING AVENA LEAVES

M. Kato, Y. Ozeki and M. Yamada

Department of Biology, The University of Tokyo, Komaba, Japan

In etiolated *Avena* leaves under the light condition, the etioplast rapidly develops into the chloroplast, proliferating thylakoid membranes during 24 hour. In such a drastic membrane genesis, the synthesis of galactolipids, the main component of thylakoid membranes, markedly increased. Our pulse-chase experiments showed that the galactolipid synthesis is driven by *de novo* synthesis of fatty acids in developing chloroplasts(Kato et al. 1987). This paper represents the development of fatty acid synthesizing system in greening *Avena* leaves, in connection with the activity level of the enzymes such as ribulose bisphosphate carboxylase (RuBPC), phosphoenolpyruvate carboxylase (PEPC), acetyl-CoA carboxylase (ACC), malonyl-CoA:ACP S-malonyltransferase (MT) and acetyl-CoA:ACP S-acetyltransferase (AT), the level of acyl carrier protein (ACP) in greening *Avena* leaves is also examined.

Materials and Methods

Seeds of oat, *Avena sativa* L.var Victory I(The General Swedish Company, Limited), were germinated at 25 °C on vermiculite and grown in the dark for 8 days. Etiolated seedlings were exposed to light(2400 lux, 25°C). The first leaves from greening *Avena* seedlings were excised and 2 cm sections located between 4 cm and 6 cm from the leaf base were used for the analysis.

Results and Discussion

There were two increases in *de novo* synthesis of fatty acids during greening of *Avena* leaves (Fig.1A). The first rise occurred 3 hour after illumination in which the prolamellar bodies are degenerated. The second rise corresponded to the formation of the thylakoids. However, the first rise was not observed in fatty acid synthesizing activity by the plastids (Fig. 1B), suggesting the involvement of cytosolic compartment in fatty acid synthesis by the chloroplast at this stage. No responsible increase to fatty acid synthesis was found in PEPC activity. The increase in ACC activity was responsible to the first rise, whereas the increases in AT and RuBPC activities were responsible to the second rise of de novo synthesis of fatty acids. This suggests that the cooperation of photosynthetic CO_2 fixation with fatty acid synthesis is initiated at this stage.The high level of MT activity throughout greening suggests that this enzyme does not control the operation of fatty acid synthesis. On the other hand, the extremely low level of AT activity and the increase in AT activity with the secnd rise of de novo fatty acid synthesis suggests that AT is a rate-limiting enzyme of the initial reactions in fatty acid synthesis in greening Avena leaves, contrast to the case of developing seeds (Turnham and Northcote 1983). The two forms of ACP (Avena-ACP I and II) were found in greening Avena leaves by immuno-blott analysis. Avena-ACP I, smaller in molecular weight., markedly increased with greening, whereas Avena-ACP II, larger in molecular weight, a little increased. This suggests that Avena-ACP I is the main ACP and light-inducible, as is ACP I in spinach leaves (Ohlrogge and Kuo 1985).

Table 1. Relationship between fatty acid synthesizing activity and the related enzyme activities

	First rise (Prolamellar body degeneration)	Second rise (Thylakoid formation)
Fatty acid synthesis		
(leaves)	+	+
(plastids)	−	+
RuBPC	−	+
PEPC	−	−
ACC	+	−
MT	−	−
AT	−	+
ACP I	−	+
ACP II	−	−

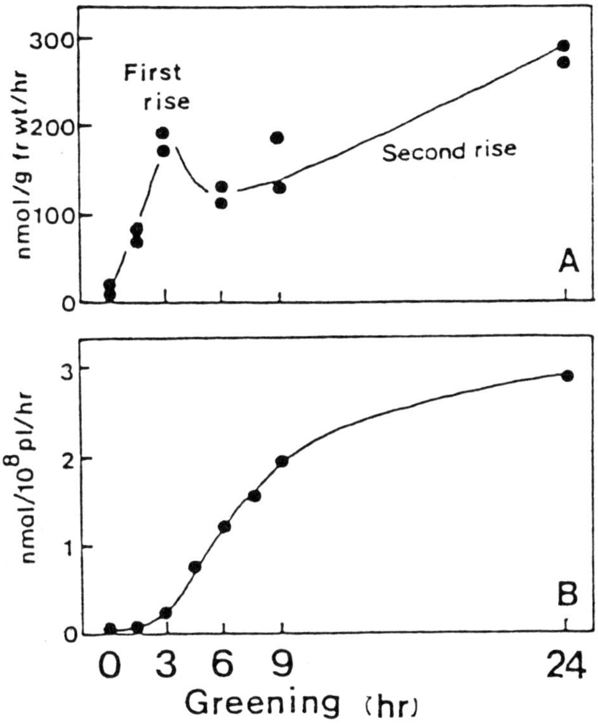

Fig. 1 (A) Level of fatty acid synthesizing activity in etiolated and greening *Avena* leaves. (B) Level of fatty acid synthesizing activity in the crude plastids isolated from etiolated and greening *Avena* leaves (Ohnishi and Yamada 1980).

REFERENCES

KATO, M., EHARA, T., KAWAGUCHI, A. and YAMADA, M. (1987) Plant Cell Physiol. 28:867-865.

OHLROGGE, J.B. and KUO, T. (1985) J. Biol. Chem. 260:8032-8037.

OHNISHI, J. and YAMADA, M. (1980) Plant Cell Physiol. 21: 1595-1606.

TURNHAM, E. and NORTHCOTE, D.H. (1983) Biochem. J. 212:223-229.

BIOLOGICAL ROLE OF PLANT LIPIDS
P.A. BIACS, K. GRUIZ, T. KREMMER (eds)
Akadémiai Kiadó, Budapest and Plenum Publishing
Corporation, New York and London, 1989

A NEW BETAINE LIPID FROM OCHROMONAS DANICA : DIACYLGLYCERYL-O-2-(HYDROXYMETHYL) (N,N,N-TRIMETHYL)-ß-ALANINE (DGTA)

G. Vogel and W. Eichenberger

Institut für Biochemie der Universität Bern, Switzerland

Ochromonas danica (Chrysophyceae) contains two major membrane lipids first described as lipids A and B (Nichols and Appleby 1969). Lipid A was subsequently identified as the betaine lipid diacylglyceryl-(N,N,N-trimethyl)homoserine (DGTS) which is widely distributed among cryptogamic green plants, while the structure of lipid B had not been elucidated so far.

Ochromonas danica, strain 933/2b (The Culture Centre of Algae and Protozoa, Cambridge) was cultivated at 26°C under sterile conditions and continuous light in 3 liter flasks using the medium of Pringsheim (1955) supplemented with glucose, thiamine-HCl, biotin and vit B_{12}. Lipids were extracted with methanol and separated on a medium pressure liquid chromatography column B-681 (Büchi). Lipid B was purified on silicagel plates (Merck 5715) with chloroform-methanol-H_2O 65:25:4 (by vol.). Fatty acid methyl esters were prepared by hydrolysis with methanolic KOH and methylation with diazomethane and were then analyzed on Carbowax 20 M (fused silica capillary column 25 m x 0.25 mm). NMR spectra were obtained on a Bruker AM 400 NMR Spectrometer (Institut für organische Chemie der Universität Bern).

Lipid B, like DGTS, which does contain neither carbohydrate nor phosphate, gives however, a positive Dragendorff test indicating the presence of a polar group of the betaine type. In pure form, lipid B is spontaneously deaminated producing trimethylamine and the deaminated compound B_D which, on hydrogenation, gives the deaminated and hydrogenated compound B_{DH}. NMR

Table 1. ^{1}H-NMR (400 MHz) and ^{13}C-NMR (100 MHz) of B, B_D and B_{DH} (, ppm in $CDCl_3$)

B		B_D			B_{DH}	
^{1}H	group	^{1}H	^{13}C	group	^{1}H	group
0.8-2.8	1	0.8-2.8	14-35	1	0.85-2.3	1"
5.35		5.35	127-131			
			169,173			
4.2	2	4.35	62.5	2'	4.14,4.3	2"
5.15	3	5.25	70-71	3'	5.28	3"
3.5-3.6	4	3.6	69	4'	3.45-3.5	4"
3.85	5	4.2	69.4	5'	3.35	5"
			136.7	6'	2.47	6"
3.4	7	5.9,6.4	127.1	7'	1.0	7"
3.25	8	2.85	44.9	8'		
			172.3	9'		

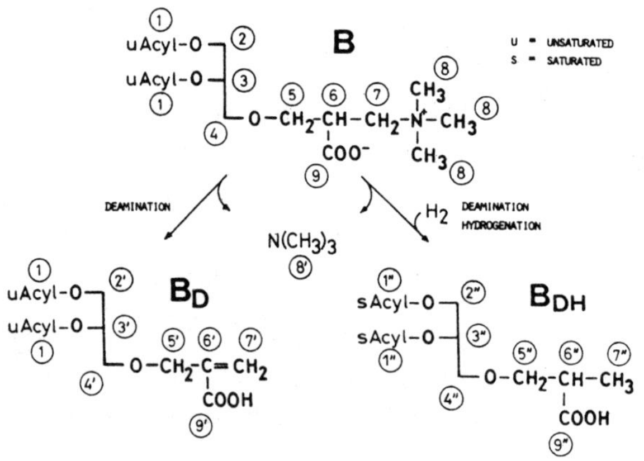

Fig. 1. Tentative structure of lipid B (B) and its derivatives obtained by deamination (B_D) and deamination/hydrogenation (B_{DH}).

data of B, B_D and B_{DH} could be attributed to different groups as shown in Table 1 and Fig. 1. The structure of diacylglyceryl-O-2-(hydroxymethyl)(N,N,N-trimethyl)-β-alanine could tentatively be assigned to lipid B. The fatty acids consist of 14:0 (5%), 16:0 (6%), 18:0 (16%), 18:1 (9%), 18:2 (10%), 20:2 (5%), 20:3 (5%), 20:4 (15%) and 22:5 (29%). The structure of derivatives of 2-(hydroxymethyl)-2-propenoic acid and 2-(hydroxymethyl)-propanoic acid could be attributed to B_D and B_{DH}, respectively.

Nichols, B.W. and Appleby, R.S. (1969) The distribution and biosynthesis of arachidonic acid in algae. Phytochemistry 8: 1907-1915.

Pringsheim, E.G. (1955) Ueber Ochromonas danica n.sp. und andere Arten der Gattung. Arch. Mikrobiol. 23: 181-194.

CHAPTER 2

STRUCTURAL AND FUNCTIONAL ORGANIZATION OF LIPIDS

BIOLOGICAL ROLE OF PLANT LIPIDS
P.A. BIACS, K. GRUIZ, T. KREMMER (eds)
Akadémiai Kiadó, Budapest and Plenum Publishing
Corporation, New York and London, 1989

STRUCTURAL AND FUNCTIONAL ASPECTS OF ACYL LIPIDS IN THYLAKOID MEMBRANES FROM HIGHER PLANTS

P.A. Siegenthaler, A. Rawyler and J.-P. Mayor

Laboratoire de Physiologie Végétale, Université de Neuchâtel, 20 Chemin de Chantemerle, CH-2000 Neuchâtel, Switzerland

Introduction

It is now well established that the protein complexes which are involved in photosynthetic electron, proton transport and phosphorylation activities are distributed asymmetrically, both in the plane of and across the thylakoid membrane. Such an asymmetric arrangement of proteins supports the vectorial properties of the membrane which are necessary for energy conservation. During these past few years, considerable efforts have been made to determine whether a similar heterogeneity existed for acyl lipids which are the second major components of this membrane [1,2].

Lateral heterogeneity of acyl lipids in photosynthetic lamellae has been assessed by several approaches such as fractionation of thylakoids into subchloroplast particles enriched in PSI or PSII activities, separation of appressed and non-appressed regions of thylakoids or purification of one of the complexes found in the membrane [2-4]. However, the use of detergent to obtaining subchloroplast fractions may result in differential displacements of certain acyl lipids, thereby changing the final lipid composition of these fractions. Thus, there are still some doubts about the localization and the degree of lateral heterogeneity of acyl lipids in photosynthetic lamellae.

In contrast, the situation concerning the transmembrane distribution of acyl lipids is more clear-cut. Based essentially on enzymatic and labeling techniques it has been found that both galactolipids and phospholipids are asymmetrically distributed between the two monolayers of the thylakoid membrane in several plant species [2,3,5]. For instance, it has been shown that (1) the outer leaflet is mainly enriched in monogalactosyldiacylglycerol (MGDG) and phosphatidylglycerol (PG) whilst digalactosyldiacylglycerol (DGDG) is confined essentially in the inner leaflet; (2) MGDG constitutes about 70%

of the total acyl lipids in the outer monolayer and only 38% in the inner one; (3) total galactolipids (MGDG + DGDG) are equally distributed in both leaflets; (4) the molar ratios MGDG/DGDG and non-bilayer/bilayer forming lipids are much higher in the outer than in the inner leaflet of the thylakoid membrane.

Thus, one can expect that, due to their different composition in acyl lipids, the two monolayers will be assigned different roles in the thylakoid membrane function. One of the approaches to test this hypothesis has been to use the acyl lipid depletion technique in which special precautions have to be taken to remove all hydrolysis products (free fatty acids and lysoderivatives) which could alter membrane function [6]. Recent evidence concerning the role of phospholipids in sustaining the uncoupled non-cyclic electron flow activity is that several distinct populations of PG and PC have to be considered [3,6]. A first one, which is easily accessible to phospholipase A_2, supports only about 15% of the activity. A second phospholipid population which is less accessible to phospholipase A_2 sustains the remaining 85% of the activity. Finally, a third population of phospholipids does not seem to be involved in the uncoupled non-cyclic electron flow activity. These results and several other reports using mainly reconstitution procedures [4,5] point to the fact that acyl lipids may sustain the photosynthetic membrane function.

The purpose of this presentation is (1) to compare the transmembrane distribution of galactolipids in intact thylakoids and in thylakoid inside-out vesicles from spinach; (2) to explore the possible role of certain membrane proteins in inducing and/or maintaining the asymmetric distribution of acyl lipids; to this aim, prothylakoids and thylakoids from oat were used; (3) finally, to define more exactly the localization, in the outer and inner monolayers, of those phospholipid populations which are necessary for electron flow activity.

Transmembrane distribution of galactolipids in spinach thylakoid inside-out vesicles

In our studies, the analysis of galactolipid hydrolysis kinetics of the thylakoid membrane in the presence of the lipase from *Rhizopus arrhizus* is the basis for the determination of galactolipid transmembrane distribution. These kinetics have been studied under a variety of conditions : temperature, ionic strength, surface potential, stacked or unstacked thylakoids,

TABLE 1. Distribution of galactolipids in the two monolayers of thylakoid inside-out vesicles and intact thylakoids from spinach. Results are expressed in mol % for each class of lipids.

	Molar outside/inside ratio	
	MGDG	DGDG
Thylakoid inside-out vesicles (n=5)	42±1/58±1	82±1/18±1
Intact thylakoids (n=3)	62±3/38±3	20±2/80±2

concentration of enzyme, presence or absence of hydrolysis products, etc. Under appropriate conditions, it is possible to hydrolyze only those galactolipids which are located in the outer monolayer of the membrane [7]. However, there is some controversy concerning the relevance of using the lipolytic approach to determine the transmembrane distribution of acyl lipids in thylakoids. The main objections are : (1) lipases do not have ready access to substrates which are localized in the outer monolayer of two adjacent stacked thylakoid membranes; (2) fatty acids and lysoderivatives released by lipase activity may influence the rate and amplitude of lipid hydrolysis and, consequently, may lead to a wrong estimation of the real transmembrane distribution; (3) the greater affinity of the lipase from *Rhizopus arrhizus* for MGDG than for DGDG may bias the estimation of the transmembrane distribution of DGDG; (4) the membrane bilayer integrity does not remain intact during lipid hydrolysis; (5) hydrolysis of one population within a given lipid class can result in a reorganization (e.g. by transbilayer movement) of the other populations of the same lipid class and then bias the native transmembrane distribution of this lipid. Objections (1) and (2) have been discussed and refuted elsewhere [7].

Results presented in Table 1 show unambiguously that, within experimental variations, the transmembrane distribution of both MGDG and DGDG in intact thylakoids and in inside-out vesicles is opposite [8]. Furthermore, it can be calculated that the molar MGDG/DGDG ratio is about 0.95 in the outer monolayer of the inside-out vesicles as well as in the inner monolayer of intact thylakoids and about 6.0 in the other homologous leaflets. These results are of particular interest since they show that inverted thylakoids

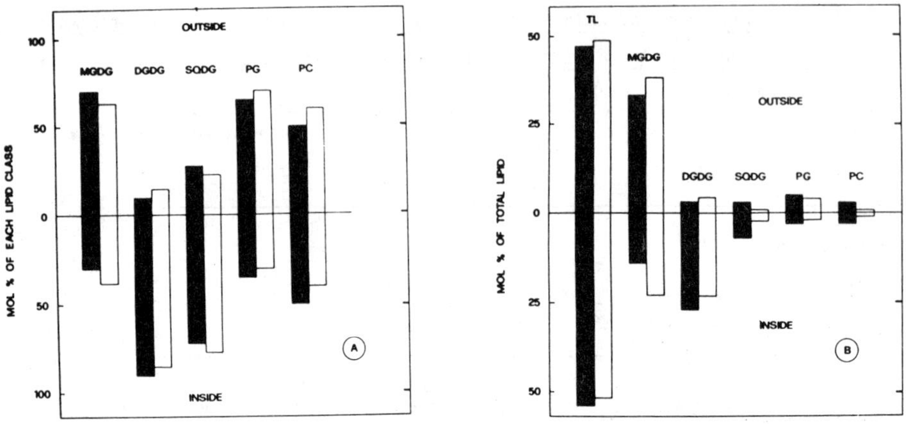

Fig. 1. Distribution of acyl lipids between the outer and inner monolayers of oat prothylakoid (■) and thylakoid (□) membranes, expressed as mol % of each lipid class (A) or mol % of total lipids (B).

which display an inverted vectoriality (light-induced extrusion of protons) are characterized by an opposite transverse galactolipid distribution compared to normal thylakoids. These results show also the validity of the lipolytic approach to determine the transmembrane distribution of acyl lipids in thylakoids and refute objections 3 to 5.

Acyl lipid asymmetry during oat thylakoid development

The transmembrane asymmetry of acyl lipids raises several interesting questions. The most intriguing ones concern the origin of this asymmetry and the reciprocal influence of proteins and acyl lipids on their respective organization. Another way to put it is to ask whether the insertion of new proteins in the thylakoid membrane induces a new or maintains the same acyl lipid asymmetry. The knowledge of the distribution of acyl lipids in prothylakoids may provide answers to the above questions [9,10]. Prothylakoids are the precursors of mature thylakoids but their function and protein composition are quite different, e.g. prothylakoids are devoid of chlorophyll-protein complexes but contain a large amount of CF_1-ATPase and of NADPH : protochlorophyllide oxidoreductase. Thus, the profound reorganization of the membrane due to insertion of new chlorophyll-protein complexes during greening of prothylakoids [11] appears to be a judicious system for our study.

Various lipolytic enzymes (phospholipases A2 from porcine pancreas and Vipera russelli, phospholipase D from Streptomyces chromofuscus, the lipase from Rhizopus arrhizus and the lipolytic acyl hydrolase from potato tubers) were used to assess the transmembrane distribution of phospholipids and galactolipids in both prothylakoids and thylakoids from oat. As summarized in Fig. 1A all acyl lipids were found to be asymmetrically distributed. The molar outside/inside distribution was 70±5/30±5 for PG in both types of membranes. Concerning phosphatidylcholine (PC), this ratio was 50±10/50±10 and 65±10/35±10 for prothylakoid and thylakoid membranes, respectively [9]. The molar outside/inside distribution was 70±8/30±8 for MGDG and 10±4/90±4 for DGDG in the prothylakoid membrane. Mature thylakoids presented a similar distribution, i.e. 63±4/37±4 for MGDG and 12±3/88±3 for DGDG. When acyl lipids are expressed in mol % of total lipids (Fig. 1B), MGDG (in the outer monolayer) and DGDG plus MGDG (in the inner monolayer) are prominent in the two types of membranes. However, both monolayers contained almost identical amounts of polar lipids (TL in Fig. 1B) [10].

In conclusion, the molar outside/inside distributions of galactolipids and phospholipids are identical (within standard deviation) in both prothylakoid and thylakoid membranes from oat. It is most remarkable and unexpected that during the biogenesis of the thylakoid in the light, which is accompanied by a significant enrichement in its relative protein amount [10] and by the insertion of new proteins [11], the distribution of acyl lipids between the two monolayers of the membrane remains the same. Although our results do not allow us to draw conclusions about the mechanism by which lipid asymmetry takes place in the thylakoid membrane, they show, however, that asymmetry appears very early during the biogenesis of this membrane and has to be looked for during the formation of the prothylakoid itself.

The above results raise an intriguing question : how can the thylakoid membrane be a bilayer in spite of the fact that its outer monolayer is highly enriched in MGDG, a typical non-bilayer forming lipid. In order to evaluate the potential of acyl lipids to adopt lamellar or hexagonal configuration, we have calculated the non-bilayer (MGDG) / bilayer forming lipids (DGDG+SQDG+PG+PC) molar ratios in each monolayer of prothylakoid and thylakoid membranes (see Table 2). It is worth mentioning that Sprague and Staehelin [12] have demonstrated that for a MGDG/DGDG mixture the transition of bilayer to non-bilayer structures starts at a molar ratio of about 2.5. Taking this ratio as being critical for the formation of non-lamellar phases means that a mixture of pure acyl lipids of similar composition to that of prothylakoid

TABLE 2. Non-bilayer forming/bilayer forming acyl lipid molar ratios in the two monolayers of prothylakoids and thylakoids from oat. Non-bilayer forming lipid : MGDG; bilayer forming lipids : DGDG, SQDG, PG and PC.

	Prothylakoid		Thylakoid	
	outer leaflet	inner leaflet	outer leaflet	inner leaflet
Non-bilayer / bilayer forming lipids [mol:mol]	2.38	0.36	4.16	0.77

and thylakoid monolayers would display lamellar structures in all monolayers (e.g. ratio < 2.5) except in the outer monolayer of the thylakoid membrane (ratio > 2.5). Although this way of reasoning is appropriate for a mixture of acyl lipids, one should keep in mind that the potential of acyl lipids to form lamellar or non-lamellar structures is modulated by several other factors such as the degree of unsaturation of the lipid acyl chains, the water content, the temperature, the pH, the cation type and concentration [2,13] and also by the proteins themselves [14].

As discussed above, the biogenesis of oat thylakoids is accompanied by a remarkable increase in the potential of the outer monolayer to form non-bilayer structures (Table 2). Thus, incorporation into the greening membrane of large amount of chlorophyll-protein complexes, which are known to interact with MGDG and PG molecules, may be necessary to preserve lamellar structures [10]. In contrast, acyl lipids alone in both prothylakoid monolayers are able to form by themselves stable lamellar structures. In conclusion, incorporation of chlorophyll-protein complexes into the nascent thylakoid membrane modifies neither the galactolipid nor the phospholipid transmembrane distribution. However, these complexes appear to be crucial to preserve a bilayer configuration to the greening membrane which, otherwise, would adopt non-lamellar structures.

Localization and function of distinct phosphatidylglycerol populations in the thylakoid membrane

As mentioned in the introduction, there is good evidence that phospholipids are organized in several distinct and discrete pools or populations within the two monolayers of the thylakoid membrane [3,5,6]. To get further information about the different PG populations of the membrane, we have designed experimental conditions, under which phospholipid depletion (occurring in the presence of pancreatic phospholipase A_2, PLA_2) and/or delocalization (via transbilayer movement) took place successively. Simultaneously, the uncoupled non-cyclic electron flow activity was measured and related to the content and localization of phospholipid molecules in either the outer or inner monolayer of the membrane. In these experiments, the content of glycolipids remained constant and complete removal of free fatty acids and lyso-phospholipids was achieved in the presence of properly defatted bovine serum albumin [6]. In addition, it has been checked that the osmotic properties of the membrane (toward sorbitol) are preserved in spite of extensive phospholipid removal [3]. The experiment involved the following steps (see Fig. 2, A-E) : (A) the outer membrane monolayer of the thylakoid was completely depleted in both phospholipids (PLA_2 at 2°C, 60 min) corresponding to a loss of 69% of PG and 60% of PC. Under these conditions, only 18% of the initial uncoupled non-cyclic electron flow activity was inhibited; (B) then, the PLA_2 activity was blocked by EGTA addition and thylakoids were post-incubated for another 90 min at 2°C. Under these new conditions, the total PG content of the membrane remained constant but some of the inner PG molecules (3% of the initial membrane PG) underwent an outward transbilayer movement (delocalization) which altered 12% of the activity; (C) a similar experiment, carried out under the same conditions but at 20°C, resulted in no change in the total PG content of the membrane but, due to increased outward transbilayer movement (delocalization) of an additional amount of inner PG molecules (3% of the initial membrane PG), to 40% inhibition of the activity; (D) in a parallel experiment, thylakoids were post-incubated for 90 min at 20°C but in the presence of PLA_2. The level of PG content dropped to 12% of the initial value and the activity was completely inhibited; (E) the remaining 12% PG which were located entirely in the inner monolayer were not involved in non-cyclic electron flow activity.

These results show quite clearly that it is not the amount of PG which is important for sustaining the electron flow activity but rather its localization. For instance, complete depletion of PG in the outer monolayer

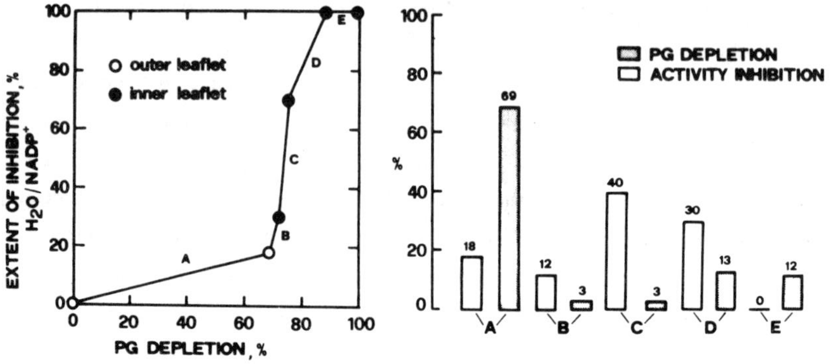

Fig. 2. Dependency of uncoupled non-cyclic electron flow activity on various PG populations in the outer and inner thylakoid membrane leaflets. PG depletion refers to either depletion *per se* (steps A and D) or to delocalization of inner PG to the outer leaflet (steps B and C). See explanations in the text.

(69% of the initial membrane PG) affected only 18% of the activity (condition A, in Fig. 2) whereas depletion of 13% in the inner monolayer impaired 30% of the activity (condition D). On the other hand, delocalization of 3% PG from the inner to the outer leaflet without lipid loss in the membrane resulted in 12% (condition B) and 40% (condition C) inhibition of the activity. Finally, there is a PG population representing 12% of the initial level which appeared to be tightly bound to some components in the inner monolayer and which did not sustain any activity.

These results suggest that as far as the uncoupled non-cyclic electron flow activity ($H_2O/NADP^+$) is dependent on PG, one can distinguish between two PG populations : the first one, weakly involved, located in the outer and the second one, strongly involved, located in the inner monolayer of the thylakoid membrane. In addition, the inner PG population is likely to contain several subpopulations of PG molecules, some of them being discrete but very efficient in sustaining electron flow activity, in agreement with earlier suggestions [3,5,6].

From these results, one may wonder which role(s) acyl lipids may play in the outer monolayer of thylakoid membranes. Two lines of research have been recently pursued. In the first one, phospholipid or galactolipid depletion in the outer leaflet was found to alter markedly both the binding and the inhibitory properties of DCMU and atrazine in thylakoid membranes isolated from susceptible and resistant biotypes of Black Nightshade (*Solanum nigrum*).

This suggests that the native conformation of the herbicide-binding protein (32 kD) may be maintained in its functional state, at least in part, by interactions with phospho- and galactolipids located in the outer monolayer [Mayor et al., these Proceedings]. The second type of studies is more specifically dealing with MGDG in the outer monolayer of spinach thylakoid membranes. The packing of MGDG (estimated from its initial hydrolysis rate by the lipase from Rhizopus arrhizus) was found to depend upon the functional state of these membranes. Closer investigations showed that MGDG packing was apparently controlled by CF_o/CF_1 being in a functioning, a resting or a dissociated/inhibited state [Rawyler and Siegenthaler, these Proceedings]. These observations open new ways for a better understanding of the role of lipids in thylakoid membranes. It may well be that the inner monolayer, because of the high bilayer-forming potential of its lipid complement, is implied in the "lipid dependency" of electron transport activities whereas the outer monolayer, through the polymorphic character of its lipid moiety, may rather be involved in some kind of "fine tuning" of membrane functions.

Acknowledgements

The authors thank Mrs Jana Smutny for her excellent technical assistance and Mrs Christiane Bettinelli for typing and arranging the manuscript. Part of this study was supported by the Swiss National Science Foundation (Grant no. 3.346-0.86 to P.A.S.).

References

[1] D.J. Murphy (1986) The molecular organisation of the photosynthetic membranes of higher plants. Biochim. Biophys. Acta 864, 33-94.

[2] K. Gounaris, J.Barber and J.L. Harwood (1986) The thylakoid membranes of higher plant chloroplasts. Biochem. J. 237, 313-326.

[3] P.A. Siegenthaler, A. Rawyler and C. Giroud (1987) Spatial organization and functional roles of acyl lipids in thylakoid membranes. In: The metabolism, structure, and function of plant lipids (P.K. Stumpf, J.B. Mudd and W.D. Nes, eds), Plenum Publishing Corporation, pp. 161-168.

[4] S.G. Sprague (1987) Structural and functional consequences of galactolipids on thylakoid membrane organization. J. Bioenerg. Biomembr. 19, 691-703.

[5] P.A. Siegenthaler and A. Rawyler (1986) Acyl lipids in thylakoid membranes : distribution and involvement in photosynthetic functions. In: Encyclopedia of Plant Physiology, New Series, vol. 19, Photosynthesis III (L.A. Staehelin and C.J. Arntzen, eds), Springer Verlag Berlin, Heidelberg, pp. 693-705.

[6] P.A. Siegenthaler, J. Smutny and A. Rawyler (1987) Involvement of distinct populations of phosphatidylglycerol and phosphatidylcholine molecules in photosynthetic electron flow activities. Biochim. Biophys. Acta 891, 85-93.

[7] A. Rawyler and P.A. Siegenthaler (1985) Transversal localization of monogalactosyldiacylglycerol and digalactosyldiacylglycerol in spinach thylakoid membranes. Biochim. Biophys. Acta 815, 287-298.
[8] P.A. Siegenthaler, J. Sutter and A. Rawyler (1988) The transmembrane distribution of galactolipids in spinach thylakoid inside-out vesicles is opposite to that found in intact thylakoids. FEBS Lett. 228, 94-98.
[9] P.A. Siegenthaler and C. Giroud (1986) Transversal distribution of phospholipids in prothylakoid and thylakoid membranes from oat. FEBS Lett. 201, 215-220.
[10] C. Giroud and P.A. Siegenthaler (1988) Development of oat prothylakoids into thylakoids during greening does not change transmembrane galactolipid asymmetry but preserves the thylakoid bilayer. Plant Physiol., in the press.
[11] J.N. Mathis and K.O. Burkey (1987) Regulation of light-harvesting chlorophyll protein biosynthesis in greening seedlings. Plant Physiol. 85, 971-977.
[12] S.G. Sprague and L.A. Staehelin (1984) Effect of reconstitution method on the structural organization of isolated chloroplast membrane lipids. Biochim. Biophys. Acta 777, 306-322.
[13] W.P. Williams and P.J. Quinn (1987) The phase behavior of lipids in photosynthetic membranes. J. Bioenerg. Biomembr. 19, 605-624.
[14] T.F. Taraschi, B. de Kruijff, A. Verkleij and C.J.A. van Echteld (1982) Effect of glycophorin on lipid polymorphism. A ^{31}P-NMR study. Biochim. Biophys. Acta 685, 153-161.

BIOLOGICAL ROLE OF PLANT LIPIDS
P.A. BIACS, K. GRUIZ, T. KREMMER (eds)
Akadémiai Kiadó, Budapest and Plenum Publishing
Corporation, New York and London, 1989

MONOGALACTOSYLDIACYLGLYCEROL DESATURATION IN SPINACH CHLOROPLASTS

Jaen Andrews*, H. Schmidt and E. Heinz

Institut für Allgemeine Botanik, Universität Hamburg, Hamburg, FRG
*currently: Sungene Technologies, San Jose, CA, USA

Chloroplasts play a central role in fatty acid metabolism in the leaf cell; not only do they synthesize fatty acids--and may in fact be the sole site of this synthesis--but they are also able to desaturate fatty acids. Desaturation of 18:0 to 18:1 occurs while the acyl group is esterified to ACP in the stroma; the desaturase is a soluble protein, and requires both a source of electrons--reduced ferredoxin is sufficient--and oxygen (Jacobson et al, 1974; Mckeon and Stumpf, 1982). All other fatty acid desaturation in plant cells, in both the plastid and elsewhere, appears to occur while the fatty acids are esterified to glycerolipids (Roughan and Slack, 1982). This is in stark contrast to animal systems in which fatty acids esterified to CoA are the predominant substrates for desaturation.

All chloroplasts so far examined are able to assemble PA from the products of fatty acid synthesis, 16:0 and 18:1. PA may then be directed either into PG, which appears to be universally synthesized by plastids, or into the glycolipids MGDG and SL, which in only certain types of plants are assembled in the plastid. Thus, when isolated intact spinach chloroplasts are incubated with labelled acetate in the light under fatty acid and glycerolipid synthesizing conditions, MGDG, PG, and SL are synthesized and their fatty acids desaturated to some extent (Roughan et al, 1979; Heinz and Roughan, 1983; Sparace and Mudd, 1982; Joyard et al, 1986). However, desaturation of the fatty acids occurs only after the appropriate headgroup is attached; it does not occur in either PA or DG. This desaturation does not require light as it is observed after incubations of the chloroplasts in the dark.

We have confirmed and extended these observations with chloroplasts isolated from both spinach and pea leaves (table 1). The degree of desaturation of PG synthesized in both types of chloroplasts is about the same. Although pea chloroplasts normally synthesize MGDG and SL to only very low levels, the C 18 fatty acids in both lipids can be desaturated. However, pea chloroplasts differ from spinach chloroplasts in that they are unable to desaturate 16:0 in MGDG. Interestingly, they are able to desaturate C 18 fatty acids found at both positions 1 and 2 of MGDG (unpublished observations).

TABLE I. ***Fatty acid desaturation in newly-synthesized lipids in isolated chloroplasts from spinach and pea leaves.*** Acetate incorporation (nmoles/mg chl· hr) was 328 for spinach and was 394 for pea.

Plant	Lipid	% Total Lipid Label	Fatty Acids (% total fatty acid label) 18:1	18:2	18:3	16:0	16:1	16:2	16:3
Spinach	MGDG	66.7	15.2	26.0	18.5	21.4	5.5	9.7	3.5
	PG	6.9	46.9	6.9	3.3	41.2	--	--	--
	SL	9.5	20.4	12.0	2.1	64.4	--	--	--
Pea	MGDG	6.7	38.4	15.8	9.1	28.6	--	--	--
	PG	15.9	41.5	11.8	1.8	38.2	--	--	--
	SL	3.4	61.4	6.4	--	26.6	--	--	--

Further investigations were made of fatty acid desaturation in MGDG in spinach chloroplasts, as this lipid is a major product and its fatty acids are desaturated to a relatively high degree. The reactions of fatty acid synthesis were separated from those of subsequent desaturation by employing pulse-chase experiments. During the pulse, chloroplasts were incubated for short periods of time (3-5 min) under fatty acid and glycerolipid synthesizing conditions with ^{14}C-acetate. They were then diluted, sedimented, resuspended, and reincubated for the chase, which was typically 60 min. The lipid products were extracted and MGDG separated by TLC. The fatty acids in MGDG were transmethylated, and the resulting FAMEs analyzed by either argentation TLC or HPLC. Analysis of both the fatty acids and molecular species of MGDG indicated that 18:1/16:0 was the major species after the pulse period (fig. 1). As reported previously, desaturation occurred in both the light and the dark, although it was slightly higher in the light, and in the C 16 fatty acids (table 2). In order to eliminate the effects of light, most of the subsequent chase incubations were carried out in the dark.

We were unable to significantly increase desaturation by adding either NADH or NADPH exogenously, or reducing equivalents capable of entering the chloroplast. Additional compounds which stimulated desaturation in other systems were without effect, and a systematic evaluation of the compounds present in the pulse resulted in the eventual inclusion of sorbitol, bicarbonate, phosphate and EDTA in the chase medium. Efforts to decrease internal reducing equivalents were without effect, so it is not known whether NAD(P)H is required. However, oxygen is required, as demonstrated by the addition of glucose, glucose oxidase and catalse to the chase (table 3). When added to intact chloroplasts evolving oxygen in light- and CO_2-dependent reactions, these

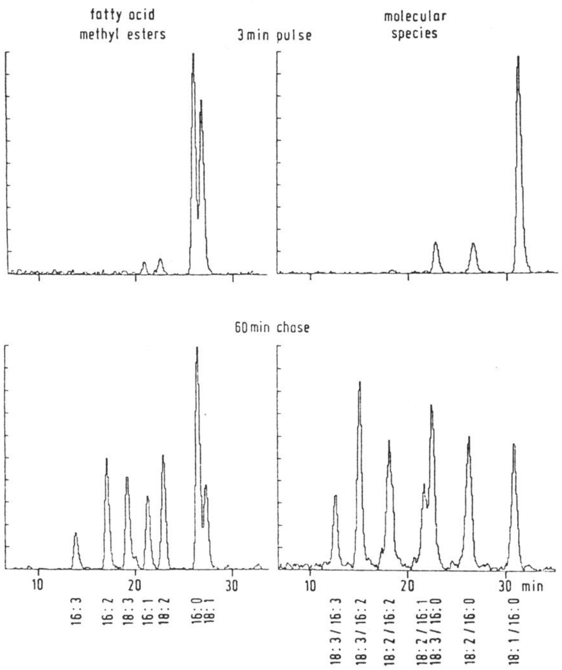

FIGURE 1 ***Radioactivity traces of MGDG fatty acid esters and molecular species separated by HPLC after a 3 min pulse and a 60 second chase***

TABLE 2. ***Fatty acid desaturation in newly-synthesized MGDG in spinach chloroplasts***

Fatty Acid	Amount Synthesized (nmoles/mg chl · hr)	
	Light	Dark
18:2	1.20	.93 ± .20
18:3	.96	.59 ± .15
C 18	**2.15**	**1.52 ± .35**
16:1	1.02	.82 ± .15
16:2	1.04	.78 ± .14
16:3	.55	.28 ± .05
C 16	**2.61**	**1.88 ± .34**
TOTAL	**4.77**	**3.39 ± .49**

compounds resulted in a rapid decrease in the oxygen concentration to about 15 μM within one min; in the dark, the concentration further decreased to zero. These values may be compared to a concentration of about 5 μM at the start of the experiment. When added to the chase, they completely blocked desaturation (table 3).

The location of desaturation within plastids is unknown. Its occurrence in the envelope membranes would be consistent with the assembly of glycerolipids there; however, neither cytochromes nor flavoproteins have been detected in these membranes. Desaturation in the thylakoid membranes would require transport of newly-synthesized lipids from the envelopes to the thylakoids; on the other hand, these membranes do contain components which catalyse electron transport in both the light and the dark.

Localization studies usually involve first disrupting the chloroplasts and then subfractionating them; breakage methods include osmotic shock, mechanical shearing, and freezing and thawing. However, breaking the chloroplasts by any of these three methods resulted in complete inactivation of desaturation; all efforts to protect the activity during disruption were unsuccessful. Desaturation was sensitive to freezing and thawing, as it was not observed in chloroplasts which survived intact afterward. Finally, additional experiments indicated that desaturation was also osmotically sensitive. Suspension of chloroplasts in media which contained decreasing concentrations of sorbitol (from 0.3 to 0.06 M) resulted in increasing breakage during the chase. However, desaturation observed in those chloroplasts which survived the osmotic shock intact and which were subsequently reisolated also decreased, and in proportion to the decrease in the sorbitol concentration. This decrease was at least partially reversible by subsequently increasing the sorbitol levels to the normal concentration of 0.3 M during the chase, though the extent of reversibility decreased as the duration of the hypotonic treatment increased from 1 - 10 min. The observed inhibition might be due to dilution of cofactors or enzymes, as the osmotic volume of chloroplasts in 0.06 M sorbitol is 2-3 times that in 0.3 M (Kaiser et al, 1971).

It is interesting to note that assembly of glycerolipids was unaffected by any of the breakage methods employed. On the other hand, fatty acid synthesis from acetate did not occur in broken chloroplasts, but did occur in intact chloroplasts which were osmotically stressed or which survived freezing and thawing. Desaturation did not occur in any broken chloroplasts, nor in chloroplasts stressed either osmotically or by freezing and thawing. One possible explanation of the inhibition of desaturation during stress is that intraorganellar lipid transport was disrupted, thus preventing lipids from moving from their site of synthesis in the evelopes to their site of desaturation. However, when chloroplasts were first subjected to the stress either before (freezing and thawing) or during (osmotic) incubation under lipid synthesizing conditions, and then fractionated into envelope and thylakoid membrane fractions, both fractions contained MGDG, though to varying extents.

TABLE 3. ***Requirement of oxygen for desaturation, as demonstrated by adding glucose, catalase and glucose oxidase to the chase incubation***

Fatty Acid	Addition	Pulse	Chase	Chase	Chase
	glucose	+	+	-	+
	catalase	+	+	-	+
	glucose oxidase	-	-	+	+
18:1		26.8	11.3	12.1	27.4
18:2		12.3	8.9	10.0	12.2
18:3		3.1	22.4	23.1	4.8
16:0		45.9	25.1	24.1	43.7
16:1		7.9	6.8	6.4	6.6
16:2		4.0	11.1	12.0	4.0
16:3		nd*	14.5	12.2	1.2

* nd = not detected

An alternative approach to localizing desaturation within the chloroplasts is to observe the distribution of newly desaturated fatty acids between the envelope and thylakoid membranes during the chase. In one such experiment, chloroplasts were broken (which then stops desaturation) after a chase of either 0, 5, or 30 minutes, and then subfractionated. The envelope fraction contained about 25 % of the total newly synthesized lipids, and the distribution of lipid products was about the same in both membranes, at all times (table 4). In addition, the degree of fatty acid desaturation was almost identical in both membrane fractions throughout the chase (table 4). These results indicate that lipid transfer occurred rapidly within the chloroplast. In addition, as the amounts of desaturated fatty acids increased during the chase, but remained evenly distributed between the envelope and thylakoid membrane fractions, either desaturation occurred in both membrane systems, or the lipids were transported several times before achieving their final desaturated form.

Desaturation in chloroplasts requires oxygen, and probably an electron donor as well. By analogy with animal systems, the electrons would be transferred from the donor to the desaturase via electron transport components. In fact, NAD(P)H can serve as a donor for "reverse" electron transport in thylakoids. In Chlamydomonas reduction of PQ by NADH through a flavoprotein has been reported (Godde and Trebst, 1980). In a similar fashion, added NADPH is able to reduce the primary acceptor Q of PS II in lysed spinach chloroplasts (Mills et al, 1979).

TABLE 4. ***Distribution of newly-synthesized lipids, MGDG, and MGDG fatty acids between envelope and thylakoid membranes during a chase***

Chase	0 min		5 min		30 min	
Membrane	Envel	Thyl	Envel	Thyl	Envel	Thyl
%Total Lipid	26	71	26	72	23	76
%MGDG	19	24	21	30	17	30
Fatty Acid in MGDG	(%)	(%)	(%)	(%)	(%)	(%)
18:1	31	29	28	25	22	19
18:2	5	7	12	12	18	20
18:3	1	1	3	3	7	8
16:0	55	53	46	45	34	35
16:1	4	6	7	8	7	8
16:2	1	2	4	3	8	8
16:3	1	1	1	0.5	4	2

This reduction is thought to occur in the dark via FNR, Fd, cytochrome b and PQ, and thus includes portions of both cyclic and non-cyclic carrier sequences.

Therefore, in an effort to determine whether any of these components might be involved in desaturation, the effects of a wide range of inhibitors normally used to block light- or respiratory-driven electron transport in thylakoids or mitochondrial membranes were examined. The inhibitors were added to pulsed, resuspended chloroplasts just before beginning the chase incubation. Many of these inhibitors have been used only with thylakoid preparations, whereas desaturation is of necessity investigated in intact chloroplasts. In order to ensure that these chemicals can cross the chloroplast envelope, their effects on oxygen evolution in intact chloroplasts was examined. Each inhibitor could completely block this activity, though at widely varying concentrations (table 5).

Neither photosystem I nor II were implicated in desaturation, as light is not required. Supporting this concept are the observations that neither DCMU nor atrazine (which prevent reduction of PQ at the Q_b site of PS II, fig. 4) inhibited desaturation (fig. 2). Because PQ is involved in "reverse" electron transport, the effects of inhibitors of its function were determined. DBMBIB, UHDBT, BPA, stigmatellin (STG), and DNP-INT interfere with oxidation of reduced PQ by blocking the Qz site (fig. 4), and in fact the first three were able to inhibit desaturation (fig. 3 and table 5). Both DNP-INT and STG may bind at an overlapping but more

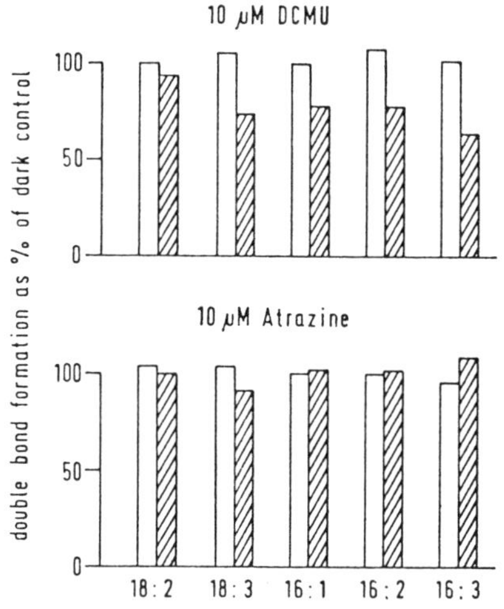

FIGURE 2 *Effects of DCMU and atrazine (both at 100 μ M) when added to chloroplasts during a chase period in either the light (clear bars) or the dark (hatched bars).*

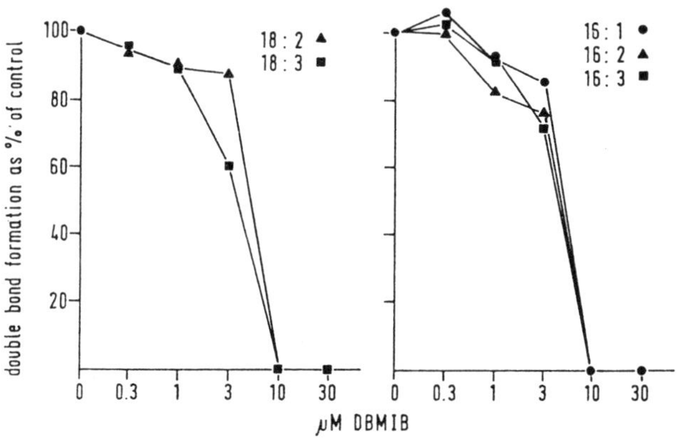

FIGURE 3 *Effect of DBMIB when added to chloroplasts during a chase period in the dark*

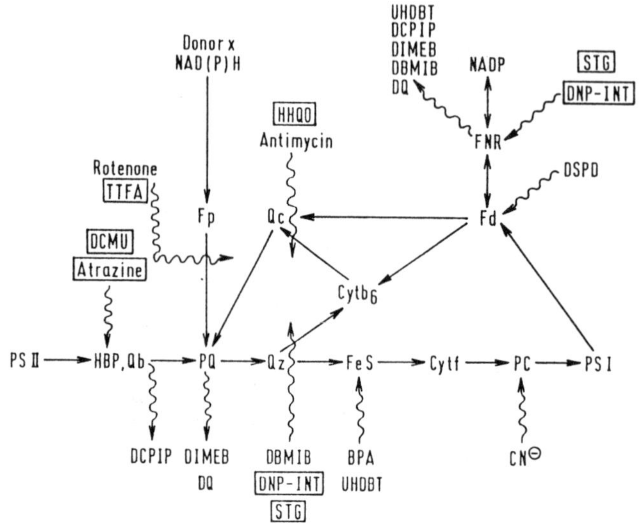

FIGURE 4 ***Sites and modes of action on photosynthetic electron transport of inhibitors examined in fatty acid desaturation.*** Those which remove electrons are indicated by arrows directed away from the affected sites; those which prevent electron flow are indicated by arrows directed toward the sites affected. Compounds which did not inhibit desaturation are boxed.

distant site than DBMIB (Malkin, 1986), which may explain their lack of inhibition. The next set of inhibitors, DIMEB and DQ, withdraw electrons directly from PQH_2 (fig 4) and these too can inhibit desaturation (table 5). Attempts were made to circumvent the inhibition by DBMIB by adding the electron donors DCPIP, DAD, and TMPD; when reduced by the presence of ascorbate, these reagents donate electrons to PC. However, their addition did not restore desaturation, suggesting that the electron transport components from PC and beyond were not involved. Cyanide also blocks electron transport at plastocyanin (PC, fig. 4), and inhibited desaturation as well (table 5); however, the effective concentration for desaturation was so much lower than that for oxygen evolution that it appears that cyanide does not act at PC but perhaps at a desaturase itself.

Taken together, these results suggest that reduced PQ may be involved in desaturation. However, all of these inhibitors except DNP-INT and STG also act as electron acceptors for FNR (fig 4), and thus could inhibit elsewhere in desaturation by withdrawing electrons from a similar type of flavoprotein.

TABLE 5. ***Inhibitor concentrations required for 50% reduction of MGDG desaturation or oxygen evolution.*** Assays contained 70 - 110 μg of chlorophyll.

Site	Compound	Concentration (μM) for 50% Inhibition of: MGDG Desaturation	Oxygen Evolution
QB	Atrazine	no inhibition at 10 uM	3
	DCMU	slight inhibition at 10 uM	0.5
QZ	DBMIB	5	2
	UHDBT	10	1
	BPA	150	45
	DNP-INT	slight inhibition at 100 uM	0.5
	STG	250	7
PQH2	DIMEB	3	16
	DQ	10	4
	DQH2	10	not measured
QC	Antimycin	200	40
	HHQO	no inhibition at 1000 uM	60
Fp[1]	Rotenone	1000	530
	TTFA	1000	50
Misc	DSPD	100	140
	DTE	1000	1450
	KCN	300 for 16:1 25-70 for all others	1100

[1]Fp = flavoprotein which reduces ubiquinone

The next group of inhibitors block reduction of quinones in thylakoids and/or mitochondria by different mechanisms. Both Antimycin A and HHQO inhibit cyclic electron transport in thylakoids at apparently the same site (fig. 4); Antimycin A is thought to prevent the reduction of PQ by cytochrome b at the Qc site. Antimycin A did inhibit desaturation, but only at relatively high concentrations, and HHQO was without effect up to 1 mM (table 5). Thus, it may be that this site is not involved in desaturation. Ubiquinone reduction in mitochondria is inhibited by both rotenone (via NADH) and TTFA (via succinate), and both compounds inhibit the NADH-PQ oxidoreductase activity in Chlamydomonas (Godde and Trebst, 1980). Both compounds also partially inhibited desaturation at 1 mM (table 5). While these concentrations are high, it suggests that electrons destined for desaturation might be donated via a flavoprotein similar to that operating in Chlamydomonas, and one which could be inhibited by the dyes listed previously.

Additional inhibitors are included under miscellaneous (table 5). Inhibition by DTE was surprising, as 1 mM DTE appears to enhance 18:0 desaturation by the stromal desaturase (Jaworski and Stumpf, 1974); however, at 5 mM it strongly inhibits microsomal desaturation (Slack et al, 1976). Cyanide inhibits desaturation of both 18:1 to 18:2 in animal ER (Oshino et al, 1971) and plant microsomes (Stymne and Stobart, 1978) and 18:0 to 18:1 in plant chloroplasts (Jaworski and Stumpf, 1974). This further supports the idea that cyanide acts at a desaturase itself. Interestingly, desaturation of 16:0 was less sensitive towards cyanide than desaturation of 16:1, perhaps pointing to distinct properties of the desaturases acting on MGDG. The 16:1 desaturase enzyme is also missing from pea chloroplasts (see above). John Browse et al (1988) have isolated a mutant from Arabadopsis in which the formation of double bonds in all chloroplast lipids appears missing. This results in glycerolipids with only 16:1 and 18:1, and suggests that desaturation of both these fatty acids to 16:2 and 18:2 may occur with a single enzyme.

In summary, the results of these inhibitors studies suggest that reduced PQ could be involved in desaturation. However, the Q_Z site is unlikely to be involved, as DNP-INT and STG do not block desaturation. Alternatively, these compounds may inhibit desaturation by removing electrons from other unidentified sites of the desaturation pathway, and the involvement of FNR or a similar activity does not necessarily include participation of PQ, and vice versa.

REFERENCES

Browse, JB (1988) Reduced levels of lipid unsaturation in mutants of Arabidopsis. Plant Physiol 4: 39

Godde, D and Trebst, A (1980) NADH as electron donor for the photosynthetic membrane of *Chlamydomonas reinhardii.* Arch Microbiol 127: 245 - 252

Heinz, E and Roughan, PG (1983) Similarities and differences in lipid metabolism of chloroplasts isolated from 18:3 and 16:3 plants. Plant physiol 72: 273 - 279

Jacobson, BS, Jaworski, JG, and Stumpf PK (1974) Fat metabolism in higher plants. LXII. Stearyl-acyl carrier protein desaturase from spinach chloroplasts. Plant Physiol 54: 484 - 486

Joyard, JE, Blee, E and Douce, R (1986) Suofolipid synthesis from $^{35}SO_4^{2-}$ and [1-^{14}C]acetate in isolated intact spinach chloroplasts. Biochim Biophys Acta 879: 78 - 87

Kaiser, WM, Stepper, W and Urbach, W (1981) Photosynthesis of isolated chloroplasts and protoplasts under osmotic stress: Reversible swelling of chloroplasts by hypotonic treatment and its effect on photosynthesis. Planta 151: 375 - 380

Malkin, R (1986) Interaction of stigmatellin and DNP-INT with the Rieske iron-sulfur center of the chloroplast cytochrome *b6-f* complex. FEBS Lett 208: 317 - 320

Mckeon, TA and Stumpf, PK (1982) Purification and characterization of the stearoyl-acyl carrier protein desaturase and the acyl-acyl carrier protein thioesterase from maturing seeds of safflower. J Biol Chem 257: 12141 - 12147

Mills, JD, Crowther, D, Slovacek, RE, Hind, G and McCarty, RE (1979) Electron transport pathways in spinach chloroplasts. Reduction of the primary acceptor of photosystem II by reduced nicotinamide adenine dinucleotide phosphate in the dark. Biochim Biophys Acta 547: 127 - 137

Oshino, N, Imai, Y and Sato, R (1971) A function of cytochrome b-5 in fatty-acid desaturase by rat liver cytochromes. J Biochem 69: 155 - 167

Roughan, PG, Mudd, JB, McManus, TT and Slack, CR (1979) Linoleate and α-linolenate synthesis by isolated spinach (*Spinacia oleracea*) chloroplasts. Biochem J 184: 571 - 574

Roughan, PG and Slack, CR (1982) Cellular organisation of glycerolipid metabolism. Ann Rev Plant Physiol 33: 97 - 132

Slack, CR, Roughan, PG and Terpstra, J (1976) Some properties of a microsomal oleate desaturase from leaves. Biochem J 155: 71 - 80

Sparace, SA and Mudd, JB (1982) Phosphatidylglycerol synthesis in spinach chloroplasts: Characterization of the newly synthesized molecule. Plant Physiol 70: 1260 - 1264

Stymne, S and Appelqvist, LA (1978) The biosynthesis of linoleate from oleoyl-CoA via oleoyl-phosphatidylcholine in microsomes of developing safflower seeds. Eur J Biochem 90: 223 - 229

BIOLOGICAL ROLE OF PLANT LIPIDS
P.A. BIACS, K. GRUIZ, T. KREMMER (eds)
Akadémiai Kiadó, Budapest and Plenum Publishing
Corporation, New York and London, 1989

THE FUNCTION OF MEMBRANE GLYCEROLIPID FATTY ACYL UNSATURATION IN SIGNAL TRANSDUCTION AND CELL CYCLE CONTROL: A BIOPHYSICAL APPROACH

Y.Y. Leshem

Departmemt of Life Sciences, Bar Ilan University, Ramat Gan 52100, Israel

INTRODUCTION

When describing the movement of suspended molecules of submicroscopic size in liquids, Einstein [1] has stated that classical thermodynamics cannot be looked upon as applicable with precision. Such molecules include amphiphilic membrane glycerolipids. Visual observation of the above mentioned submicroscopic particles was originally made by the botanist Robert Brown in 1828 who noted that movement of pollen grains suspended in a drop of water was "uninterrupted and swarming". Such Brownian movement was ascribed by Gouy in 1888 to the effect of the thermal molecular motions of the suspending liquid and formulae enabling calculation of these motions which were found to be both *rotational* and *lateral* (or as commonly designated, *translational*) were worked out by Einstein [2]. With certain modifications his work is still regarded as a basis for behaviour of amphiphiles and similar molecules in liquids [3].

Relating the above to plant membranes, the present paper endeavors to indicate that the physico-chemical nature of the membrane bilayer, in particular interfacial surface tensions or pressures could markedly affect biological activity. We shall attempt to show that anisotropic rotational and translational diffusion rates of membrane amphiphiles –phospholipids (PL) and galactolipids (GL) – in each individual monolayer leaflet of the bilayer depends to a marked degree on the degree of unsaturation of fatty acyl chains. A second factor influencing these parameters, that of the intensity of the negative charge of anionic PL headgroups has been dealt with elsewhere [4].

It is hypothesized that the cytosol being essentially an aqueous phase, were it alone, would possess a surface tension approximating that of water which at 24°C is 72.5 milliNewtons (mN)/m^{-1}, thus value being relatively high. As in seastorms "pouring oil on troubled (cytosolic) waters" would have a marked calming effect by reducing surface tension by 40–50 percent. In biological membranes this enables greater mobility, not only of surface lipids but also of embedded proteins, which metaphorically float like icebergs or buoys in a lipid sea. Degree of movement of such proteins which are part and parcel of the signal transduction process and include enzymes, hormone receptors and ionic channels [5–10] is to a major degree linked to surface tension parameters. Their freedom of movement and hence biological activity could be enhanced or limited by surface tension changes which depends not only on glycerolipid molecular areas as a function of fatty acid unsaturation (i.e. a **tail** function), but also as shown by Landau and Leshem [11] and Leshem [12], on degree of electrostatic cross-linking between the components themselves (i.e. a **head** function). This crosslinking which *in vivo* is effected by divalent cations – $Ca^{2+} > Mg^{2+}$, by forming multimolecular aggregates, markedly hinders motility and has an essentially detrimental and rigidifying effect on membranes.

METHODS

a) Effect of fatty acyl tails on diffusional motion

Basing upon the elegant research on fatty acid structure conducted by de Jong [13] – see **Fig. 1**, scale models of the most probable biological conformations, were constructed and molecular radii-measured.

These values were entered in the Einstein equation to enable calculation of rotational and translational diffusion (D_r and D_T). The rationale, in terms of surface parameters was that in a system seen as in **Fig. 2**, surface tension in A should markedly differ from B.

We point out that Merril and Nichols [14] have shown that each *cis* C=C bond induces a *ca* 30° bend in the molecule, thus according to de Jong's models 18 : 3 > 18 : 2 > 18 : 1 in degree of 'scorpion tail' bending. Arachidonic acid, 18 : 4, in his

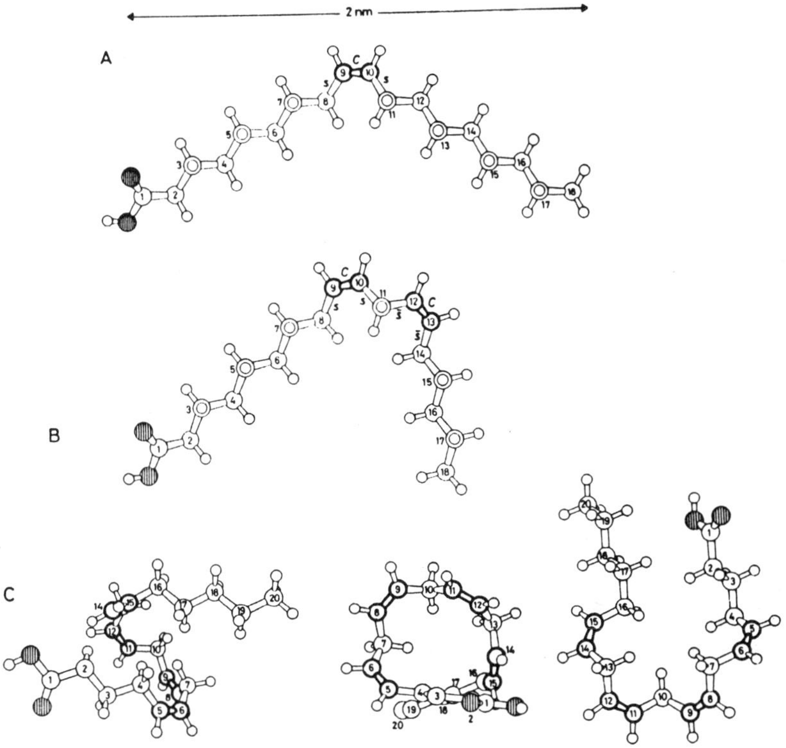

Fig. 1: *Probable crystal structure biological conformations of prevalent unsaturated fatty acids as suggested by de Jong [13]. A – oleic acid; B – linoleic acid; C – arachidonic acid: left and middle – side and top views of spiral conformation, right – U conformation.*

model, contrary to what is commonly believed [15], is *not* U shaped in biological systems but assumes a spiral conformation and thus while being more unsaturated, arachidonic acid possesses a radius approximating that of 18 : 2.

The Einstein formulae employed were as follows: –

a) Rotational diffusion – D_r (being an indication of no. of revolutions per sec.)

$$D_r = \frac{K_B T}{8\pi\eta r^3} \quad \frac{1}{\text{sec}} \tag{1}$$

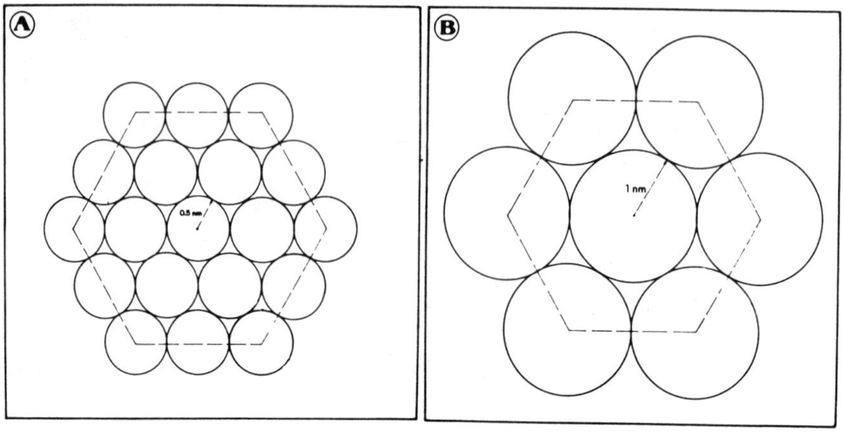

Fig. 2: *Given area spacing in a hexagonal lattice of liquid-suspended glycerolipid molecules possessing different radii. In A and B total interstitial free space is equal, but total molecular periphery in A in greater (top view).*

b) Lateral (translational) diffusion D_T – being an indication of area traversed per sec:–

$$D_T = \frac{K_B T}{6\pi\eta r} \quad cm^2/sec \qquad (2)$$

In the equations K_B is the Boltzmann Constant; T, the absolute temperature; η – the microviscosity; and r, the particle radius.

b. Effect of PL acyl tail unsaturation on membrane rigidification

In this trail the experimental premise was that if taking a common membrane PL such as phosphatidylcholine (PC) and comparing one species with an *oleoyl* (18 : 1) *sn*–2 acyl tail to another, with a *linoleoyl* (18 : 2) one, the latter would possess a greater area. Contrariwise, within a given field, more molecules of an oleoyl PC species would be able to be packed and thus in keeping with the experimental surmise detailed in 'a' above, should have a rigidifying effect on the membrane.

This aim was achieved by spreading monolayers of either of the two PC species, – *sn*–2 oleoyl or *sn*–2 linoleoyl (the rest of the molecule being indentical and having a *sn*–1 palmitoyl tail) over an aqueous subphase in a Langmuir-Blodget trough and then employing the movable teflon barrier to gradually compact the

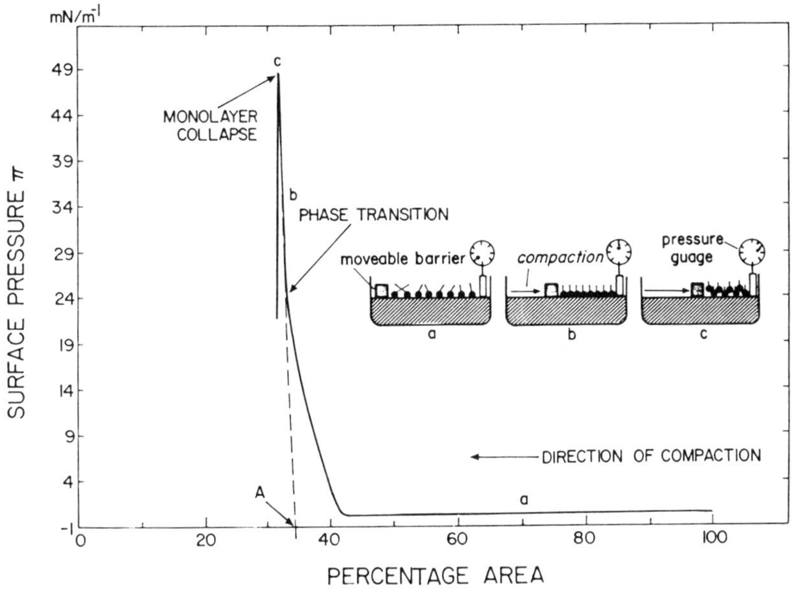

Fig. 3: *Langmuir film balance: schematic.* ***a*** *– monolayer of lipid in a quasi gas state,* ***b*** *– state of maximal compaction indicative of molecular area,* **c** *– monolayer collapse point reflecting monolayer rigidity. Tested monolayer was palmitic acid over* H_2O *to produce a typical surface pressure/molecular area isotherm.*

layer until collapse point (*cf* Landau and Leshem [11] for details of experimental procedure). By taking readings on a recording Wilhelmy balance, surface pressure/molecular area isotherms were drawn, and by drawing a tangential intercept to the exponential section of the isotherm and employing the Langmuir equation (detailed henceforth), molecular areas were determined. Compaction was continued until the monolayer collapsed. **Fig. 3** indicates the principle employing a palmitic acid monolayer.

The percentage area (A) at which the tangential intercept cuts the y axis is entered to the Langmuir formula to calculate molecular area of the tested molecular species comprising the monolayer:

$$\text{nm}^2/\text{molecule} = \frac{A \bullet 10^{-18}}{\text{Avogadro number} \bullet M \bullet V} \qquad (3)$$

In Eq. (3) A is % compaction at phase transition; Avogadro's no. $= 6.02 \cdot 10^{-23}$; M, molarity of the applied monolayer film; V, volume of applied monolayer droplet. In the above example where palmitic acid was dissolved in hexane and spread over a water subphase, entering an A value of 34% to Eq. (3) yielded a molecular area of *ca* 0.215 nm^2. The experiment was performed on a Lauda Langmuir trough fitted with a computerized recorder programmed with Eq. (3) to present direct readings of molecular areas.

RESULTS AND DISCUSSION

Table 1 indicates results of mobility calculations obtained by entering r values obtained from the de Jong models to Equations (1) and (2).

This table shows that the greater the radii, the less diffusional freedom. Increasing degree of *cis* unsaturation from 1 to 3, decreases D_r by about 80%, and D_T by 40%. Arachidonic acid is the exception that proves both the rule and the de Jong assumption that it turns back upon itself and is not U shaped and that diffusional values are radius linked.

In terms of physiological implications, in a tetragonal lattice which is the current concept of glycerolipid packing [16] a simple geometrical calculation indicates that in both constellations (Fig. 2 A and B) while total interstitial free space is equal, more individual molecules fill the area when radius of each decreases; this could lead to closer packing and hence to greater intermolecular friction. This in an *in vivo* membrane could cause rigidification, and this was the premise of the next experimental step. Bamberger *et al.* [17] have shown that treatment of a model membrane with the superoxide ($O_2^{\bar{\bullet}}$) free radical, stabilized in crownether has a similar rigidfying effect. This assumedly was caused by decrease of membrane PL radius as brought about by $O_2^{\bar{\bullet}}$ attack on the *cis* double bonds in the fatty acyl chains.

Table 2 presents results of employing similar Langmuir methodology to compare the two above mentioned phosphatidyl species.

These results show that inclusion of an 18 : 2 instead of an 18 : 1 *sn*–2 acyl chain:– a) increases molecular area from 0.62 to 0.69 nm^2 b) considerably reduces collapse point, from 46 to 33 mN/m^{-1}. This clearly indicates that a more saturated PC species, here the oleoyl PC, resists compaction more efficiently and is more rigid,

Table 1: *Relative rotational and translational motilities of a series of increasingly* cis *unsaturated fatty acids – 18 : 1, 18 : 2, 18 : 3 and 20 : 4.*

Values of $8.79 \cdot 10^6$ and $3.808 \cdot 10^{-8}$ respectively, for D_r and D_T served a basis of comparison.

No. of *cis* double bonds	r (nm)	D_r	D_T
1	0.57	100	100
2	0.83	32	69
3	0.94	22	61
4	0.77	41	74

Table 2: *Effect of unsaturation of the sn–2 acyl chain in sn–1 palmitoyl PC on molecular area and collapse points. Tabulated parameters were calculated from molecular area/surface area isotherms as outlined in methods.*

Monolayer species tested	Molecular area	Collapse Point
	nm^2	mN/m^{-1}
sn–2 **oleoyl** (18 : 1) PC	0.62	46.0
sn–2 **linoleoyl** (18 : 2) PC	0.69	33.0

and on the contrary the more unsaturated linoleoyl species derigidifies the monolayer, a phenomenon which as pointed out in the Introduction is regarded as physiologically desirable.

Taken together, experimental results indicate that surface parameters of model membranes are markedly effected by the nature of the unsaturated acyl chains which *in vivo* could occur at the *sn*–1 site in the plasma membrane [7] and at either the *sn*–1 and –2 sites in chloroplastic membranes [18]. It therefore is suggested that these biophysical parameters are contributing factors in determination of mobility and hence availability of the biochemical factors involved in cell cycle signal transduction. Moreover other processes e.g. pollen tube penetration into stylar tissue may also be facilitated by decrease of membrane rigidity and in this context in almond flowers it has been reported [19] that with progress of pollen grain germination, there is an increase in 18 : 3, accompanied by a decrease in 18 : 1 fatty acyl species.

ACKNOWLEDGMENTS

The author wishes to thank Dr. M. Deutsch of the Department of Physics, Bar Ilan University, Dr. E. M. Landau of the Department of Structural Chemistry, the Weizmann Institute of Science, Rehovoth, Israel and Dr. A. Berçzi of the Institute of Biophysics, Szeged, Hungary, for their aid and helpful discussions.

REFERENCES

1. Einstein A. 1956. Investigations on the theory of the Brownian movement. Ed. with notes R. Fürth. Dover, N.Y. 119 pp.
2. Einstein, A. 1906. Zur Theorie der Brownschen Bewegung. Annal. der Physik, **14**, 371–81.
3. Wiegel, F.W. 1980. Fluid flow through porous macromolecular systems. In Lecture Notes in Physics. Eds. J. Ehlers, K. Hepp, H.A. Weidenmuller, R. Kippenhahn, and F. Zittartz. Springer Verlag, Berlin, pp. 1–102.
4. Landau E.M. and Leshem Y. 1989. A monolayer model study of surface tension – associated parameters of membrane phospholipids II. Anionic headgroup interactions with pH, Ca^{2+} and auxins. Jour. Exp. Bot. (in press).
5. Blowers, D.P., Hetherington, A. and Trewavas, A. 1985. Isolation of plasma membrane bound calcium/calmodulin regulated protein kinase from pea using western blotting. Planta **166**, 208–15.
6. Paliyath, G. and Poovaiah, B.W. 1985. Calcium and calmodulin promoted phosphorylation of membrane proteins during senescence in apples. Pl. Cell. Physiol. **26**, 977–86.

7. Leshem, Y. 1987. Membrane phospholipid catabolism and Ca^{2+} activity in control of senescence. Physiol. Plant **69**, 551–59.
8. Pfaffman, H., Hartman, E., Brightman, A.O. and Moreé, D.J. 1987. Phosphatidylinositol specific phospholipase C of plant systems. Plant Physiol. **85**, 1151–5.
9. Zbell B. and Walter, G. 1987. About the research of the molecular action of auxin binding sites on membrane localized rapid phosphoinositide metabolism in plant cells. In Plant Hormone Receptors. Ed. D. Klambt, Springer Verlag – Berlin. pp. 143–53.
10. Sommarin, M. and Sandelius, A.S. 1988. Phosphatidyl inositol and phosphatidyl inositolphosphate kinases in plant plasma membranes. Biochem. Biophys Acta, **958**, 268–786.
11. Leshem, Y., Landau, E.M. and Deutsch, M. 1989. A monolayer model study of surface tension associated parameters of membrane phospholipids. I. Effect of unsaturation of fatty acyl tails. Exp. Bot. (in press).
12. Leshem, Y.Y. 1987. Ca^{2+} and inter molecular bridging of membranal phospholipids and proteins. In The Metabolism Structure and Function of Plant Lipids. Eds. P.K. Stumpf, J.B. Mudd, and U.D. Ness. Plenum Press, Calif. pp. 225–7.
13. De Jong, S. 1980. Diacylglycerol structures and fatty acid conformations. A theoretical approach. Ph.D. thesis University of Ulrecht. The Netherlands 171 pp.
14. Merril, A.H. and Nichols, J.W. 1986. Techniques for studying phospholipid membranes. In Phospholipids and Cellular Regulation. Ed. J.F. Kuo. Vol. I. CRC Press, Boca Raton, Florida, pp. 69–96.
15. Holman, R.T. 1951. Die rolle der Fettsauren in der Ernahrung. Fette, Seifen Anstrichmit. **53**, 332–6.
16. Kjaer, K., Als-Nielsen, Helm, C.A., Laxhuber, L.A.A. and Möhwald 1987. Order in lipid monolayer studied by synchroton X ray diffraction and fluorescence microscopy. Phys. Rev. Let. **58**, 2224–6.
17. Bamberger, E.S., Alter, M., Landau, E.M. and Leshem, Y. 1989. Biophysical effects of superoxide on surface parameters of a model membrane. In Medical, Biochemical and Chemical Aspects of Free Radicals. Eds. Hayaishi, O., Niki, E., Kondo, M., and Yoshikawa, T. Elsevier Amsterdam (in press).

18. Gounaris, K., Barber, J. and Harwood, J. L. 1986. The thylakoid membrane of higher plant chloroplasts Biochem. Jour. **237**, 313–26.

19. Rikhter, A.A. 1987. Fatty acid composition of lipids in almond leaves and generative organs. Fiziol. Rastenii, **34**, 949–55.

BIOLOGICAL ROLE OF PLANT LIPIDS
P.A. BIACS, K. GRUIZ, T. KREMMER (eds)
Akadémiai Kiadó, Budapest and Plenum Publishing
Corporation, New York and London, 1989

ROLE OF PHOSPHATIDYLGLYCEROL IN PHOTOSYNTHETIC MEMBRANE: STUDY WITH MUTANTS OF CHLAMYDOMONAS REINHARDTII

A. Trémolières, J. Garnier, D. Guyon, J. Maroc and B. Wu

Laboratoire de Biochimie Fonctionnelle des Membranes Végétales, Bâtiment 9, UPR 0054-CNRS
91198 Gif sur Yvette, France

INTRODUCTION

Studies of interactions between PG-containing C16:1-*trans* and LHCP have been first performed in chloroplasts of higher plants (1, 2, 3, 4, 5). It was concluded that this lipid could play a role in the oligomeric organization of the LHCP and its efficiency for the light energy collection (6). But the discovery by Somerville group (7, 8) of a mutant of *Arabidopsis thaliana* lacking C16:1-*trans* in its PG re-opened the question of the exact function of this fatty acid. This question was approached using mutants of *Chlamydomonas reinhardtii* showing photosynthetic and lipid deficiencies (9).

MATERIAL AND METHODS

The characteristics of the wild type of *C. reinhardtii*, of the high fluorescent mutant *F l* 39 and of the low fluorescent mutants *mf* 1, *mf* 2 and *mf* 3 have been indicated (10, 11). Algae were grown in light, in Tris-acetate medium (12). Lipids and chlorophyll-protein complexes were analyzed as previously described (9). The fluorescence induction kinetics and the low temperature emission spectra of whole cells were measured as in (9).

RESULTS

Four mutants were compared with the wild type. As seen in Fig. 1, the wild type evidently possesses all the protein-chlorophyll complexes usually observed in the photosynthetic membrane of *C. reinhardtii* : the PS I system

---

Abbreviations : C16:1-*trans*, Δ_3-*trans*-hexadecenoic acid ; CP, chlorophyll-protein complex ; LHCP, light-harvesting chlorophyll-protein complex ; PG, phosphatidylglycerol ; PS, photosystem.

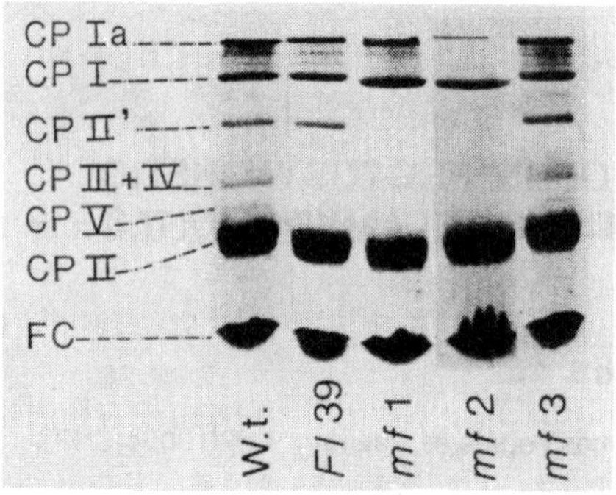

Fig. 1 - Chlorophyll-protein complexes of the different strains of *C. reinhardtii*.

complex (CP Ia and CP I), the two bands of the PS II center (CP III + IV), the main light-harvesting antenna (CP II) and its oligomeric form (CP II') recently pointed out for the first time in *C. reinhardtii* (9). The three non-photosynthetic mutants *Fl* 39, *mf* 1 and *mf* 2 are lacking CP III + IV (the PS II center). *mf* 3 which is a photosynthetic mutant showing a low level of fluorescence, shows all the complexes found in the wild type. One of the three non-photosynthetic mutants, *Fl* 39, contains an oligomeric form of the main light-harvesting antenna ; the two weakly fluorescent mutants are lacking this oligomeric CP II' form, but have a normal monomeric antenna (CP II).

Fig. 2 indicates that, despite the lack of PS II center for *Fl* 39 and an abnormally low level of fluorescence for *mf* 3, these two mutants are able to carry out reversible state I $\rightleftarrows$ state II transition, as shown by the

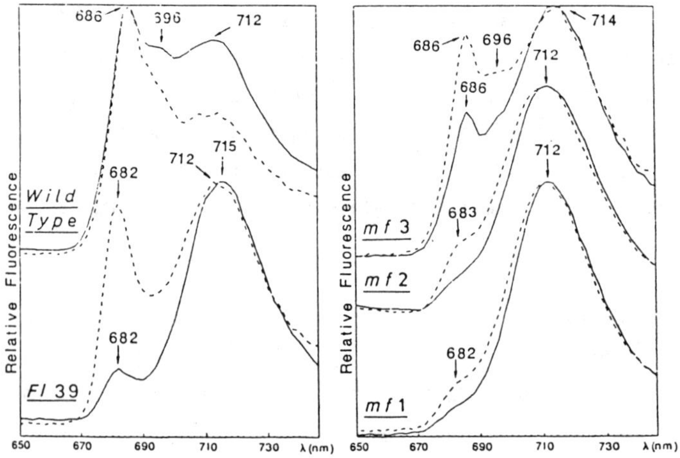

Fig. 2 - 77 K fluorescence emission spectra of cells incubated with NaN_3 (———) or pre-illuminated in far red light (----) before freezing.

TABLE I - Main characteristics of the different strains

Components or ability	W.t.	*Fl* 39	*mf* 1	*mf* 2	*mf* 3
PG-C16:1-*trans*	+	+	-	-	+
PS II	+	-	-	-	+
CP II'	+	+	-	-	+
State I $\rightleftarrows$ state II transition	+	+	-	-	+

changes in the low temperature fluorescence emission spectra induced by variations of the redox state of the plastoquinone pool. On the other hand, the mutants *mf* 1 and *mf* 2 always show a main emission peak in the region of the fluorescence of the PS I center and they appear unable to perform a significant state I $\rightleftarrows$ state II transition.

As indicated in Table I, if the wild type and the mutants *Fl* 39 and *mf* 3 contain a high percentage of C16:1-*trans* in their PG, *mf* 1 and *mf* 2 are totally lacking this fatty acid (and in addition *mf* 2 was shown to be lacking the Δ_3-transdesaturase).

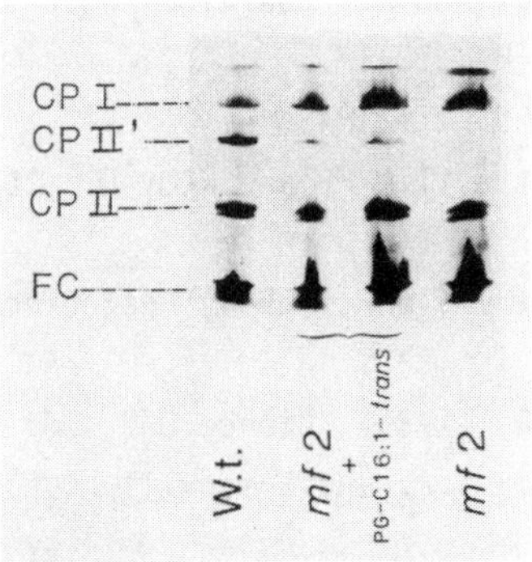

Fig 3 - Reconstitution of oligomeric CP II' by incubation of *mf* 2 cells with liposomes of PG-C16:1-*trans*

In Fig. 3 is reported the result of an experiment where *mf* 2 cells were incubated with liposomes of PG-C16:1-*trans* for 36 h. Lipid analysis of the washed chloroplasts showed that PG-C16:1-*trans* has been incorporated in weak amounts in the photosythetic membranes. We can see that the apparition of a weak green band exactly at the same place as the oligomeric PC II' is observed in the gel, when *mf* 2 cells have been incubated with liposomes.

CONCLUSIONS

In this work, a correlation is observed between the lack of C16:1-*trans* in the PG, the absence of the oligomeric form CP' and the unability to carry out state I $\leftrightarrows$ state II transition. On the other hand, incorporation of PG-C16:1-*trans* in *mf* 2 cells by incubation with liposomes allows to reconstitute an oligomeric form of the main light-harvesting antenna. Among ten mutants of *C. reinhardtii* affected at various steps of the light-harvesting and/or photosynthetic processes, six were found to be able to perform a state I $\leftrightarrows$ state II transition : all these six mutants show both oligomeric CP II' and C16:1-*trans* in their PG.

We propose, as a working hypothesis, that the oligomeric form CP II' could represent the part of the mobile antenna linked to the PS II core-antenna through an interaction with PC-C16:1-*trans*, which is the lipid showing the higher affinity for antenna-complexes.

REFERENCES

1) Dubacq J.P. and Trémolières A. (1983) Physiol. Vég. 21, 293-312.
2) Rémy R., Trémolières A., Duval J.C., Ambard-Bretteville F. and Dubacq J.P. (1982) FEBS Lett. 137, 271-275.
3) Huner N.A.P., Krol M., Williams J.P., Maissan E., Low P.S., Roberts D. and Thompson J.E. (1987) Plant Physiol. 84, 12-18.
4) Krupa Z. (1984) Photosynthesis Research 5, 177-184.
5) Trémolières A., Dubacq J.P., Ambard-Bretteville F. and Rémy R. (1981) FEBS Lett. 130, 27-31.
6) Rémy R., Trémolières A. and Ambard-Bretteville F. (1984) Photobiochem. Photobiophys. 7, 267-276.
7) Browse J., Mc Court P. and Somerville C.R. (1985) Science 227, 763-765.
8) Mc Court P., Browse J., Watson J., Arntzen C.J. and Somerville C.R. (1985) Plant Physiol. 78, 853-858.
9) Maroc J., Trémolières A., Garnier J. and Guyon D. (1987) Biochim. Biophys. Acta 893, 91-99.
10) Garnier J., Maroc J. and Guyon D. (1986) Biochim. Biophys. Acta 851, 395-406.
11) Garnier J., Maroc J. and Guyon D. (1987) Plant Cell Physiol. 28, 1117-1131.
12) Gorman D.S. and Levine R.P. (1965) Proc. Natl. Acad. Sci. US 54, 1665-1669.

BIOLOGICAL ROLE OF PLANT LIPIDS
P.A. BIACS, K. GRUIZ, T. KREMMER (eds)
Akadémiai Kiadó, Budapest and Plenum Publishing
Corporation, New York and London, 1989

ESR STUDIES OF LIPID-PROTEIN INTERACTIONS IN PHOTOSYNTHETIC MEMBRANES

D.J. Murphy[1], L. Gang[2], I. Nishida[3] and P.F. Knowles[2]

[1]Department of Biological Sciences, University of Durham, UK
[2]Department of Biophysics, University of Leeds, UK
[3]National Institute for Basic Biology, Okazaki, Japan

INTRODUCTION

The nature of lipid-protein interactions in photosynthetic membranes has been examined by means of electron spin resonance (ESR) and saturation transfer ESR (STESR) spectroscopy. Nitroxide-labelled acyl lipids were introduced into thylakoid membranes and sub-membrane preparations and the resultant ESR spectra were analysed under various conditions.

LIPID SPECIFICITY FOR THYLAKOID PROTEINS

Spin-labelled membrane preparations gave rise to spectra consisting of two major components. One component was similar to the spectrum of lipid spin-labels in aqueous dispersions and corresponds to that of a fluid lipid bilayer. The other component was characteristic of motionally restricted lipid spin-labels which were associated with membrane proteins. The two components were quantitated using computational spectral subtractions. All spectra were normalised to the same double integrals (to equate the number of spins). A single-component spectrum was then subtracted from a two-component spectrum to give the other component (Marsh, 1985). The results show that in whole thylakoid membranes and sub-membrane fractions from pea chloroplasts, phosphatidylglycerol (PG) spin-labels have a higher specificity than phosphatidylcholine (PC) and monogalactosyldiacylglycerol (MGD) spin-labels for association with thylakoid proteins. Using the above spin-labels in freshly-prepared thylakoid membranes, it was calculated that the proportion of motionally restricted lipids of each class was: 14-PGSL, 43%, 14-PCSL, 31%; 12-PCSL, 34%; 12-MGDSL, 26%. SL = spin-label, numbers (12 or 14) before each lipid spin-label refer to the acyl carbon number to which the nitroxide moiety was attached. This indicates the following selectivity of thylakoid proteins for lipid: PG < PC < MGD.

TABLE 1

Spin-labelled lipid	14 PGSL	14 PCSL	12 PCSL	12 MGDSL
T_C	1.5×10^{-5}s	3.6×10^{-5}s	2.8×10^{-3}s	1.5×10^{-5}s
$2A_{max}$	60.1	59.8	60.3	60.3

SATURATION-TRANSFER EST STUDIES

The nature of the lipid-protein interactions can be further studied by the use of ST-ESR techniques. ST-ESR extends the useful timescale of ESR to > 10^{-5}s and can be used for the analysis of the amplitude and rate of the rotation of spin-labelled lipids. The rotational correlation time (T_C) which can readily be deduced from ST-ESR spectra, is inversely proportional to the rate of rotation of the spin probe. In the case of the PSII preparations the values for T_C are given in Table 1. In general all of these rotation rates are very slow indeed. A T_C of the order of 10^{-5}s corresponds to that of a large protein in a model membrane, or to lipid molecules in the gel phase. Although each of the lipid classes is undergoing a very slow rate of rotation, there is some selectivity, with PC exhibiting appreciably slower motion than either PG or MGD. Interestingly, the hyperfine splitting constants ($2A_{max}$) of all four spin probes are almost identical. Since this constant is determined by the amplitude and the rate of motion, we can say that although PC rotates more slowly than PG or MGD, it has a correspondingly higher amplitude of rotation.

These data are potentially important since they indicate that the different spin-labels bind at different sites, the site for PCSL binding being the most motionally restricted. The most attractive explanation for these effects is that PCSL is associated with a larger protein complex than either PGSL or MGDSL and therefore tumbles more slowly. However, location of the spin-labels between protein aggregates or in gel-phase lipid pools cannot be ruled out as alternative explanations.

ACKNOWLEDGEMENTS

We gratefully thank Drs D. Marsh and L.J. Horvath for the use of STESR facilities in Göttingen and for valuable discussions.

REFERENCE

Marsh, D. (1985) in Progress in Lipid-Protein Interactions (Watts, A. & du Pont, J.H.M. eds) Elsevier, Amsterdam.

BIOLOGICAL ROLE OF PLANT LIPIDS
P.A. BIACS, K. GRUIZ, T. KREMMER (eds)
Akadémiai Kiadó, Budapest and Plenum Publishing
Corporation, New York and London, 1989

THE PHASE BEHAVIOUR OF MEMBRANE LIPIDS AND THE ORGANISATION OF THE PHOTOSYNTHETIC MEMBRANE

P.J. Quinn

Department of Biochemistry, King's College London, Campden Hill, London W8 7AH, UK

INTRODUCTION

The structure of the photosynthetic membrane in common with all biological membranes is considered to be a lipid bilayer matrix to which the various functional membrane proteins are associated. The form of the association can be intrinsic to the lipid component in which case there is presumed to be contact between the polypeptide chains of the proteins and the hydrophobic domain of the lipid. Other proteins are believed to be peripheral to the lipid bilayer and associate mainly through electrostatic interactions.

The lipids comprising the photosynthetic membrane of higher plant chloroplasts are dominated by galactolipids of which monogalactosyldiacylglyercol represents about half the total polar lipid fraction. Of the remainder, digalactosyldiacylglycerol comprises approximately half and the other lipid classes consist mainly of sulphoquinovosyldiacylglycerol and the phospholipids, phosphatidylcholine and phosphatidylglycerol. The metabolic pathways involved in synthesis of these lipids and their biochemical modification in response to environmental factors has been the subject of a recent review [1]. In this report the phase behaviour of lipids of the photosynthetic membrane will be considered in the context of the structural organisation of the thylakoid.

PHASE BEHAVIOUR OF MEMBRANE LIPIDS

The molecular species of membrane lipids including that of the photosynthetic membrane is highly complex with members of each of the major lipid classes containing fatty acyl substituents that differ in length and extent of unsaturation and position of attachment to the glycerol backbone. The need for this molecular complexity and the biochemical mechanisms whereby it is maintained within relatively narrow limits, is not clearly understood. Nevertheless, it is possible to categorise membrane lipids into a few groups on the basis of their physical properties when dispersed in excess physiological salt solutions. Thus in all biological membranes it is possible to identify molecular species of lipids that when dispersed alone will form bilayer phases in which the hydrocarbon chains are arranged in a gel or liquid-crystalline configuration at the growth temperature or hexagonal-II structure.

A number of detailed biophysical studies [2,3] have shown that with the exception of monogalactosyldiacylglycerols of the photosynthetic membrane of higher plant chloroplasts most of the remaining lipids will form liquid-crystalline bilayers in aqueous systems at the growth temperature but some saturated molecular species of phosphatidylglycerols may form lamellar gel phases under these conditions. The monogalactolipid which dominates the lipids of this membrane, by contrast, generally assumes an hexagonal-II phase.

The phase structure of total polar lipid extracts of chloroplast thylakoid membranes when dispersed in physiological salt solutions is characterised by non-lamellar arrangements of lipid [4]. This contrasts markedly with the general belief that the dominant form of the lipids in the photosynthetic membrane is in a bilayer configuration. The conclusion that must be drawn from this is that the interation of the different polar lipids with the other membrane components imposes a bilayer structure and prevents phase separation of non-bilayer forming lipids into separate domains within the membrane.

The factors that effect the mixing or phase separation of polar membrane lipids is illustrated schematically in Figure 1. In this figure

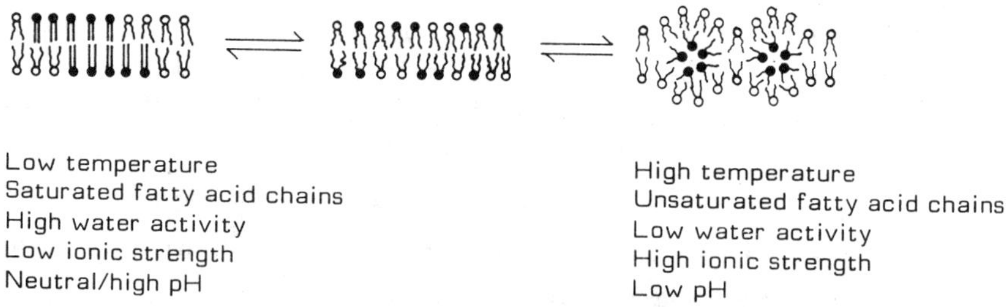

Fig.1 Factors affecting lipid phase separations.

the bilayer forming lipids are represented by open polar groups and the non-bilayer lipids by filled circles. It is noteworthy that screening the charges of the acidic lipids by inorganic ions or pH can have a profound effect on the phase behaviour of the dispersion pointing to an important function of these lipids in the membrane.

MEMBRANE STABILITY

It is clear from the above discussion that the stability of biological membranes and the photosynthetic membrane in particular relies largely on the interaction between the various constituents. To probe the nature of these interactions we have manipulated the environmental conditions of intact chloroplasts and examined the consequences on the structure of the thylakoid membrane as well as photosynthetic functions.

In experiments in which chloroplasts were exposed to different temperatures for 5 minutes and examined for both morphological and functional changes at 20°C it was found that a normal morphology with stacked grana and function was preserved with treatments up to 35°C. With incubation at temperatures between 35 and 45°C a complete destacking of the grana was observed and higher temperatures caused a phase separation of non-bilayer lipids into stable aggregates of cylindrical inverted micelles [5]. Interpretation of the effects of temperatures greater than 45°C is based on phase conditions that result in a release of constraints imposed by interaction of the monogalactodiacylglycerol with other membrane components and its segregation into domains of non-bilayer lipid structure. Gross phase separations of this type require that the shift in thermal stability of the stacked membrane is relatively large because three-dimensional aggregates of lipids are not observed if the chloroplast membrane is destacked by manipulation of the ionic environment before heat treatment.

The explanation of the effect of heat treatment on destacking the native membrane is not straightforward. The factors responsible for membrane stacking have been postulated in a model proposed by Barber [6]. According to this model thylakoid stacking and related phenomena are explained in terms of the effect of cations on electrostatic charges on the membrane surface. A difference in surface charges of light-harvesting chlorophyll a/b protein complexes associated with photosystem II (believed to carry little or no charge) and P-700-chlorophyll a protein complexes associated with photosystem I (possessing a relative abundance of negative charge) is said to result in a randomisation of the complexes laterally in the membrane in conditions of low salt concentration. It is argued that addition of cations screens the charges and reduces electrostatic repulsion between photosystem-II complexes, allowing a reorganisation of the membrane system in which the two photosytems become spatially segregated. The formation of the grana stacks is explained by a reduction in coloumbic repulsion forces between opposing membrane surfaces [7].

A detailed freeze-fracture analysis of the structural changes associated with heat destacking [8] revealed that dramatic changes in particle size and distribution occurs during the destacking process. One explanation for these changes was that as a result of a change in phase behaviour of the monogalactolipid there was a dissociation of core complexes of photosytem-II and their associated light-harvesting protein complexes which aggregated into domains of focal point attachment between the membrane leaving the core complexes free to diffuse into the unstacked regions of the membrane.

These structural rearrangements are consistent with functional changes in photosystem-II observed in heat stressed chloroplasts [9,10]. On the basis of these functional and structural changes it has been suggested that the function of non-bilayer lipid of the photosynthetic membrane is to package the light-harvesting chlorophyll _a_/_b_ protein complexes together with the photosystem-II core protein complex into efficient functional units located within the grana stack [3].

Other factors that can cause bulk phase separations of non-bilayer structures include high Mg^{2+} concentrations, low pH, treatment with 6-M guanidine thiocyanate or digestion with phospholipase A_2. Stabilisation of thylakoid membranes to heat stress can be achieved by reducing the tendency of monogalactosyldiacylglycerol to form non-bilayer arrangements. This has been demonstrated by hydrogenating the fatty acyl residues of the membrane lipids _in situ_ in the presence of homogeneous catalysts [10].

CONCLUSIONS

The above discussion indicates that a knowledge of the phase behaviour of individual molecular species of lipid comprising the matrix of the thylakoid membrane of higher plant chloroplasts can be informative of the factors governing the stability of the membrane. Perhaps the most important conclusion is that the presence of non-bilayer forming lipids is not simply required to facilitate the dynamic functions such as membrane fusion etc., but also to play a role in the creation of oligomeric functional complexes of the different membrane proteins.

REFERENCES

[1] Quinn, P.J. (1988). Regulation of membrane fluidity in plants. In: Advances in Membrane Fluidity, (Aloia, R.C., Curtain, C.C. and Gordon, L.M., eds.) Vol. 3, pp 293-321, Alan R. Lis, New York.

[2] Shipley, G.G., Green, J.P. and Nichols, B.W. (1973). The phase behaviour of monogalactosyl, digalactosyl, and sulphoquinovosyl diglycerides, Biochim. Biophys. Acta, 311, 531-544.

[3] Quinn, P.J. and Williams, W.P. (1983). The structural role of lipids in photosynthetic membranes, Biochim. Biophys. Acta, 737, 223-266.

[4] Gounaris, K., Sen, A., Brain, A.P.R., Quinn, P.J. and Williams, W.P. (1983a). The formation of non-bilayer structures in total polar lipid extracts of chloroplast membranes, Biochim. Biophys. Acta, 728, 129-139.

[5] Gounaris, K., Brain, A.P.R., Quinn, P.J. and Williams, W.P. (1983b). Structural and functional changes associated with heat-induced phase-separations of non-bilayer lipids in chloroplast thylakoid membranes, FEBS Lett., 153: 47-52.

[6] Barber, J. (1980). An explanation for the relationship between salt-induced thylakoid stacking and the chlorophyll fluorescence changes associated with changes in spillover of energy from photosystem II to photosystem I. FEBS Lett., 118: 1-10.

[7] Mullet, J.E. and Arntzen, C.J. (1980). Simulation of grana stacking in a model membrane system. Mediation by a purified light-harvesting pigment-protein complex from chloroplasts, Biochim. Biophys. Acta, 589: 100-117.

[8] Gounaris, K., Brain, A.P.R., Quinn, P.J. and Williams, W.P. (1984). Structural reorganisation of chloroplast membranes in response to heat stress, Biochim. Biophys. Acta, 766: 198-208.

[9] Schreiber, U. and Armond, P.A. (1978). Heat-induced changes of chlorophyll fluorescence in isolated chloroplasts and related heat-damage at the pigment level, Biochim. Biophys. Acta, 502: 138-151.

[10] Thomas, P.G., Dominy, P.J., Vigh, L., Mansourian, A.R., Quinn, P.J and Williams, W.P. (1986). Increased thermal stability of pigment-protein complexes of pea thylakoids following catalytic hydrogenation of membrane lipids, Biochim. Biophys. Acta, 849: 131-140.

BIOLOGICAL ROLE OF PLANT LIPIDS
P.A. BIACS, K. GRUIZ, T. KREMMER (eds)
Akadémiai Kiadó, Budapest and Plenum Publishing
Corporation, New York and London, 1989

LIPID COMPOSITION OF THYLAKOID MEMBRANES FROM TOBACCO MUTANT CHLOROPLASTS EXHIBITING EITHER ONLY PHOTOSYSTEM I OR PHOTOSYSTEM I AND II ACTIVITY

J. Bednarz, A. Radunz and G.H. Schmid

Universität Bielefeld, Fakultät für Biologie, Lehrstuhl Zellphysiologie, 4800 Bielefeld 1, FRG

Serological studies have shown that the lipid distribution in the lamellar system of chloroplasts shows a lateral and transversal asymmetry (Radunz, 1980). We have analysed the lipid composition of chloroplasts from the variegated tobacco mutant Nicotiana tabacum NC 95.

Chloroplasts from green leaf areas of this mutant show grana stacking and exhibit photosysten I and photosystem II activity, whereas chloroplasts from yellow-green leaf areas exhibit no grana stacking and no photosystem II activity (Homann and Schmid, 1967). This structural relationship permits without using detergent treatment the comparison of the lipid composition of the intergrana region with that of a lamellar system having grana and intergrana regions.

From chloroplasts of green leaf areas we have isolated photosystem II particles according to Berthold et al. (1981) using the unpolar detergent Triton X-100. A portion of these particles was used to separate the light-harvesting and the reaction center complex according to Ghanotakis and Yocum (1986) using the nonionic detergent octyl-ß-D-glucopyranoside. The lipid composition of chloroplasts from green and yellow-green leaf areas as well as that of photosystem II particles, reaction center and light-harvesting complexes is shown in table 1.

In chloroplasts of yellow-green leaf areas we observe a much higher ratio of monogalactosyldiglyceride (MGDG) to digalactosyldiglyceride (DGDG) and an increased amount of phospholipids than in chloroplasts of green leaf areas. On the other hand the amount of carotenoids and chlorophyll is decreased in

Table 1: Per Cent of Lipids Referred to the Total Lipid Content of Chloroplasts from Green (Chl_{gr}) and Yellow-Green (Chl_{ye}) Leaf Areas and of Photosystem II Particles (PS II), Reaction Center Complexes (RCC) and Light-Harvesting Complexes (LHC) from Green Leaf Areas of the Tobacco Mutant *Nicotiana tabacum* NC 95

	Chl_{gr}	Chl_{ye}	PS II	RCC	LHC
MGDG	42.6±7.5	62.7±8.9	21.6±6.2	22.5±3.0	18.5±1.6
DGDG	21.6±3.0	11.7±2.5	9.0±2.4	8.9±1.1	8.3±0.9
SQDG	10.7±2.3	8.4±0.7	6.0±2.7	2.3±1.3	4.1±1.7
PC	3.0±0.5	8.7±0.2	3.7±0.7	0.44±0.04	1.14±0.09
PG	2.0±0.6	1.2±0.2	0.14±0.06	0.06±0.03	0.7 ±0.2
PI	1.4±0.2	3.2±0.2	0.1	0.17±0.09	0.08±0.01
PE	0.5±0.2	1.2±0.1	n.d.	n.d.	n.d.
ß-Car.	0.54±0.03	0.10±0.01	0.90±0.16	2.55±0.17	1.55±0.17
Lut.	0.90±0.08	0.30±0.01	2.90±0.19	1.08±0.06	3.97±0.52
Viol.	0.36±0.06	0.14±0.01	0.53±0.07	0.15±0.06	
Neox.	0.30±0.05	0.10±0.01	0.83±0.13	0.21±0.02	1.38±0.09
Chl a	10.7±0.5	1.54±0.02	37.6±4.3	51.5±1.3	42.1±0.3
Chl b	5.4±0.5	0.72±0.02	16.7±1.7	10.1±0.1	18.2±0.1

Abbreviations: MGDG monogalactosyldiglyceride; DGDG digalactosyldiglyceride; SQDG sulfoquinovosyldiglyceride; PC phosphatidylcholin; PG phosphatidylglycerol; PI phosphatidylinositol; PE phosphatidylethanolamin; ß-Car.ß-Carotene; Lut. Lutein; Viol. violaxanthin; Neox. neoxanthin; Chl chlorophyll
n.d. not detected

chloroplasts of yellow-green leaf areas. In photosystem II particles we find an increase of carotenoids and chlorophyll and a decrease of glycolipids and phospholipids compared to chloroplasts of green leaf areas. These results are discussed in more detail in an earlier publication (Bednarz et al.,1988). The isolated photosystem II particles consist of polypeptides with molecular weights of 11, 14, 16, 18, 23, 25, 26, 28, 33, 34, 42 and 48 kDa. The 18 and 25 kDa peptides are absent in the reaction center complex. The 26 and 28 kDa peptides which are compounds of the light-harvesting complex are present only in traces. The light-harvesting complex is enriched in the 26,

28 and also in the 18 and 25 kDa peptides. But qualitatively all peptides present in the photosystem II particles are found in addition, except the 16 kDa peptide which is lost during the isolation procedure.
In relation to the protein content the lipid content of the reaction center complex is half as high as that of the photosystem II particles and only a quarter as high as that of chloroplast preparations (table 2).

Table 2: Ratio of Protein to Chlorophyll and Ratio of Protein to Lipid in Chloroplasts, Photosystem II Preparations, Reaction Center Complexes and Light-Harvesting Complexes from the Tobacco Mutant <u>Nicotiana</u> <u>tabacum</u> <u>NC95</u>

	Ratio of Protein/Chlorophyll	Ratio of Protein/Lipid
Chloroplast Preparations of Green Leaf Areas	6.1 ± 1.0	1.0 ± 0.2
Chloroplast Preparations of Yellow-Green Leaf Areas	36.4 ± 5.8	0.8 ± 0.1
Photosystem II Preparations	3.8 ± 0.5	2.1 ± 0.4
Reaction Center Complexes	6.8 ± 0.4	4.2 ± 0.2
Light-Harvesting Complexes	3.3 ± 0.7	2.0 ± 0.4

After treatment of photosystem II particles with octyl-ß-D-glucopyranoside only 80% of the MGDG and DGDG, 60% of the sulfoquinovosyldiglyceride (SQDG) and 40% of the phospholipids are recovered in the light-harvesting and reaction center complexes. The recovery of carotenoids and chlorophyll is more than 95%.
Compared to photosystem II particles the reaction center complex is enriched in ß-carotene and chlorophyll a, whereas the light-harvesting complex is enriched in all carotenoids and chlorophyll b. The ratio of MGDG to DGDG being greater than 5 in the intergrana regions, is with a value of approximatly 2 nearly identical in photosystem II particles, reaction center

and light-harvesting complexes. This result is also obtained with the tobacco mutant Xanthi (see contribution of Radunz et al.).

Our results show, that the membranes of the intergrana region showing only photosystem I activity are enriched in MGDG and phospholipids whereas the amount of carotenoids and chlorophyll is decreased in comparison to the entire lamellar system of chloroplasts.

In highly purified photosystem II reaction center complexes we find only traces of phospholipids. The ratios of the glycolipids MGDG, DGDG and SQDG to each other are nearly the same as in the whole lamellar system. It appears that a high amount of ß-carotene and chlorophyll a is found whereas the other carotenoids and chlorophyll b are main components of the light-harvesting complex.

References

Bednarz, J., Radunz, A. and Schmid, G. H. (1988)
Lipid composition of photosystem I and II in the tobacco mutant Nicotiana tabacum NC 95
Z. Naturforsch. 43c, 423-430

Berthold, D. A., Babcock, G. T. and Yocum, C. F. (1981)
A higly resolved, oxygen-evolving photosystem II preparation from spinach thylakoid membranes
FEBS Letters 134, 231-234

Ghanotakis, D. F. and Yocum, C. F. (1986)
Purification and properties of an oxygen-evolving reaction center complex from photosystem II membranes
FEBS Letters 197, 244-248

Homann, P. H. and Schmid, G. H. (1967)
Photosynthetic reactions of chloroplasts with unusual structures
Plant Physiology 42, 1619-1632

Radunz, A. (1980)
Binding of antibodies onto the thylakoid membrane
VI. Asymmetric distribution of lipids and proteins in the thylakoid membrane
Z. Naturforsch. 35c, 1024-1031

BIOLOGICAL ROLE OF PLANT LIPIDS
P.A. BIACS, K. GRUIZ, T. KREMMER (eds)
Akadémiai Kiadó, Budapest and Plenum Publishing
Corporation, New York and London, 1989

STUDY ON THE LIPID COMPOSITION OF CHLOROPLASTS OF THE TOBACCO MUTANT N. TABACUM VAR. XANTHI EXHIBITING AN INCREASED PHOTOSYSTEM I ACTIVITY

A. Radunz, J. Bednarz and G.H. Schmid

Univesität Bielefeld, Lehrsthuhl Zellphysiologie, 4800 Bielefeld 1, FRG

Introduction

Chloroplasts of the sponge parenchyme of whitish-yellow leaf areas of the plastome mutant Nicotiana tabacum var. Xanthi exhibit in comparison to chloroplasts of green leaf tissue of this mutant a 10-12 fold increased photosystem I activity (2160-4040 µmoles O_2 x mg chlorophyll^{-1} x h^{-1} in chloroplasts of whitish-yellow leaf areas and 220-340 in chloroplasts of green leaf areas). These chloroplasts are a suitable object for comparative lipid and protein studies with chloroplasts of green leaf areas of the same mutant which exhibit photosystem I as well as photosystem II activity and with chloroplasts of yellow-green leaf areas of N. tabacum mutant NC 95 which exhibit only photosystem I reactions (Homann and Schmid, 1967, Schmid, 1967, Bednarz et al., 1988) above all since these chloroplasts can be isolated in aqueous buffer media without the use of structure destroying detergents.

Results

In the above described chloroplasts with increased photosystem I activity the lamellar system consists, just as in chloroplasts of yellow-green leaf areas of the N. tabacum mutant NC 95 which also only exhibit photosystem I activity, of single isolated thylakoids. Grana stacking of the thylakoids is completely lacking. With respect to the polypeptide composition these chloroplasts are characterized by the fact that polypeptides with the molecular weights of 9-18 kDa are completely lacking, or occur in only very small amounts, whereas in the molecular weight region of 30-75 kDa some polypeptides occur in higher concentrations than in chloroplasts of the green leaf areas which have developed a normal ratio of thylakoids of the grana regions to those of the intergrana region.

A quantitative lipid analysis by thin layer chromatography (Radunz, 1969, Radunz, 1976) leads to the result (Table 1), that lipids of chloroplasts with increased photosystem I activity consist as the chloroplasts of green leaf areas by 2/3 of glycolipids. However, a difference exists in the distribution of the individual glycolipids. Whereas chloroplasts with photosystem I and photosystem II activity contain relatively more digalactolipid, it appears that chloroplasts with increased photosystem I activity have a higher content in monogalactolipid and sulfolipid. In this respect these chloroplasts fully correspond to those of the variegated *N. tabacum* mutant NC 95 which also only exhibit photosystem I activity.

A substantial difference between chloroplasts of different photosynthetic activities appears to exist with respect to the phospholipid content. Whereas in normally developed chloroplasts of green leaf areas the phospholipid content is approximately 7%, this content increases in chloroplasts with exclusively photosystem I activity to 14% and increases further in chloroplasts with increased photosystem I activity to 28% of the total lipid content whereas it appears that the ratio of phosphatidylglycerol to phosphatidylcholine and to phosphatidylinositol is the same in the two types of chloroplasts of the plastome mutant. Chloroplasts of the tobacco mutant NC 95 with only photosystem I activity seem to contain besides phosphatidylinositol essentially only phosphatidylcholine as the main component.

Moreover, chloroplasts exhibiting different photosynthetic activities differed, as expected, with respect to their chlorophyll content. Whereas chloroplasts with photosystem I and photosystem II activity and normal grana stacking contained a very high chlorophyll portion in their lipids,

Table 1: Comparison of the lipid composition of chloroplasts and leaves of the plastome mutant *Nicotiana tabacum* var. Xanthi with the lipid composition of chloroplasts of *N. tabacum* var.NC 95

	Per cent composition of lipids			
	N. tabacum var. Xanthi			N. tabacum var. NC 95
	whitish-yellow leaf areas	chloroplasts[1] of whitish-yellow leaf areas (sponge parenchyme)	chloroplasts[3] of green leaf areas	chloroplasts[2] of yellow-green leaf areas
MGD	22.0	44.8	38.6	63.2
DGD	14.0	10.8	19.4	11.8
SL	15.0	9.8	6.0	8.4
SG	+			
DPG	13.1			
PE	8.6			1.2
PG	10.7	15.2	3.8	1.2
PC	6.6	7.6	1.8	8.8
PI	9.5	5.0	1.1	3.2
Chl a	0.4	5.1	22.3	1.5
Chl b	0.1	1.7	7.0	0.7

MGD, Monogalactosyldiglyceride; DGD, Digalactosyldiglyceride; Sl, Sulfoquinovosyldiglyceride; SG, Sterylglycoside; DPG,Diphosphatidylglycerol; PE, Phosphatidylethanolamine; PG, Phosphatidylglycerol; PC, Phosphatidylcholine; PI, Phosphatidylinositol, Chl a, Chlorophyll a, Chl b, Chlorophyll b. Total lipids without carotinoids were normalized to 100. 1, chloroplasts with high photosystem I activity; 2, chloroplasts with only photosystem I activity (Bednarz et al., 1988); 3, chloroplasts with photosystem I and II activity.

Table 2: Fatty acids composition of leaf and chloroplast lipids of the plastome mutant of *Nicotiana tabacum* var. Xanthi and lipids of chloroplasts of yellow-green leaf areas of *N. tabacum* var. NC 95.

	Per cent composition of fatty acids of			
	N. tabacum var. Xanthi			*N.t.* var. NC 95
	chloroplasts of green leaf areas	chloroplasts of whitish-yellow leaf areas	MGD of chloroplasts of whitish-yellow leaf areas	chloroplasts of yellow-green leaf areas
$C_{12:0}$	-	2.4	-	-
$C_{14:0}$	0.5	2.1	-	2.2
$C_{16:0}$	10.0	24.9	2.0	21.9
$C_{16:1}$ (trans)	3.0	5.1	-	3.9
$C_{16:3}$ } $C_{18:0}$	13.5	2.4	20.0	6.0
$C_{18:1}$	2.1	3.1	2.5	6.4
$C_{18:2}$	5.2	15.5	3.5	5.9
$C_{18:3}$	65.7	44.5	72.0	53.6

the chlorophyll a and b portion is very low in photosystem I chloroplasts of the two N. tabacum mutants.

A comparison of the fatty acid composition clearly shows (Table 2) that the lipids of chloroplasts with photosystem I activity are composed by 25-29% of saturated fatty acids, whereas chloroplasts having a normal lamellar system with grana and intergrana regions exhibiting and normal activity of photosystem I and photosystem II are composed by only 11% of lipids that have saturated fatty acids. In this respect these chloroplasts correspond to chloroplasts of other higher plants (Radunz, 1976). Hence, the degree of saturation of lipids of the "photosystem I chloroplasts" is very high. The monogalactolipid which contributes to 20% of the membrane composition in these photosystem I chloroplasts, corresponds with respect to its fatty acid composition to photosystem I active chloroplasts of the variegated N. tabacum mutant NC 95, as well as to photosystem II particles prepared from green leaves or green leaf patches of N. tabacum NC 95 (Bednarz et al., 1988). These fatty acids are composed by 98% of unsaturated fatty acids with the two main components octadecatrienoic and hexadecatrienoic acid. Hence, the monogalactolipid is not different with respect to its fatty acid composition regardless whether it participates in the composition of membranes active in photosystem I or photosystem II.

References

Bednarz, J., Radunz, A. and Schmid, G.H., 1988, Z. Naturforsch. 43c, in press

Homann, P. and Schmid, G.H., 1967, Plant Physiol. 42, 1619-1632

Radunz, A., 1969, Hoppe-Seyler's Z. Physiol. Chem. 350, 411-417

Radunz, A., 1976, Z. Naturforsch. 31c, 589-593

Schmid, G.H., 1967, J. de Microscopie 6, 485-497

BIOLOGICAL ROLE OF PLANT LIPIDS
P.A. BIACS, K. GRUIZ, T. KREMMER (eds)
Akadémiai Kiadó, Budapest and Plenum Publishing
Corporation, New York and London, 1989

CHANGES IN THE MOLECULAR ORGANIZATION OF MONOGALACTOSYLDIACYL-GLYCEROL BETWEEN RESTING AND FUNCTIONAL STATES OF THYLAKOID MEMBRANES

A. Rawyler and P.A. Siegenthaler

Laboratoire de Physiologie Végétale, Université de Neuchâtel, 20, Ch. de Chantemerle
2000 Neuchâtel, Switzerland

Considerable progress in the knowledge of the molecular organization of thylakoid membranes (TM) has been made during the last decade. Pronounced lateral heterogeneity and transversal asymmetry have been recognized for protein complexes and, to a lesser extent, for lipids, thus leading to the widely accepted concept that transversal and lateral heterogeneities in TM are the structural counterpart of its vectorial functionality [1]. However, this concept entirely relies upon the basic assumption that the molecular organization of TM is similar under "resting" conditions (mostly used to study structural aspects) and "functioning" conditions. A few works have indeed suggested that protein rearrangments occur upon a dark-to-light transition [2]. Is it also the case for lipids? We show here that such changes effectively occur, at least for monogalactosyldiacylglycerol (MGDG).

Two parameters were studied : (a) transversal distribution and (b) initial hydrolysis rates (reflecting the packing) of MGDG in TM treated with the lipase from Rhizopus arrhizus (LRa). Spinach TM were submitted to various pretreatments and conditions (see Table 1). MGDG hydrolysis was measured together with electron flow activity (when required), the LRa treatments being short enough so as to keep TM fully active.

As shown previously [1], the transversal distribution of MGDG in dark- incubated TM was such that the outer monolayer is enriched in MGDG. Allowing TM to perform either uncoupled electron flow alone or photophosphorylation during LRa treatment did not change this asymmetry (Table 1). Under our conditions, activities were hardly altered by the complete removal of MGDG and of its split products from the outer monolayer and no other lipid was degraded by LRa. During these short incubations (12 min, 15°C), the inner MGDG pool (reached after 6 min) was very stable, due to a negligible outwards transbilayer movement of inner MGDG molecules.

When TM were put in steady-state before LRa treatment, initial hydrolysis rates were markedly affected (Table 1). Since initial rates were strongly decreased by cholesterol incorporation in TM (not shown), they effectively reflected the packing of MGDG within the

TABLE 1. Effects of various pretreatments and conditions on the transversal distribution and initial hydrolysis rate of MGDG in LRa-treated TM

Condition(s) applied	Outer MGDG (mol %)	Inner MGDG (mol %)
Dark	62	38
Light + NH_4Cl (10 mM)	61	39
Light + ADP/Pi (3/3 mM)	62	38
	Initial hydrolysis rate (relative)	Hydrolysis extent (% degraded in 1 min)
Light + ADP/Pi	133	50.8
Light + FCCP (10 µM)	100	39.0
Light, basal	100	41.8
Light + DCMU (20 µM)	100	36.1
Light + ATP (3 mM)	100	39.4
Dark	100	40.0
Dark + ATP	100	40.9
Light, ADP/Pi + phlorizin (3 mM)	43	15.9
Light, ADP/Pi +DCCD (2 mM)	43	20.2
Light, ADP/Pi, NaBr-TM	67	27.6

TM. Experiments with liposomes made of total thylakoid lipids revealed that DCMU, FCCP, DCCD and phlorizin had negligible effects on the initial hydrolysis rate of MGDG by LRa. A membrane structure is thus required to obtain the results of Table 1. These results show that, compared to dark (resting) condition, MGDG packing in the outer monolayer is: (A) influenced neither by light alone, nor by the electron flow rate and ΔpH amplitude, nor by the presence of ATP (be it in the dark or in the light). (B) increased upon CF_1 removal (NaBr-treated TM) and even more by inhibition of either CF_o (by DCCD) or of CF_1 (by phlorizin). (C) decreased only when TM are synthesizing ATP, (D) closely controlled by the functional status of CF_o/CF_1, likely via MGDG-CF_o, MGDG-CF_1 and CF_o-CF_1 interactions. Thus, a decreased MGDG packing in photophosphorylating TM may facilitate lateral diffusion processes, thereby enhancing the overall capacity of chloroplasts.

Technical assistance of Mr. D. Monney and partial financial support of the Swiss National Science Foundation are gratefully acknowledged.

REFERENCES

[1] P.A. Siegenthaler & A. Rawyler (1986) Acyl lipids in thylakoid membranes: distribution and involvement in photosynthetic functions. In: Encyclopedia of Plant Physiology, New Series, vol. 19 (L.A. Staehelin and C.J. Arntzen, eds), Springer-Verlag, Berlin, pp. 693-705.

[2] J.L. Ellenson, D.J. Pheasant & R.P. Levine (1978) Light/dark labeling differences in chloroplast membrane polypeptides associated with chloroplast coupling factor 0. Biochim. Biophys. Acta 504, 123-135.

BIOLOGICAL ROLE OF PLANT LIPIDS
P.A. BIACS, K. GRUIZ, T. KREMMER (eds)
Akadémiai Kiadó, Budapest and Plenum Publishing
Corporation, New York and London, 1989

ASSOCIATIONS OF PIGMENT-PROTEIN COMPLEXES IN PHOSPHOLIPID ENRICHED BACTERIAL PHOTOSYNTHETIC MEMBRANES

W.H.J. Westerhuis[1], M. Vos[2], R.J. van Dorssen[2], R. van Grondelle[3], J. Amesz[2] and R.A. Niederman[1]

[1]Department of Biochemistry, Rutgers University, P.O. Box 1059, Piscataway, NJ 08854-1059, USA
[2]Dept. of Biophysics, Huygens Lab. of the State Univ., P.O. Box 9504, 2300RA Leiden, Holland
[3]Dept. of Biophysics, Physics Lab. of the Free Univ., De Boelelaan 1081, HV Amsterdam, Holland

The intracytoplasmic membrane (ICM) of the nonsulfur-purple bacterium *Rhodobacter sphaeroides* contains both peripheral and core light-harvesting pigment-protein complexes designated as B800-850 and B875, respectively. It has been proposed [1] that B800-850 is arranged in large "lakes" and transfers collected light energy to B875 which surrounds and interconnects photosynthetic reaction centers and focuses these excitations on the bacteriochlorophyll *a* (BChl) "special pair". Recently, this model for supramolecular associations within these antenna arrays has been refined by singlet-singlet annihilation measurements [2] and with membrane preparations with incremental increases in bilayer surface area [3,4]. Here, these procedures, together with detailed fluorescence yield studies, are used to demonstrate that B800-850 multimers can be detached from B875-core particles, thus providing additional support for a model in which reaction centers embedded within the B875 arrays are interconnected by these B800-850 clusters.

EXPERIMENTAL PROCEDURE

ICM vesicles (chromatophores) and the upper pigmented fractions were purified from *R. sphaeroides* NCIB 8253 as described previously [5]. Small unilamellar liposomes were prepared from soybean phosphatidylcholine (type IV-S, Sigma) that was purified by column chromatography on silicic acid in chloroform-methanol (30:70, vol/vol). To obtain phospholipid-enriched preparations in high yield, a procedure was developed in which fixed concentrations of pigmented membranes (1 mM BChl) and varying levels of liposomes were fused by repeated cycles of freeze-thaw-sonication [6] followed by isopycnic centrifugation. Low-temperature absorption and fluorescence spectra were obtained as described previously [3]; singlet-singlet annihilation was measured from the rate of quenching of the integrated fluorescence yield induced by picosecond laser excitation pulses of varying intensity as in [2].

RESULTS AND DISCUSSION

In this study, the efficiency of chromatophore-liposome fusion was improved by several cycles of freeze-thaw-sonication; 5 cycles gave optimal results as judged from the disappearance of the chromatophore band in sucrose gradients. With this procedure, a set of fused preparations was obtained in which the buoyant densities varied between that of liposomes and chromatophores (1.030 and 1.156 g/cm^3, respectively) (Table 1). The relative phospholipid contents of the lipid-enriched membranes increased up to 15 fold and indicated that 50 to 80% of added lipid was incorporated into the pigmented membranes. The specific BChl values showed that relative pigment and protein contents remained essentially unchanged during fusion.

Table 1. Composition and energy transfer properties of phospholipid-enriched chromatophores.

Preparation	Buoyant density (g/cm^3)	BChl/protein (μg/mg)	Phospholipid/protein (w/w)	Phospholipid/protein (-fold increase)	Energy transfer efficiencies[a] Carotenoid →B875	Q_x band →B875	B800 →B875	B850 →B875
Chromatophores	1.156	62.5	0.30	-	0.78	0.75	0.90	1.00
2	1.137	67.0	0.65	2.2	0.70	0.77	0.85	0.90
3	1.112	63.0	1.1	3.7	0.67	0.76	0.86	0.73
4	1.077	64.5	1.9	6.3	0.61	0.68	0.86	0.56
5	1.059	61.0	4.0	13	0.56	0.67	0.86	0.45
6	1.055	61.5	4.4	15	0.49	0.63	0.86	0.38

[a] Calculated from excitation and fractional absorption spectra at 4 K. The spectra were normalized at the B875 absorption maxima (885-887 nm). B875 and B850 emission were detected at 915 and 875 nm, respectively.

The visible and near-IR absorption spectra measured at 4 K confirmed that the chromophores were retained (not shown); however, the B850 absorption band showed a small shift to shorter wavelengths (≤ 5 nm) in preparations of higher lipid content. Fluorescence excitation spectra (4 K) of emission from B875 revealed a gradual decrease in energy transfer efficiency between B850 and B875 from 100% down to ~40% (Table 1). The excitation spectra of B850 emission were unaltered indicating that the B850 complex remained intact. Together, these results suggest that dilution of the phospholipid bilayer causes a physical detachment of B800-850 multimers from the B875 complexes.

The decrease in energy transfer efficiency between the peripheral and core antennae was also reflected in the emission spectra at 4 K (Fig. 1). Marked increases in the minor B850 emission band near 880 nm were observed which correlated with the level of phospholipid enrichment; this showed a dependence upon excitation wavelength and was especially pronounced when the B800-850 complex was excited directly into the B800 band. It is also

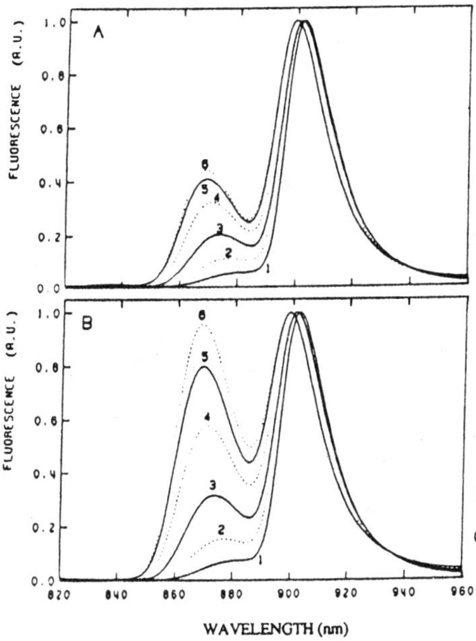

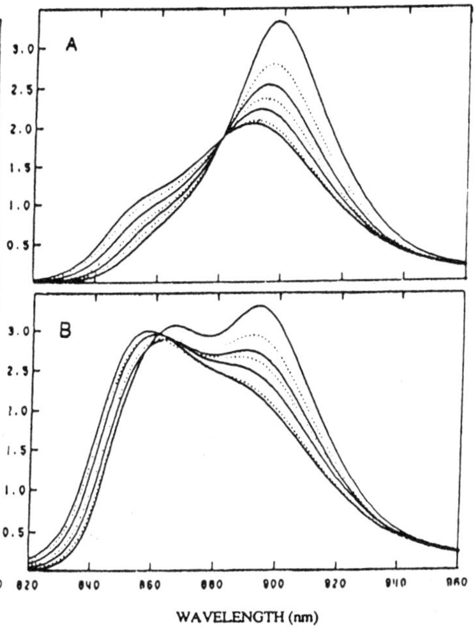

Fig. 1 (left). Fluorescence emission spectra of the phospholipid-enriched chromatophore preparations at 4 K, excited at 590 nm (A) and 800 nm (B). The spectra are normalized at the emission maxima and were corrected for the response of the measuring system. Both emission bands exhibit a blue shift related to the level of lipid enrichment with the greatest shift in the B850 band.

Fig. 2 (right). Temperature dependence of fluorescence emission in phospholipid-enriched chromatophores excited at 800 nm. (A), chromatophore control. (B), preparation #5. The temperatures were 180 K, 200 K, 220 K, 240 K, 260 K, 280 K and 300 K and are identified with respective decreased B875 and increased B850 emission bands.

noteworthy that upon excitation in this band, the fluorescence yield from B875 did not diminish as a result of the reduced B850→B875 energy transfer. Furthermore, upon preferential excitation of B875 at 295 K, the overall fluorescence yield in enriched preparation #5 was ~2 fold higher than in the control. Calculations of the loss yield of B875 with all the reaction center traps closed suggested some decrease in energy transfer from B875 to the reaction center (not shown).

Energy transfer between the B800-850 and the B875 complexes was examined further from the relative fluorescence yields as a function of temperature in the range 180-300 K (Fig. 2). Analysis of these data as in [4] indicated that in the control membranes, the emission spectra reflected a Boltzmann equilibrium distribution of the excitation densities in B850 and B875. A decrease in energy transfer between B850 and B875 in the enriched preparation is evident in Fig. 2 from smaller variations with temperature and higher emission yields from B850. As a

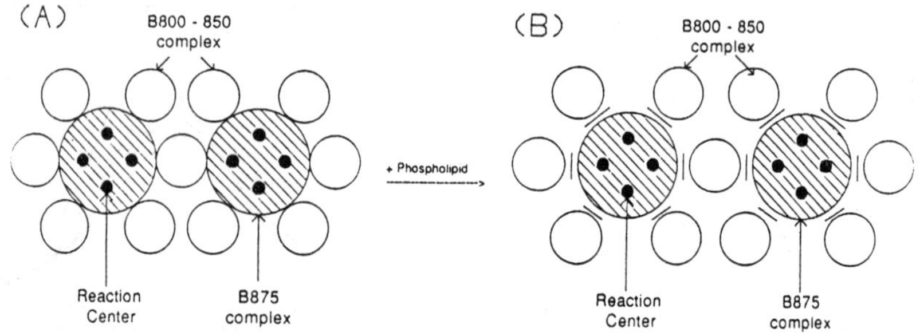

Fig. 3. Model for the organization of photosynthetic units in *R. sphaeroides* based upon excitation annihilation and energy transfer measurements. (A), Organization in control chromatophores [2]; the diameters of the circles correspond approximately to domain sizes of 45 interconnected BChl850 for the B800-850 complex and ≤ 100 BChl875 in the B875-reaction center core units. (B), Organization in phospholipid-enriched membranes. The B800-850 units have dissociated from B875 such that the interchromophore BChl850-BChl875 distances, calculated as in [2], have increased by ~1.75 fold and energy is no longer transferred efficiently to B875.

result, the excitation densities in the peripheral and core antenna did not reach a thermal equilibrium within the fluorescence lifetime.

Singlet-singlet annihilation measurements were also performed on these preparations to determine domain sizes (N_D), i.e. the number of functionally interconnected BChl molecules of each antenna component [2]. N_D values for B875 in the control chromatophores subjected to the fusion procedure decreased from ≥225 at 295 K to ≤100 at 4 K. The larger domain sizes at the higher temperature were explained by transfer among B875 arrays via multimeric B800-850 units interspersed between the B875-reaction center core particles (Fig. 3A). Because back transfer to B800-850 does not occur at 4 K, excitations are confined to single core units which consist of ~100 BChl875 in which 4 reaction centers are embedded. After lipid bilayer dilution, N_D values for B875 were largely independent of temperature and varied from ~100 at 295 K to ~60-80 at 4 K. B850 domains were determined with the upper pigmented fraction which has a high B850 fluorescence yield; N_D of ~40-55 at 4 K were obtained for both control and lipid-enriched preparations, although formation of smaller clusters cannot be excluded. Overall these effects of phospholipid dilution on domain sizes provide important confirmation for the model of Vos et. al. [2] (Fig. 3A) and permit measurement of the true N_D value for the cores (i.e., the actual homogeneous B875 connectivity) at room temperature. The results are consistent with the model shown in Fig. 3B in which the different pigment-protein particles have moved apart upon lipid dilution. As a consequence of strongly reduced energy transfer between B850 and B875, the core clusters are no longer interconnected by B800-850. Experiments are currently in progress to further assess possible dissociations within these B875-reaction center units.

(Supported by U.S. National Science Foundation grant DMB85-12587. W.H.J.W. was supported in part by a predoctoral fellowship award from the Busch Memorial Fund by the Bureau of Biological Research).

REFERENCES

[1] Monger, T. G. and Parson, W. W. (1977) Biochim. Biophys. Acta 460, 393-407

[2] Vos, M., van Grondelle, R., van der Kooij, F. W., van de Poll, D., Amesz, J. and Duysens, L. N. M. (1986) Biochim. Biophys. Acta 850, 501-512

[3] Pennoyer, J. D., Kramer, H. J. M., van Grondelle, R., Westerhuis, W. H. J., Amesz, J. and Niederman, R. A. (1985) FEBS Lett. 182, 145-150

[4] Westerhuis, W. H. J., Vos, M., van Dorssen, R. J., van Grondelle, R., Amesz, J. and Niederman, R. A. (1987) in: Progress in Photosynthesis Research, vol. I (Biggins, J. ed.) pp. 29-32, Martinus Nijhoff, Dordrecht.

[5] Reilly, P. A. and Niederman, R. A. (1986) J. Bacteriol. 167, 153-159

[6] Casadio, R., Venturoli, G., Di Gioia, A., Castellani, P., Leonardi, L. and Melandri, B. A. (1984) J. Biol. Chem. 259, 9149-9157

BIOLOGICAL ROLE OF PLANT LIPIDS
P.A. BIACS, K. GRUIZ, T. KREMMER (eds)
Akadémiai Kiadó, Budapest and Plenum Publishing
Corporation, New York and London, 1989

EFFECTS OF CATALYTICAL HYDROGENATION IN SITU OF PHOSPHATIDYLGLYCEROL-ASSOCIATED TRANS-Δ^3-HEXADECENOATE ON THE STABILITY OF THE THYLAKOID LIGHT HARVESTING COMPLEX

Ibolya Horváth[1], L. Vigh[1], S.H. Cho[2] and G.A. Thompson Jr.[2]

[1]Inst. of Biochemistry, Biological Research Center, Hung. Academy of Sciences, Szeged, Hungary
[2]Department of Botany, University of Tewas, Austin, Texas 78713, USA

INTRODUCTION

Over the past few years steadily increasing evidence (reviewed by Dubacq and Trémolières 1983) has pointed to a very specific role for phosphatidylglycerol (PG) containing trans- Δ^3-hexadecenoic acid (t-16:1) in stabilizing the oligomeric form of the chlorophyll-containing light harvesting complex proteins (LHCP) of chloroplast. Among the more recent findings supporting this concept are 1) the observation that LHCP of cold hardened rye had a 54% lower level of PG-bound t-16:1 and a 60% lower LHCP oligomer/monomer ratio, as measured on non-denatuing electrophoretic gels, than did LHCP from non-hardened rye (Krupa et al. 1987), 2) correlations between the lack of t-16:1 in Chlamidomonas mutants and their lack of an oligomeric form of LHCP on electrophoretic gels (Maroc et al. 1987) and 3) the finding that hydrolyzing most of the t-16:1-containing PG of tobacco thylakoids with exogenous phospholipase A_2 resulted in a substantial loss of the LHCP oligomer band on gels (Remy et al. 1982). In agreement with the above observations, McCourt et al. (1985) noted a disappearance of the LHCP oligomer band from electrophoresis gels of an Arabidopsis thaliana mutant lacking t-16:1. Surprisingly, however, the mutant was no less efficient than wild type Arabidopsis in photosynthetic energy transfer, raising doubts as to the need for the apparent t-16:1 PG stabilisation.
In the experiments described below we have extended the use of the $Pd(QS)_2$ catalyst and specifically correlated the content of PG-bound t-16:1 in hydrogenated thylakoids with oligomeric LHCP stability on SDS gels.

METHODS

Dunaliella salina (UTEX 1644) was grown to the late logarithmic phase and harvested as previously described (Lynch and Thompson 1982). The cells were resuspended in ice cold 0.4 mannitol, 25 mM TRIS, 2 mM EDTA, 1 mM $MgCl_2$, pH 8.2, 1mM benzamidine, and 5mM ε-amino-n-caproic acid and were disrupted in a Parr bomb. Chloroplasts were sedimented at 2000 x g for 5 min, and the pellet was resuspended in the same medium and sedimented again. The chloroplast were then lysed by resuspension in 20 mM Na TRICINE, 15 mM NaCl, 5 mM $MgCl_2$, 2mM EDTA, 1 mM benzamidine, and 5 mM ε-aminocaproic acid, pH 8,0, (to give a chlorophyll concentration of about 200 µg/ml) and stirring in the dark on ice for 10 min.
Thylakoids were sedimented at 12000 x g for 10 min, and the pellet was taken up in hydrogenation buffer (2 mM EDTA, 2mM $MgCl_2$, 10mM NaCl, 1 mM benzamidine, 5 mM ε-aminocaproc acid in 15 mM Na phosphate buffer, pH 6.0) to give a chlorophyll concentration of 20 µg/ml. Hydrogenation with $Pd(QS)_2$ catalyst (0.1 mg/ml) proceeded under 1 atm. H_2 for either 10 or 40 min. Control preparations contained catalyst but no H_2.
After hydrogenation the thylakoids were washed with 20 mM Na TRICINE, 1 mM benzamidine, 5 mM ε-aminocaproic acid, pH 8. Half of each sample (600 µg chlorophyll) was solubilised at once in darkness at 0-2°C using 0.88% (w/v)octyl-β-D-glucopyranoside, 0.22% (w/v)SDS, 15% (v/v)glycerol in 20 mM Na TRICINE, pH 8, and subjected to PAGE either on 1.3 mm thick standard slab gels or 0.7 mm thick minigels. Elecrophoresis was done at 4 °C in darkness by the procedure of Waldron and Anderson (1979) except that no sodium deoxycholate was in the gels. Scanning of the gels was done at 671 nm on a Beckman DU-8 spectrophotometer equipped with a gel scanning attachment.
In some cases, strips containing individual protein complexes were cut from the gels, and lipids were extracted. PG was separated from other lipids by TLC using chloroform:acetic acid:methanol:water (75:25:5:2.2, by vol.), and fatty acid methyl esters were prepared as described below.
Lipids were extracted from the other half of each sample and separated on TLC. The individual lipid spots were scrapped off and transmethylating using 5% HCL in methanol. Fatty acid methyl esters were analyzed by GC.

RESULTS.

Analysis of PG fatty acids isolated from thylakoids of 15°C grown and 30°C grown D. salina revealed a decreased level of t-16:1 at low temperature (Table 1). This change in t-16:1 was correlated with a marked decrease in the LHCP oligomer/monomer ratio (Fig.1), which declined from a value of 0.80 in 30°C-cells to 0.24 in 15°C-cells.

Table 1
Fatty acid composition of PG isolated from thylakoids of 15°C and 30°C adapted D. salina

Fatty acid	at 15°C	at 30°C
	%	
16:0	10	8
t-16:1	42	49
18:0	2	3
18:1	3	2
18:2	8	11
18:3	35	27

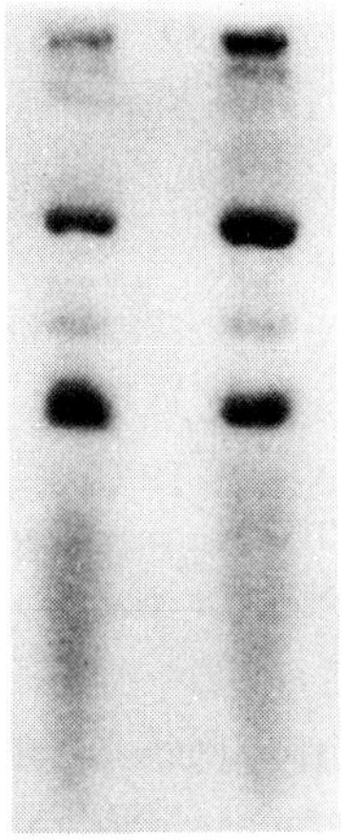

Fig.1 Unstained electrophoretic gel of thylakoids from 15°C-grown cells (left) and 30°C-grown cells (right).

In order to investigate further the cause and effect relationship between the content of t-16:1 and the stability of LHCP oligomer supported by the above observation with D. salina, thylakoids freshly isolated from 30°C-grown cells were hydrogenated with the $Pd(QS)_2$ catalyst for either 10 or 40 min. There was a rapid and pronounced decrease in the degree of fatty acid unsaturation in all lipid classis, as illustrated by the examples shown in Table 2.

Table 2

Hydrogenation of selected thylakoid lipids during treatment with $Pd(QS)_2$

Time of hydrogenation (min)	Phosphatidylglycerol t-16:1/16:0	Phosphatidylglycerol 18:3/18:0	Monogalactosyldiacylglycerol 16:4/16:0
0	6.5	9.2	26.2
10	6.4	0.7	1.5
40	3.4	0.3	1.7

But, the PG-bound t-16:1 was unexpectedly resistant to hydrogenation in comparison with other unsaturated fatty acids bound to PG or other lipids. After 10 and 40 min only 1% and 16%, respectively, of the PG-bound t-16:1 had been reduced to 16:0. Apart from the increising hydrogenation of t-16:1, few additional changes in unsaturation occured between 10 and 40 min. of hydrogenation; therefore, any differences between these two samples would largely reflect the declining t-16:1 level.

To determine whether all the regions of the thylakoids were affected by the catalyst, lipids were extracted from individual protein complexes cut from preparative gels, and PG was recovered from each extract by TLC. In each of the compexes analysed, namely, LHCP oligomer, monomer, and photosystem I (CPIa), t-16:1, as a percentage of total PG fatty acids, was reduced by more than 15%.

By determining the LHCP oligomer stability in these thylakoids it was possible to test directly the hypothesis that a lower content of PG t-16:1 is the key factor which promotes dissociation of oligomeric to monomeric LHCP on the diagnostic gel. Therefore, control and hydrogenated thylakoids were solubilised in a variety of detergent mixtures and electrophoresed under many different conditions. Under the mild conditions of thylakoid solubilization, the degree of oligomeric LHCP dissociation was dependent upon the conditions of electrophoresis. When electrophoresed on standard slab gels , which require 4 hr per run, both the non-hydrogenated and the hydrogenated thylakoids exhibited a considerable dissociation, yielding in each case an O/M ratio of 0.36 (Fig. 2a and b). Considerably less dissociation occured when samples were run on the minigels routinly used in our laboratory. Probably because of the shorther time (45 min) on these small gels, the O/M ratios were 1.1 for both the control and the 40 min-hydrogenated samples (see fig.2c and d for examples).

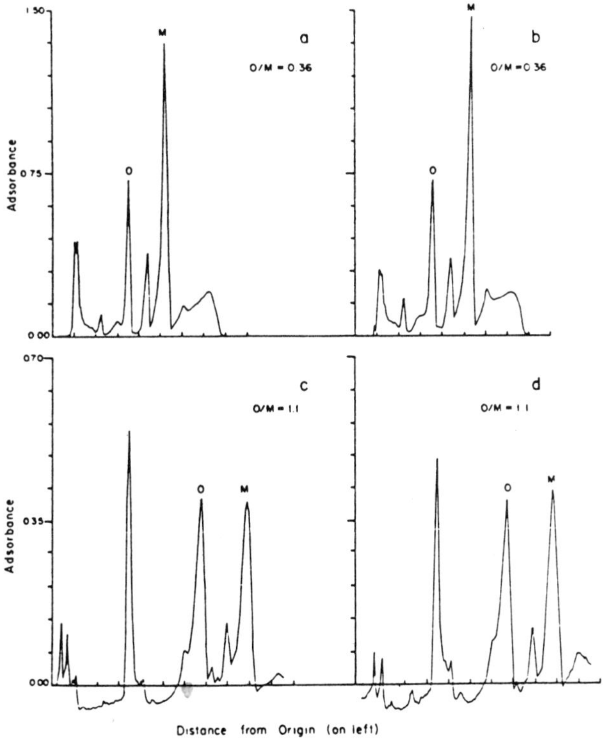

Fig.2. Spectrophotometer scan at 671 nm of normal size electophoretic gels of control(a) and 40 min. hydrogenated thylakoids (b), and contrasting minigels of the same control (c) and 40 min. hydrogenated (d) preparations.

DISCUSSION

These findings constitute an apparent exception to the frequently noted correlation between the concentration of t-16:1 in thylakoid PG and the tendency of LHCP to remain associated in the form of oligomers on non-denaturing electrophoretic gels. Even though prolonged catalytic hydrogenation reduced t-16:1 to a level below that found in low temperature-acclimated D. salina cells, it did not depress LHCP oligomer association to a degree found in thylakoids of cold-hardened cells. Because the PG-associated t-16:1 was very slow to be hydrogenated, it seemed possible that a subpopulation of t-16:1-containing PG molecules bound tightly to the LHCP oligomers was protected against hydrogenation and, for that reason, still able to bind LHCP monomers together. That possibility appears unlikely in view of the finding that both LHCP monomers and oligomers electrophoretically isolated from hydrogenated thylakoids contained significantly decreased level of t-16:1.

It will be informative to repeat these hydrogenation studies using thylakoids from other plants in which a direct cause-and-effect relationship between LHCP oligomer stability and t-16:1 content has been postulated. But for the present time we conclude that high levels of t-16:1 in PG are not themselves responsible for preserving the stability of LHCP oligomers on electrophoretic gels.

REFERENCES

Dubacq, J.P. and Trèmoliéres, A. (1983) Physiol. Veg. 21, 293-312

Krupa, Z., Huner, N.P.A, Williams, J.P., Maissan, E. and James, D.R. (1987) Plant Physiol. 84, 19-24

Lynch, D.V. and Thompson, G.A., Jr. (1982) Plant Physiol. 69, 1369-1375

Maroc, J., Trèmoliéres, A., Garnier, J., and Guyon, D. (1987) Biochim. Biophis. Acta 893, 91-99

McCourt, P., Browse, J., Watson, J., Arntzen, C.J. and Sommerville, C.R. (1985) Plant Physiol. 78, 853-858

Remy, R., Trèmoliéres, A., Duval, J.C., Ambard-Bretteville, F. and Dubacq, J.P. (1982) FEBS Letters 137, 271-275

Waldron, J.C. and Anderson, J.M. (1979) Eur. J. Biochem. 102, 357-362

BIOLOGICAL ROLE OF PLANT LIPIDS
P.A. BIACS, K. GRUIZ, T. KREMMER (eds)
Akadémiai Kiadó, Budapest and Plenum Publishing
Corporation, New York and London, 1989

THE EFFECTS OF PACLOBUTRAZOL ON STEROL AND ACYL LIPID COMPOSITION OF MEMBRANES IN APIUM GRAVEOLENS AND RHODOTORULA GRACILIS

Penny A. Haughan, Carole E. Rolph, J.R. Lenton* and L.J. Goad

Department of Biochemistry, University of Liverpool, Liverpool, UK
*Long Ashton Research Station, Long Ashton, Bristol, UK

Paclobutrazol is a triazole plant growth retardant which also has fungicidal properties [1]. It is an inhibitor of cytochrome P-450 dependant enzymes such as *ent*-kaurene oxidase and sterol 14α–methyl-demethylase [1]. The fungicidal activity of azole compounds is believed to result from the inhibition of sterol 14α-demethylation . This causes an accumulation of 14α-methylsterols and loss of ergosterol which may have adverse effects on membrane properties [2]. However, there is evidence for other essential requirements for sterol in cell proliferation in addition to a membrane structural role [3-6].For example, in yeast mutants traces of ergosterol stimulate phospholipid formation [7] and protein kinase activity [8]. This report describes the effects of paclobutrazol on the sterol and phospholipid compositions of membranes from plant and yeast cultures.

METHODS

Apium graveolens and *Rhodotorula gracilis* were grown (see [6] & [9]) in the presence or absence of paclobutrazol (50 and 75µM respectively). Plant and yeast cells were harvested after 7 days and 2 days growth, respectively. Membranes were isolated (see [10] & [11]) and sterol and phospholipid compositions determined (see [6] & [9]) Membrane fluidity was measured by fluorescence polarization using DPH as a probe [12].

RESULTS

Plasma membranes isolated from paclobutrazol treated *A.graveolens* cells exhibited a reduced membrane fluidity and an accumulation of 14α-methylsterols(mainly obtusifoliol) accompanied by a decrease in phospholipid:sterol ratio. In addition, the relative proportions of phosphatidylethanolamine was reduced and that of phosphatidylserine was increased. A direct carboxylation of existing phosphatidylethanolamine molecules could account for this latter observation [13]. In contrast, mitochondrial membranes exhibited an increased fluidity and there was an accumulation of both the 14α-methylsterols and phosphatidylglycerol. Mitochondrial phosphatidylglycerol accumulation has been reported to occur as the consequence of the inhibition of cell division (13).

Treatment of *R.gracilis* cultures with paclobutrazol resulted in the accumulation of phosphatidate and 14α-methylsterols (particularly lanosterol and 24-methylenedihydrolanosterol) in both plasma and mitochondrial membranes. These changes may be directly related to the observed increase in membrane fluidity. Whether the accumulation of phosphatidate is a direct result of phospholipid retailoring mechanisms via phospholipase D activity is not yet known.

CONCLUSIONS

Paclobutrazol treatment of *A.graveolens* and *R.gracilis* resulted in changes in the phospholipid and sterol compositions of both the plasma and mitochondrial membranes. These changes led to alterations in the physical properties of these membranes.The results suggest that sterol production and sterol type may in some way influence phospholipid biosynthesis.

ACKNOWLEDGEMENTS: The authors gratefully acknowledge the help of Dr. A.R.Cossins in the determination of membrane fluidity and the assistance of Mr David Cooke in the preparation of membrane fractions.

REFERENCES

[1]Burden, R.S., Clark, T. & Hollloway, P.J. (1987) Pestic. Biochem. Physiol. **27**, 289
[2]Vanden Bosche, H. (1985) in " Current Topics in Medical Mycology. Vol.1" pp313, McGinnis, M.R. ed., Springer-Verlag
[3]Rodriguez, R.J., Low, C., Bottema, C.D.K. & Parks, L.W. (1985) Biochim. Biophys. Acta **837**,336
[4]Ramgopal, M. & Bloch, K. (1983) Proc. Nat. Acad. Sci. **80**, 712
[5]Pinto, W.J. & Nes, W.R. (1983) J. Biol. Chem. **258,** 4472
[6]Haughan, P.A., Lenton, J.R. & Goad, L.J. (1988) Phytochemistry **27**, 2491
[7]Kawasaki,, S., Ramgopal, M., Chin, J. & Bloch, K. (1985) Proc. Nat. Acad. Sci. **82**, 5715
[8]Dahl, C. , Biemann, H.B. & Dahl, J. (1987) Proc. Nat. Acad. Sci. **84**, 4012
[9]Rolph, C.E. (1988) Ph.D. Thesis, University of Wales
[10]Larsson, C.H. (1985) in "Modern Methods of Plant Analysis. New Series, Vol.!: Cell Components" Linskins, H.F. & Jackson, J.F., eds. Springer-Verlag, Berlin
[11]Bottema, C.D.K., McLean-Bowen, C.A. & Parks, L.W. (1983) Biochim. Biophys. Acta **734**, 235
[12]Cossins, A.R (1977) Biochim. Biophys. Acta **470**, 395
[13]Mudd, J.B. (1980) in "The Biochemistry of Plants: a comprehensive treatise, vol. 4", Stumpf, P.K., ed., Academic Press.

BIOLOGICAL ROLE OF PLANT LIPIDS
P.A. BIACS, K. GRUIZ, T. KREMMER (eds)
Akadémiai Kiadó, Budapest and Plenum Publishing
Corporation, New York and London, 1989

EFFECTS OF PHOSPHOLIPASE A_2 DIGESTION ON THE ELECTRIC-FIELD SENSING CAROTENOIDS IN PHOTOSYNTHETIC MEMBRANES OF RHODOBACTER SPHAEROIDES

Leticia M. Olivera and R.A. Niederman

Department of Biochemistry, Rutgers University, P.O. Box 1059, Piscataway
New Jersey 08855-1059, USA

The intracytoplasmic membrane (ICM) of the photosynthetic bacterium *Rhodobacter sphaeroides* contains two light-harvesting bacteriochlorophyll *a* (BChl)-carotenoid-protein complexes designated as B800-850 and B875 which function as peripheral and core antennae, respectively. Phospholipase digestion has proved useful in studying the structural organization of these integral membrane proteins [1]. When ICM vesicles (chromatophores) were treated with phospholipase A_2, the spectral response of the carotenoids to electric field alterations disappeared [1]; these field-sensing pigments are associated exclusively with B800-850 [2]. It is shown here that this effect of the phospholipase does not merely reflect a loss of the ability of the treated chromatophores to sustain a field, but instead, is exerted upon the electrochromically active carotenoid species and is correlated with degradation of phosphatidylethanolamine (PE) at the membrane surface.

RESULTS AND DISCUSSION

The formation of lysophospholipids and fatty acids by phospholipase A_2 digestion results in alterations in the surface charge and fluidity of the membrane and could also affect permeability barrier properties. It was therefore possible that the observed phospholipase-induced disappearance of the electrochromic response [1] reflected the loss of the ability of the membrane to sustain a potential, rather than an effect on the field-sensing carotenoids. Alternatively, alterations in activities of field-generating components could be manifested in a disappearance of the carotenoid absorption change. These components were bypassed by examining the effects of phospholipase treatment on changes in the carotenoid spectrum induced by a valinomycin-K^+ diffusion potential (Fig. 1). The amplitude of the absorption change was decreased by >50% over the course of a 60 min digestion; however the rate of decay of the absorption transients remained essentially constant at $0.12 \pm 0.01\ s^{-1}$. This suggests that the effects of phospholipase digestion are exerted upon the field-sensing chromophores and do not reflect an increase in the permeability of the membrane to cations. In contrast , a preliminary examination of the biphasic decay of the flash-induced carotenoid bandshift indicated that some increase in proton permeability occurred after 15 min of digestion (not shown).

The phospholipase-treated chromatophore preparations were extracted with butanol to assess the extent of degradation of individual phospholipid species (Table 1). After 10 min of

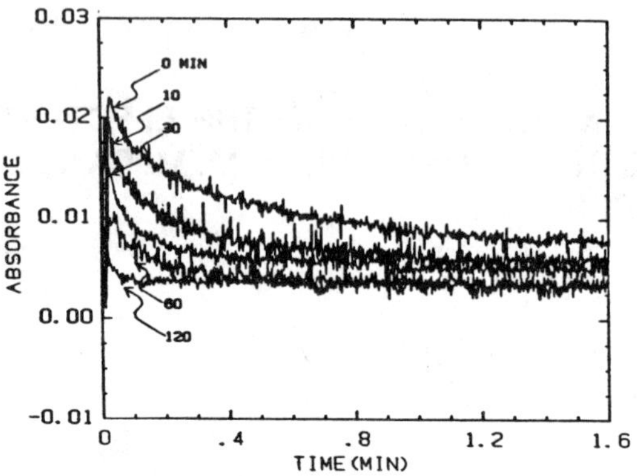

Fig. 1. Effect of phospholipase on K^+-valinomycin-induced carotenoid bandshift. Chromatophores (10 µg BChl/ml) were incubated with 0.01 units phospholipase A_2 (*Naja naja,* Sigma)/ml in 20 mM MOPS (pH 7.0) containing 1 mM $CaCl_2$, 0.1 M NaCl and 1 µM antimycin A at 37° C. To each sample cuvette, valinomycin and KCl were added to final concentrations of 1 µM and 12.5 mM, respectively. Measurements were made at 523-507 nm on a Johnson Research Foundation DBS-3 spectrophotometer.

digestion in which ~25% of PE was degraded to lysophosphatidylethanolamine (LPE), the amplitude of the salt-induced spectral response was reduced by nearly one-third; no breakdown of the other phospholipid species was observed over this period. The hydrolysis of PE was increased to 40% by 2 hr; at this time, 10% of the phosphatidylglycerol (PG) was degraded to lysophosphatidylglycerol (LPG). No lysophosphatidylcholine (LPC) was detected, suggesting that phosphatidylcholine (PC) is either confined to the inner leaflet of the chromatophore bilayer or protected from attack by other components.

Overall, these results confirm that intact phospholipids are necessary for the electrochromic response and suggest that the effect of phospholipase treatment on the electrochromically active carotenoids is due to the degradation of PE at the surface of the membrane. This may result in structural changes which alter local field interactions with these chromophores, possibly through dislocation or shielding of polarizing point charges [5]. (Supported by U.S. National Science Foundation grant DMB85-12587 and the Busch Memorial Fund Award).

Table 1
Lipid composition of chromatophores after phospholipase digestion (as percent total lipid)[a]

Min	PE	LPE	PG	LPG	PC	LPC
0	31	0	38	0	31	0
10	26	8(24)[b]	36	0	31	0
120	17	11(39)[b]	33	4(11)[b]	35	0

[a] Incubated as described in Fig. 1 except that antimycin A was omitted and NaCl was replaced by KCl. Lipids were extracted in butanol [3] and composition was determined by integration of ^{31}P-NMR peaks [4].
[b]Lysophospholipids as % of total respective species.

REFERENCES

[1] Symons, M. and Swysen, C. (1983) On the location of the carotenoids in the light-harvesting pigment-protein complexes of the photosynthetic bacterium *Rhodopseudomonas capsulata*. Biochim. Biophys. Acta 723,454-457.

[2] Holmes, N.G., Hunter, C.N., Niederman, R.A. and Crofts, A.R. (1980) Identification of the pigment pool responsible for the flash-induced carotenoid band shift in *Rhodopseudomonas sphaeroides*. FEBS Lett. 115,43-48.

[3] Bjerve, K.S., Daae, L.N.W. and Bremer, J., (1974) The selective loss of lysophospholipids in some commonly used lipid-extraction procedures. Anal. Biochem. 58,238-245.

[4] Sotirhos, N., Herslof, B. and Kenne, L. (1986) Quantitative analysis of phospholipids by ^{31}P-NMR. J. Lipid Res. 27,386-392.

[5] Kakitani, T., Honig, B. and Crofts, A.R. (1982) Theoretical studies of the electrochromic response of carotenoids in photosynthetic membranes. Biophys. J. 39,57-63.

BIOLOGICAL ROLE OF PLANT LIPIDS
P.A. BIACS, K. GRUIZ, T. KREMMER (eds)
Akadémiai Kiadó, Budapest and Plenum Publishing
Corporation, New York and London, 1989

UPD-GALACTOSE-INDEPENDENT SYNTHESIS OF MONO-GALACTOSYLDIACYLGLYCEROL (MGDG). AN ENZYMATIC ACTIVITY OF THE SPINACH CHLOROPLAST ENVELOPE

J.W.M. Heemskerk[1], F.H.H. Jacobs[2] and J.F.G.M. Wintermans[2]

[1]Correspondence: Dept. of Biochemistry, BMC, University of Limburg, P.O. Box 616
6200 MD Maastricht, The Netherlands
[2]Dept. of Botany, University of Nijmegen, Nijmegen, The Netherlands

INTRODUCTION

Various authors have noted that isolated chloroplasts, incubated under appropriate conditions, can synthesize small amounts of monogalactosyldiacylglycerol (MGDG) in the absence of UDPgalactose, either from labeled glycerol 3-phosphate [1] or from labeled acetate [2]. Van Besouw and Wintermans [3] tried to explain this phenomenon by a mechanism involving UDPgalactose:diacylglycerol galactosyltransferase (UDGT). In this paper we present evidence that UDPgalactose-independent MGDG synthesis is an enzymatic activity of the chloroplast envelope and that galactolipid:galactolipid galactosyltransferase (GGGT) is involved in the synthesis. Thus, GGGT catalyzes the reversible reaction of Eq.(1), permitted by an expected, low reaction enthalphy:

$$\text{DGDG} + \text{diacylglycerol} <=> \text{MGDG} + \text{MGDG} \qquad (1).$$

Since UDGT and GGGT activities are tightly coupled under most experimental conditions [4,5], proof of the exact mechanism of UDPgalactose-independent MGDG synthesis is made possible only after the development of assays sufficiently specific for either galactosyltransferase [4,6].

EXPERIMENTS AND RESULTS

In a typical experiment in which intact spinach chloroplasts were incubated under illumination with [2-^{14}C]acetate in the absence of UDPgalactose, 44% of the label was found in diacylglycerol and 4% in MGDG (fatty acids and phospholipids accounting for the rest).

When isolated spinach envelope membranes briefly were sonicated with liposomes of di[1-^{14}C]oleoyl-PC or [*glycerol*-^{3}H]-PC and subsequently treated with phospholipase C (from *Bacillus cereus*), the resulting labeled diacylglycerol was partially transformed into MGDG (in the absence of UDPgalactose) (see Fig. 1).

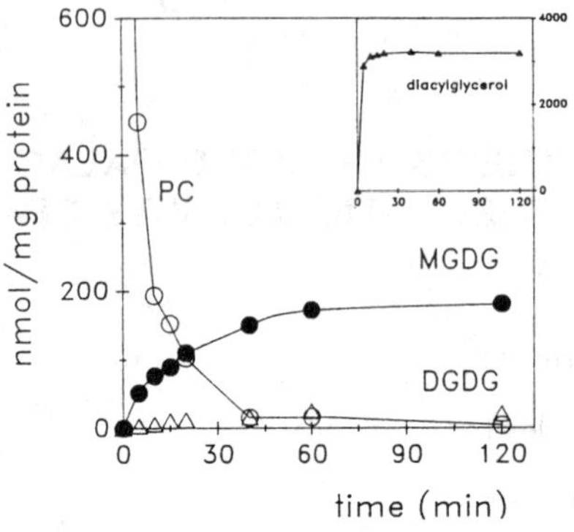

Fig. 1. Time-course of UDPgalactose-independent MGDG synthesis. Spinach envelope membranes (150 ug protein) were sonicated with 520 nmol di[^{14}C]oleoyl-PC and incubated with phospholipase C for various times at pH 7.2 (30 °C). Indicated is the distribution of radioactive lipids.

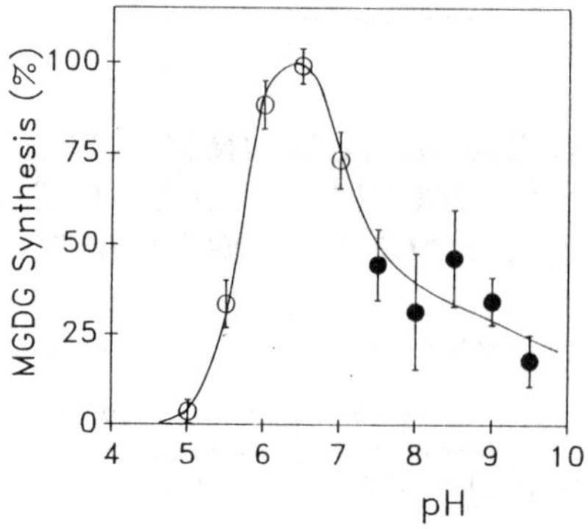

Fig. 2. UDP-Galactose-independent MGDG synthesis: dependence of pH. Envelope membranes were sonicated with di[^{14}C]oleoyl-PC and incubated for 20 min with phospholipase C at pH 7.2. Then the mixture was resonicated and samples were incubated for another 20 min at the indicated pH values. The MGDG produced during the second incubation is expressed in percentages of maximum (S.D., n = 3).

Production of labeled MGDG was no longer found, when the envelope membranes were heat-treated before the experiment, or when envelopes were mixed with liposomes without sonication. We therefore attribute the MGDG formation to an enzymatic activity of the envelope membranes, not to the phospholipase preparation.

Fig. 1 shows that the synthesis rate of MGDG proceeds over 2 hours but declines already after 10 min, when only a few percent of the labeled diacylglycerol substrate is converted. This is probably due to local exhaustion of substrate, since resonication after 20 min caused a new wave of MGDG production (not shown). The rate of MGDG synthesis depends strongly on the diacylglycerol concentration. A Lineweaver-Burk plot gives a straight line, indicating an V_{max} of 275 nmol/mg envelope protein.h at a diacylglycerol concentration of 4000 nmol/mg protein (0.6 mM) (not shown).

The optimal pH for UDPgalactose-independent MGDG synthesis is around pH 6.5 (Fig. 2). The reaction rate is not affected by Mg^{2+}, Ca^{2+}, PP_i or EDTA. Similarly, the reaction is not modified by the presence of either UDPgalactose or UDP, which are the substrate and a competitive inhibitor of UDGT, respectively [3]. Double-labeling experiments with di[^{14}C]oleoylglycerol, UDP-[^{3}H]galactose and various concentrations of UDP gave evidence that MGDG can be simultaneously synthesized by two enzymatic reactions, one of which is catalyzed by UDGT and the other one being UDP- galactose-independent (not shown).

Table 1. Effect of galactolipid concentrations on UDPgalactose-independent MGDG synthesis. Envelope membranes (255 ug protein) were sonicated with di[^{14}C]oleoyl-PC (850 nmol) and various amounts of [*galactose*-^{3}H]-MGDG or -DGDG (see below). Phospholipase C was added to the mixtures and incubations lasted for 30 min at pH 7.2. The enzymic reactions represent: GGGT activity (MGDG -> DGDG), reversed GGGT activity (DGDG -> MGDG), and UDPgalactose-independent MGDG synthesis (DG -> MGDG).

Galactolipid concentration (uM)		Rate of enzymatic reaction (nmol/mg protein/h)		
MGDG	DGDG	[^{3}H]MGDG -> DGDG	[^{3}H]DGDG -> MGDG	[^{14}C]DG -> MGDG
75	25	218	–	87
390	25	236	–	63
910	25	252	–	63
19	80	–	33	39
19	380	–	50	42
19	880	–	110	99

The galactolipids MGDG and DGDG are abundantly present in envelope membranes. They can be expected to donate their galactosyl groups to diacylglycerol, thus yielding MGDG. This was tested by the inclusion of various amounts of [*galactose*-^{3}H]-MGDG or -DGDG into envelope membranes, sonication of these galactolipid enriched envelopes with di[^{14}C]oleoyl-PC, and followed by phospholipase C treatment. Table 1 shows that the synthesis rate of [^{14}C]MGDG only increases with higher concentrations of DGDG, but decreases with higher MGDG concentrations. Further, the consumption rates of DGDG and of diacylglycerol are approximately in a 1:1 ratio. This stoichiometry is well described by Eq.(1), a reversible reaction that can be catalyzed by GGGT. GGGT is known to be inhibited by treatment of intact chloroplasts with the proteinase thermolysin [4,5]. Indeed, we found that envelope membranes isolated from thermolysin-treated chloroplasts were almost devoid of both GGGT activity and UDPgalactose-independent MGDG synthesis, whereas UDGT activity was not inhibited at all (data not shown).

DISCUSSION

The present data indicate that chloroplast envelope membranes are able to synthesize MGDG in the absence of UDPgalactose, and that UDGT is probably not involved in this process. The thermolysin experiments point to a localization of the responsible enzyme in the outer envelope membrane, known as site of GGGT [4,5]. Further evidence for involvement of GGGT is found in the pH dependence of the reaction and in its independence of UDPgalactose and UDP. However, the effects of divalent cations (Mg^{2+} and Ca^{2+}) and of chelating agents (PP_i and EDTA) deviate. GGGT activity (i.e., the reverse of Eq.(1)) is strongly stimulated by divalent cations and is inhibited by EDTA [4,6], whereas UDPgalactose-independent MGDG synthesis is not affected by these compounds. This, however, is not necessary an argument against involvement of GGGT. Gounaris et al. [7] could induce in chloroplast lipid extracts a phase separation of the non-bilayer forming MGDG upon the addition of divalent cations. We suggest that, similarly, in the envelope membrane, cation-dependent phase separation of MGDG can force GGGT in the direction of DGDG synthesis (where MGDG is a substrate). On the other hand, the reverse GGGT reaction with DGDG and diacylglycerol as substrates may be less affected by MGDG phase segregation.

A physiological role of Eq.(1) might be the synthesis of eukaryotic MGDG in the outer envelope membrane. Many data in literature [2] indicate that eukaryotic MGDG is derived from a diacylglycerol group supplied by cytoplasmic PC. The experiments reported here can be seen as an *in vitro* reconstruction of such process. Alternatively, Eq.(1) may provide a mechanism for regulating the diacylglycerol concentration in the envelope membranes. This concentration is naturally quite low, 0.25% of total acyl-lipids, comparable to that present in mammalian cell membranes. The diacylglycerol level in the chloroplast may be regulated, among other processes, by the equilibrium of the forward and reverse reaction of GGGT.

REFERENCES

[1] Douce, R. and Guillot-Salomon, T. (1970) Sur l'incorporation de la radioactivité de *sn*-glycérol 3-phosphate-^{14}C dans le monogalactosyldiglycéride des plastes isolées. FEBS Lett. 11, 21-124.

[2] Heinz, E. and Roughan R. (1983) Similarities and differences in lipid metabolism of chloroplasts isolated from 18:3 and 16:3 plants. Plant Physiol. 72, 273-279.

[3] Van Besouw, A. and Wintermans, J.F.G.M. (1979) The synthesis of galactosyldiacylglycerols by chloroplast envelopes. FEBS Lett. 102, 33-37.

[4] Heemskerk, J.W.M., Jacobs, F.H.H., Scheijen, M.A.M., Helsper, J.P.F.G. and Wintermans, J.F.G.M. (1987) Characterization of galactosyltransferases in spinach chloroplast envelopes. Biochim. Biophys. Acta 918, 189-203.

[5] Heemskerk, J.W.M., Bögemann, G., Helsper, J.P.F.G. and Wintermans, J.F.G.M. (1988) Synthesis of mono- and digalactosyldiacylglycerol in isolated spinach chloroplasts. Plant Physiol. 86, 971-977.

[6] Heemskerk, J.W.M., Bögemann, G. and Wintermans, J.F.G.M. (1983) Turnover of galactolipids incorporated into chloroplast envelopes. An assay for galactolipid:galactolipid galactosyltransferase. Biochim. Biophys. Acta 754, 181-189.

[7] Gounaris, K., Sen, A., Brain, A.P.R., Quinn, P.J. and Williams, W.P. (1983) The formation of non-bilayer structures in total polar lipid extracts of chloroplast membranes. Biochim. Biophys. Acta 728, 129-139.

BIOLOGICAL ROLE OF PLANT LIPIDS
P.A. BIACS, K. GRUIZ, T. KREMMER (eds)
Akadémiai Kiadó, Budapest and Plenum Publishing
Corporation, New York and London, 1989

INTERACTIONS OF THE CHLOROPLAST ATP SYNTHASE WITH GLYCOLIPIDS

U. Pick[1], K. Gounaris[2] and J. Barber[2]

[1]The WIS, Dept. of Biochemistry, Rehovot 76100, Israel
[2]Imperial College of Science and Technology, Dept. of Pure and Applied Biology
London SW7 2BB, UK

Introduction

We have chosen to study the chloroplast ATP synthase (CF_0CF_1) as a system to find out whether interactions with specific glycolipids have a functional role in the catalytic activity of this enzyme. The reasons for selecting this enzyme are:
(a) CF_0CF_1 can be easily and reversibly delipidated, whereby it looses most of it catalytic activity. The activity is regained upon reconstitution with phospholipids.
(b) CF_0CF_1 is specificaly activated by octylglucoside, a glycodetergent, suggesting that octylglucoside mimics specific interactions between CF_0CF_1 and glycolipids in thylakoid membranes (1). In this paper we summarize the results of several approaches aimed to identify interactions between CF_0CF_1 and chloroplast glycolipids.

Results

(a) Copurification of glycolipids with CF_0CF_1

Solubilization of of CF_0CF_1 with octylglucoside + Na-cholate, and purification on a triton X-100 containing sucrose gradient results in a progressive delipidation of the enzyme down to 5-10 mol equivalents in both higher plants, and the green alga *Dunaliella salina* (Table 1). The bound lipids are almost exclusively sulphoquinovosyldiacylglycerol (SQDG). The fatty acid composition of the bound sulpholipids are considerably more saturated than the bulk sulpholipids in thylakoids (mainly 16:0, 18:0 and 18:1). There is no exchange between the bound and bulk SQDG (2).

The results indicate that there is a distinct population of sulpholipids which are tightly-bound to CF_0CF_1 which may be important for its structural and functional integrity.

Table 1

Lipid composition of CF_0CF_1 preparations

Preparation	mol % of lipid			
	MGDG	DGDG	SQDG	PG
spinach thylakoids	2,000	1,000	320	540
spinach crude CF_0CF_1	320	104	224	136
spinach purified CF_0CF_1	0	<1	5	0
D. salina purified CF_0CF_1	2	2	12	2

Crude and purified CF_0CF_1 preparations were obtained before and after purification of triton-containing sucrose gradients. lipids were extracted, separated on TLC plates, scraped off and analysed by GLC (spinach) or radioactivity *(D. salina)*.

(b) <u>Interference of fatty acids with the catalytic activity of CF_0CF_1</u>

Fatty acids uncouple ATP formation in chloroplasts (3) by a unique mechanism, which presumably involves interaction with the ATPase complex. Fig. 1 demonstrates that the inhibition of photophosphorylation by fatty acids depends both on the chain length and on the degree of unsaturation - only fatty acids having at least 16 carbons effectively inhibit, and saturated fatty acids are more efficient than unsaturated fatty acids.

(c) <u>The role of different thylakoikd glycolipids on catalytical properties of CF_0CF_1</u>

The extensive delipidation of CF_0CF_1 during purification causes a progressive inactivation but the activity can be recovered by reconstitution of the purified enzyme with different lipids (Table 2, ref. 4). This provides, therefore, an experimental system to analyse the role of specific lipids in ATP synthesis and hydrolysis.

The results in Table 2 demonstrate that:

a) CF_0CF_1 requires lipids for catalytic activity.
b) Thylakoid glycolipids stimulate ATP hydrolysis far better than phospholipids.
c) MGDG - the major chloroplast glycolipid by itself, but none of the other thylakoid lipids, stimulates ATP hydrolysis when reconstituted with CF_0CF_1.
d) Unsaturated fatty acids, in either MGDG or PC, seem to be essential for activity.

The results suggest that MGDG specifically activates CF_0CF_1
$\Delta \mu_{H}^{+}$ will affect the interactions between different CF_0CF_1 subunits and MGDG. It may be noted that this energization treatment stimulates both ATP synthesis and hydrolysis of proteoliposomes by 3-10 fold.

Fig 1:

Fatty acid specificity for inhibition of photophosphorylation.

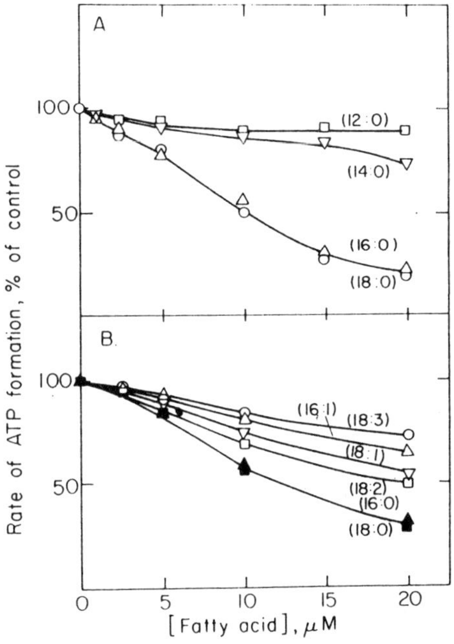

ATP formation was measured with lettuce thylakoids (20μg Chl/ml) in the presence of methyl viologen, ADP and ^{32}Pi and different fatty acids (indicated by the carbon chain length and number of double bonds) at the indicated concentrations. The observation that palmitic (16:0) and stearic (18:0) acids are the most efficient inhibitors of photophosphorylation and also the major constituents of bound-sulpholipids, is consistent with the idea that they may inhibit by displacing tightly bound lipids from CF_0CF_1.

(d) Cross linking between MGDG and specific CF_0CF_1 subunits.

In order to find out which of the 9 CF_0CF_1 polypeptide subunits comes in contact with MGDG we attemped crosslinking the lipid to the protein. ^{14}C-MGDG was derivatized by periodate oxidation and the aldehyde derivative was cross-linked to the protein, reduced by $NaCNBH_3$ (5) and the labeled subunits have been analysed by SDS-PAGE. Table 3 summarizes the results of cross-linking experiments which

Table 2

Effects of different lipids and lipid mixtures on catalytical activity of CF_0CF_1 proteoliposomes.

Lipid composition	ATPase nmols/mg protein/min.
none	38
Soybean phospholipids	190
PC (egg yolk)	60
PC (soybean)	180
Thylakoid glycolipids	420
MGDG	295
MGDG (hydrogenated)	175
DGDG	50
SQDG/PG (1:1)	20
MGDG/DGDG/SQDG (6:3:1)	390

CF_0CF_1 was reconstituted with different lipid mixtures by the cholate-dilution technique and analysed for the ATPase activity.

Table 3

Effect of energization on crosslin king of CF_0CF_1 subunits with MGDG

Subunit		MW (kD)	^{14}C incorporation (cpm/band) control	$+\Delta\mu H^+$
CF_1	α	58	950	1,900
	β	55	250	550
	γ	37	100	180
	δ	18	150	100
	ϵ	13	-	-
CF_0	I	20	650	500
	II	15.5	1,900	1,650
	III	13	-	-
	IV	8	2,450	2,200

CF_0CF_1 reconstituted with MGDG:DGDG:SQDG (6:3:1) mixture, including oxidized ^{14}C-MGDG were incubated for 30 min at 21°C. Energization ($\Delta\mu_H{}^+$) was induced by 2 min incubation in 30mM Na succinate at pH 5.0 followed by alkalinization to pH 8.2 by addition of K^+-tricine. The proteoliposomes were treated with $Na\text{-}CNBH_3$ and the polypeptide subunits separated by SDS-PAGE. Each band was cut, and analysed for ^{14}C counts.

The experiment demonstrates that there is a significant interaction between MGDG and the α subunit of CF_1 which is increased upon energization. The incorporation of ^{14}C into hydrophobic subunits of CF_o II and IV is hardly affected by energization. This result suggest that energization induces a conformational change in CF_0CF_1 which brings the catalytic subunits, particularly α, into close contact with the membrane interface. It also supports the idea that MGDG has a role in the activation of CF_0CF_1 by $\Delta\mu_{H}{}^{+}$

Summary

Two thylakoid glycolipids play different roles in the catalytic activity of CF_0CF_1: MGDG specifically activates ATP hydrolysis. Energization increase cross linking between MGDG and the α subunit of CF_1 suggesting a possible role of MGDG in catalytic activation. SQDG co-purifies with CF_0CF_1 (5-10mol equivalents/enzyme). The bound sulpholipids differ from bulk sulpholipids in having mostly saturated fatty acids. Free palmitic and stearic acids uncouple photophosphorylation in thylakoids presumably by interaction with CF_0CF_1. It is suggested that fatty acids inhibit by displacement of tightly-bound sulpholipids.

References

1. Pick, U. and Bassilian, S. (1982) Biochemistry *24*, 6144-52.
2. Pick, U., Gounaris, K., Weiss, M. and Barber, J. (1985) Biochim. Biophys. Acta *808*, 415-420.
3. Pick, U., Weiss, M. and Rottenberg, H. (1987) Biochemistry *26*, 8295-302.
4. Pick, U., Weiss, M., Gounaris, K. and Barber, J. (1987) Biochim. Biophys. Acta, *891*, 28-39.
5. Pattela, U. (1984) Biochem. International *8*, 77-82.

BIOLOGICAL ROLE OF PLANT LIPIDS
P.A. BIACS, K. GRUIZ, T. KREMMER (eds)
Akadémiai Kiadó, Budapest and Plenum Publishing
Corporation, New York and London, 1989

DUAL LOCALIZATION OF GALACTOSYL TRANSFERASE ACTIVITY IN ISOLATED PEA CHLOROPLASTS ?

R.O. Mackender

Biology Department, The Queen's University of Belfast, Belfast BT7 1NN, N. Ireland, UK

INTRODUCTION

UDPgalactosyltransferase (UDPgalT) catalyses the synthesis of monogalactolipid (MGDG) from UDPgalactose (UDPgal) and diglyceride (DAG). This enzyme is assayed by incubating plastids or plastid membrane fractions with UDP-^{14}C (^{3}H) gal and measuring the incorporation of ^{14}C or ^{3}H into lipid. It is assumed that the fraction being assayed contains sufficient DAG to satisfy the reaction. The question is "Is this a reasonable assumption?" With fractions derived from "16:3 plants" it may well be, but with those derived from "18:3 plants" it is probably not since preincubation with precursors of DAG synthesis can stimulate UDPgalT activity when UDPgal is added subsequently (1; Mackender unpublished).

This paper reports briefly other methods for enhancing the endogenous DAG content of pea (18:3 plant) plastids and the effects these have on UDPgalT activity in isolated envelope membrane fractions.

MATERIALS AND METHODS

Chloroplasts were isolated and purified as described previously (2) from destarched 11-12 day old pea seedlings (16 h photoperiod; 21C day 16C night). Inner (IEM) and outer (OEM) envelope membranes were isolated as described by Nguyen _et al_ (3).

Reactions were terminated by the addition of C:M and either aqueous KCl or KCl in dilute phosphoric acid (4) so that C:M:aqueous = 1:1:1 v/v/v. Lipids were separated by TLC using either C:Acet:M:AcH:H_2O 50:20:10:10:5 by volume or pet ether:diethyl ether:formic acid 80:20:2. Lipids were quantified by GLC using an internal standard of heptadecanoic acid. Radioactivity was measured by liquid scintillation counting and counts were quench corrected when necessary using the channels ratio method. Chlorophyll was calculated using the equations of Lichtenthaler *et al* 1982 (5).

RESULTS AND DISCUSSION

The eukaryotic pathway for MGDG synthesis (6) predicts that the DAG for galactosylation is generated from phosphatidylcholine (PC). If this involves phospholipases it must be highly regulated since the envelope membranes which are the site of DAG synthesis contain phospholipids (7). The aims of the experiments reported here were to ascertain (i) whether it was possible to stimulate DAG synthesis at the expense of PC and (ii) how an enhanced DAG content affects UDPgalT activity.

Phospholipases require Ca^{2++} for full activity. In preliminary experiments it was found that the effects of adding Ca^{2+} and Mg^{2+} on the incorporation of ^{14}C-galactose into lipid, were synergystic, and that the time course of the incorporation tended to be sigmoidal. When pea chloroplasts were incubated with UDPgal (isoosmotic conditions) at either pH 5.5 or 7.4 $\pm$ 10 mM Ca^{2+} or Mg^{2+} for 40 minutes and then for a further 25 minutes at the opposite pH $\pm$ the other cation (Fig. 1) it was found that the incorporation of ^{14}C-galactose prior to adjustment was usually greatest at pH 5.5 + 10 mM Ca^{2+}, but greatest at pH 7.4 + 10 mM Mg^{2+} (+ 10 mM Ca^{2+}) following it. The pH and cation requirements for these two phases of ^{14}C-galactose incorporation were optimized and found to be pH 6.5 + 10 mM Ca^{2+} for the pretreatment and for UDPgalT activity pH 8 + 4 mM Mg^{2+}.

The incubation of chloroplasts with phospholipase C (PlC) resulted in DAG accumulation (Fig. 2) and enhanced UDPgalT activity. DAG also accumulated in response to the pH 6.5/Ca^{2+} treatment, an effect which could be inhibited 50% by the addition of 1 mM $ZnCl_2$ or 5 mM NEM and 100% by prior treatment of chloroplasts with heat, pronase or trypsin. This suggests enzyme involvement. However since preliminary analyses indicate MGDG as the source of DAG, the enzyme is not a phospholipase but is in all probability the intergalactolipid transferase (IGT) (8). This enzyme is localised in the OEM (7; 9). Therefore since its activity stimulates UDPgalT activity, either this latter enzyme is in the OEM or if it is in the IEM the DAG must be able to migrate to it. In experiments comparing the UDPgalT activities of OEM and IEM isolated from treated and non treated chloroplasts (Table 1) it has been found that (i) UDPgalT activity is apparently present in both membrane fractions but is severely

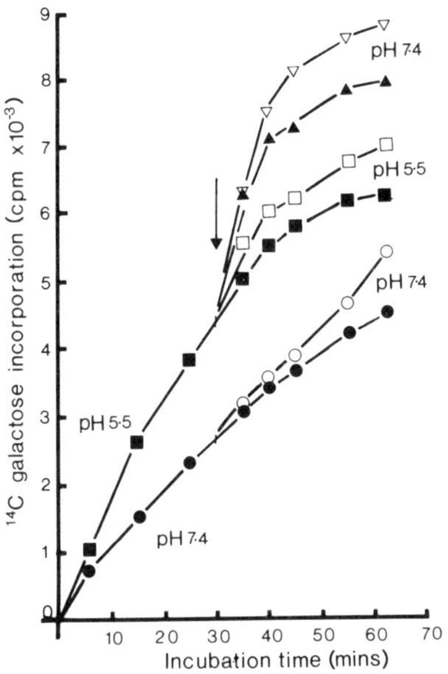

Fig. 1. The effect of pH and cations on the incorporation of ^{14}C galactose into lipid by isolated pea chloroplasts. Prior to pH adjustment (↑) the incubation contained 10 mM Ca^{2+}; following it 10 mM Mg^{2+} was added to some incubations (closed symbols) but not to others (open symbols).

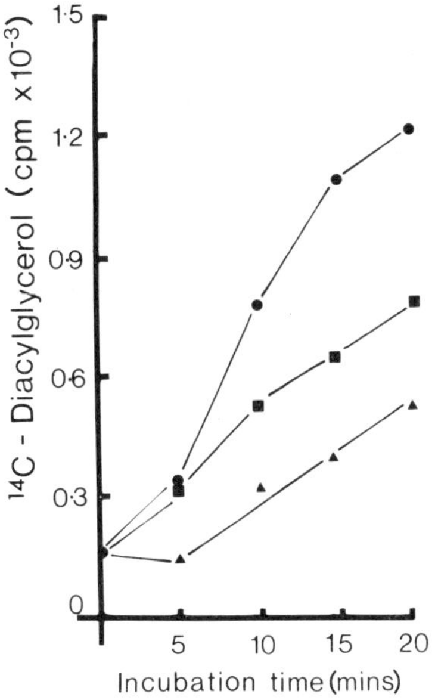

Fig. 2. The time course of ^{14}C-DAG accumulation in pea chloroplasts isolated from ^{14}C acetate labelled leaves, following their incubations (1 mg chl/ml) at either pH 5.5 (▲) or pH 6.5 + 10 mM Ca^{2+} (●) or with phospholipase C - 0.1 μg/ml (■).

Table 1. UDPgalT activity (cpm/nmole MGDG) in OEM and IEM isolated from non treated chloroplasts (control) or from chloroplasts incubated at pH 6.5 + 10 mM Ca^{2+} or with phospholipase C.

Chloroplast treatment*	UDPgalT activity (cpm/nmole MGDG) OEM	IEM
None (control)	129	200
pH 6.5 + 10 mM Ca^{2+}	1227	584
plC buffer pH 8	108	171
+ 0.025 μg/ml plC	75	247
+ 0.100 μg/ml plC	46	413

* Chloroplasts were incubated under the specified conditions for 20 minutesand then recentrifuged through percoll before isolating the envelope membranes.

substrate limited (ii) the pH 6.5/Ca^{2+} pretreatment preferentially enhances activity in the OEM (x10) and (iii) plC treatment stimulates the activity in the IEM (x3) but abolishes it in the OEM. This data is taken as further evidence (Dubacq, Mackender and Mazliak - in preparation) for two UDPgalT activities in pea chloroplast envelope membranes - one in the OEM utilizing DAG generated via the eukaryotic pathway and one in the IEM utilizing DAG generated via the prokaryotic pathway. Cross examination of each fraction by the other cannot explain these results.

There may be several reasons for the failure to detect more than one UDPgalT activity in envelope membranes; two of these may be a low endogenous DAG content or the use of chloroplasts from fully expanded leaves where it is possible that only one pathway of DAG synthesis is operating to any great extent.

REFERENCES

[1] Gillanders B and Mackender RO. Galactosyltransferase activity in developing plastids from naturally greening oat leaves in Advances in Photosynthesis Research Vol VI. Ed Sybesma C. Publ Martinus Nijhoff and Dr W Junk, The Hague 1984. pp 603-606.

[2] Mills WR and Joy KW. A rapid method for the isolation of purified physiologically active chloroplasts used to study the intracellular distribution of amino acids in pea leaves. Planta, 148 75-83. 1980.

[3] Nguyen TD, Miguel M, Dubacq J-P and Sigenthaler P-A. Localization and some properties of a Mg^{2+}-dependent ATPase in the inner membrane of pea chloroplast envelopes. Plant Sci Letts, 50(1) 57-63. 1987.

[4] Bertrams M and Heinz E. Positional specificity and fatty acid selectivity of purified sn-glycererol 3-phosphate acyltransferases from chloroplasts. Plant Physiol, 68 653-657. 1981.

[5] Lichtenthaler HK, Bach TJ and Wellburn AR. Cytoplasmic and plastid isoprenoid compounds of oat seedlings and their distinct labelling from ^{14}C-mevalonate in Biochemistry and Metabolism of Plant Lipids Eds Wintermans JGF and Kuiper PJC. Publ Elsevier Biomedical Press, Amsterdam 1982. pp 489-500.

[6] Roughan PG, Holland R and Slack CR. The role of chloroplasts and microsomal fractions in polar lipid synthesis from [1-^{14}C]acetate by cell free preparations from spinach (Spinacea oleracea) leaves. Biochem J, 188 17-24. 1980.

[7] Dorne A-J, Block MA, Joyard J and Douce R. Studies on the localization of enzymes involved in galactolipid metabolism in chloroplast envelope membranes in see ref 5. pp 153-164. 1982.

[8] Van Besouw, Wintermans JFGM and Bogemann G. Galactolipid formation in chloroplast envelopes III Some observations on galactose incorporation by envelopes with high and low content of diacylglycerol. Biochim et Biophys Acta, 663 108-120. 1981.

[9] Cline K and Keegstra K. Galactosyltransferases involved in galactolipid biosynthesis are located in the outer membrane of pea chloroplast envelopes. Plant Physiol, 71 366-372. 1983.

BIOLOGICAL ROLE OF PLANT LIPIDS
P.A. BIACS, K. GRUIZ, T. KREMMER (eds)
Akadémiai Kiadó, Budapest and Plenum Publishing
Corporation, New York and London, 1989

LIPID SYNTHESIS IN ISOLATED CHLOROPLASTS FROM ACETABULARIA MEDITERRANEA

R. Bäuerle, F. Lütke-Brinkhaus, H. Kleinig

Institut für Biologie II, Zellbiologie, Schänzlestr. 1, D-7800 Freiburg, FRG

As it was already shown by Shephard & Bidwell (1973) and Moore & Shephard (1977) isolated chloroplasts from the unicellular, siphonous green alga Acetabularia mediterranea incorporate $NaH^{14}CO_3$ in chlorophylls and carotenoids. We have confirmed and extended these investigations.

Chloroplasts from A. mediterranea were obtained by a very gentle isolation method combining both membrane filtration and a Percoll density gradient step. These chloroplasts exhibit unique long-term biosynthetic activities with a CO_2-fixation rate of 80 µmoles CO_2 fixed per mg chl.$^{-1}$ x h^{-1} and a photosynthetic activity of up to 65 µmoles O_2 evolved per mg chl.$^{-1}$ x h^{-1} (5 h after isolation). For comparison, whole Acetabularia cells show a photosynthetic capacity of 105 µmoles O_2 evolved per mg chl.$^{-1}$ x h^{-1}.

Results

Isolated chloroplasts incorporate $NaH^{14}CO_3$ as well as ^{14}C-labeled isopentenyl diphosphate (IPP), acetate and pyruvate into chloroform-methanol soluble products. It should be mentioned, that with $NaH^{14}CO_3$ as substrate, the bulk of the radiolabel (95%) is found in nonlipid material.

Precursor	Acyllipids	Chlorophyll/Prenyllipids	Others
	(% of total incorporation into lipids)		
$NaH^{14}CO_3$	45.7	48.6	4.8
[1-^{14}C] IPP	--	97.5	2.5
[1-^{14}C]acetate	100	--	--
[2-^{14}C]pyruvate	100	--	--
[2-^{14}C]mevalonate	--	--	--
[5-^{14}C]MevPP	--	--	--

Table 1: Incorporation of various precursors into lipids of isolated intact chloroplasts from Acetabularia mediterranea.

The incorporation of CO_2 on the one hand and IPP on the other hand shows a striking peculiarity: by lowering sorbitol concentration in the incubation medium, i.e. impairing the functional integrity of the plastids, incorporation of CO_2 , of course, decreases dramatically. However, IPP incorporation markedly increases and may even be enhanced by disrupting the chloroplasts mechanically, thus showing the chloroplast membrane to be a barrier for IPP permeation. Radioactivity of $NaH^{14}CO_3$ appears in acyllipids as well as in the chlorophyll/prenyllipid fraction (Tables 1 and 2), whereas the label derived from the precursor IPP can only be found in the chlorophyll/prenyllipid fraction. Mevalonate and 5-phosphomevalonate (MevPP) don't seem to be metabolized at all, although they are the direct precursors in IPP biosynthesis.

Acetate and pyruvate which are converted to acetyl-CoA via acetyl-CoA synthetase and pyruvate dehydrogenase complex, respectively, are exclusively channelled to fatty acid synthesis by building up the acyl moieties of the acyllipids. No labeling of prenyllipids could be observed.

Radioactive Precursor	Acyl-lipids	Polyprenyl-diphoshates	Polyprenyl-alcohols	Chlorophyll	Carotenes	Xanthophylls	Others
	Distribution of labeled Lipids (%)						
$[1\text{-}^{14}C]$IPP	--	16.4	44.7	12.9	23.8	--	2.1
$NaH^{14}CO_2$	45.7	--	10.8	17.0	19.5	1.3	5.6

Table 2: Lipid labeling pattern of isolated chloroplasts from Acetabularia mediterranea. (With ^{14}C-IPP 100% correspond to 580 000 cpm, with $NaH^{14}CO_2$ 100 % correspond to 600 000 cpm).

Concluding Remarks

Isolated chloroplasts from Acetabularia mediterranea have been shown to incorporate $NaH^{14}CO_3$ and IPP into prenyllipids in high yields. However, the pathway from Calvin cycle intermediates to isoprenoid end products remains obscure since acetate and pyruvate were not utilized in the carotenogenic pathway. Furthermore, mevalonate and 5-phosphomevalonate as direct precursors of IPP synthesis were not accepted as substrates. This means, that the pathway is not accessible for intermediates applied in vitro. The question of the molecular organization of this pathway arises.

Literature

D.C. Shephard & R.G.S. Bidwell, 1973: Photosynthesis and Carbon Metabolism in a Chloroplast Preparation from Acetabularia. Protoplasma **76**, 289-307.

F.D. Moore & D.C. Shephard, 1977: Biosynthesis in Isolated Acetabularia Chloroplasts. II. Plastid Pigments. Protoplasma **92**, 167-175.

BIOLOGICAL ROLE OF PLANT LIPIDS
P.A. BIACS, K. GRUIZ, T. KREMMER (eds)
Akadémiai Kiadó, Budapest and Plenum Publishing
Corporation, New York and London, 1989

INCORPORATION OF INORGANIC PHOSPHATE INTO THE PHOSPHOLIPIDS OF ISOLATED CHLOROPLASTS

P.G. Roughan[1] and J.E. Cronan Jr.[2]

[1]Division of Horticulture and Processing, DSIR, Mt. Albert Research Centre, Private Bag, Auckland, New Zealand
[2]Department of Microbiology, University of Illinois, Urbana, USA

An apparent discrepancy in rates of phospholipid synthesis from glycerol 3-phosphate labelled with ^{3}H or ^{32}P suggested the possibility that chloroplasts might contain a diacylglycerol kinase. Chloroplasts isolated from leaves of spinach, pea and Amaranthus incorporated inorganic orthophosphate (Pi) into lyso-phosphatidic acid (LPA), phosphatidic acid (PA), and phosphatidylglycerol (PG) when incubated in the light with glycerol 3-phosphate (G3P), CTP and acetate. In 15 min assays, the incorporation was 90% stimulated by light (Table 1) and was, therefore, presumed to be dependent upon ATP synthesis. However, 3-phosphoglycerate (PGA), which supports electron transport and ATP synthesis in Amaranthus chloroplasts (Roughan, 1986), strongly inhibited the incorporation (Table

TABLE 1. EFFECT OF DIFFERENT TREATMENTS ON THE INCORPORATION OF ^{32}Pi INTO PHOSPHOLIPIDS OF CHLOROPLASTS ISOLATED FROM AMARANTHUS.

P-lipid	Treatment					
	Control	Dark	- Ac	- G3P	+ PGA	+ CTP
	10^{-3}x cpm/assay					
LPA	8.1	1.2	11.9	3.9	2.3	3.7
PG	5.2	0.5	2.9	5.5	1.7	10.1
PA	12.8	0.9	6.1	13.6	3.3	3.6
Sum	26.1	2.6	20.9	23.0	7.3	17.4
% of control	-	10.0	80.1	88.1	28.0	66.7

Amaranthus chloroplasts equivalent to 46 μg of chlorophyll were incubated in the light for 15 min at 25° with 6.8 μCi of ^{32}Pi in 0.25 ml of the basal medium (Roughan, 1986) containing cold acetate and 0.4 mM G3P. Additions (4 mM PGA, 1 mM CTP) or deletions were as shown. Phospholipids were separated by tlc and detected by autoradiography.

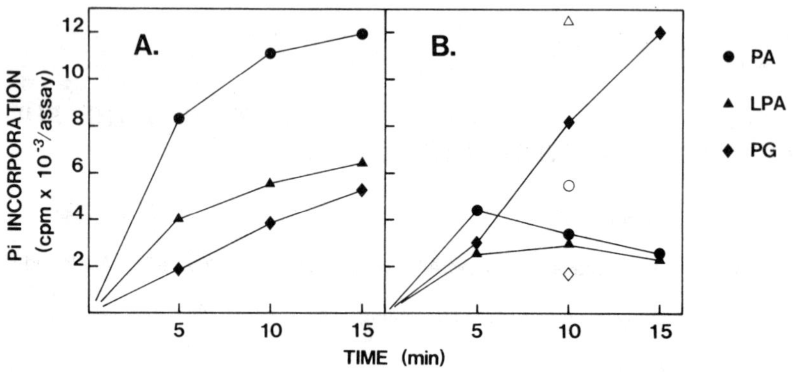

Figure 1. Time-course of Pi incorporation into phospholipids of Amaranthus chloroplasts in the absence (A) and presence (B) of 0.5 mM CTP. The assays also contained 0.4 mM G3P, 0.2 mM acetate, chloroplasts equivalent to 61 μg chlorophyll and 5 μCi ^{32}P. Open symbols; acetate omitted.

1). Increased concentrations of G3P in the medium only slightly affected Pi incorporation which was about 10% higher at 0.4 mM compared with no added G3P (Table 1). However, adding G3P increased the proportion of label accumulated in LPA . In the absence of added acetate, and high rates of fatty acid synthesis, Pi was incorporated primarily into LPA, and total incorporation was reduced by about 20% compared with when 0.2 mM acetate was added (Table 1). Added CTP reduced total Pi incorporation into Amaranthus lipids by about 30% whilst concomitantly doubling labelling of PG (Table 1). About 0.2% of the supplied label was incorporated into phospholipids of chloroplasts equivalent to 50 μg of chlorophyll in 15 min. A time-and CTP-dependent incorporation into PG of Amaranthus chloroplasts is shown in Figure 1. Also shown (Fig.1) is the accumulation of Pi within LPA when acetate was omitted from the incubation medium.

A DAG kinase which also phosphorylates monoacylglycerol (MAG) might explain the present data. Such activities may serve to retrieve LPA and PA which had received the unwelcomed attentions of a phosphatase. Incorporation of Pi into phospholipids of peanut mitochondria was dependent upon ATP synthesis and was attributed to the activity of a diacylglycerol (DAG) kinase (Bradbeer and Stumpf, 1960).

References

Bradbeer C, Stumpf PK 1960. J Lipid Res 1 214-220

Roughan PG 1986 Biochim Biophys Acta 878 371-379.

BIOLOGICAL ROLE OF PLANT LIPIDS
P.A. BIACS, K. GRUIZ, T. KREMMER (eds)
Akadémiai Kiadó, Budapest and Plenum Publishing
Corporation, New York and London, 1989

AN AUXIN-MEDIATED CONTROL OF A RAPID PHOSPHOINOSITIDE RESPONSE IN ISOLATED PLANT CELL MEMBRANES

B. Zbell

Botanical Institute, Ruprecht-Karls-University, Heidelberg, FRG

INTRODUCTION

In many types of animal cells the so-called phosphoinositide (PI) response is involved in the signal transduction of light, hormones or neurotransmitters [1]. Extracellular signals are percepted by their specific receptors at the outside of the plasma membrane, and these informations are transduced across the membrane by the operation of high-affinity GTP-binding proteins (G proteins) [2]. The activated G proteins stimulate obviously the phosphoinositidase C on the cytoplasmic side of the plasma membrane, which hydrolyzes membrane-bound phosphatidyl (4,5) bisphosphate (PIP_2) to membrane-bound diacylglycerol and soluble inositol (1,4,5) trisphosphate (IP_3) [1]. The compounds act as second messengers for the stimulation of the protein kinase C and the release of Ca^{2+} from endomembrane stores, respectively [1,3]. The protein phosphorylation and the Ca^{2+} mobilization are suggested to cooperate as persistent and transient signals for the control of cellular processes, respectively [3]. The PI response is discussed now also for plants, since phosphatidylinositol (4) monophosphate (PIP) and PIP_2 as lipids [4], their specific lipid kinases [5] as well as the phosphoinositidase C [6] were found to be localized at the plasma membrane, and high-affinity GTP binding proteins of as yet unknown function [7,8] were detected on membranes of plant cells. The phytohormone auxin was found to function as one putative signal for the initiation of the PI response in plants [8,9,10,11]. This report summarizes briefly some important aspects concerning the auxin action on microsomal membranes prepared from carrot suspension cells.

MATERIAL AND METHODS

<u>Chemicals and Plant Material.</u> [γ-^{32}P]ATP (0.55 or 1.07 TBq $mmol^{-1}$) were purchased from the Radiochemical Centre Amersham (Buckinghamshire,UK). All other chemicals used are analytical grade and were purchased from Merck (Darmstadt, FRG) and

Boehringer (Mannheim, FRG). Microsomal membranes were prepared from carrot suspension cells of the log-phase of the culture cycle as described previously [9,11].

In vitro -Assay of PI Response. The assay contained 93 kBq [γ-^{32}P]ATP, 100 μM Na_2·ATP, 10 μM GTP, 10 mM $MgSO_4$, 25 mM LiCl , 125 μg membrane protein in a final volume of 500 μl buffer (25 mM Hepes·KOH, pH 7.5, 250 mM sucrose). The reaction was started by the addition of an aliquot of the ATP/GTP mixture, incubated at room temperature, and terminated by the addition of 1 ml ice-cold stop solution containing 2-propanol/conc. HCl (100/1; v/v). The lipids were extracted from the acidified propanolic solutions with a solvent system using n-hexane as organic solvent. Following the lipid extraction the inositolphosphates of the aqueous phases were separated by anion exchange chromatography as described previously [9,11]. [^{32}P]-label of the extracts was measured via the Cerenkov-radiation in a liquid scintillation counter.

RESULTS

Membranes prepared from carrot suspension cells were found to have the capacity for a rapid *in vitro* phosphorylation of their lipids. This lipid phosphorylation depends absolutely on the presence of Mg^{2+} and can be saturated with increasing ATP concentration indicating Mg·ATP as the true enzymatic substrate [9,11]. The products of these reactions were identified by thin-layer chromatography and subsequent visualization of the labelled lipids by autoradiography as phosphatidic acid (PA), PIP, lysoPIP, and PIP_2 [11].

The detection of labelled polyphosphoinositides were the experimental prerequisite for the search of an auxin effect on the membrane-bound inositol phospholipids [9]. Actually, the kinetics of the lipid phosphorylation in the absence or presence of 1 μM IAA were analyzed. In comparison with the control a very rapid reduction of the [^{32}P]-label was observed in the phospholipid fraction in the presence of auxin (fig.1). This decrease of [^{32}P]-label was not originated from an auxin-mediated inhibition of the lipid phosphorylation reaction but it was really caused by an auxin-mediated hydrolysis of [^{32}P]-labelled lipids, since [^{32}P]-labelled inositol polyphosphates (IP_x) could be detected in the aqueous extracts of the same assays after anion exchange chromatography (fig.1). The kinetics of the IAA-mediated release of the [^{32}P]-labelled inositol compounds IP_3 and IP_2 correspond well with the kinetics of the loss of the [^{32}P]-label in the lipid fraction. The auxin-dependent reactions exhibit also a dose response relationship either by a continuous loss of [^{32}P]-label of the phospholipids or by a stimulated release of [^{32}P]-labelled IP_x with increasing IAA concentrations as demonstrated previously [9,11].

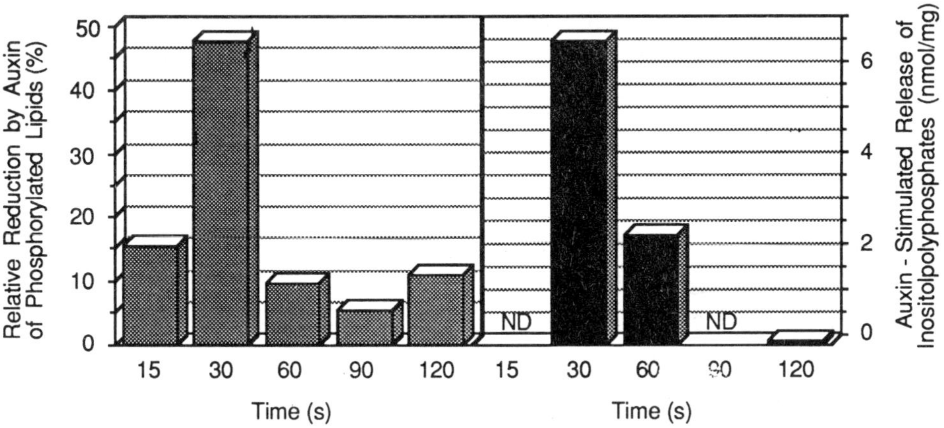

Fig. 1 Kinetics of the effect of 1μM IAA on the lipid phosphorylation of microsomal membranes (left panel) and on the corresponding release of the pooled fractions of IP_2 and IP_3 (right panel).

DISCUSSION

At the moment the data presented can be evaluated only as a further indication for the occurrence of a hormone-controlled PI response in plants, but it is still open if it is a similar or a distinct signal transduction mechanism in comparison to the well-known animal process [1,3]. The hypothesis concerning an auxin-mediated PI response in plant cells was postulated some years ago [12]. The first direct experimental indication for an auxin action on membrane-bound PI turnover was found in isolated membranes of *Glycine max* [13]. However, this auxin effect observed was suggested to be caused not by the action of a phosphoinositidase C (EC 3.1.4.10) but by the coordinated actions of a phospholipase D (EC 3.1.4.4) and a PA phosphatase (EC 3.1.3.4) [13,14]. In contrast to this early finding recent investigations *in vivo* on suspension cells of *Catharanthus roseus* [10] and *in vitro* on microsomal membranes from carrot suspension cells [9,11] point to an auxin-mediated stimulation of a phosphoinositidase C-like reaction. There are many open questions concerning the precise mechanism of the auxin-stimulated PI response. First, the localization of the reactions at the plasma membrane is as yet not proven. In respect to the putative involvement of a G protein [12] the auxin action must be carefully analyzed, since the hormonal effect on carrot microsomes was also observed in the absence of GTP, which can be caused by the known contamination with GTP of the commercial ATP used for the lipid phosphorylation *in vitro*. Moreover, the auxin specificity of the hormone-stimulated PI response must be also elucidated. The positions of the phosphate groups of the phosphoinositides as well as of the IP_x should be analyzed by chemical methods and not only by their chromatographic properties, that the substrates and products of the phosphoinositidase C-like reaction could be identified.

REFERENCES

[1] Berridge MJ (1987) Inositol trisphosphate and diacylglycerol: two interacting second messengers. Ann Rev Biochem 56: 159-193

[2] Gilman AG (1987) G proteins: transducers of receptor-generated signals. Ann Rev Biochem 56: 615-649

[3] Berridge MJ (1987) Inositol lipids and cell proliferation. Biochim Biophys Acta 907: 33-45

[4] WheelerJJ, Boss WF (1987) Polyphosphoinositides are present in plasma membranes isolated from fusogenic carrot cells. Plant Physiol 85: 389-392

[5] Sommarin M, Sandelius AS (1988) Phosphatidylinositol and phosphatidylinositolphosphate kinases in plant plasma membranes. Biochim Biophys Acta 958: 268-278

[6] Melin PM, Sommarin M, Sandelius AS, Jergil B (1987) Identification of Ca^{2+}-stimulated polyphosphoinositide phospholipase C in isolated plant plasma membranes. FEBS Lett 223: 87-91

[7] Drøbak BK, Allan EF, Comerford JG, Roberts K, Dawson AP (1988) Presence of a guanine nucleotide-binding protein in a plant hypocotyl microsomal fraction. Biochem Biophys Res Commun 150: 899-903

[8] Zbell B, Walter-Back C, Hohenadel H, Schwendemann I (in press): Polyphosphoinositide turnover and signal transduction of auxin on isolated membranes of *Daucus carota* L. In: Plant Growth Substances 1988 (Pharis RP, Rood, SB) Springer-Verlag, Berlin Heidelberg

[9] Zbell B, Walter C (1987) About the search for the molecular action of high-affinity auxin binding sites on membrane-localized rapid phosphoinositide metabolism in plant cells. In: NATO ASI Series, Vol. H10, Plant Hormone Receptors (Klämbt D), Springer-Verlag, Berlin Heidelberg, pp 141-153.

[10] Ettlinger C, Lehle L (1988) Auxin induces rapid changes in phosphatidylinositol metabolites. Nature 331: 176-178

[11] Zbell B, Walter-Back C (in press): Signal transduction of auxin on isolated plant cell membranes: Indications for a rapid polyphosphoinositide response stimulated by indole acetic acid. J Plant Physiol

[12] Zbell B (1983) Über die molekulare Wirkung von Auxin. Thesis, The Free University, Berlin

[13] Morré DJ, Gripshover B, Monroe A, Morré JT (1984) Phosphatidylinositol turnover in isolated soybean membranes stimulated by the synthetic growth hormone 2,4-dichlorophenoxyacetic acid. J Biol Chem 259: 15364-15368

[14] Zbell B, Morré DJ (1987) The molecular action of auxin. In: Models in Plant Physiology and Biochemistry, Vol. II (Newman DW, Wilson KG), CRC Press, Boca Raton, FL, pp 87-90

BIOLOGICAL ROLE OF PLANT LIPIDS
P.A. BIACS, K. GRUIZ, T. KREMMER (eds)
Akadémiai Kiadó, Budapest and Plenum Publishing
Corporation, New York and London, 1989

POSITIONAL DISTRIBUTION OF POLYUNSATURATED FATTY ACIDS IN GALACTOLIPIDS FROM SOME ALGAE OF RHODOPHYTA, PHAEOPHYTA, BACILLARIOPHYTA AND CHLOROPHYTA

T. Arao and M. Yamada

Department of Biology, Univesity of Tokyo, Komaba, Tokyo 153, Japan

Polyunsaturated fatty acids such as octadecatetraenoic acid(18:4) and eicosapentaenoic acid(20:5) are characteristic of marine algae. Previously, we reported that 20:5 in the diatom Phaeodactylum tricornutum was exclusively located in the sn-1 position of monogalactosyldiacylglycerol(MGDG) and digalactosyldiacylglycerol(DGDG)[Arao et al. 1987]. In higher plants, biosynthetic path for fatty acids depends on the molecular species of lipids, as classified into eukaryotic and prokaryotic lipids[Roughan and Slack 1982].

This paper represents the positional distribution of fatty acids in galactolipids from Gloiopeltis complanata, Grateloupia filicina and Gymnogongrus flabelliformis (Rhodophyta), Ishige okamurai, Padina arborescens and Sargassum ringgoldianum (Phaeophyta), Heterosigma akashiwo (Raphydophyta) and Monostroma nitidum (Chlorophyta) to find a clue for the biosynthesis of these acids.

Material and Methods

Each alga collected was washed with sea water, frozen in liquid nitrogen, ground and extracted by Bligh Dyer method. Positional distribution of fatty acids in lipids were determined by Arao et al.(1987).

Results and Discussion

Table 1 summarized the positional distribution of fatty acids in algal galactolipids. From the viewpoint of fatty acid synthesis and desaturation, location of long chain

Table 1. Positional distribution of the major fatty acids from some algae of Rhodophyta, Phaeophyta, Bacillariophyta, Raphydophyta and Chlorophyta.

Fatty acid	Positional distribution of fatty acids			
	MGDG		DGDG	
16:0	I,II(Rh,Ph),	I(Ra)	I,II(Rh,Ph,Ba),II(Ch,Ra)	
16:1(n−7)	I,II(Ba)	II(Ra)	I,II(Ba)	
16:3		II(Ba)		II(Ch)
16:4		II(Ch)	I,II(Ch)	
18:1		II(Rh,Ph)		II(Rh,Ph,Ra)
18:2		II(Ph)		II(Ph)
18:3(n−3)	I,II(Ph)	I(Ch)	I,II(Ph,Ch)	
18:4	I,II(Ph),I(Ch),II(Ra)		I,II(Ph,Ra)	I(Ch)
18:5?		I(Ra)		I(Ra)
20:4(n−6)	I,II(Rh)			
20:5(n−3)	I,II(Rh),I(Ph,Ba,Ra)			I(Rh,Ph,Ba,Ra)

I,II represent the Sn−1 and sn−2 positions,respectively
Rh, Rhodophyta;Ph, Phaeophyta;Ba, Bacillariophyta;
Ra, Raphydophyta;Ch, Chlorophyta.

fatty acids such as C16− and C18−acids at the sn−2 position, in contrast to that of very long chain fatty acids such as C20− acids at the sn−1 position, suggests the occurrence of acyltransferases specific for chain length.

In galactolipids containing unsaturated C18−acids, 18:1 and 18:2 occur only in the sn−2 position, but 18:3 and 18:4 in both the sn−1 and sn−2 positions, suggesting that the 18:1 desaturase is different from 18:2 and 18:3 desaturases. The occurrence of 16:3 and 16:4 at the sn−2 position and that of 18:3 and 18:4 at the sn−1 position in MGDG and DGDG of the green alga suggests that successive unsaturation of 18:3/16:3−galactolipid to 18:4/16:4−galactolipids in the green alga. Similarly, the occurrence of 20:5 at the sn−1 position and that of 18:2 at the sn−2 position in MGDG and DGDG of the brown algae suggest that desaturation of 20:4/18:1−galactolipid to 20:5/18:2−galactolipid in the brown algae.

References

Arao,T., Kawaguchi,A. and Yamada,M. (1987) Positional distribution of fatty acids in lipids of the marine diatom Phaeodactylum tricornutum Phytochemistry 26;2573−2576.

Roughan, P.G. and Slack, C.R. (1982) Cellular Organization of Glycerolipid Metabolism Ann. Rev. Plant Physiol. 33; 97−132.

BIOLOGICAL ROLE OF PLANT LIPIDS
P.A. BIACS, K. GRUIZ, T. KREMMER (eds)
Akadémiai Kiadó, Budapest and Plenum Publishing
Corporation, New York and London, 1989

EFFECT OF WATER STRESS ON FATTY ACIDS OF CHLOROPLASTS IN SARATOVSKAYA 55 AND LUTESCENCE 1848 VAR. WHEAT SEEDLING

T.V. Shigalova, O.B. Chivkunova and M.N. Merzlyak

Faculty of Biology, Moscow State University, Moscow, USSR

Water stress causes numerous damages to photosynthetic apparatus of higher plants: decrease in net photosynthetic activity, inhibition of several photosynthetic reactions, structural changes in membranes and etc. |1,2|. Under these conditions the changes in total lipid contents, disturbances in the ratios between individual lipids and fatty acids in isolated chloroplasts take place |3|. In this paper we report data on the contents of chloroplast fatty acids and the level of thiobarbituric acid - reactive products (TBARP) in the leaves in two (17-d old) wheat varieties of different drought resistance exposed to water deficit stress by withholding irrigation.

Wheat seedlings were grown at 25°C, 60% humidity and 2 klx illumination. The irrigation of plants was stopped 3 days before analysis. Water deficit in the leaves was determined according to |4|. Methods of lipid extraction, determination of fatty acid composition and TBARP were described in ref. |5|.

It was found that at low water deficit in the leaves drought-resistant wheat var. Saratovskaya 55 contained higher level of linolenic acid in chloroplast as compared with drought-sensitive var. Lutescence 1848 (Table). At moderate water deficit the insaturation of chloroplast fatty acids tended to increase that probably reflects the adaptation of plants to stress. Higher water deficit in Lutescence 1848 seedlings induced considerable decrease in chloroplast linolenate. In addition in this drought-sensitive plants water stress caused the accumulation of TBARP in the leaves. The contents of

Table. The effects of water stress on fatty acid composition in chloroplasts of two wheat varieties

Var.	Water deficit, %	Fatty acid content, mol% 16:0	16:1	18:0	18:1	18:2	18:3
Saratovskaya	0.1	17.5	2.1	2.6	5.3	6.9	65.6
55	6.8	17.1	2.4	1.5	2.2	6.1	70.6
	7.9	14.8	2.4	3.6	2.8	6.2	70.2
	27.4	18.4	1.8	3.9	4.3	8.5	63.2
Lutescence	0.1	3.2	2.5	4.7	7.1	10.8	51.8
1848	3.2	20.0	4.2	4.0	3.7	8.2	59.9
	6.0	20.7	2.7	3.2	4.4	7.9	61.1
	42.7	34.9	1.8	6.6	7.7	8.8	40.2

the products were increased with increasing of water deficit from 10 to 30%, coefficient of linear correlation (Kcor) 0.99. These changes were not observed in Saratovskaya 55 leaves and the level of TBARP was decreased in water-stressed plants (water deficit up to 45%, Kcor = -0.95). Thus the data presented suggest that the loss of linolenate in chloroplast lipids of drought-sensitive wheat seedlings is probably related to peroxidation of membrane lipids. The stability of chloroplast fatty acids in Saratovskaya 55 seedlings suggest about tolerance of photosynthetic apparatus of drought-resistant plants to water deficit.

REFERENCES

1. Boyer J.S. 1971. Recovery of photosynthesis in sunflower after a period of low leaf water potential. Plant Phisiol., *47*, 816-820.
2. Keck R.W., Boger J.S. 1974. Chloroplast response to low leaf water potentials. III. Differing inhibition on electron transport and photophosphorilation. Ibib., *53*, 474-479.
3. Martin B.A., Schoper J.B., Rinne R.W. 1986. Changes in soybean (*Glycine max*/L./Merr.) glycolipids in response to water stress. Ibid, *81*, 798-801.

4. Patel J.A., Vora A.B. 1985. Free proline accumulation in drought-stressed plants. Plant Soil, *84*, 427-429.
5. Bolychevsteva Yu.V., Chivkunova O.V., Merzlyak M.N., Karapetyan N.V. 1987. Effect of norflurazone on chlorophyll, fatty acid and lipid peroxidation products content in barly seedlings grown under different illumination conditions. Sov. Biochemistry, *51*, 160-167.

BIOLOGICAL ROLE OF PLANT LIPIDS
P.A. BIACS, K. GRUIZ, T. KREMMER (eds)
Akadémiai Kiadó, Budapest and Plenum Publishing
Corporation, New York and London, 1989

PHOTOACUSTIC SPECTRA OF CHLOROPHYLLS AND CAROTENOIDS IN FRUITS AND IN PLANT OILS

E.M. Nagel[1], H.K. Lichtenthaler[1], L. Kocsányi[2] and P.A. Biacs[3]

[1]Botanical Institute II, Plant Physiology and Plant Biochemistry, University of Karlsruhe, Kaiserstr. 12. 7500 Karlsruhe, FRG
[2]Technical University of Budapest, Institute of Physics, Dept. of Atomic Physics, Budafoki út 8 1111 Budapest, Hungary
[3]Central Food Research Institute (KEKI), Herman Otto út 15, 1022 Budapest, Hungary

INTRODUCTION

Photoacoustic (PA) spectroscopy is a non-destructive method to detect the heat production (yield of non-radiative de-excitation processes) induced by modulated (chopped) light absorbed in the sample (Rosencwaig 1980).
The sample is placed in a PA-cell, in which modulated heat creates an acoustic signal. This PA-signal is influenced only by the heat generated within the so-called thermal diffusion length of the probe, because the heat waves from deeper layers are damped. This thermal diffusion length decreases with increasing chopping frequency of the excitation light. By recording the PA-signal during wavelength scans one receives PA-spectra. Since most pigments loose absorbed light energy via non-radiative de-excitations, they can be measured by the PA-method. The PA-spectra are in most cases similar to absorption spectra. This fact permits to determine the absorption characteristics of pigments in vivo both in highly or weakly absorbing materials (Tam 1986). Another advantage of the PA-method is the fact that it is not influenced by scattered light. A unique possibility is the estimation of the pigment distribution within a sample (depth-profiling) by a systematic change of the thermal diffusion length by varying the chopping frequency.

MATERIALS AND METHODS

The PA-spectra were taken with a spectrometer described by Nagel et al. (1987) equipped with a commercial available PA-cell (EG&G, 6003).The pigments were isolated by thin layer chromatography (Lichtenthaler 1987) and measured photoacoustically after evaporation of the developing solvent. The plant oils of sunflower (Helianthus annuus), olive (Olea europea) and St. John´s Wort (Hypericum perforatum) were measured directly without diluting. Red and green paprika (Capsicuum annuum) fruits were investigated with different chopping frequencies (22, 210 and 515 Hz).

RESULTS

Photoacoustic and absorption spectra of isolated pigments.
PA-spectra of chlorophyll a and b as well as ß-carotene are similar to their absorption spectra (Fig. 1A and 1B). The chlorophylls possess maxima in the blue-light part and the red-light part of the spectra, whereas the ß-carotene only possess a maximum in the blue-light region.

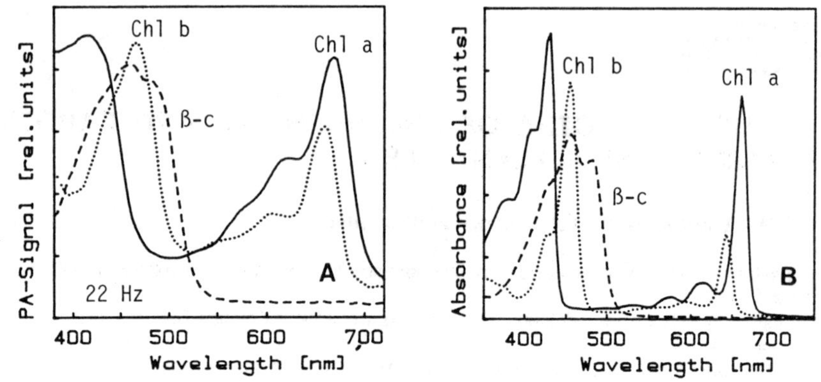

Fig. 1. (A) Absorption spectra (in diethyl ether) and (B) PA-spectra of chlorophyll a, chlorophyll b and ß-carotene. The PA-spectra were measured with a chopping frequency of 22 Hz.

Photoacoustic spectra of plant oils.

The PA-spectra of plant oils show a broad ground absorption in the visible range (Fig. 2), which makes it difficult to measure them with an absorption spectrometer. The pigment absorption is superimposed on this ground absorption. The PA-spectrum of the oil of the sun flower possess a small maximum in the blue-light region. In addition to this maximum the spectrum of the olive oil possess a maximum in the red-light region. The spectrum of the oil of the St. John´s Wort shows maxima in the blue, the green and the red light region.

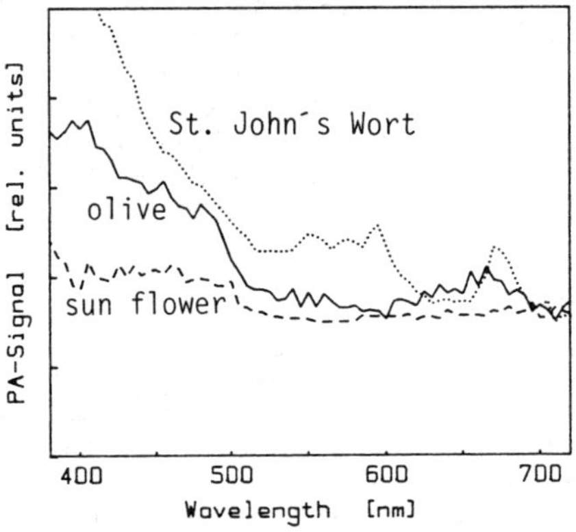

Fig. 2. Photoacoustic spectra of oils of sun flower, olive and St. John´s Wort measured directly with low chopping frequencies.

Photoacoustic spectra of paprika fruits.

The PA-spectrum of a red paprika has a broad maximum between 420 and 550 nm which becomes smaller with increasing chopping frequency (Fig. 3A), whereas the signal above 650 nm and below 420 nm rises. The spectrum of the green fruit taken at 22 Hz has also a maximum in the blue light region and in addition it has a lower, but characteristic maximum at 675 nm (Fig. 3B). The high-frequency PA-spectrum (515 Hz) of the green fruit between 420 and 720 nm shows also an inversion of the maxima and minima. However below 420 nm the PA-signal rises in both the high and low frequency measurements.

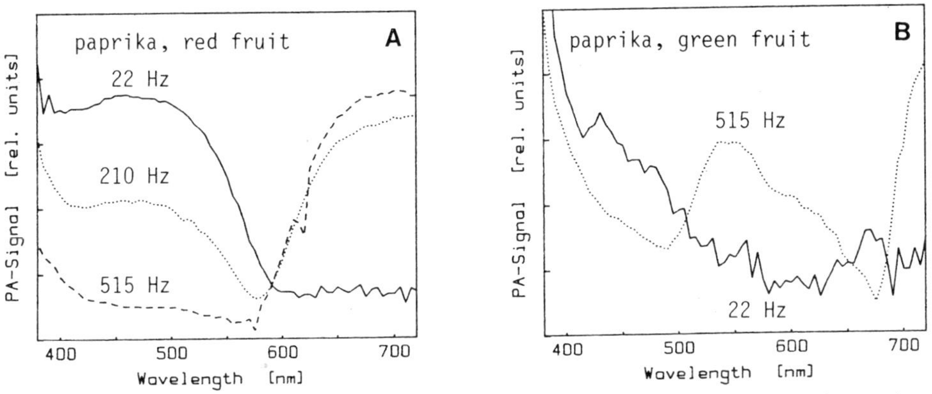

Fig. 3. Photoacoustic spectra of red (A) and green (B) paprika fruits taken with different modulation frequencies.

DISCUSSION

The similarity of PA-spectra and absorption spectra allows non-destructive absorption measurements of pigments with the PA-method. The spectra of the plant oils show that the method is sensitive to detect low pigment concentrations in strongly absorbing material without diluting. The oil of the sun flower only contains a small amount of carotenoids, whereas the olive oil also possesses traces of chlorophylls. In addition to these pigments the oil of St. Johns Wort contains hypericin (Robinson 1963) which absorbs green light. PA-spectra of fruits measured at a low modulation frequency (22 Hz) show their absorption characteristics. The red fruit contains only carotenoids, whereas in the green fruit chlorophylls are also present. With increasing chopping frequency (decreasing thermal active layer) the absorption maxima of chlorophylls and carotenoids decrease and at 515 Hz they are no more visible. At very high frequencies the PA-signal arises only from the epidermal layer which is colourless. The characteristic maxima of the 515 Hz PA-spectra are not only influenced by the direct light absorption but also by the absorption of the light reflected from deeper layers (Nagel et al. 1987). This is the reason for the spectral inversion between 420 and 720 nm. Below 420 nm the PA-spectra are determined by the high absorption of the flavonoids located inside the epidermal cells.

ACKNOWLEDGEMENTS: The authors express their sincere thank to Dr. C. Buschmann for helpfull discussion.

REFERENCES

NAGEL, E.M., BUSCHMANN, C. & LICHTENTHALER, H.K. (1987): Photoacoustic spectra of needles as an inicator of the activity of the photosynthetic apparatus of healthy and damaged conifers. Physiol. Plantarum 70, 427-437.

LICHTENTHALER, H.K. (1987): Chlorophylls and Carotenoids: Pigments of Photosynthetic Biomembranes. Methods in Enzymology 148, 350-382.

ROBINSON, T. (1963): The organic constituents of higher plants, First edition, Burgess Publishing company, Minneapolis, pp 109-110.

ROSENCWAIG, A. (1980): Photoacoustics and Photoacoustic Spectroscopy; in Chemical Analysis, Volume 57, John Wiley & Sons, New York.

TAM, A.C. (1986): Applications of photoacoustic sensing techniques. Rev. Mod. Phys. 58, 381- 431.

BIOLOGICAL ROLE OF PLANT LIPIDS
P.A. BIACS, K. GRUIZ, T. KREMMER (eds)
Akadémiai Kiadó, Budapest and Plenum Publishing
Corporation, New York and London, 1989

EPICUTICULAR WAXES OF ZEA MAYS ssp. MAYS AND RELATED SPECIES

P. Avato[1], G. Bianchi[2] and N. Pogna[3]

[1]Dipartimento Farmaco-Chimico, Univesità Bari, Italy
[2]Dipartimento di Chimica Organica, Università Pavia, Italy
[3]Instituto Sperimentale per la Cerealicoltura, Sant'Angelo Lodigiano, Italy

INTRODUCTION

Previous studies have shown that chemical composition of the leaf lipid compounds provide a useful tool to elucidate phylogenetic affinities (1-4). In fact, wax constituents have frequently been chosen as chemosystematic markers for the classification of several plant families.

Objective of the present contribution is to establish possible phylogenetic relationships among modern corn (*Zea mays* ssp. *mays*) and related species from the analysis of their epicuticular waxes. Chemical composition of waxes from fourth-fifth leaf seedlings of modern corn and its putative ancestors annual teosinte (*Zea mays* ssp. *mexicana* and ssp. *parviglumis*) and diploid perennial teosinte (*Zea diploperennis*) is reported. Details of the experimental procedures are given in reference 5.

RESULTS AND DISCUSSION

There are competing theories about the origin of corn based on genetic studies, breeding experiments and archeological evidence(6-8). One is that annual teosinte was the direct ancestor of modern corn. In particular, recent results would indicate that the most likely ancestor is *Z. mays parviglumis*. According to another current hypotesis modern corn and annual teosinte are instead both descended from the perennial teosinte, *Z. diploperennis*.

Some general comments can be made considering the distribution of the epicuticular wax components shown in Table 1.

Table 1. Composition (%) of epicuticular waxes from maize and related species

Components	Maize	Teosinte Chalco 1)	Teosinte El Salado 2)	Teosinte Manantlan 3)
Alkanes	1	19	6	8
Esters	16	24	20	21
Aldehydes	20	10	10	19
Alcohols	63	42	61	38
Acids	tr	5	3	14

1) *Z. mays* ssp. *mexicana*; 2) *Z. mays* ssp. *parviglumis*; 3) *Z. diploperennis*

A close similarity of wax composition between maize and Z. mays mexicana, race Chalco, and Z.mays parviglumis from El Salado, is evident. Alcohols are the main component (42 to 63%), followed by esters (16 to 24%) and aldehydes (10 to 20%). However higher amounts of alkanes (6,19% vs 1%) and acids (3,5% vs traces) are synthesized by the two subspecies of annual teosinte. The perennial teosinte from Sierra de Manantlan, with the same chromosome number (2n=20) of Z.mays and annual teosintes, produces surface lipids in which alcohols(38%)are still dominant over the other components but to a lesser extent and acids(14%) have much increased their importance.

Being generally accepted that chemical evolution proceeds by expansion of metabolic routes and that aldehydes and alcohols originate from acids through successive reductive steps, it seem reasonable to consider more advanced those species which accumulate alcohols in their waxes. Thus the primitive perennial teosinte fits well into the classification based on wax composition. The same criterion applied to the two subspecies of annual teosinte seems instead to support that they are as much advanced as maize.

No clear trend of evolution can be instead discerned from comparison of the alcohol and aldehyde chain lenghts of maize and teosinte plants prevailing the C_{32} homologue (73 to 99%) in all the species (complete data in press). When composition of alkanes is compared maize and teosinte seem to diverge (complete data in press). The 29 and 31 carbon chain lenghts are common components, the latter being the major alkane (41 to 59%). As a difference, maize is characterized by the presence of C_{27} (13%)as further important chain, whereas annual and perennial teosintes present high amounts of the C_{33} homologue (11 to 23%).

Reasonably supposing that shorter carbon chains precede longer ones on evolution, the presence of high amounts of the C_{33} chain in teosinte makes them chemically advanced species. This assumption, however, is tempered with the indications derived from the wax class composition. Taking into account that in maize the amount of alkanes appears strongly reduced, we like to propose that, meanwhile adaptive selection has favoured the formation of alcohols, the synthetic machinery giving rise to the alkanes has become less specialized.

REFERENCES

1. Eglinton,G. and Hamilton,R.J. (1967). Science, 156:1332.
2. Tulloch,A.P. (1976). Phytochemistry, 15:1145.
3. Salatino,M.L.F. and Salatino,A. (1983). Revta Brasil.Bot., 6:23.
4. Bianchi,G. (1985). Genet.Agr., 33:471.
5. Avato,P., Bianchi,G., Nayak,A., Salamini,F. and Gentinetta,E. (1987). Lipids, 22:11.
6. Beadle,G.W. (1980). Sci.Amer., 242:96.
7. Mangelsdorf,P.C. (1986). Sci.Amer., 255:72.
8. Ilts,H.H. and Doebly, J.F. (1984). in : "Plant Biosystematics", Grant W.F., ed., Academic Press, London, 587.

ACKNOWLEDGEMENT

We thank CNR (Rome) for financial aid.

CHAPTER 3

BIOSYNTHESIS AND FUNCTION OF PRENYLLIPIDS

BIOLOGICAL ROLE OF PLANT LIPIDS
P.A. BIACS, K. GRUIZ, T. KREMMER (eds)
Akadémiai Kiadó, Budapest and Plenum Publishing
Corporation, New York and London, 1989

ENZYMIC SYNTHESIS OF MEVALONIC ACID IN PLANTS*

J. Bach and T. Weber

Botanisches Institut II, Universität Karlsruhe, Kaiserstr. 12, 7500 Kalsruhe 1, FRG

INTRODUCTION

The biosynthesis of mevalonic acid ("MVA"), starting from acetyl-CoA ("Ac-CoA"), requires the action of three enzymes: a) **a**ceto**a**cetyl-**C**oA **t**hiolase ("AACT", EC 2.3.1.9), b) 3-**h**ydroxy-3-**m**ethyl**g**lutaryl-CoA **s**ynthase ("HMGS", EC 4.1.3.5), and c) HMG-CoA **r**eductase ("HMGR", EC 1.1.1.34). Our recent research has mainly centered around the purification and characterization of membrane-bound HMGR from radish seedlings (1-3). HMGR activity is commonly regarded as playing an important role in the regulation of substrate flux from acetate to the various isoprenoid endproducts, mainly of sterols (cf. 2,6). It is well documented that regulation of mammalian HMGR is mediated by rapid processes both at the translational and post-translational level (cf. 10). Regulation of mammalian HMGS appears to parallel that of HMGR (cf. 5). Similar processes can be expected to occur in plant cells. A thorough study in this direction requires a profound knowledge of the properties of the enzymes involved in the formation of HMG-CoA. This substrate, besides being reduced by HMGR to yield MVA, can also be utilized by HMG-CoA **l**yase ("HMGL", EC 4.1.3.4) to form acetoacetate, Ac-CoA and HS-CoA, an important reaction which in mammalian cells (mitochondria) contributes to the HMG-CoA cycle and hence to the formation of ketone bodies. None of these enzymes has ever been further characterized from plant materials, a major drawback in any study of the enzymic regulation of the isoprenoid pathway in plants (6). If there is some rudimentary information available, those data are difficult to compare because of the variety of plants used. Thus we have started to characterize all enzymes involved in the formation and metabolism of HMG-CoA in radish seedlings.

METHODS

HMGR was assayed as described (3). The assay for AACT/HMGS was largely based on the radioctive method described by Clinkenbeard et al. (4) as well as that one of HMGL. Details will be reported elsewhere. HMGR was purified from yeast applying the protocol of Kirtley and Rudney (8) up to the heat precipitation step, with little modifications. As a final step chromatography on HMG-CoA-agarose was added.

* Dedicated to the late Professor William R. Nes, Philadelphia, in recognition of his great scientific contributions in isoprenoid biochemistry.

Radish seedlings were grown in the dark and membranes were isolated as described (3). AACT and HMGS activity was solubilized using 2% polyoxyethlene ether (Brij W-1) and incubation at 30°C. Details will be given elsewhere.

Liquid chromatography columns were equilibrated with buffer system "A" (3) containing 1% Brij W-1, however, if emphasis was given on separation of HMGL activity, EDTA was ommitted. In kinetic studies substrate concentrations were varied as was appropriate. Data were plotted as described (3).

RESULTS AND DISCUSSION

Membranes isolated from etiolated radish seedlings and pelletable at 16,000*g contain considerable part of the total cellular enzyme activity, comprising AACT and HMGS, which are capable of synthesizing HMG-CoA from Ac-CoA **(table 1)**. They also contain a HMG-CoA utilizing activity, apparently HMGL. Since this membrane fraction is the major source of HMGR activity found in radish plants, we primalily focused our attention on the further characterization of those membrane-associated enzymes. If enzymes competing for the same substrate or channelling substrate flow are sharing the same intracellular distribution, some way of coarse control as well as of fine tuning must exist. Although the coupled enzyme system AACT/HMGS could be solubilized from the membranes under the same conditions as developed for HMGR (3), milder conditions, e.g. incubation at lower temperature (30° instead of 37°C) have turned out to be favorable to the enzymes' stability. However, all attempts to further purify the proteins have failed so far, with the exception of gel filtration and treatment with polyethlenimine **(table 2)**. The activity eluted at an apparent molecular mass of ca. 56 kDA. The enzyme system proved to be quite unstable in the presence of high salt concentrations or in specific buffer solutions, e.g. used in chromatofocusing. Nevertheless, the results presented in **table** 2 give the first example of any substantial purification of this enzyme system from a higher plant source. However, we have not succeeded in separating AACT from HMGS activity. Usually, each fraction was assayed twice, once only in presence of ^{14}C-Ac-CoA, and once being additionaly supplied with unlabeled AcAc-CoA. If we had a fraction exclusively containing HMGS enzyme, we should observe formation of ^{14}C-HMG-CoA only in the second case. There were some slight variations in the corresponding assays **(table 2)** in the presence or absence of AcAc-CoA, which itself was revealed to be a strong inhibitor of the enzyme activity (not shown). This was certainly not due to any substrate dilution effect. With this partially purified enzyme a K_m of 15 µM for AcCoA was determined. The observation that these two enzyme activities somehow conspire was further confirmed through the experiment outlined in **table 3.** Here we could also prove that the enzyme system produced ^{14}C-HMG-CoA, because highly purified yeast HMGR could completely convert the product into ^{14}C-MVA. This reaction was not affected by the addition of NADH, the cosubstrate for liver ß-hydroxyacyl-CoA dehydrogenase ("ß-HAC-DH"). High levels of this enzyme in the assay system should have been capable of trapping any intermediate AcAc-CoA, converting it into ß-hydroxybutyryl-CoA, and thereby preventing any formation of ^{14}C-HMG-CoA. However, the apparent inability of this enzyme to affect HMG-CoA (or MVA) formation led us to conclude that there might exist a very close co-operation through formation of a tightly fitting enzyme complex. The product arising from the

Table 1: Intracellular distribution of AACT/HMGS and HMGL activities.

Enzyme system / Spec. activity (pmol/mg/min)	Homogenate[a]	P 16000	S 16000	P 140000	S 140000[b] $(NH_4)_2SO_4$ prec. 59%	95%	100%
AACT/HMGS	747	944	654	221	165	104	98
(total activity, nmol/min)	(1872)	(242)	(1195)	(28)	(99)	(43)	(18)
HMGL	143	255	33	97	13	-	-
(total activity, nmol/min)	(358)	(66)	(60)	(12)	(8)	-	-

a) Homogenates were prepared from 4-day-old etiolated radish seedlings. Cell debris and PVP-particles had been removed by 10 min centrifugation at 1000*g. **b)** Cytosolic proteins were precipitated by ammonium sulfate at the saturations indicated.

Table 2: Solubilization and partial purification of AACT/HMGS.

Fraction	specific activity (nmol/mg/min)		purification factor	yield
P 16000	0.89[a]	(0.62)[b]	1	100 %
P 16000+Brij	1.6	(0.79)	1.8	179 %
S 100000+Brij	4.5	(1.4)	5.1	222 %
P 100000+Brij	0.15	(0.02)	0.17	17 %
S 0.3% Imine[c]	26.3	(7.6)	29.4	55 %
Gel filtration (peak fraction)	114.9	(31.4)	129	5.4 %

The solubilization was achieved through incubation with 2% (w/v) Brij for 30 min at 30°C. **a)** Substrate: 82 µM ^{14}C-Ac-CoA (= 27,470 dpm); **b)** in presence of 50 µM AcAc-CoA; **c)** supernatant 10,000*g.

Table 3: Co-operation between AACT and HMGS solubilized from membranes (P 16,000) of 4-day-old etiolated radish seedlings.

Treatment	Product formation (dpm)	% of control	
Control	10,789[a]	100	a) assayed as ^{14}C-HMG-CoA
+ NADPH + HMGR (yeast)	10,790[b]	100	b) assayed as ^{14}C-MVA
+ NADH	10,606[a]	98.3	
+ NADPH + HMGR (yeast) + NADH + ß-HA-CoA-DH (liver)	10,680[b]	99.0	**Conclusion:** No free substrate exchange of AcAc-CoA intermediate with the solvent system.

AACT reaction could be immediately transferred to the consecutive enzyme without any substantial exchange with the medium. In view of the numerous enzymes competing for Ac-CoA, a directed flux of carbon units into the isoprenoid pathway makes sense. However, the final proof of the formation of tightly bound enzyme complexes, even if they occur only temporarily in vivo (7), requires careful in vitro NMR studies.

Another enzyme activity that has to be considered in this regard is HMGL, competing for the same substrate as HMGR. We found this enzyme largely associated with heavy membranes (**table 1**). Under certain conditions its activity can interfere with the HMGR radioassay, viz. pretend HMGR activity. However, since HMGL activity is dependent on the presence of Mg^{++}, in contrast to HMGR as well as AACT/HMGS, its interfering activity can be suppressed. The enzyme is rather unstable at > 30° and ceases to be active in the presence of EDTA, but it is apparently not sensitive to increased salt concentrations. The enzyme which also precipitates at > 38% ammonium sulfate (HMGR activity precipitates at 0-38%), does survive this treatment, using detergent-solubilized proteins. Gel filtration of partially purified enzyme yielded an apparent molecular mass of ca. 72 kDa. Recent experiments indicated that the enzyme has a pI of about 6.8. The physiological role this enzyme plays in the plant cell's metabolism remains to be elucidated. To our knowledge no clear data are available as yet which would indicate the presence of the so-called HMG-CoA cycle and ketone body formation in plants. If the MVA shunt mechanism is operating in plants (cf. 9), HMGL might be involved in the process of shunting carbon units away from their inclusion into the isoprenoid pathway.

ACKNOWLEDGEMENTS

This work has been supported by the Deutsche Forschungsgemeinschaft.

LITERATURE

1) **Bach, T.J.**- Hydroxymethylglutaryl-CoA reductase, a key enzyme in phytosterol synthesis? Lipids 21: 82-88 (1986).
2) **Bach,T.J.**- Synthesis and metabolism of mevalonic acid in plants. Plant Physiol. Biochem. 25: 163-178 (1987).
3) **Bach, T.J.**, D.H. Rogers, and H. Rudney.- Detergent-solubilization, purification and characterization of 3-hydroxy-3-methylglutaryl-CoA reductase from radish seedlings. Eur. J. Biochem. 154: 103-111 (1986).
4) **Clinkenbeard, K.D.**, D.W. Reed, R.A. Mooney, and M.D. Lane.- Intracellular localization of the 3-hydroxy-3-methylglutaryl coenzyme A cycle enzymes in liver: Separate cytoplasmic and mitochondrial 3-hydroxy-3-methylglutaryl coenzyme A generating systems for cholesterogenesis and ketogenesis. J. Biol. Chem. 250: 3108-3116 (1975).
5) **Gil,G.**, J.L.Goldstein, C.A. Slaughter, and M.S. Brown.- Cytoplasmic 3-hydroxy-3-methylglutaryl coenzyme A synthase from hamster: I. Isolation and sequencing of a full-length cDNA. J. Biol. Chem. 261: 3710-3716 (1986).
6) **Gray, J.C.**- Control of isoprenoid biosynthesis in higher plants. Adv. Bot. Res. 14: 25-91 (1987).
7) **Kaprelyant, A.S.**- Dynamic spatial distribution of proteins in the cell. Trends Biochem Sci. 13: 43-46 (1988).
8) **Kirtley, M.E.**, and H. Rudney.- Some properties and mechanisms of action of the ß-hydroxy-ß-methylglutaryl coenzyme A reductase of yeast. Biochemistry 6: 230-238 (1967).
9) **Nes, W.D.**, and T.J. Bach.- Evidence for a mevalonate shunt in a tracheophyte. Proc. R. Soc. Lond. B 225: 425-444 (1985).
10) **Sabine,J.R.**- Monographs on enzyme biology: HMG-CoA reductase, J.R. Sabine ed., CRC Press, Boca Raton 1983.

BIOLOGICAL ROLE OF PLANT LIPIDS
P.A. BIACS, K. GRUIZ, T. KREMMER (eds)
Akadémiai Kiadó, Budapest and Plenum Publishing
Corporation, New York and London, 1989

PHYTOENE DESATURASE, A TARGET FOR HERBICIDAL INHIBITORS

G. Sandmann, A. Schmidt, H. Linden, J. Hirschberger* and P. Böger

Lehrstuhl für Physiologie und Biochemie der Pflanzen, Universität Konstanz, FRG
*Department of Genetics, Hebrew University, Jerusalem, Israel

Phytoene desaturase catalyzes the insertion of two double bonds into the first carotene in a series leading to α- and ß-carotene. In several fungi this key enzyme is responsible for light/dark regulation of carotenogenesis (1). Furthermore, it is the most important target for many bleaching herbicides (2).

Herbicides interferring with the carotenogenic pathway can be characterized by in-vitro carotene biosynthesis (3). Table 1 shows the interference of three selected herbicidal inhibitors with phytoene desaturase, ζ-carotene desaturase and lycopene cyclase in Aphanocapsa thylakoids and Chenopodium chloroplasts. In the control, ^{14}C-phytoene was converted into ζ-carotene, lycopene and ß-carotene with a certain rate. In the presence of difunone, less phytoene was desaturated to the subsequent carotenes. With J 852, a pyrimidine derivative, we obtained an increased accumulation of ζ-carotene at the expense of lycopene and ß-carotene. When CPTA was added, more lycopene was formed and less ß-carotene. Phytoene, ζ-carotene and lycopene are the substrates of the three enzymes in the carotenogenic pathway from phytoene to ß-carotene. Therefore, we can conclude that difunone inhibits phytoene desaturase, J 852 inhibits ζ-carotene desaturase and CPTA lycopene cyclase. As we observed the same inhibition pattern with the blue-green alga Aphanocapsa and with Chenopodium, we can use the cell-free system from Aphanocapsa as a very efficient model system to study carotene biosynthesis of photosynthetic organisms.

Several resistant Anacystis mutants have been selected against norflurazon, another herbicide directed against phytoene desaturase. This herbicide

TABLE 1:

IN VITRO CONVERSION OF ^{14}C-PHYTOENE INTO SUBSEQUENT CAROTENES IN THE PRESENCE OF DIFFERENT INHIBITORS OF CAROTENOGENIC ENZYMES

	Incorporation of radioactivity (dpm) into			
	Phytoene	ζ-Carotene	Lycopene	ß-Carotene
A. Aphanocapsa				
Control	3924	1013	816	1070
Difunon, 1 µM	7354	397	227	377
J 852, 50 µM	4035	2180	399	561
CPTA, 50 µM	3649	941	1432	411
B. Chenopodium				
Control	4125	712	316	355
Difunon, 1 µM	5331	276	147	123
J 852, 50 µM	4291	1688	171	231
CPTA, 50 µM	3846	744	622	200

^{14}C-phytoene generated from 0.5 µCi D,L- ^{14}C-mevalonic acid, membranes equivalent to 120 µg chlororphyll, incubation time 2 hrs at room temperature.

blocks carotenogenesis at the level of phytoene. After treatment, accumulation of phytoene and epoxydated and hydroxylated phytoene derivatives have been observed. The strains NFZ 4 and NFZ 18 showed the strongest resistance against norflurazon (Table 2). The resistance factors (I_{50} mutant/I_{50} wild strain) were about 70 and 40, respectively.
The degree of resistance was the same in intact cells as in the enzyme assay. This indicates that the mutants contain a phytoene-desaturase enzyme which is modified in a way that herbicide interference is less effective. Further studies have excluded that the resistance in both mutants is caused by overexpression of the phytoene desaturase gene leading to an overproduction of phytoene desaturase. We observed cross resistance with some other hercicides interfering with phytoene desaturase like fluorochloridone. In contrast, our mutants were not resistant against fluridone.

In several photosynthetic species we determined the molecular weights of phytoene desaturase by SDS polyacryl-amide gel electrophoresis and identification by Western blotting with a phytoene-desaturase antibody (Table 3). This antibody was obtained after fusion of the phytoene desaturase gene from Rhodobacter with the lac Z gene and expression of this fusion protein in E. coli.

We found that the molecular weights of phytoene desaturase from the prokaryotic organisms, the photosynthetic bacterium Rhodobacter (MW = 64 kd) and the cyanobacterium Aphanocapsa (MW = 65 kd), are very similar to the MW of the protein from several higher plants (all at 64 kd). The MW of the phytoene desaturase determined here is identical to the value calculated from the DNA sequence. The eukaryotic alga Bumilleriopsis contains a phytoene desaturase with a significant lower molecular weight of 55 kd. This result does not fit in the general picture that phytoene desaturase is highly conserved in the evolution of photosynthesis.

TABLE 2:
INHIBITION OF PHYTOENE DESATURATION BY NORFLURAZON IN ANACYSTIS MUTANTS

Strain	Cellular I_{50} value (μM)	Factor of resistance	in vitro I_{50} value (μM)	F-R
wild type (R2)	0.11	-	0.03	-
NFZ 4	8.11	73.6	1.92	76.8
NFZ 18	4.57	41.6	1.01	40.0

TABLE 3:
MOLECULAR WEIGHTS OF PHYTOENE DESATURASES FROM PHOTOSYNTHETIC ORGANISMS

Species	MW (kd)
Rhodobacter	64
Aphanocapsa	65
Bumilleriopsis	55
Chenopodium	64
Rape	64
Spinach	64

ACKNOWLEDGMENT

This work was supported by the Deutsche Forschungsgemeinschaft and by the "Stiftung Umwelt und Wohnen".

REFERENCES

1. Rau W. (1976) Pure and Appl. Chem. 47, 237-243.
2. Sandmann G., Clarke I.E., Bramley P.M. and Böger P. (1984) Z. Naturforsch. 39c, 443-449.
3. Sandmann G., Bramley P.M. and Böger P. (1985) J. Pestic. Sci. 10, 19-24.

BIOLOGICAL ROLE OF PLANT LIPIDS
P.A. BIACS, K. GRUIZ, T. KREMMER (eds)
Akadémiai Kiadó, Budapest and Plenum Publishing
Corporation, New York and London, 1989

DEHYDROGENATION AND CYCLIZATION REACTIONS IN CAROTENE BIOSYNTHESIS

P. Beyer and H. Kleinig

Institut für Biologie II, Universität Freiburg im Breisgau, Schänzlerstr. 1, D-7800 Freiburg i. Brsg., FRG

In vitro, daffodil chromoplasts (Narcissus pseudonarcissus L.) synthesize α-and β-carotene in high yields from [1-^{14}C]isopentenyl diphosphate as precursor. The associated enzymes are distributed between two spatially and functionally distinct sites: whereas the synthesis of 15-cis phytoene is mediated by peripheral membrane proteins (the phytoene synthase complex), the desaturase reactions leading to the polyene chromophor, together with the cyclization reactions which form the α-and β-ionone end groups are catalyzed by integral membrane enzymes (Kreuz et al. 1982). Owing largely to this topological separation, the reaction sequence phytoene - α/β-carotene, in particular, is poorly understood at present.

Some time ago we developed conditions to investigate this system. Conditions were defined for the successful solubilization of integral membrane enzymes, using CHAPS, a zwitterionic detergent, and for reconstituting these enzymes into proteoliposomes (Beyer et al. 1985, Beyer 1985). As it is possible to separate proteins in micellar solution under certain circumstances, it should also be possible to characterize individual enzymes of the sequence by use of the reconstituted system. Problems, however, arise in such a system when a reaction is mediated by more than one enzyme and it will be shown that indeed carotene desaturation proceeds by cooperation of the membrane-integral desaturases with peripheral protein factors. In addition to these problems there are two other crucial points as to the nature of the terminal electron acceptor and - as has also been found - the state of geometrical isomerism of the carotene intermediates.

These relations become evident by use of a dissected system: intact chromoplasts synthesize α-and β-carotene from [1-^{14}C]isopentenyl diphosphate as a precursor without significant accumulation of the carotene inter-

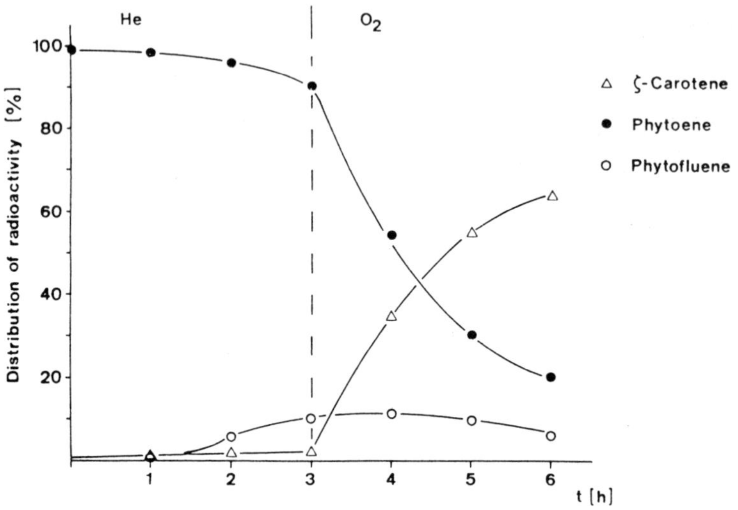

(Fig. 1) Time course of the desaturation of radiolabelled phytoene in the absence and presence of O_2. Substrate: [^{14}C]phytoene, 100 000 cpm ml^{-1}

mediates. Disrupted chromoplasts, on the other hand, often show [^{14}C]ζ-carotene as an intermediate and, in particular when membranes were washed, [^{14}C]ζ-carotene was the only product formed from radiolabelled 15-cis phytoene. When such a dissected system was investigated more closely, it was found that it worked independently of usual cofactors, FAD and $NADP^+$, so that the nature of the electron acceptor became a matter of interest. Decisive evidence in this respect came from strictly anaerobic experiments, made in the presence of helium (Fig. 1). For as long as O_2 was absent, no formation of [^{14}C]ζ-carotene was observed, and the substrate 15-cis[^{14}C]phytoene was not metabolized. On the addition of O_2, the reaction accelerated to normal values.

This desaturation reaction is in many ways analogous to the desaturation of ACP-bound stearate to oleate as is known from plastids and bacteria. It was demonstrated (Jaworski and Stumpf, 1974) that this reaction was dependent on oxygen which serves as an electron acceptor and is reduced to H_2O, with NADPH being used as a coreductant (see also Kikuchi and Kusaka, 1986). The reaction was mediated by an enzyme, ferredoxine: $NADP^+$-oxidoreductase (EC 1.18.1.2.).

An oxidoreductase of this type, exhibiting the same diaphorase activity ($NADPH \cdot H^+ + DCPIP \rightarrow NADP^+ + DCPIPH_2$) was also found in chromoplasts as a peripheral membrane protein. This enzyme, when purified, stimulated the

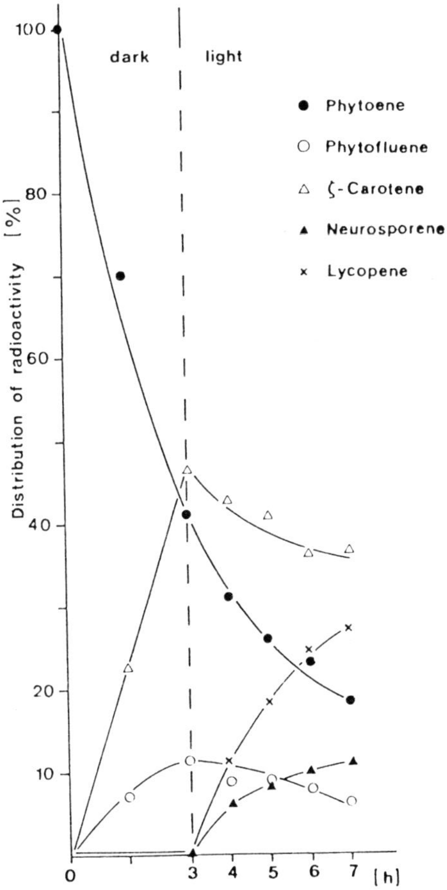

(Fig.2) Time course of the formation of ζ-carotene in the dark and the desaturation of ζ-carotene upon illumination. Substrate: 15-cis [^{14}C]phytoene, 1500 000 cpm ml^{-1}

desaturation of [^{14}C]phytoene so that it seems reasonable to conclude that the O_2-dependence of this carotene desaturase reaction is mediated by this enzyme and proceeds in a manner analogous to the desaturation of ACP-stearate.

(Table 1) Activation of ζ-carotene synthesis by a purified oxidoreductase preparation from daffodil chromoplasts Substr.: phytoene (100 000 cpm = 100%)

	phytoene	phytofluene	ζ-carotene
washed membranes	67.6%	9.3%	23.2%
+ oxidored. (2 nkat)	36.0%	14.3%	50.0%

The accumulation of ζ-carotene can be explained in terms of geometrical isomerism. This is based on the following observation: [^{14}C]ζ-carotene which accumulated in the dark on incubation with radiolabelled 15-cis phytoene as a substrate, is desaturated to lycopene upon illumination (Fig. 2). It was concluded that a photoisomerization of ζ-carotene is the crucial point in this sequence. This can be demonstrated by analytical HPLC (Fig. 3). The enzymatically formed ζ-carotene species is in the 15-cis configuration and this central cis double bond, deriving from 15-cis phytoene, is photoisomerized to trans. The resulting all-trans ζ-carotene is then accepted as a substrate in the lycopene-forming enzyme system.

The accumulation of lycopene which is now observed, is again due to O_2 which acts in an inverse way in the lycopene cyclization reactions (Fig. 4). When homogenates of chromoplasts are incubated in the presence of radioactive all-trans ζ-carotene, then in the presence of O_2, the formation of lycopene is observed as this reaction is dependent on O_2 as was found for the formation of ζ-carotene. Under anaerobic conditions (effected here by an enzymatic oxygen trap; glucose, glucose oxidase, catalase) no more desaturation is observed, but a rapid cyclization of the accumulated lycopene and formation of α-and β-carotene takes place.

Such a dual role of O_2, with desaturation reactions being absolutely dependent on O_2 and being coupled to cyclization reactions which, in contrast, function only in the absence of O_2 is apparently contradictory and only explicable under the assumption of a protein complex of high topological order in the plane of the lipid bilayer. Then, however, a working hypothesis can be made based on the assumption that the desaturation reactions, which proceed by the reduction of $O_2 \rightarrow H_2O$ mediated by the oxidoreductase, could create an anaerobic microenvironment around the active center of the cyclase(s), thus enabeling it to function. The regulatory implications of such a model that cyclization occurs only when desaturation reactions actively proceed might provide a first experimental access. The daffodil chromoplast system, as presented here, is disturbed in its topological order and, as a consequence, aerobic and anaerobic conditions had to be administered artificially.

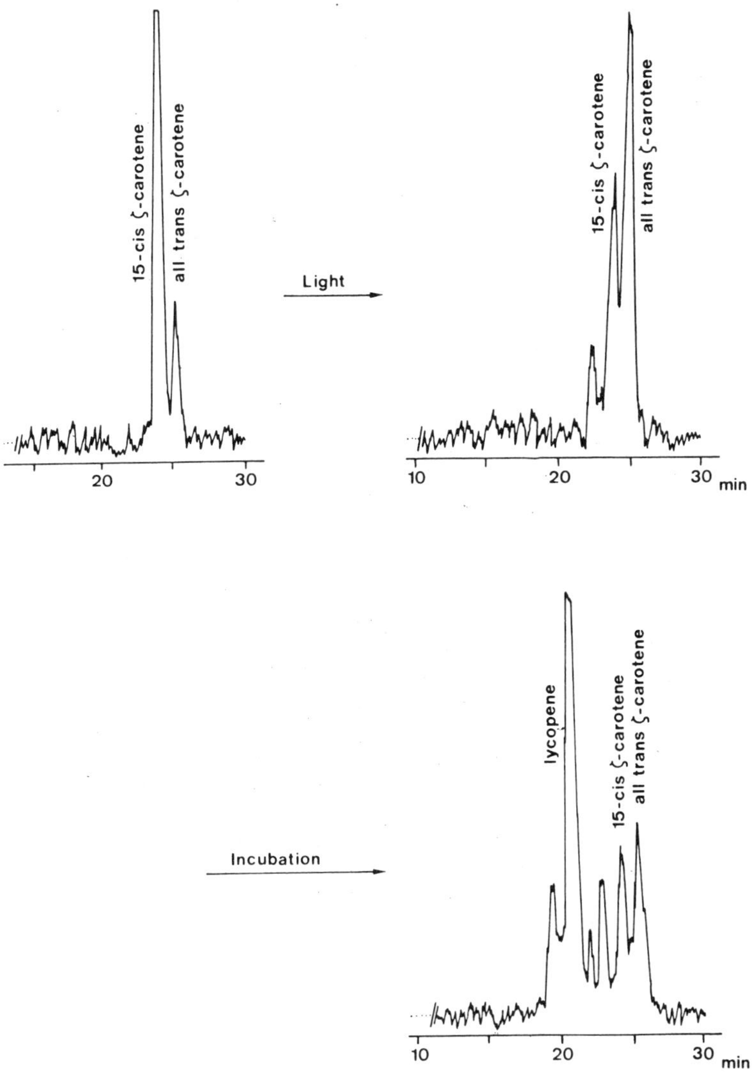

(Fig.3) HPLC sepatation of the enzymatically-formed 15-cis $|^{14}C|$ζ-carotene (upper left), its photoisomerization to all trans ζ-carotene and the conversion of the latter into lycopene

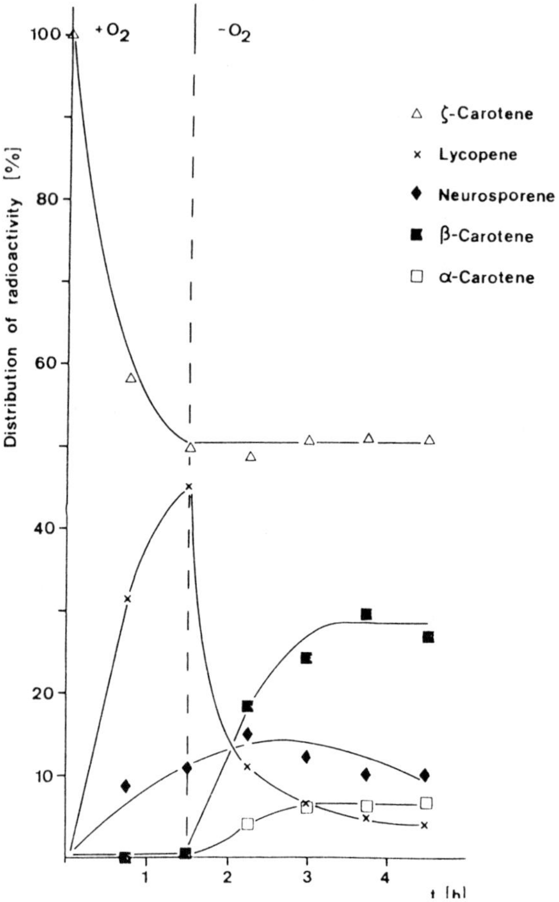

(Fig.4) Time course of the formation of [^{14}C]lycopene from all-trans [^{14}C]ζ-carotene in the presence of O_2 and of the cyclization of [^{14}C]licopene under anaerobic conditions. Substrate: all trans[^{14}C]ζ-carotene,120 000 cpm m

Beyer, P., G. Weiss and H. Kleing (1985) Solubilization and reconstitution of the membrane-bound carotenogenic enzymes from daffodil chromoplasts. Eur. J. Biochem. 153, 341-346.

Beyer, P. (1987) Solubilization and reconstitution of carotenogenic enzymes from daffodil chromoplast membranes using 3[(3-cholamidopropyl)dimethyl-ammonio]-1-propane sulfonate. in Meth. in Enzymol. 148, 392-400.

Kreuz, K., P. Beyer and H. Kleinig (1982) The site of carotenogenic enzymes in chromoplasts from Narcissus pseudonarcissus L.. Planta 154, 66-69.

Kikuchi, S. and T. Kusaka (1986) Isolation and partial characterization of a very long-chain fatty acid desaturation system from the cytosol of Mycobacterium smegmatis. J. Biochem. 99, 723-731.

BIOLOGICAL ROLE OF PLANT LIPIDS
P.A. BIACS, K. GRUIZ, T. KREMMER (eds)
Akadémiai Kiadó, Budapest and Plenum Publishing
Corporation, New York and London, 1989

THE ROLE OF ISOPENTENYL-DIPHOSPHATE Δ-ISOMERASE IN PHYTOENE SYNTHESIS OF DAFFODIL CHROMOPLASTS

M. Lützow, P. Beyer and H. Kleinig

Institut für Biologie II, Universität Freiburg, Schänzlestr. 1, 7800 Freiburg i. Brsg, FRG

The synthesis of phytoene constitutes the first part of the pathway leading to the formation of carotenes and xanthophylls in prokaryots, fungi and plastids of algae and higher plants. In six successive reactions which are performed by the putative phytoene synthase complex the watersoluble substrate isopentenyl diphosphate (IPP) is converted to the hydrophobic hydrocarbon phytoene, which is metabolized further by integral and peripheral membrane proteins. The complex in plastids is partially soluble after rupturing the isolated organells, a property which led to the opinion that it is located in the stroma. Several attempts to purify the intact phytoene synthase complex failed, usually the activity during purification was lost due to the dissociation into subunits. For that reason we decided for our system, the chromoplast of the daffodil flower an other strategy: First the characterisation of the individual enzymes, afterwards the study of their relation to the whole complex.

The isopentenyl diphosphate isomerase, which catalyses the equilibrium between IPP and dimethylallyl diphosphate (DMAPP) was purified 200 fold to near homogenity |1|. Analysis of the preparation by SDS-PAGE revealed two protein bands with M_r values of 32,000 and 28,000, which are similar to those reported from other plant sources. Other properties of the partially purified enzyme as the dependence of the activity on Mg^{2+}/Mn^{2+} and pH are the same as reported for different isomerases of plant and animal sources. The enzyme displays a remarkable high specific activity in the stroma of daffodil chromoplasts, fifty

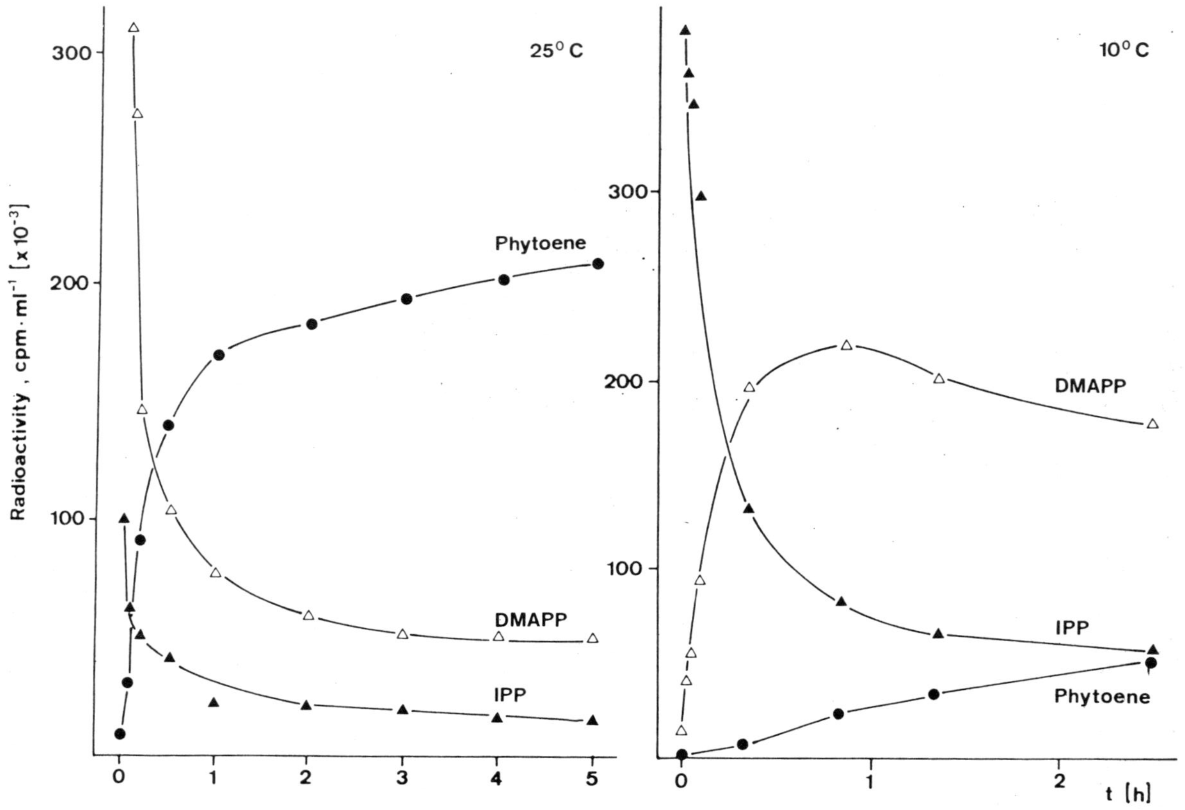

Figure 1. Time dependence of the isomerisation of isopentyl diphosphate and the formation of phytoene in the stroma of daffodil chromoplasts. A: at 25°C. B: at 10°C.

times higher than in comparable systems. Detailed analyses of time-dependent incubations in the stroma showed that in vitro the isomerase works independently of the consecutive enzymes of the phytoene synthase complex. Exogenous supplied radioactively labeled IPP is converted at room temperature rapidly to DMAPP (Fig. 1a). After two minutes 70% of the radioactivity is found in DMAPP. This ratio of free DMAPP/IPP with an equilibrium constant of 3.1 is maintained during the whole incubation, although the synthesized phytoene contains the inverse ratio of the two C5-diphosphates. A similar experiment at 10°C reveals that first the isomerase forms DMAPP and that the synthesis of phytoene follows (Fig. 1b). Isomerisation and phytoene synthesis proceed independently. This view is supported by inhibitor studies. The IPP/DMAPP isomerase is strictly inhibited by the alkylating agent iodo acetamide, in this case only traces of phytoene are synthetized (Figure 2). Phytoene synthesis is restored to the control level when DMAPP is added as a second substrate. This kind of experiment was repeated with several other putative and reported inhibitors |2,3| (Table). In the case of dimethylaminoethyl diphosphate, the transition-state analogue of the isomerisation, only the isomerase was inhibited, whereas the synthesis of phytoene was not affected. The fluorinated derivative of IPP blocked not only the isomerase but the prenyl-transferase too. In the third case, the herbicide Command, no effect on isomerase or subsequent enzymes was detected. This finding is somewhat in contrast to recently reported results for in vivo systems, which postulated a target for this effective pre-emergence herbicide |3| in the metabolical pathway from IPP to geranylgeranyl diphosphate.

However, our different incubation data show, that the first enzyme of the phytoene synthase complex in daffodil chromoplasts, the isopentenyl diphosphate isomerase, works independently and maintains with high efficiency a pool of free DMAPP/IPP. On the other hand the consecutive enzymes, the prenyltransferase and the phytoene-synthase utilized this pool in an effective way for the formation of phytoene. Whether this conclusions can be extrapolated to the in vivo situation remains uncertain. The complicated situation is daffodil chromo-

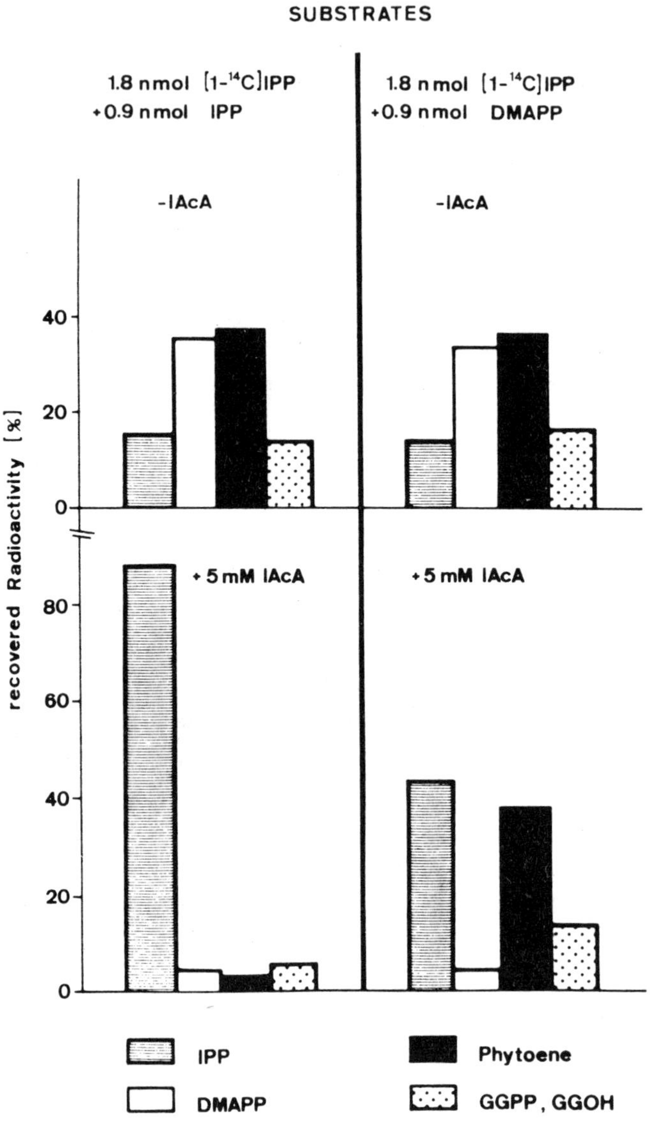

Figure 2. Inhibition of the IPP/DMAPP isomerase by iodo acetamide. Phytoene synthesis is restored by additio of DMAPP.

Table. Relative inhibition of isopentenyl-diphosphate Δ-isomerase and phytoene synthase by several inhibitors. The inhibitor concentrations used are stated in brackets. The isomerase was incubated with |1-^{14}C|IPP, the phytoene synthase dito plus DMAPP.

	Isopentenyl-diphosphate Δ-isomerase Enzyme Activity	Phytoene synthesis
Control	100%	100%
2-(Dimethylamino)-ethyl diphosphate	0% (1 μM)	95% (0.1 mM)
3-(Fluoromethyl)-3-butenyl diphosphate	10% (1 μM)	6% (0.1 mM)
Command (FMC 57020)	97% (2 mM)	107% (2 mM)

plasts where two pathways leading to mono- and tetraterpenes respectively |4| have to be supplied with IPP and DMAPP may be reflected by the situation we found in vitro.

REFERENCES

1. Lützow, M., Beyer, P.: The isopentenyl diphosphate Δ-isomerase and its relation to the phytoene synthase complex in daffodil chromoplasts BBA *959* (1988) 118-126
2. Muehlbacher, M., Poulter, C.D.: Isopentenyl Diphosphate: Dimethylalyl Diphosphate Isomerase. Irreversible Inhibition of the Enzyme by Active-Site-Directed Covalent Attachment J. Am. Chem. Soc. *107* (1985) 8307-8308.
3. Sandmann, G., Böger, P.: Interconversion of Prenyl Pyrophosphates and subsequent Reactions in the Presence of FMC 57020 Z. Naturforsch. *42c* (1987) 803-807
4. Mettal, U. et al.: Biosynthesis of monoterpene hydrocarbons by isolated chromoplasts from daffodil flowers Eur. J. Biochem. *170* (1988) 613-616

BIOLOGICAL ROLE OF PLANT LIPIDS
P.A. BIACS, K. GRUIZ, T. KREMMER (eds)
Akadémiai Kiadó, Budapest and Plenum Publishing
Corporation, New York and London, 1989

MONOTERPENE BIOSYNTHESIS BY CHROMOPLASTS FROM DAFFODIL FLOWERS

U. Mettal, W. Boland, P. Beyer and H. Kleinig

Institut für Biologie II, Universität Freiburg im Breisgau, Schänzlestr. 1, 7800 Freiburg i. Brsg, FRG

Chromoplasts from the daffodil flower (Narcissus pseudonarcissus) are well suited for in vitro investigations on the biosynthesis of tetraterpenoid carotenes. Soluble enzymes of the chromoplast stroma convert the hydrophilic substrate isopentenyl diphosphate (IPP) to the liphophilic and membrane associated phytoene, further carotenes are then formed by integrale membrane enzymes (Beyer et al. 1980).

In the course of our work on carotene biosynthesis we noticed that certain amounts of the radioactively labeled precursor IPP were incorporated into volatile compounds. These compounds turned out to be monoterpene hydrocarbons which also occur in the volatile pattern of daffodil flowers. These findings enabled us to make the first report on chromoplasts being the site of monoterpene biosynthesis (Mettal et al. 1988). Another plastid type, i.e. leucoplasts from Citrofortunella mitis fruit exocarps, has been shown to synthesize monoterpene hydrocarbons from IPP (Gleizes et al. 1983).

This presentation will go into some details concerning analysis, identification and characteristics of the in vitro formed monoterpenes.

Results and discussion

It turned out that the enzymes which are responsible for monoterpene biosynthesis are present as soluble proteins in the chromoplast stroma. For qualitative and quantitative analysis of the volatile compounds incubations of chromoplast stroma in

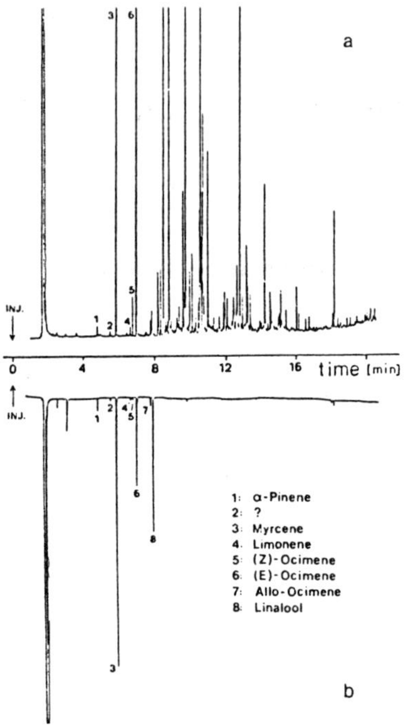

Fig. 1. Gaschromatographical separation of the natural volatile pattern from daffodil flowers (a) and the in vitro pattern (b). The range of about 8 - 11 min comprises monoterpene ketones and non-terpenoid volatiles, the range above 11 min sesquiterpenes and other compounds.

the presence of [1-^{14}C]IPP are performed in a closed system, a "closed-loop-stripping" apparatus (Boland et al. 1984). After the stripping procedure the volatile constituents are quantitatively separated from diterpenoid, tetraterpenoid and phosphorylated products of the incubation assay and are ready for analysis.

The gaschromatographical analysis of the volatile compounds in comparison with the volatile pattern synthesized by whole daffodil flowers (Fig. 1) revealed that the compounds formed in vitro were monoterpene hydrocarbons plus linalool, mainproducts being: myrcene (56.4%), ocimene (18%) and linalool (22.9%). Moreover the bicyclic α-pinene (2.8%) and the monocyclic limonene (1%) were formed. The low portion in cyclic monoterpenes is in good agreement with the natural pattern. Monoterpene ketones as well as sesquiterpenes which both occur in the natural pattern, were not obtained in chromoplast stroma incubations. On the other hand, linalool was not present in the in vivo pattern.

That is to say, the chromoplasts' constribution to the volatile isoprenoids of whole daffodil flowers comprises only monoterpene hydrocarbons.

The identification of the volatile products was based on combined gaschromatographical and mass spectroscopical analysis. The additional use of the deuterated precursor IPP as substrate for the incubation experiments allowed an unequivocal assignment of the in vitro formed compounds to those of the in vivo

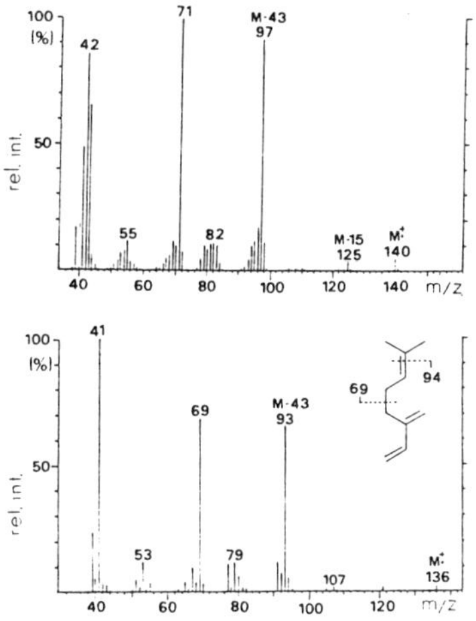

Fig. 2. Mass spectra of [H]myrcene (below) and $[^2H_4]$myrcene (above) using $[1-^2H_2]$IPP as substrate.

pattern (in Fig. 2 the spectra for myrcene are shown as an example.).

First attempts to purify the enzymes responsible for monoterpene biosynthesis have been made. The following methods were used: a first ion exchange chromatography on DEAE-cellulose, gelfiltration on Sephadex G-100 and a second ion exchange chromatography on a FPLC column. Upon all three steps all the enzymatic activities for monoterpene synthesis coeluted with geranylgeranyl diphosphate synthase so that a branch point on the stage of geranylgeranyl diphosphate synthase must be assumed.

At the moment our work concentrates on the characterisation of this branch point of monoterpene synthesis and phytoene synthesis, respectively. Both ways can be activated by different means. Phytoene synthesis is stimulated by 3 mM ATP and liposomes whereas optimal monoterpene synthesis occurs in the absence of ATP and liposomes.

Another access to the enzymes responsible for monoterpene synthesis is provided by dimethylaminoethyl diphosphate, an analogon of dimethylallyl diphosphate which inhibits the isopentenyl diphosphate Δ-isomerase at very low concentrations (Mühlbacher and Poulter 1985). This inhibitor was tested for its influence on monoterpene biosynthesis using radioactively labeled IPP and nonlabeled dimethylallyl diphosphate as substrates and optimized monoterpene conditions as mentioned above. Monoterpene synthesis was inhibited up to 50% at an inhibitor concentration of 2 µM whereas the synthesis of di- and tetraterpenes and geranylgeranyl diphosphate was not impaired (Fig. 3).

In recent experiments unlabeled geranyl diphosphate was used

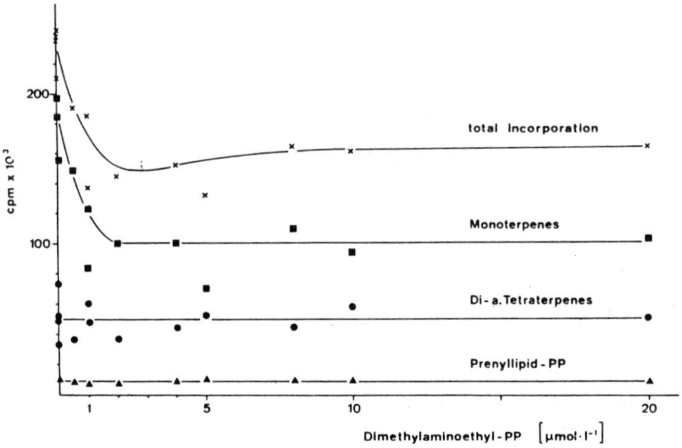

Fig. 3. Dependence of incorporation of [1-^{14}C]IPP into monoterpenes, di- and tetraterpenes and geranylgeranyl diphosphate on the concentration of dimethylaminoethyl diphosphate. Incubation assay: 0.5 ml chromoplast stroma, 100 mM Tris/HCl, pH 7.2, 2 mM DTE, 10 mM Mg^{2+}, 1 mM Mn^{2+}, 9 μM [1-^{14}C]IPP (2.07 GBq/mmol), 9 μM dimethylallyl diphosphate.

as substrate and turned out to be very well accepted by the monoterpene synthesizing enzymes.

References

Beyer, P., Kreuz, K., Kleinig, H. (1980) β-carotene synthesis in isolated chromoplasts from Narcissus pseudonarcissus. Planta 150, 435-438.

Boland, W., Ney, P., Jaenicke, L., Gassmann, G. (1984) A "closed-loop-stripping" technique as a versatile tool for metabolic studies of volatiles, in Analysis of volatiles (Schreiber, P., ed.) pp 371-380, Walter de Gruyter, Berlin.

Gleizes, M., Pauly, G., Carde, J.P., Marpeau, A., Bernard-Dagan, M. (1983) Monoterpene hydrocarbon biosynthesis by isolated leucoplasts of Citrofortunella mitis. Planta 159, 373-381.

Mettal, U., Boland, W., Beyer, P., Kleinig, H. (1988) Biosynthesis of monoterpene hydrocarbons by isolated chromoplasts from daffodil flowers. Eur. J. Biochem. 170, 613-616.

Mühlbacher, M., Poulter, C.D. (1985) Isopentenyl diphosphate: dimethylallyl diphosphate isomerase. Irreversible inhibition of the enzyme by active-site-directed covalent attachment. J. Am. Chem. Soc. 107, 8307-8308.

BIOLOGICAL ROLE OF PLANT LIPIDS
P.A. BIACS, K. GRUIZ, T. KREMMER (eds)
Akadémiai Kiadó, Budapest and Plenum Publishing
Corporation, New York and London, 1989

ISOLATION OF CELL COMPARTMENTS INVOLVED IN THE BIOSYNTHESIS OF LOWER TERPENOIDS OF CITROFORTUNELLA MITIS FRUITS

L. Belingheri, M. Gleizes, G. Pauly, J.P. Carde and A. Marpeau

Laboratoire de Physiologie Cellulaire Végétal (U.A. CNRS 568), Université Bordeaux I,
Avenue des Facultés, 33405 Talence Cédex, France

INTRODUCTION

The biosynthesis of mono- and sesquiterpene hydrocarbons takes place in different cell compartments of plant producing essential oils (Croteau and Loomis, 1972 ; Bernard-Dagan et al., 1982). Gleizes et al.(1983) have isolated the organelles involved in monoterpene hydrocarbon synthesis from the epicarp of young fruits of calamondin (*Citrofortunella mitis*). These colorless organelles devoid of plastoribosomes (Carde, 1984) are leucoplasts characterized by a dense stroma surrounded by a double membrane. Their inner membrane system is a peripheral reticulum connected with the envelope. In the presence of [1-^{14}C] isopentenyl pyrophosphate (IPP), leucoplasts elaborate limonene, the main monoterpene hydrocarbon, with some traces of α- and β-pinene. They are also able to realize the isoprenoid chain elongation with formation of C_{15} and C_{20} pyrophosphates which, however, do not lead to the corresponding terpenoid derivatives (Gleizes et al., 1987). Nevertheless sesquiterpene hydrocarbons are present in small amounts in the epicarp of fruits and it was established recently (Belingheri et al., 1988) that they are synthesized by endoplasmic reticulum. In the present report this result is completed by electron microscope investigations after zinc iodide-osmium tetroxide impregnation made *in situ* and on membranes isolated from the epicarp of calamondin.

MATERIAL AND METHODS

The young fruits used in our experiments were collected on plants growing in the greenhouse of the University Bordeaux I.

All the procedures of membrane isolation, incubation, terpene extraction, enzymic reactions and analytical methods are described in our previous paper (Belingheri et al., 1988).

The electron microscopy study was carried out according to Carde (1987) for sequential fixation with glutaraldehyde, osmium tetroxide and tannic acid and osmium-iodide impregnation.

RESULTS AND DISCUSSION

When crude pellets obtained from 4,500 g, 12,000g or 100,000g centrifugations are incubated with [1-^{14}C] IPP, sesquiterpene hydrocarbons are synthesized with a maximum of activity in the cell compartments sedimenting at 100,000g (Table I). A complete separation of cell compartments involved respectively in sesquiterpene and monoterpene biosynthesis is obtained by the use of partition in aqueous two phase polymer systems with dextran T500 and polyethylene glycol (PEG) 4000. As it is shown in Table II, the membranes partitioning in the PEG-rich upper phase elaborate only sesquiterpene hydrocarbons and the synthesis of monoterpenes is associated with organelle vesicles of the dextran-rich lower phase in which is also recovered part of the membranes associated with sesquiterpene biosynthesis. The PEG-rich upper phase synthesizing the C_{15} hydrocarbons is pelleted down at 100,000g and the obtained membranes are separated on sucrose gradient. Four bands are recovered called A,B,C,D from the top to the base of the gradient respectively. The four membrane fractions are incubated with [1-^{14}C] IPP and assayed for 5 enzymic activities. Figure 1 shows that the membranes sedimenting in the upper band A exhibit the greatest activity in the synthesis of sesquiterpene hydrocarbons particularly β-selinene and germacrene D. Lower amounts of radioactivity are also present in B and C whereas only traces of synthesis of C_{15} hydrocarbons are found in D.

The distribution of enzyme activities in the different membrane systems (Fig. 1) shows a high activity of antimycin-A insensitive NADH cytochrome c reductase, cinnamic acid 4 - hydroxylase and phosphorylcholine-glyceride transferase, all markers of endoplasmic reticulum (ER), in the upper band A and, to a lesser extent, in B and C. The enzymes characteristic of plastid envelopes and plasma membranes are poorly represented in band A, they are active mainly in band C for galactosyl transferase and in band D for glucan synthetase II. No markers of mitochondria are represented here because this cell compartment was found exclusively in the dextran-rich lower phase of the partition.

Table 1 - Incorporation of [1-^{14}C] isopentenyl pyrophosphate (IPP) into monoterpene and sesquiterpene hydrocarbons by the 4,500 g, 12,000 g and 100,000 g pellets.

Pellets	IPP incorporated into			
	C_{10} Hc (nmol)	%	C_{10} Hc (nmol)	%
4,500	0.44	94	0.03	6
12,000	0.17	63	0.07	37
100,000	0.07	28	0.18	72

C_{10} Hc : monoterpene hydrocarbons ; C_{15} Hc : sesquiterpene hydrocarbons

Table 2 - Incorporation of [1-^{14}C] isopentenyl pyrophosphate (IPP) into terpene hydrocarbons after incubation of the two pellets obtained after two-polymer phase partition of 100,000 g pellet.

	IPP incorporated into			
	C_{10} Hc (nmol)	%	C_{10} Hc (nmol)	%
Upper phase	-	-	0.15	100
Lower phase	0.02	64	0.01	36

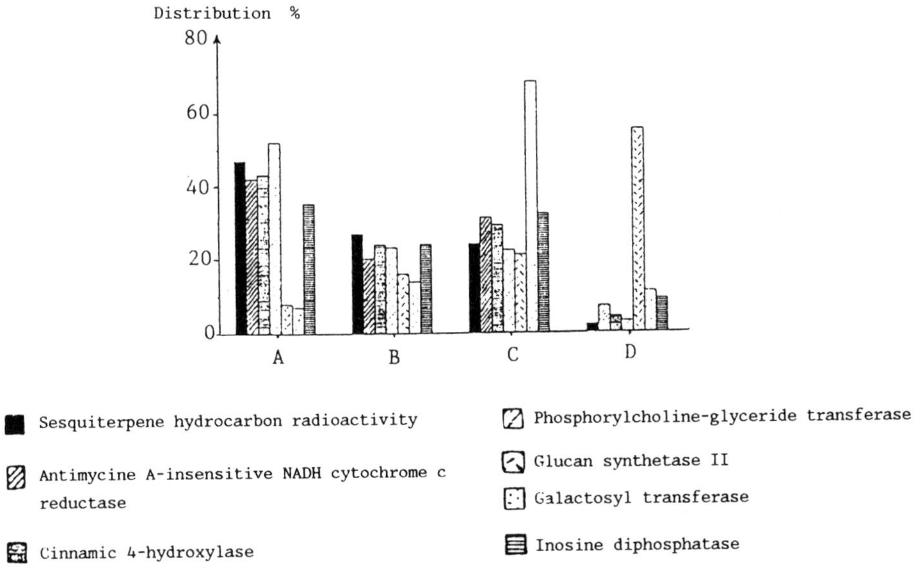

Fig. 1 - Distribution of enzymic marker specific activities on a discontinuous sucrose density gradient after centrifugation of the PEG upper phase obtained after phase partition of the microsomal pellet.

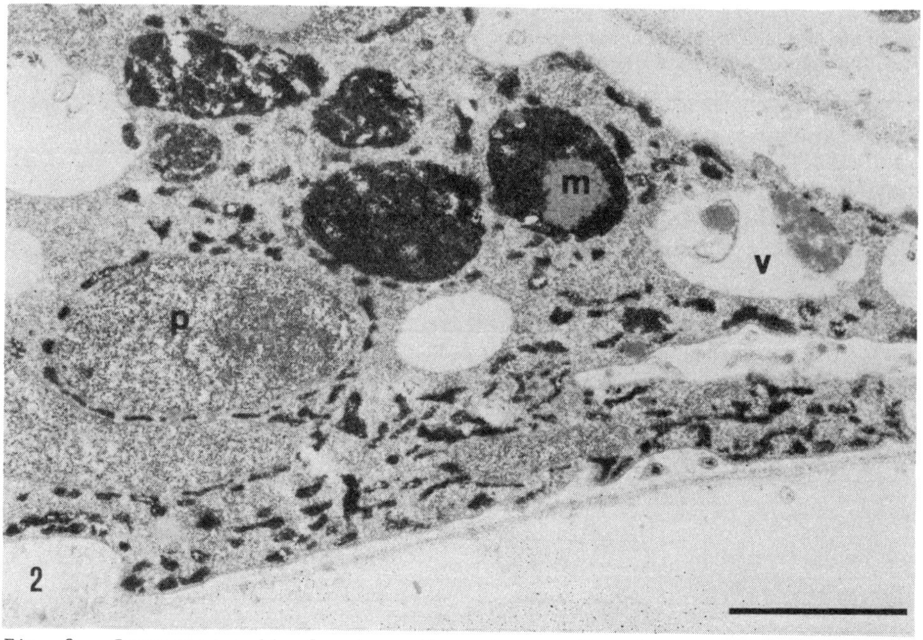

Fig. 2 - Secretory cell of calamondin epicarp after impregnation with zinc iodide - osmium tetroxide. Bar = 1 μm.

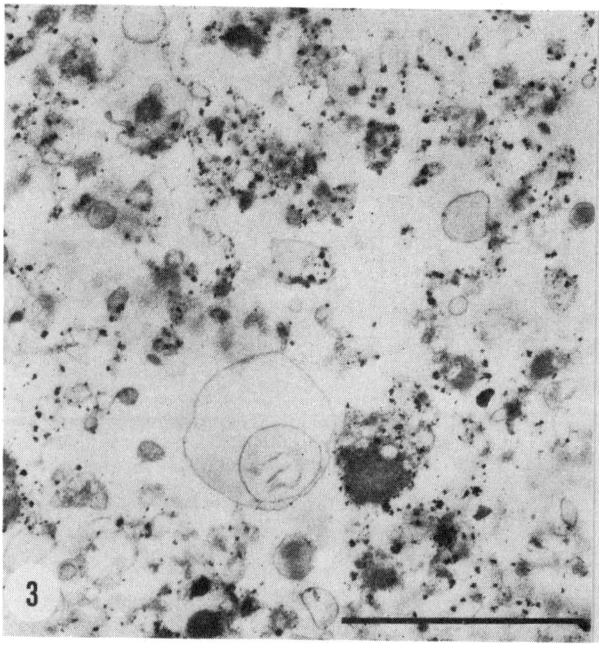

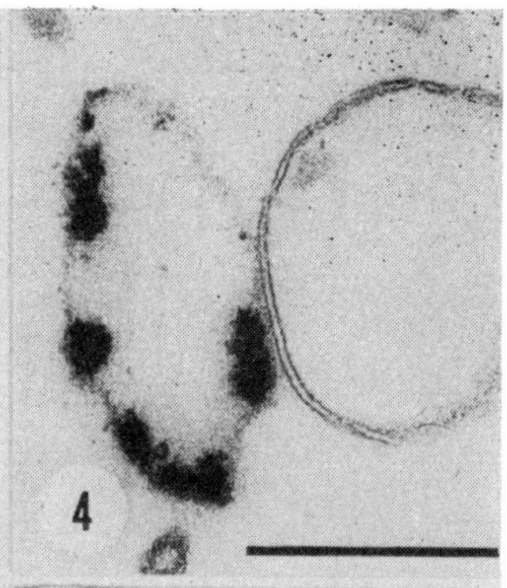

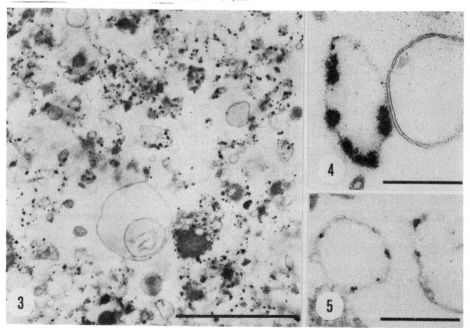

Fig. 3 - Fraction A treated with zinc iodide-osmium tetroxide. Bar = 1 μm.
Fig. 4 and 5 - Details of Fig. 3. Only thin-walled vesicles are reactive and bring heavy (Fig. 4) or light (Fig. 5) precipitates. Bar = 0.2 μm.

The inosine diphosphatase activity is high in bands A, B and C. However its activity is nearly the same in the 3 bands whereas the radioactivity in sesquiterpene decreases from A to C.

The membranes recovered in the four bands obtained from the sucrose gradient were also examined by electron microscopy after tannic acid postfixation or impregnation with zinc iodide osmium tetroxide mixtures. A secretory cell impregnated with the zinc-osmium mixture is shown on Figure 2. The staining concerns mainly the nuclear envelope, ER membranes, Golgi apparatus and mitochondria. On the contrary, the plasma membrane, the tonoplast and the plastid envelopes are not contrasted. On the basis of membrane thickness, it was concluded that the membranes collected in band A are enriched in ER with a contamination by tonoplast. Band B is also rich in ER and more contaminated by tonoplast whereas in band C, ER vesicles are still present with thylacoïds and various membranes.

Band D presents a greater homogeneity with small vesicles originating in the fragmentation of the plasma membrane, an observation which is in good agreement with the high activity of glucan synthetase II. The observation of band A after treatment by zinc-osmium shows (Fig. 3) that the dense precipitates are not scattered randomly but specifically associated to distinct membrane profiles. Only thin-walled vesicles are reactive and bring heavy (Fig. 4) or light (Fig. 5) precipitates. It is likely that the same membranes are contrasted *in situ* and after fragmentation and that ER vesicles are the main membranes of the fraction A in good agreement with enzyme activities shown in Fig. 1. Both the results of electron microscopy and enzyme markers indicate that the endoplasmic reticulum is involved in the sesquiterpene biosynthesis, in good agreement with the data previously obtained on pine seedling leaves (Gleizes et al., 1980). This cell compartment contains all the enzyme systems able to elongate the isoprenoid chain from isopentenyl pyrophosphate to geranyland farnesyl pyrophosphate (prenyltransferase) and to cyclise the FPP to sesquiterpenes such as β-selinene and germacrene D.

REFERENCES

BELINGHERI L., PAULY G., GLEIZES M., and MARPEAU A. (1988) - Isolation by an aqueous two-polymer phase system and identification of endomembranes from *Citrofortunella mitis* fruits for sesquiterpene hydrocarbon synthesis. J. Plant Physiol., 132 , 80-85.

BERNARD-DAGAN C., PAULY G., MARPEAU A., GLEIZES M., CARDE J.P. and BARADAT P. (1982) - Control and compartmentation of terpene biosynthesis in leaves of *Pinus pinaster*. Physiol. Vég., 20, 775- 795.

CARDE J.P. (1984) - Leucoplasts : a distinct kind of organelles lacking typical 70 S ribosomes and free thylakoids. Eur. J. Cell Biol., 34, 18-26.

CARDE J.P. (1987) - Electron microscopy of plant cell membranes. In : L. Packer and R. Douce, eds., Plant cell membranes, Methods in Enzymology, 148, Academic Press, New-York, 599-622.

CROTEAU R. and LOOMIS W.D. (1972) - Biosynthesis of mono- and sesquiterpenes in peppermint from mevalonate-2-^{14}C. Phytochemistry, 11, 1055-1066.

GLEIZES M., CARDE J.P., PAULY G. and BERNARD-DAGAN C. (1980) - *In vivo* formation of sesquiterpene hydrocarbons in the endoplasmic reticulum. Plant Sci. Lett., 20, 79-80.

GLEIZES M., PAULY G., CARDE J.P., MARPEAU A. and BERNARD-DAGAN C. - 1983-- Monoterpene hydrocarbon biosynthesis by isolated leucoplasts of *Citrofotunella mitis*. Planta, 159, 373-381.

GLEIZES M., CAMARA B. and WALTER J. (1987) - Some characteristics of terpenoid biosynthesis by leucoplasts of *Citrofortunella mitis*. Planta, 170, 138-140.

BIOLOGICAL ROLE OF PLANT LIPIDS
P.A. BIACS, K. GRUIZ, T. KREMMER (eds)
Akadémiai Kiadó, Budapest and Plenum Publishing
Corporation, New York and London, 1989

PARTIAL PURIFICATION OF THE ENZYME SYSTEMS INVOLVED IN MONO-, SESQUI- AND DITERPENE BIOSYNTHESIS

J. Walter, L. Belingheri, A. Cartayrade, G. Pauly and M. Gleizes

Laboratoire de Physiologie Cellulaire Végétal (U.A. CNRS 568), Université Bordeaux I, Avenue des Facultés, 33405 Talence Cédex, France

INTRODUCTION

Mono-, sesqui- and diterpenes are 10-, 15- and 20- carbon compounds derived biogenetically from an acyclic precursor containing 2, 3, 4 isoprenoid units linked together. It was already shown that, in the epicarp of young fruits of calamondin (*Citrofortunella mitis*), the site of biosynthesis of C_{10} hydrocarbons is located in leucoplasts present in epithelial cells surrounding the secretory cavities (Gleizes *et al.*, 1983), whereas the C_{15} hydrocarbon biosynthesis is bound to endoplasmic reticulum (Belingheri *et al.*, 1988). Nevertheless it appeared that all the enzymes implied in these biosynthetic processes (prenyltransferases and cyclases) are partly solubilized during the cell fractionation. They are presumably, either peripheral membrane proteins or, as recently demonstrated (Mettal *et al.*, 1988), located in the plastid stroma. In the present report, we describe the data from our first experiments of purification of the enzymes extracted from 100,000 g supernatants and involved in the biosynthesis of mono-, sesqui- and diterpene hydrocarbons present in the essential oils of calamondin fruits and maritime pine (*Pinus pinaster* Ait.) needles.

MATERIAL AND METHODS

The young fruits of calamondin were collected on plants growing in the greenhouse of University Bordeaux I. The maritime pine needles were harvested during their growing period (June-July) at the Station d'Amélioration des Arbres Forestiers INRA, Pierroton -Cestas, France.

The steps of enzyme purification were carried out first by polyethylene glycol 6000 (PEG) concentrations of the 100,000 g supernatants according to Chayet *et al.*, (1977). The PEG precipitates (30-40%) were separated on DEAE-Sephacel. This separation was followed by a chromatography on an affinity column using Sepharose-Aminophenethyl pyrophosphate (Dogbo and Camara, 1987). Electrophoresis on polyacrylamide gel was adapted from the data of Davis (1964) and SDS gel electrophoresis was carried out according to Laemmli (1970).

RESULTS AND DISCUSSION

The PEG precipitate (30-40%) obtained from the 100,000 g supernatant of the epicarp of immature fruits of calamondin was submitted to chromatography on a column of DEAE-Sephacel and developed with a linear gradient of NaCl (0.1-0.2 M). As shown in fig. 1, the limonene and selinene cyclase activities were resolved. The electrophoresis of the fractions involved in mono- and sesquiterpene biosynthesis showed the purification obtained after the DEAE-Sephacel column. Incubations of small gel pieces with [^{3}H]-GPP and [^{3}H]-FPP led to the localization of the cyclases involved in both limonene and selinene synthesis and indicated that the two cyclases are well separated on the gel. In order to improve the purification of these cyclases, the active fractions eluted from the DEAE-Sephacel column were submitted to affinity chromatography using Sepharose APP. This procedure contributed to a 90-fold and 70-fold purification for limonene and selinene cyclases respectively. An isoelectric point of 5.4 ± 0.2 was found for limonene-cyclase whereas β-selinene-cyclase presented a value of 6 ± 0.2. After gel permeation with Sephadex G 200, the same molecular weight of 67 ± 5 kD was attributed to the two cyclases. In addition, prenyltransferases obtained from a DEAE-Sephacel column developed with a step gradient (0.1, 0.2, 0.3 M NaCl) were also isolated on the same affinity column. These prenyltransferases were 160 fold purified. Their molecular weight was about 70 kD and they exhibited an isoelectric point of 5.2.

In pine needles, the enzymes present in 100,000 g supernatants were able to synthesize sesquiterpene and diterpene hydrocarbons from [1-^{14}C] IPP whereas the synthesis of monoterpene hydrocarbons was never observed. However, it appeared that the enzymes, extracted from needles originating from Landes area, were able to convert IPP into sesquiterpene hydrocarbons with small amounts of diterpenes. In contrast, the enzymes from needles of maritime pines from Corsican origin used IPP to form mainly diterpene hydrocarbons with small amounts of C_{15} compounds. The diluted proteins extracted from 100,000 g supernatants obtained from needles of Corsican pines, were also precipitated with PEG 6000 (30-40%). After purification by ion exchange column chromatography developed with a NaCl step gradient (0.1, 0.2, 0.3 M), an enzyme system, able to carry out all the steps leading to C_{15} and C_{20} hydrocarbons from [1-^{14}C] IPP, was obtained at 0.2 M NaCl (fig. 2). [^{3}H]-GGPP was also used by this enzyme complex and was cyclised into diterpene hydrocarbons, mainly 7,13-abietadiene (Walter, unpublished data). This complex was

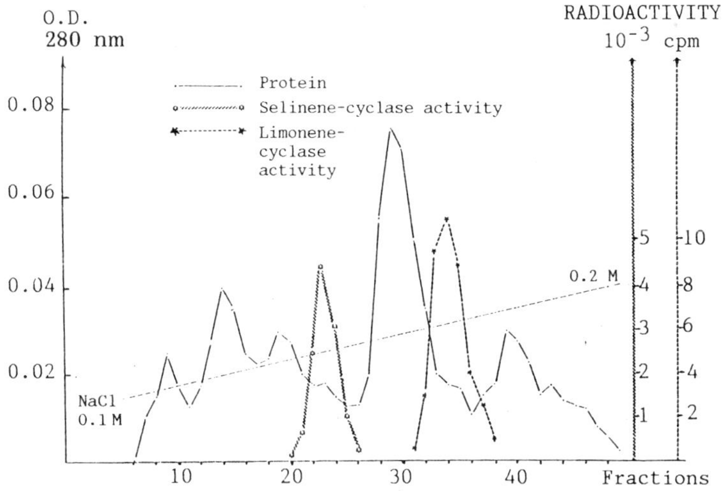

Fig. 1 - Separation on DEAE-Sephacel chromatography column of cyclase activities involved in biosynthesis of mono- and sesquiterpene hydrocarbons, with a linear gradient of NaCl (0.1 - 0.2 M). Limonene-cyclase and selinene-cyclase were detected after incubation with [^{3}H] GPP and [^{3}H] FPP respectively.

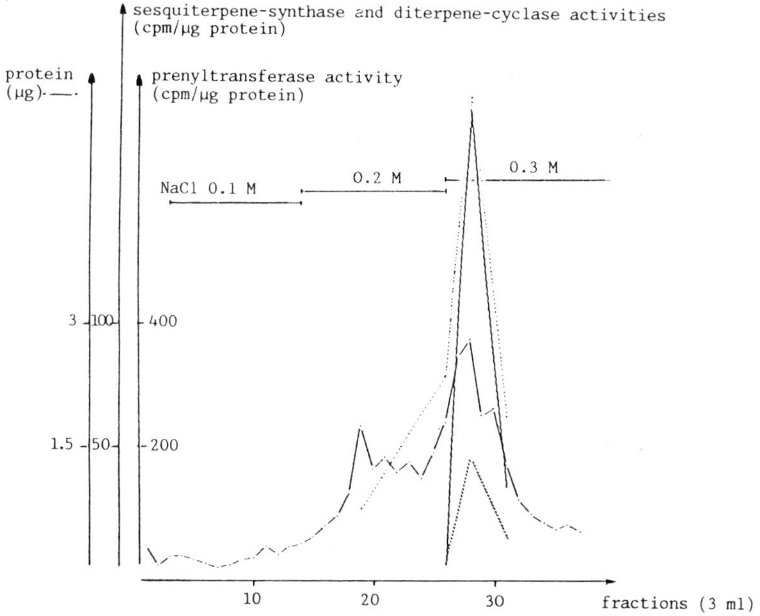

Fig. 2 - Ion-exchange chromatography column of the polyethyleneglycol 6000 fraction on DEAE-Sephacel. Fractions (3 ml) were collected and incubated with [1-^{14}C] IPP. For all the active fractions, prenyltransferase, sesquiterpene-synthase and diterpene-cyclase were eluted as an enzyme complex.

sesquiterpene-synthase activity. diterpene-cyclase activity.

prenyltransferase activity.

submitted to affinity chromatography. Prenyltransferases involved in the elongation of the isoprenoid chain from IPP to C_{10}, C_{15}, C_{20} phosphorylated intermediates were obtained with a 70-fold purification. A SDS-electrophoresis of these purified prenyltransferases was followed by Western-blotting using anti-GGPP-synthase antibodies from chromoplasts of *Capsicum* (generous gift of Dr. B. Camara). These antibodies cross-reacted with GGPP-synthase from pine needles and a molecular weight of 37 ± 2 kD was attributed to a monomer of this protein.

At this time, we are improving the purification of all the enzymes involved in the biosynthesis of the mono-, sesqui- and diterpene hydrocarbons present in the essential oils of calamondin fruits and pine needles. Antibodies against the purified enzymes will be further used to localize the enzymes at the cell level by immunocytochemistry.

REFERENCES

BELINGHERI L., PAULY G., GLEIZES M. and MARPEAU A. (1988) - Isolation by an aqueous two-polymer phase system and identification of endomembranes from *Citrofortunella mitis* fruits for sesquiterpene hydrocarbon synthesis. J. Plant Physiol., 132 : 80-85.

CHAYET L., ROJAS M.C., CARDEMIL E., JABALQUINTO A.M., VICUNA J.R. and CORI O. (1977) - Biosynthesis of monoterpene hydrocarbons from [1-^{3}H] neryl pyrophosphate and geranyl pyrophosphate by soluble enzymes from *Citrus limonum*. Arch. Biochem. Biophys., 180 : 318-327.

DAVIS B.J. (1964) - Disc Electrophoresis. II. Method and application to human serum proteins. Ann. N.Y. Acad. Sci., 121 : 404-427.

DOGBO O. and CAMARA B. (1987) - Purification of isopentenyl pyrophosphate isomerase and geranylgeranyl pyrophosphate synthase from *Capsicum* chromoplasts by affinity chromatography. Biochim. Biophys. Acta, Lipids and lipid metabolism, 920 : 131-139.

GLEIZES M., PAULY G., CARDE J.P., MARPEAU A. and BERNARD-DAGAN C. (1983) - Monoterpene hydrocarbon biosynthesis by isolated leucoplasts of *Citrofortunella mitis*. Planta, 159 : 373-381.

LAEMMLI U.K. (1970) - Cleavage of structural proteins during the assembly of the head of Bacteriophage T_4. Nature (London), 227 : 680-685.

METTAL U., BOLAND W., BEYER P. and KLEINIG H. (1988) - Biosynthesis of monoterpene hydrocarbons by isolated chromoplasts from daffodil flowers. Eur. J. Biochem., 17 : 613-616.

BIOLOGICAL ROLE OF PLANT LIPIDS
P.A. BIACS, K. GRUIZ, T. KREMMER (eds)
Akadémiai Kiadó, Budapest and Plenum Publishing
Corporation, New York and London, 1989

ON THE ORIGIN OF ISOPRENOID INTERMEDIATES FOR THE SYNTHESIS OF PLASTOQUINONE-9 AND ß-CAROTENE IN DEVELOPING CHLOROPLASTS

G. Schultz and D. Schulze-Siebert *

Botanisches Institut der Tierärztlichen Hochschule, Bünteweg 17d, 3000 Hannover 71, FRG

INTRODUCTION

The existence of a chloroplast mevalonate (Mev) pathway, which synthesizes isopentenyl diphosphate (IPP) to form chloroplast isoprenoids (e.g. carotenes, plastoqinone-9 (PQ), tocopherols and the phytyl moiety of chlorophyll), is still under debate /3/. In Goodwin's laboratory /5/ it was shown that 5 to 6-day germinated cereal seedlings incorporate $^{14}CO_2$ into chloroplast isoprenoids but only less into sterols. As known the latter are formed in the cytosol at the endoplasmatic reticulum (ER). Vice versa, when labeled Mev was applied it was preferentially found in sterols but not in chloroplast isoprenoids. This gave rise to the idea of two independent compartments in which IPP will be formed. It was further substantiated by later findings (e.g. carotene and PQ formation from CO_2 by isolated chloroplasts of fast growing spinach /2/, enzymes for IPP synthesis in chloroplasts of pea /10/, Nepeta and spinach /1/, and acetoacetyl-CoA formation from acetyl-CoA in spinach /6/). The origin of IPP from leucine could be excluded /6/. On the other hand, Kreuz and Kleining /4/ found that chloroplasts from market spinach synthesize isoprenoids from added IPP but not from CO_2. Therefore, the authors suggest that IPP is only formed in the cytosol at the ER and from there delivered to isoprenoid synthesizing organelles.

---

*Supported by the Deutsche Forschungsgemeinsch...

This striking differences in findings on the same object led to the suggestion that the chloroplast isoprenoid metabolism alters fluently from the developing chloroplast at the basal growing region of the leaf to the mature chloroplast at the tip of the leaf. The latter corresponds to the stage of the differentiated fully photosynthetically active chloroplast in expanded leaves. To study the alterations in chloroplast isoprenoid metabolism, spinach was useful to investigate isolated chloroplasts, showing highest intactness, whereas leaves of barley seedlings were suited best for following the fluent metabolic changes from the developing to the mature chloroplast.

METHODS

$NaH^{14}CO_3$, /U-^{14}C/3-D-phosphoglycerate (3-PGA) (4.96 GBq $mmol^{-1}$) and /2-^{14}C/D,L-Mev, DBED salt (0.981 GBq $mmol^{-1}$) were from Amersham Buchler, Braunschweig, FRG, /4-^{14}C/IPP (1.87 GBq $mmol^{-1}$) was from NEN, Dreieich, FRG. Chloroplasts from spinach (*Spinacia oleracea*) var. Butterfly were isolated as described in /9/. The reaction mixture contained in mM: $NaHCO_3$ 5, sorbitol 330, Hepes-KOH (pH 7.6) 50, $MgCl_2$ 2, $MnCl_2$ 1, NaCl 20, $NaNO_3$ 2, KH_2PO_4 0.5, ascorbate 4, EDTA 2, glutamate 0.05, oxaloacetate 1. Chloroplast suspensions were illuminated for 30 min (1 kW m^{-2}, 20 $\pm$ 2°C). In experiments for chloroplast development, first foliage leaf from 8 to 10-day germinated barley seedlings (*Hordeum vulgare*) var. Gerbel was cut into four pieces. Each of the regions was inserted with the base into a microvessel containing the labeled substance dissolved in 25 µl tap water and illuminated for 3 hrs (Osram HQI-T/400 W/DH, 10^3 lx, 24+2°C). After uptake of the substance, tap water was provided ad libidum (2-3 x 50 µl). β-Carotene, PQ, sterols and fatty acids as well as marker enzymes to check the purity of chloroplasts were assayed as described in /7-9/.

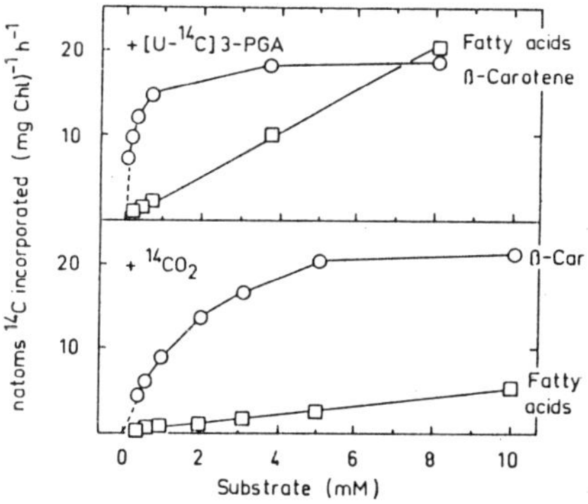

Fig. 1. ^{14}C-Incorporation from /U-^{14}C/3-PGA or $NaH^{14}CO_3$ (=CO_2) into β-carotene and fatty acids by purified intact chloroplasts from primary leaves of spinach (CO_2 fixation rate: 98 μmol (mg chlorophyll (Chl))$^{-1}$ h^{-1}). Enzyme activities of nonreversible NADP-glyceraldehyde 3-phosphate dehydrogenase and NADP-isocitrate dehydrogenase as marker for the cytosol, hydroxypyruvate reductase for peroxisomes and cytochrom c reductase for mitochondria were 0.3 to 2% compared to those of the total leaf extract.

RESULTS

β-Carotene synthesis from $NaH^{14}CO_3$ or /U-^{14}C/3-PGA by chloroplasts from primary leaves of spinach. Purified chloroplasts isolated from primary leaves of 3-4 weeks old plants are higly capable of incorporating $^{14}CO_2$ into β-carotene. Fig. 1 shows that β-carotene is more effectively formed from $^{14}CO_2$ as well as from /U-^{14}C/3-PGA than fatty acids. At this stage of chloroplast development, thus, enzymes of carotene synthesis compared to those of fatty acid synthesis possess a higher affinity to photosynthetically fixed carbon.

^{14}C-Incorporation of $NaH^{14}CO_3$ or /2-^{14}C/D,L-Mev into different regions of the first foliage leaf of barley seedlings. The basal region of barley leaves, possessing developing chloroplasts, showed a different behavior than the tip region of the leaves, having mature chloroplasts (Fig. 2). To compare the incorporation rates, the leaf fresh weight was suited best as reference base; the chlorophyll content was less useful because

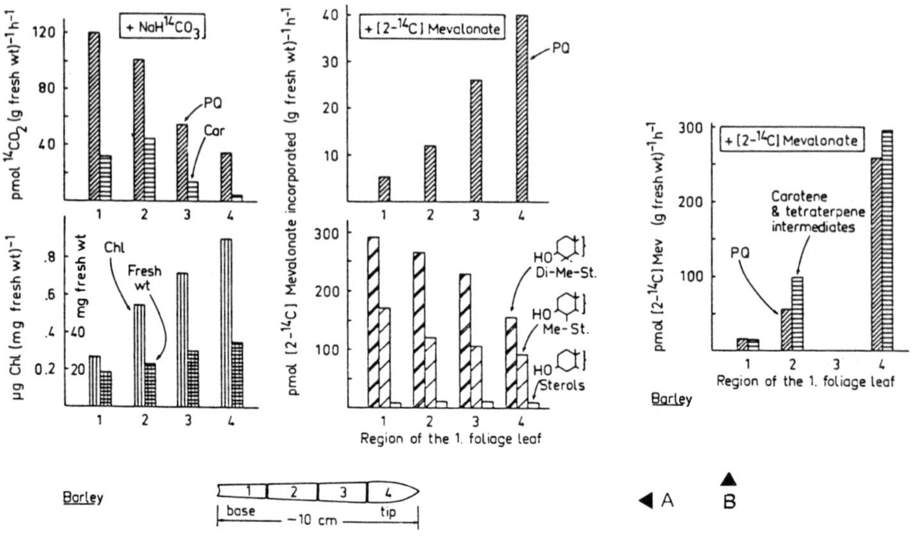

Fig. 2. ^{14}C-Incorporation of $NaH^{14}CO_3$ or /2-^{14}C/D,L-Mev into different parts cut from the first foliage leaf of barley seedlings. A - 40 Leaves of seedlings were used. Each part of the leaf (1, base; 2; 3; 4, tip) was supplied with 612 Bq /2-^{14}C/D,L-Mev or 10 KBq $NaH^{14}CO_3$ (0.337 GBq $mmol^{-1}$) and iluminated for 3 hrs. B - The same as in A but parts of 12 leaves were used (2.57 KBq /2-^{14}C/D,L-Mev per part).

it increases with chloroplast development (Fig. 2A). Highest rates of ^{14}C-incorporation from $NaH^{14}CO_3$ into PQ were found in the basal region of the leaves. The rates more or less linearly decrease towards the tip of the leaves. A similar behavior was found for β-carotene. Vice versa, the rates of ^{14}C-incorporation from labeled Mev into PQ increases from the leaf base, showing only low rates, to the tip of leaves by the factor 8 to 10. As shown in Fig. 2B, the same was found for the incorporation into β-carotene including tetraterpenoid intermediates. To encounter the objection that labeled Mev is not transferred across the plasma membrane of the cells in the basal region, the incorporation into sterols was assayed. 4-Dimethylsterols, 4-methylsterols and sterols are formed in the basal region of the leaves at the highest rates more or less decreasing towards the tip of leaves (Fig. 2A). Thus, no barrier for

Mev exists at the plasma membrane in the investigated stages of leaf development.

Formation of β-carotene from /4-^{14}C/IPP by isolated spinach chloroplasts. The uptake of IPP by isolated chloroplasts, extensively studied by Kleinig's group /4/, could be shown for all differentiated leaves of spinach. Illuminated chloroplasts incorporate labeled IPP not only into the phytol moiety of chlorophylls but also into β-carotene at considerable rates (Fig. 3).

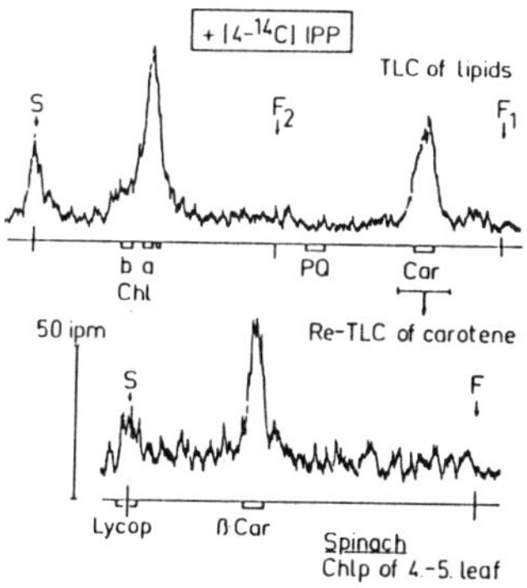

Fig. 3. Formation of β-carotene from /4-^{14}C/IPP by isolated spinach chloroplasts in the light. 24.4 KBq /4-^{14}C/IPP was added to 2 ml chloroplast suspension (100 ug Chl ml^{-1}) and illuminated for 40 min. The thinlayer chromatography (TLC) system for lipids was silicagel - light petrol/ethylether 20:1 (S, origin; F_1, front) and repeated TLC using petrol (b.p. 100°) /isopropanol 10:1 (with F_2 as front). Re-TLC of carotene was performed on silicagel - hexane/5% benezene.

DISCUSSION

The results obtained allow the interpretation that developing chloroplasts at the leaf base are capable of forming isoprenoids from photosynthetically fixed CO_2 via plastidic Mev pathway. Intermediates of Mev metabolism are not uptaken across the envelope membranes of developing chloroplasts. This situation significantly alters in the course of chloroplast matura-

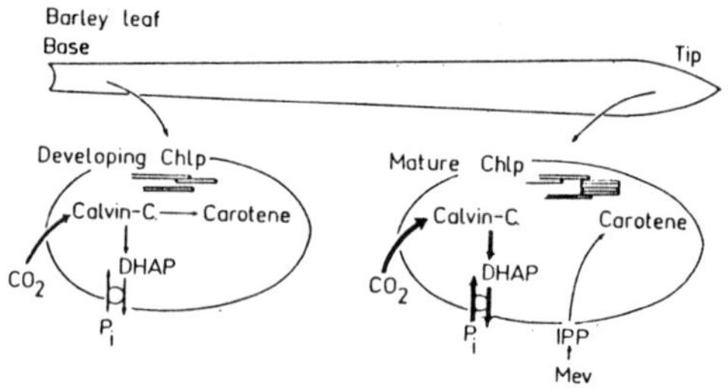

Fig. 4. Scheme to illustrate the changes of isoprenoid metabolism during maturation of chloroplasts in different regions of the leaf. (left) The formation of isoprenoids (e.g. carotene) from photosynthetically fixed CO_2 via plastidic IPP synthesis in developing chloroplasts and the basal region of the leaf. (right) Decrease of the plastidic activity for IPP synthesis and import of IPP from the cytosol into fully photosynthetically active, mature chloroplast at the tip of the leaf. DHAP, dihydroxyacetonephosphate; P_i, PO_4

tion. The mature chloroplasts at the tip of the leaf, having full photosynthetic activity, may possess only low capacities for IPP synthesis. However, IPP synthesized at the ER is effectively transferred from the cytosolic compartment across the envelope membrane into the chloroplast. Remarkably, the transfer occurrs in the form of the diphosphate. From the view of ontogeny this changes in isoprenoid metabolism may be interpreted as transition from the stage of the metabolically (not genetically) autonomic, developing chloroplast to that of the mature chloroplast, fully active in photosynthesis, but in which a division of labor takes place.

REFERENCES

1. Arebalo R.E. Mitchell E.D. (1984) Phytochemistry *23*, 13-18.
2. Bickel H. Schultz G. (1976) Phytochemistry *15*, 1253-1255.
3. Gray J.C. (1987) Adv. Bot. Res. *14*, 25-91.
4. Kreuz K. Kleinig H. (1984) Eur. J. Biochem. *141*, 531-535.
5. Rogers L.J., Shah S.P.J. Goodwin T.W. (1966) Biochem. J. *99*, 381-388.

6. Schulze-Siebert D. (1987) Zur Autonomie der Isoprenoidbiosynthese der Chloroplasten Höherrer Pflanzen. Thesis, University of Hannover (FRG).
7. Schulze-Siebert D. Schultz G. (1987) Plant Physiol. *84*, 1233-1237.
8. Schulze-Siebert D. Schultz G. (1987) Plant Physiol. Biochem. *25*, 145-153.
9. Schulze-Siebert D., Heintze A. Schultz G. (1987) Z. Naturforsch. *42c*, 570-580.
10. Wong R.J., Mc. Cormack D.K. Russel D.W. (1982) Arch. Biochem. Biophys. *216*, 631-638.

BIOLOGICAL ROLE OF PLANT LIPIDS
P.A. BIACS, K. GRUIZ, T. KREMMER (eds)
Akadémiai Kiadó, Budapest and Plenum Publishing
Corporation, New York and London, 1989

EFFECTS OF 9β,19-CYCLOPROPYLSTEROLS ON THE STRUCTURE AND FUNCTION OF THE PLASMA MEMBRANE FROM MAIZE ROOTS

M.A. Hartmann, A. Grandmougin, P. Ullmann, P. Bouvier-Navé and P. Benveniste

Laboratoire de Biochemie Végétale (UA CNRS 1182), Université Louis Pasteur, Institut de Botanique, 28 Rue Goethe, 67083 Strasbourg, France

INTRODUCTION

In most higher plants, sterols are present as a mixture of Δ^5-sterols. Sitosterol, stigmasterol and 24-methylcholesterol are usually cited as the typical plant sterols (Fig.1). These sterols are concentrated mainly in the plasma membrane (PM) (1). Using sterol biosynthesis inhibitors, we have shown that it is possible to obtain plant cell suspensions or whole plants with a completely modified sterol profile (2,3). Such plants constitute most suitable material for studying structural and functional roles of sterols, which are still largely unknown.

In the present paper, we report qualitative and quantitative effects of fenpropimorph, a systemic morpholine fungicide, on the lipid composition (free sterols, phospholipids (PL) and fatty acids (FA)) of a PM-rich fraction from maize roots. The consequences of such a treatment on the properties of the PM-bound vanadate-sensitive H^+-ATPase were investigated.

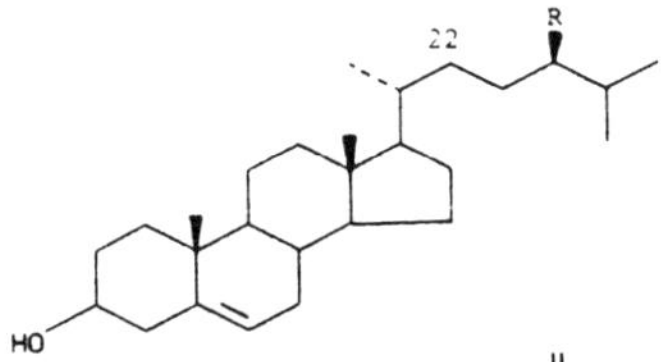

R, CH_3 : (24R)-methylcholesterol
R, C_2H_5 : sitosterol
R, C_2H_5 and Δ^{22} : stigmasterol

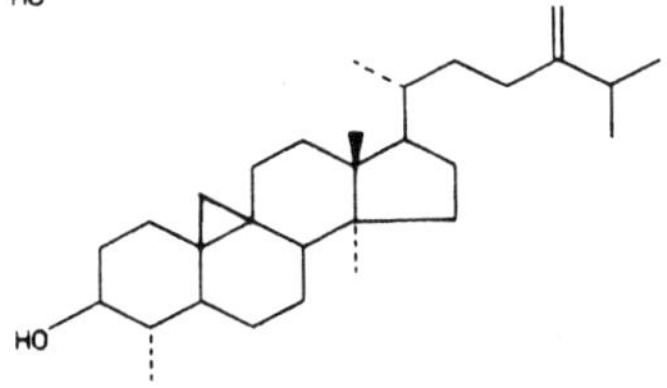

Cycloeucalenol

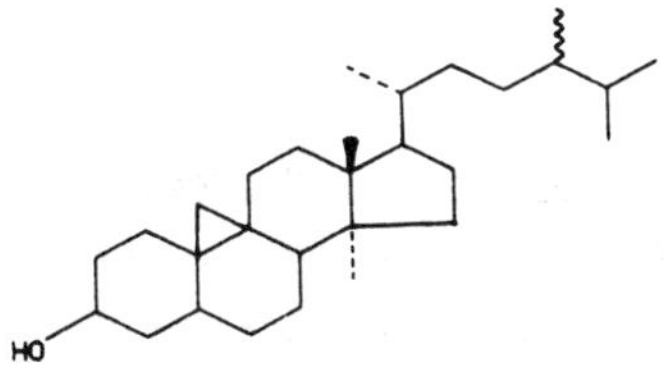

24-methylpollinastanol

Fig.1. Structures of plant sterols cited in this study.

MATERIAL AND METHODS

Plant material. Maize (*Zea mays* L., INRA cv LG 11) seeds were germinated and grown in moist vermiculite in the dark at 25°C. The vermiculite was daily soaked with 0.5 litre with or without (R,S)-fenpropimorph (20 mg/l). Roots were excised after 7 days.
Isolation of PM. PM-rich preparations from control and treated roots were isolated and characterized as described elsewhere (4,5).
Lipid analysis. Sterols, PL and total FA were determined according to Refs (1,5).

RESULTS AND DISCUSSION

Lipid analysis

Sterols. As shown in Table 1, major sterols in the control PM are the typical Δ^5-sterols : sitosterol, stigmasterol and 24-methylcholesterol. In the fenpropimorph-treated PM, the Δ^5-sterols are present only in low amounts and are replaced by 9β,19-cyclopropylsterols such as cycloeucalenol, 24-dihydrocycloeucalenol and 24-methylpollinastanol, indicating that the cycloeucalenol-obtusifoliol isomerase is the major target of the fungicide in maize roots (6). These unusual sterols exhibit a cyclopropane ring at 9β,19, an additional methyl group at C-14 and most of them (97%) have a C-9 side-chain, whereas major Δ^5-sterols (i.e. stigmasterol) have a C-10 side-chain.

TABLE 1. Sterols of PM from control and treated maize roots.

	Control	Treated
24-methylcholesterol (% sterols)	28	Tr
Stigmasterol	57	4
Sitosterol	14	1
24-methylpollinastanol[a]	0	46
Cycloeucalenol	1	26
24-dihydrocycloeucalenol	0	22
Total Δ^5-sterols	99	5
Total 9β,19-cyclopropylsterols	Tr	94
Total sterols (µg/mg prot)	60	85
Sterols/PL (molar ratio)	0.45	0.62

Tr, Traces (<1%) ;

[a]24-methylpollinastanol + 5% 24-methylenepollinastanol.

Phospholipids. PM preparations from control and treated plants were found to contain the same classes of PL, with PC and PE as a major compounds (Table 2). No quantitative change was evidenced following fenpropimorph treatment, so that an increase in the sterol to PL molar ratio of the PM was obtained.

TABLE 2 : Phospholipids of PM from control and treated maize roots.

	Control	Treated
PA (% PL)	4	10
PG	14	10
PE	35	33
PI + PS	9	12
PC	39	34
Total PL (μmol/mg prot)	0.29	0.30

Fatty acids. In control and treated PM preparations, major FA are oleic acid (18:2) and palmitic acid (16:0 (Table 3). No qualitative or quantitative change was evidence in response to the modification of the sterol profile by fenpropimorph.

TABLE 3 : Fatty acids of PM from control and treated maize roots.

	Control	Treated
16:0 (% FA)	33	30
18:0	2	4
18:1	2	3
18:2	58	57
18:3	3	5
UI	1.3	1.3
Total FA (μg/mg prot)	88	91

UI, Unsaturation index : [(% monoenes + 2 x (% dienes) + 3 x (% trienes)]/100.

ATPase activity in a fenpropimorph-treated PM preparation.

As shown in Table 4, in both PM preparations, vanadate-sensitive ATPase (pH 6.5) exhibits similar properties and hydrolyzes ATP to the same extent. The thermotropic behavior of the enzyme as evidenced by Arrhenius plots was also found to be similar in both fractions (data not shown), thus ruling out any important contribution by the modification of the sterol profile of the PM to ATP hydrolysis. Some preliminary experiments were carried out with an ATPase reconstituted into lipid vesicles containing or not sterols (Δ^5-sterols or 24-methylpollinastanol). H^+-pumping was measured by using the fluorescent probe quinacrine. Sterols were found to induce a significant inhibition (55 to 65%) of H^+-transport compared to that measured in the absence of sterols. By contrast, only low changes (10%) in ATP hydrolysis were observed by addition of sterols in liposomes.

TABLE 4. ATPase properties in control and fenpropimorph treated PM from maize roots.

Activities are expressed as the percentage of the value under standard conditions.

	Control	Treated
- 1 mM NaN_3	120	110
+ 100 μM Na_3VO_4	60	50
+ 100 mM $NaNO_3$	100	110
+ 100 μM DCCD	65	60
+ 100 μM $Na_2M_oO_4$	85	85
+ fenpropimorph (from 0.5 to 50 μM)	100	nd

Control PM : 17 $\mu mol.h^{-1}.mg^{-1}$ prot (mean value of 10 replicates);
Treated PM : 18 $\mu mol.h^{-1}.mg^{-1}$ prot (mean value of 3 replicates);
nd, not determined.

CONCLUSION

All these results taken together indicate that cyclopropylsterols can replace effectively Δ^5-sterols as membrane components in higher plants as well as in other organisms such as the GL7 yeast mutant (7), Mycoplasma capricolum (8) or Tetrahymena pyriformis (9). Further investigations are necessary to determine whether other sterol structures like Δ^8-sterols or Δ^8,14α-methylsterols are also suitable for playing a similar role.

REFERENCES

1 - Hartmann M.A. and Benveniste P. (1987) Methods Enzymol. 148, 632-650.
2 - Schmitt P., Rahier A. and Benveniste P. (1982) Physiol. Vég. 20,559-571.
3 - Bladocha M. and Benveniste P. (1983) Plant Physiol. 71, 756-762.
4 - Hartmann M.A. and Benveniste P. (1978) Phytochemistry 17, 1037-1042.
5 - Hartmann M.A., Grandmougin, Ullmann P., Bouvier-Navé P. and Benveniste P. (1988) Plant Physiol., submitted to publication.
6 - Rahier A., Schmitt P., Huss B., Benveniste P. and Pommer E.H. (1986) Pest. Biochem. Physiol. 25, 112-124.
7 - Odriozola J.M., Waitzhin E., Smith J.L. and Bloch K. (1978) Proc. Natl. Acad. Sci. USA 75, 4107-4109.
8 - Nes W.D. and Patterson G.W. (1981) J. Nat. Prod. 44, 215-220.
9 - Raederstorff D. and Rohmer M. (1988) Biochim. Biophys. Acta 960, 190-199.

BIOLOGICAL ROLE OF PLANT LIPIDS
P.A. BIACS, K. GRUIZ, T. KREMMER (eds)
Akadémiai Kiadó, Budapest and Plenum Publishing
Corporation, New York and London, 1989

TRITERPENOIDS IN EPICUTICULAR WAXES

P.-G. Gülz

Botanisches Institut der Universität zu Köln, Gyrhofstr. 15, 5000 Köln 41, FRG

INTRODUCTION

All aerial organs of higher plants are covered primarily with a thin continuous wax layer. These surface or epicuticular waxes consist of a very complex mixture of different components. In most cases these very long chained lipids are found in form of homologous series. The composition of the wax lipids shows species specific and also organ specific patterns. But numerous plants in addition contain triterpenoids,mostly pentacyclic compounds. The composition of triterpenoids from two Euphorbia species and from the leaf waxes of the trees Citrus halimii, Tilia tomentosa and Tilia x europaea will be summarized in this paper (Table 1).

RESULTS

In E.dendroides leaf wax triterpenoids were found in amounts of 26% in addition to the common wax lipids. The triterpenois ß-amyrin and lupeol were identified in this wax with the aid of TLC, GC, GC/MS and comparison with authentic samples. These alcohols were found not only free but also esterified with very long chain fatty acids. At last ß-amyrin and lupeol were present in form of their corresponding ketones,too. Δ12-Olean-en-3-one and Δ20-lupen-3-one could be isolated by TLC and were identified by chemical reactions and their mass spectra. All these triterpenoids showed the typical colour reaction with carbazole and could be checked by TLC in this manner (1).

E.aphylla wax contained triterpenoids in amounts of 31%, beside the common wax lipids. In this wax the alcohols ß-amyrin, α-amyrin, lupeol and also simiarenol were found. With the exception of simiarenol the first triterpenols occured free as well as esterified with long chain fatty acids and with acetic acid, too, and they were found also in form of their ketones. The publications of further *Euphorbia* wax triterpenoids are in preparation. All *Euphorbia* waxes show qualitative and quantitative species specific compositions of triterpenoids, which are not in agreement with triterpenoids described from the latex in most cases (2).

In *C.halimii* leaf wax triterpenoids were found in the very high amount of 72%, beside the wax lipids. The free triterpenols dominated with about 62%. Identified could be ß-amyrin, α-amyrin and lupeol. In small amounts esters of ß-amyrin and α-amyrin containing long chain fatty acids were found. Again triterpene ketones were isolated and detected with the aid of GC/MS. In this wax lupenone was present and also one other ketone named fridelanone in the remarkable amount of about 12%. They could be reduced to the corresponding alcohols. Fridelanone and fridelanol have a pentacyclic structure without any double bond. Therefore these substances showed quite different Rf values in TLC and no colour reaction with carbazole (3)(Fig.1).

Several deciduous broadleaved trees have leaf waxes containing triterpenoids in addition to the wax lipids. Epicuticular leaf waxes of *T.tomentosa* and *T.* x *europaea* showed triterpenoids in amounts of 56% and 44%. Identified could be only one triterpenol: ß-amyrin. in these leaf waxes ß-amyrin was found free as well as esterified with long chain fatty acids and also with acetic acid. ß-Amyrin acetate was the dominating wax component with about 34% of the wax. In *Tilia* waxes triterpenones could not be detected. In both *Tilia* species the same triterpenoids were analysed with one significant dominating main component: ß-amyrin acetate (4).

These results demonstrate, epicuticular triterpenoids are widespread in the plant kingdom. These substances were found primarily as alcohols, free as well as esterified and often also in form of their ketones, in a very different species specific composition.

Table 1: Composition and characterization of triterpenoids in epicuticular waxes

Triterpenoids	TLC Rf_1 a)	TLC Rf_3 b)	TLC Carbazole	GC rrt c)	GC/MS M^+	E.aphylla	E.dendroides	C.halimii	T.tomentosa
Triterpenol esters (in % wax)						1.5%	4.6%	0.8%	11.3%
ß-Amyrin esters	0.62		+			+	+	+	+
α-Amyrin esters	0.62		+			+	-	+	-
Lupeol esters	0.62		+			+	+	-	-
Triterpenol acetates (in % wax)						7.2%	--	--	34.3%
ß-Amyrin acetate	0.35	0.64	+	1.097	468	+	-	-	+
α-Amyrin acetate	0.35	0.64	+	1.126	468	+	-	-	-
Lupeol acetate	0.35	0.46	+	1.129	468	+	-	-	-
Triterpenones (in % wax)						2.1%	13.0%	14.8%	--
ß-Amyrinone	0.19	0.53	+	0.983	424	+	+	-	-
α-Amyrinone	0.19	0.53	+	1.013	424	+	-	-	-
Lupeonone	0.19	0.42	+	1.016	424	+	+	+	-
Fridelanone	0.15	0.57	-	1.095	426	-	-	+	-
Titerpenols (in % wax)						20.6%	8.6%	62.2%	10.1%
ß-Amyrin	0.06	0.30	+	1.000	426	+	+	+	+
α-Amyrin	0.06	0.30	+	1.029	426	+	-	+	-
Lupeol	0.06	0.20	+	1.034	426	+	+	+	-
Simiarenol	0.10	0.04	+	1.098	426	+	-	-	-

a) toluene; b) $AgNO_3$ + dichloromethane/ethylacetate (24:1); c) relative retention times

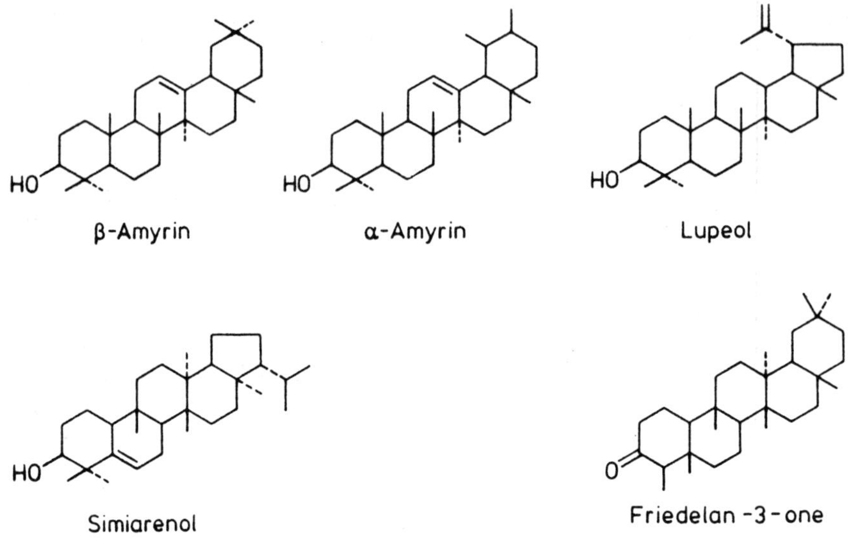

Figure 1: Chemical structures of triterpenoids

REFERENCES

1. Gülz,P.-G., Hemmers,H., Bodden,J.,and Marner,F.-J. (1987) Epicuticular Leaf Wax of Euphorbia dendroides L.Euphorbiaceae Z.Naturforsch. 42c, 191-196

2. Gülz,P.-G., Bodden,J., Müller,H.,and Marner,F.-J. (1988) Epicuticular Waxes of Euphorbia aphylla Brouss.ex Willd., Euphorbiaceae. Z.Naturforsch. 43c, 80-84

3. Gülz,P.-G., Scora,R.W., Müller,E.,and Marner,F.-J. (1987) Epicuticular Leaf Wax of Citrus halimii Stone J.Agric.Food Chem. 35, 716-720

4. Gülz,P.-G., Müller,E.,and Moog,B. (1988) Epicuticular Leaf Waxes of Tilia tomentosa Moench. and Tilia x europaea L., Tiliaceae Z.Naturforsch. 43c, 173-176

BIOLOGICAL ROLE OF PLANT LIPIDS
P.A. BIACS, K. GRUIZ, T. KREMMER (eds)
Akadémiai Kiadó, Budapest and Plenum Publishing
Corporation, New York and London, 1989

FLUORESCENCE ANISOTROPY OF DIPHENYLHEXATRIENE INCORPORATED INTO SOYBEAN PHOSPHATIDYLCHOLINE VESICLES CONTAINING PLANT STEROLS

I. Schuler, G. Duportail*, P. Benveniste and M.A. Hartmann

Laboratoire de Biochimie Végétale (UA CNRS 1182), Université Luis Pasteur, Institut de Botanique, 28 Rue Goethe, 67083 Strasbourg, France
*Laboratoire de Physique (UA CNRS 491), Université Luis Pasteur, Faculté de Pharmacie, 74 Route de Rhin, 67048 Stasbourg, France

INTRODUCTION

In most higher plants, sterols are present as a mixture of Δ^5-sterols with sitosterol, stigmasterol and 24-methylcholesterol as the main compounds. Plant sterols differ from cholesterol by the presence of an additional bulky alkyl group at C-24 of the side-chain. The high content of polyunsaturated fatty acids in plant membranes, which results in a lower packing of phospholipids (PL) acyl chains compared to that of acyl chains in animal membranes, is probably an important feature for a good fit between plant PL and plant sterols. In order to investigate whether plant sterols may regulate the plasma membrane fluidity of higher plant cells as well as cholesterol does in animal cells, steady-state fluorescence polarization measurements using 1,6-diphenyl-1,3,5-hexatriene as a probe were performed with large unilamellar vesicles (LUV) prepared from soybean phosphatidylcholine (PC) and different plant sterols in varying molar ratios. The effectiveness of 24-methylpollinastanol (a 9β,19-cyclopropylsterol) was also tested.

METHODS

Vesicle preparation. LUV were prepared in 100 mM NaCl buffered with 5 mM Hepes (pH 7.4) by the reverse-phase evaporation procedure of Szoka and Papahadjopoulos (1) from soybean PC, phosphatidylglycerol (PG) and sterol(s) in different molar ratios. PG was present at 10 mol % in all the assays. Sterol-free vesicles were also prepared. Stigmasterol, sitosterol and (24R)-methylcholesterol were obtained from commercial sources. 24-methylpollinastanol was extracted from fenpropimorph-treated maize roots as reported in Ref.2. Lipid phosphorus as well as sterol content were determined after LUV recovery.

Fluorescence measurements. Steady-state fluorescence polarization measurements were performed with a SLM 8000 spectrofluorimeter at 20°C. Vesicles were labelled with 1 μM DPH by incubation for 30 min in the dark, at room temperature. Data were expressed as plots of the anisotropy parameter $[(r_o/r - 1]^{-1}$, which provides a quantitative index of the environmental resistance to the rotational motion of the fluorophore (3), versus the sterol content (in mol %) of vesicles.

RESULTS

The anisotropy parameter $[(r_o/r - 1]^{-1}$ is an indicator of the fluidity of the bilayer core : the greater this value, the higher is the apparent microviscosity of the bilayer and therefore the stronger are the PL-sterol interactions. Fig.1 shows that, whatever the sterol incorporated into the vesicles, a linear increase of the anisotropy parameter associated with an increase of the sterol content of vesicles is observed. Sitosterol, a 24-ethylcholesterol, was found to be the most effective plant sterol in decreasing the fluidity of soybean PC bilayers and stigmasterol, a 24-ethylcholesterol with an additionak double bond at C-22, the least effective one. Campesterol or (24)-methylcholesterol exhibits an intermediate behavior.

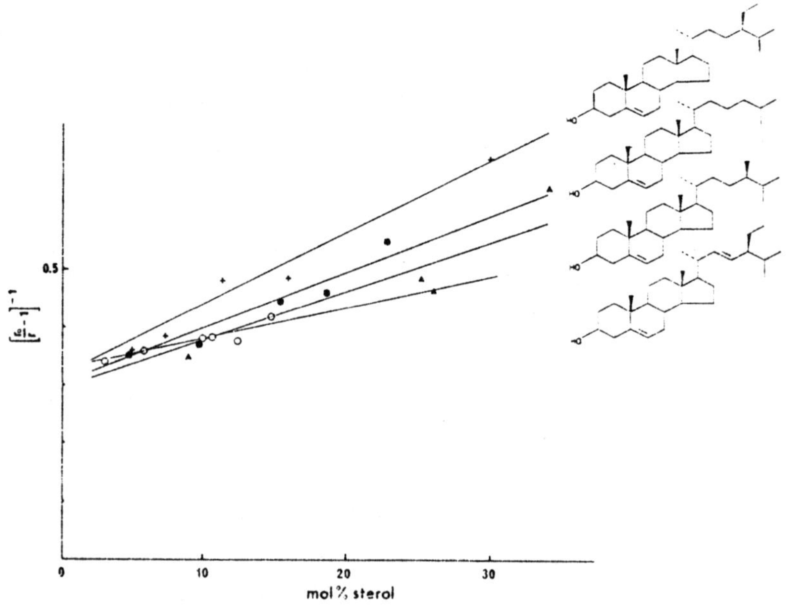

Fig.1. The effect of various plant sterols on the DPH anisotropy parameter of soybean PC vesicles at 20°C.

The effects of a mixture of plant sterols (stigmasterol/sitosterol/campesterol, 55:15:30, which is representative of the mean sterol composition of membranes from maize roots), were found to be similar to those obtained with stigmasterol, the main compound of the mixture (data not shown). Cholesterol, a minor sterol in plants, is shown to be less efficient than sitosterol. In Fig.2 is represented the comparative behavior of sitosterol- and 24-methylpollinastanol-containing LUV. The cyclopropylsterol appears to be much less efficient than sitosterol as a modulator of the fluidity of soybean PC bilayers. It behaves more like (24R)-methylcholesterol.

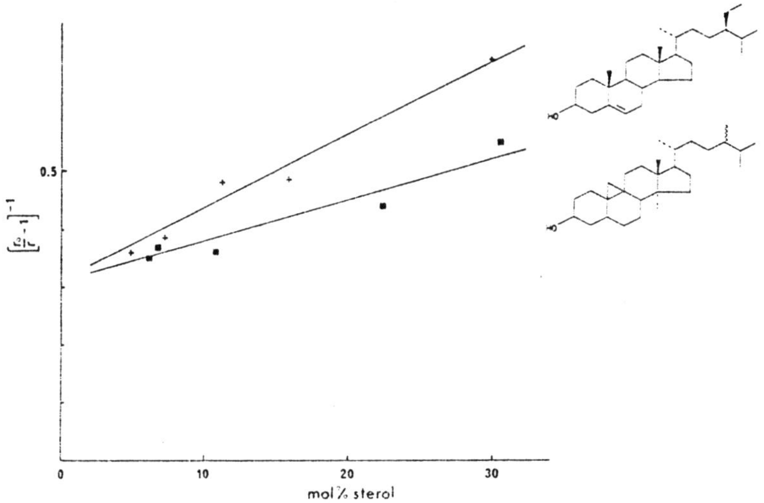

Fig.2. Comparative effects of sitosterol and 24-methylpollinastanol on the DPH anisotropy parameter of soybean PC vesicles at 20°C.

CONCLUSION

This study clearly evidences that all the plant sterols tested are able to regulate the membrane fluidity of soybean PC vesicles, with the sterol competence (above 10 mol %) following the order : stigmasterol < 24-methylpollinastanol ≃ (24R)-methylcholesterol < cholesterol < sitosterol. Further work would be to investigate if such differences in the behavior of the various sterols have a functional significance.

REFERENCES

(1) Szoka F. and Papahadjopoulos D. (1978) Proc. Natl. Acad. Sci. USA 75, 4194-4198.

(2) Bladocha M. and Benveniste P. (1983) Plant Physiol. 71, 756-762.

(3) Shinitzky M. and Barenholz Y. (1978) Biochim. Biophys. Acta 515, 367-394.

CHAPTER 4

CARRIER PROTEINS, GENETICS OF PLANT LIPIDS

BIOLOGICAL ROLE OF PLANT LIPIDS
P.A. BIACS, K. GRUIZ, T. KREMMER (eds)
Akadémiai Kiadó, Budapest and Plenum Publishing
Corporation, New York and London, 1989

MODIFICATIONS TO THE TWO PATHWAY SCHEME OF LIPID METABOLISM BASED ON STUDIES OF ARABIDOPSIS MUTANTS

J. Browse[1], L. Kunst[2], S. Hugly[2] and C. Somerville[2]

[1]Institute of Biological Chemistry, Washington State University, Pullman, Wa. 99164, USA
[2]MSU-DOE Plant Research Laboratory, Michigan State University, E. Lansing, MI 48824, USA

INTRODUCTION

In the last few years the lipid mutants of *Arabidopsis* have been very useful in elucidating details of the synthesis and desaturation of glycerolipids in plant leaves [1-3]. *Arabidopsis* is a typical 16:3-plant in which both the prokaryotic and eukaryotic pathways [4] contribute to the production of chloroplast lipids. We have investigated the pattern of lipid metabolism in wild type *Arabidopsis* and calculated the fluxes of carbon involved [5]. An abbreviated version of this analysis is shown in Fig. 1a. For every 1000 fatty acid molecules synthesized in the chloroplast 390 enter the prokaryotic pathway in the chloroplast envelope while 610 are exported as CoA esters to enter the eukaryotic pathway. Of these 340 are reimported into the chloroplast. Overall, almost equal amounts of chloroplast lipids are produced by each pathway. However, the quantities of individual lipids synthesized by the two routes are very different. All the chloroplast phosphatidylglycerol (PG) and over 70% of the monogalactosyldiacylglycerol (MGD) is derived from the prokaryotic pathway while digalactosyldiacylglycerol is synthesized mainly on the eukaryotic pathway [5]. In this paper we have outlined how four of the *Arabidopsis* mutants have changed the way we view the operation of the two pathways involved in leaf membrane lipid synthesis. More detailed information on each mutant can be found elsewhere [1-3, 5,6 and in preparation].

RESULTS AND DISCUSSION

Recently [3] we described a mutant (*act1*) which was almost completely deficient in activity of the chloroplast glycerol-3-phosphate acyltransferase the first enzyme of the prokaryotic pathway. Analysis of the fatty acid composition of leaf lipids and ^{14}C-labelling experiments indicated that PG was essentially the only lipid produced by the prokaryotic pathway in this mutant. However, this blockage in the prokaryotic pathway is compensated for by increased flux through the eukaryotic pathway and furthermore the proportions of the different chloroplast lipids made by the eukaryotic pathway are altered so that there is only a small alteration in lipid composition in the *act1* mutant compared with wild type [3]. Thus in the *act1* mutant (Fig. 1b) 950 of every 1000 fatty acids enter the eukaryotic pathway and 650 of these are subsequently returned to the chloroplast. The mutant is essentially an 18:3-plant and clearly demonstrates the existence of controls which regulate reactions in the chloroplast and endoplasmic reticulum so as to provide the complement of lipids required for correct membrane synthesis in leaf cells.

During our analysis of a mutant (*fadD*) deficient in desaturation of both 16:2 and 18:2 fatty acids [2], we noted that extrachloroplast as well as chloroplast lipids showed reduced 18:3 levels in the mutant. Assuming that the *fadD* gene product is a chloroplast *w*-3 desaturase our results indicate either that the enzyme is targeted to some other cellular compartment as well as to the chloroplast or that there is considerable exchange of lipid between the chloroplast and other cell membranes [2]. In light of the results discussed below we believe the second possibility is correct as shown in Fig. 1c.

The *fadC* mutant is deficient in a chloroplast *w*-6 desaturase which is responsible for desaturation of 18:1 and 16:1 fatty acids on the prokaryotic pathway. One effect of this mutation is a 37% decrease in the amount of prokaryotic MGD species compared with the wild type. This decrease is compensated for by increased synthesis of eukaryotic MGD so that there is only a small change in the total amount of leaf

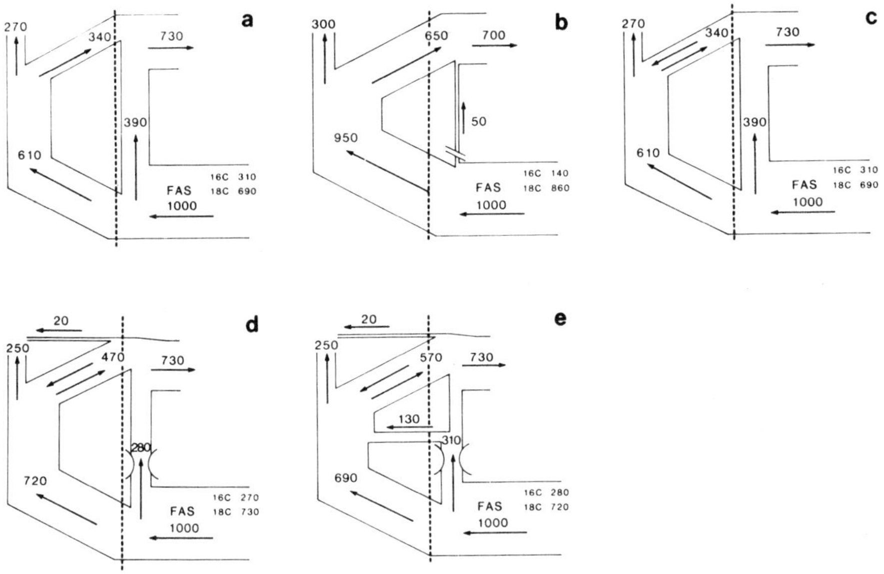

Fig.1. Fluxes of fatty acids involved in leaf lipid metabolism. The chloroplast is shown to the right of the dashed line in each scheme and the extrachloroplast compartment to the left.
a). The original conception of the two pathway hypothesis in wild type *Arabidopsis* with unidirectional flow of carbon (abbreviated from Fig. 3 in ref 2). b). A block in the prokaryotic pathway in the *act1* mutant is circumvented by increased flux through the eukaryotic pathway. c). The *fadD* mutant suggests reversible exchange of lipid between the chloroplast and endoplasmic reticulum. d). The *fadC* mutant shows reduced accumulation of prokaryotic MGD species and the possible transfer of lipid from the chloroplast to membranes other than the endoplasmic reticulum. e). In the *fadB* mutant there is considerable export of prokaryotic lipid derived from MGD to extrachloroplast membranes.

MGD[6]. In Fig. 1d this is shown by a decreased flux through the prokaryotic pathway of only 280 fatty acids compared with 390 in the wild type and an increased flux through the eukaryotic pathway . The extrachloroplast lipids- phosphatidylcholine (PC), and phosphatidylethanolamine (PE)- of the *fadC* mutant also show increased levels of 18:1. We did not expect this result since the mutant contains normal activity of the endoplasmic reticulum 18:1 desaturase. While the eukaryotic pathway very probably delivers some 18:1-containing lipids to the chloroplast, any 18:1 reexported to the endoplasmic reticulum should be efficiently desaturated. One explanation, shown in Fig. 1d, is that a proportion of the eukaryotic lipid reexported from the chloroplast goes to membrane sites which are not accessible to any 18:1 desaturase.

The *fadB* mutant is deficient in the chloroplast enzyme which desaturates 16:0 to 16:1 (*w*-9) on MGD. In this mutant the amount of prokaryotic MGD is reduced by more than 60% compared with the wild type and again there is a compensating increase in the amount of eukaryotic species of MGD. However, labelling studies with [^{14}C]-acetate demonstrated that this reduction was largely due to a steady turnover of 18:X/16:0 species of MGD rather than to a block in prokaryotic MGD synthesis. Our data indicates that these 18:X/16:0 species are not broken down to their component fatty acids but instead are exported from the chloroplast as diacylglycerol moieties and are used for the synthesis of PC and PE in the endoplasmic reticulum. Thus, in the mutant PC and PE both contain more than 15% prokaryotic molecular species (16:0 at *sn*-2 of the glycerol backbone). It is noteworthy that only the *fadB* and to a lesser extent the *fadC* mutants exhibit export of prokaryotic lipid species. 16:3 and 16:2 fatty acids cannot normally be detected in the phospholipids of wild type or *fadD* mutant plants. The fluxes of fatty acids occuring during lipid synthesis in the *fadB* mutant are shown quantitatively in Fig. 1e. Thus our characterization of the *fadB* mutant greatly extends the two pathway scheme for leaf lipid synthesis by demonstrating the considerable exchange of lipids between the chloroplast and other cell membranes.

CONCLUSION

This analysis of *Arabidopsis* mutants demonstrates that plants show considerable flexibility in operation of the two pathways of lipid synthesis. This flexibility involves both regulation of the production and turnover of different lipids and a mechanism for export of lipid molecules from the chloroplast to other organelles and membranes.

REFERENCES

1. Browse JA, McCourt PJ and Somerville CR. 1985. A mutant of *Arabidopsis* lacking a chloroplast specific lipid. Science 227,763-765.
2. Browse JA, McCourt PJ, and, Somerville CR. 1986. A mutant of *Arabidopsis* deficient in C18:3 and C16:3 leaf lipids. Plant Physiol. 81, 751-756.
3. Kunst L, Browse J and Somerville CR. 1988. Altered regulation of lipid biosynthesis in a mutant of *Arabidopsis* deficient in chloroplast glycerol phosphate acyltransferase activity. Proc. Natl. Acad. Sci. 85, 4143-4147.
4. Roughan PG and Slack CR. 1982. Cellular organization of glycerolipid metabolism. Ann. Rev. Plant Physiol. 33, 97-132.
5. Browse JA, Warwick N, Somerville CR and Slack CR. 1986. Fluxes through the prokaryotic and eukaryotic pathways of lipid synthesis in the 16:3 plant *Arabidopsis thaliana*. Biochem J. 235, 25-31.
6. Browse JA, Kunst L, Anderson S, Hugly S and Somerville CR. 1988. A mutant of *Arabidopsis* deficient in the chloroplast 16:1/18:1 desaturase. Plant Physiol. (in press).

BIOLOGICAL ROLE OF PLANT LIPIDS
P.A. BIACS, K. GRUIZ, T. KREMMER (eds)
Akadémiai Kiadó, Budapest and Plenum Publishing
Corporation, New York and London, 1989

PLANT LIPID-TRANSFER PROTEINS: STRUCTURE, FUNCTION AND MOLECULAR BIOLOGY

V. Arondel[1], F. Tchang[1], Ch. Vergnolle[1], A. Jolliot[1], M. Grosbois[1], F. Guerbette[1], Marie-Dominique Morch[2], J.-C. Pernollet[3], M. Delseny[4], P. Puigdomenech[5] and J.-C. Kader[1]

[1]Université P. et M. Curie (UA CNRS 1180), Paris, [2]Institut J. Monod, Paris, [3]INRA Versailles
[4]Université de Perpignan (UA CNRS 565), France, [5]Institut de Biologia, Barcelona, Spain

Intracellular transport of phospholipids is thought to require the participation of lipid transfer proteins (LTP) able to facilitate intermembrane lipid movements. These proteins, first detected and partially purified from potato tuber cytosol (Kader, 1985), have been purified to homogeneity from various plant tissues : spinach leaf (Kader et al., 1984), maize seedlings (Douady et al., 1982, 1985), and castor bean endosperm (Yamada et al., 1978 ; Watanabe and Yamada, 1986). These purifications opened perspectives for studying the structure, the mode of action and the molecular biology of this novel category of plant proteins involved in lipid movements.

ASSAYS FOR LIPID TRANSFER

The activity of these proteins involves the use of two categories of membranes, an artificial one (liposomes) and a natural one (mitochondria or microsomes). One of two membranes contains the lipid to be transferred (usually ^{3}H-labeled phospholipid) and a non transferable tracer ^{14}C-cholesteryl oleate. After incubation with or without added proteins, the membranes were separated by centrifugation ; the determination of the radioactive phospholipids recovered within the initially unlabeled membranes when lipid transfer protein is present, indicates the extent of the transfer. Other types of experimental mixtures have been studied : labeled castor bean microsomes and unlabeled mitochondria (Yamada et al., 1978), multilamellar vesicles and unlamellar liposomes (Guerbette, unpublished).

All these assays involved a separation of two categories of membranes . Other assays do not need such separation since they involved liposomes containing spin-labeled lipids ; the movement of these lipids towards unlabeled lipids is followed by ESR-spectroscopy.

This approach has been used to study the transfer of spin-labeled phosphatidylcholine mediated by maize protein (Kader, unpublished) or spin-labeled monogalactosyldiacylglycerol facilitated by spinach protein (Nishida and Yamada, 1986).

PURIFICATION

Lipid transfer proteins have been purified from cytosolic extracts by using classical methods of protein separation. The purification procedures for maize, spinach or castor bean proteins involved gel filtration, DEAE-Sepharose or DEAE-Trisacryl chromatography. Carboxymethyl-Sepharose chromatography and Sephadex G50 filtration (Douady et al., 1982 ; Kader et al., 1984 ; Watanabe and Yamada, 1986). In the case of maize protein, the purification procedure was simplified by adjusting the pH of the Carboxymethyl-Sepharose column to higher values (Douady et al., 1985). The purification of the castor bean protein involved final purification steps of gel filtration and hydroxyapatite chromatography (Watanabe and Yamada, 1986). Interestingly, different peaks of activity were detected after carboxymethyl Sepharose chromatography of maize, spinach or castor bean. These peaks correspond to isoforms of lipid transfer proteins (see the paper of Yamada et al. in this volume). All these purification procedures are based on the fact that lipid transfer proteins from plants are mainly basic. However, acidic proteins have been purified from castor bean endosperm by Tanaka and Yamada (1982).

OCCURRENCE OF LIPID-TRANSFER PROTEINS

In addition to maize, spinach and castor bean, lipid transfer proteins have been detected in other plant tissues : seeds of barley (Coutos-Thevenot, unpublished) and sunflower (Arondel, unpublished), wheat (Monnet, unpublished), tobacco (Gawer, unpublished), pea (Kader et al., 1985) and Avena leaves (Yamada et al., 1980).

PROPERTIES OF LIPID-TRANSFER PROTEINS

Molecular mass - The major lipid-transfer protein purified to homogeneity from plants have a low molecular mass around 9 kDa (Kader et al., 1984 ; Douady et al., 1985 ; Watanabe and Yamada, 1986 ; Nishida and Yamada, 1986). These values , determined by gel filtration or SDS-electrophoresis, have been confirmed by the determination of the amino-acid sequences (Table 1). In addition to the major protein of castor bean, some isoforms have been recently found with different molecular mass (see Yamada et al., this volume). The molecular mass of

9 kDa is lower than those of lipid transfer proteins purified from animal cells or yeasts (see Wirtz, 1982 and Kader et al., 1982 for reviews). However, polymers of plant lipid transfer proteins have been observed (dimers for maize protein (Douady et al., 1982) and for spinach protein (Kader et al., 1984)). Lower or higher values have been determined for spinach leaf proteins by Nishida and Yamada (1986) who found in the total spinach leaf extract 11 kDa - and 5.1 kDa-lipid transfer proteins and in the spinach chloroplast extract a galactolipid transfer protein of 28 kDa. Interestingly, the acidic lipid transfer proteins detected by Tanaka and Yamada (1982) present polymeric forms (from 11.1 to 69.2 kDa).

Table 1. Properties of Lipid transfer proteins from higher plants

Source	Determined[a] molecular mass	Calculated[b] molecular mass	Determined[c] pHi	Calculated[d] pHi
Maize	9 000	9 054	8.8	8.97
Spinach	9 000	8 833	9	9.36
Castor bean	9 000	9 313	10.5	8.56

[a]determined by SDS-electrophoresis
[b]calculated from amino acid sequence
[c]determined by isoelectric focusing or chromatofocusing
[d]calculated from amino acid sequence

Isoelectric point - As indicated above, the lipid transfer proteins from plants are mainly basic. The determination of pHi values by chromatofocusing or isoelectric focusing gave values of 8.8 for maize protein, higher than 10.5 for castor bean protein and around 9 for spinach protein (Table 1). The isoelectric points of the various isoform of lipid transfer proteins have not been determined. However, very low values were found for acidic proteins from castor bean (from 5.4 to 6.6) or spinach leaf (around 4.5).

Stability - Castor bean, maize and spinach proteins are remarkably stable. These proteins keep their transfer activity at 4°C for at least 1 month. Maize and castor bean proteins are resistant to elevated temperatures ; the activity was unchanged after a 30 min-heating at 90°C (Douady et al., 1982 ; Watanabe and Yamada, 1986).

SPECIFICITY FOR PHOSPHOLIPIDS

Some lipid transfer proteins isolated from castor bean (Tanaka and Yamada, 1982) (acidic ones) transfer preferentially phosphatidylcholine or phosphatidylinositol. However, the main characteristics shared by the more abundant lipid transfer proteins from plants is their non-specific character for phospholipids. As indicated by Table 2, proteins isolated from castor bean, maize or spinach are able to transfer phosphatidylcholine, phosphatidylethanolamine, phosphatidylinositol from liposomes to mitochondria. In addition, spinach protein also transfers phosphatidylglycerol and castor bean protein, phosphatidic acid. The possibility for these proteins to transfer galactolipids has been found recently by Watanabe and Yamada (1986) using ^{14}C-monogalactosyldiacylglycerol with castor bean protein and Nishida and Yamada (1986) using spin-labeled monogalactosyldiacylglycerol with spinach leaf protein. However, other lipids, like diacylglycerol or triacylglycerol are not transferred by the castor bean protein (Watanabe and Yamada, 1986).

Table 2. **Specificity of Lipid Transfer Proteins for phospholipids**

Souce	Lipid transferred	References
Maize	PC, PI, PE	Douady et al., 1982
Spinach leaf	PC, PI, PE, PG	Kader et al., 1984
Spinach leaf	PC, PI, PE, PG, MGDG	Nishida and Yamada, 1986
Castor bean	PC, PI, PE, PG, PA, MGDG	Watanabe and Yamada, 1986

BINDING OF FATTY ACIDS

The discovery of a protein in Avena leaf able to bind fatty acids (Rickers et al., 1984) opened new perspectives in the study of the properties of lipid transfer proteins. Working on Avena leaf, Rickers et al. (1984) incubated cytosolic proteins with ^{14}C-oleic acid and separated the labeled proteins on carboxymethyl-Sepharose. Three peaks of fatty acid binding were observed, the main one corresponding to a protein having the same properties as lipid transfer proteins : molecular mass of 8.7 kDa and pHi of 8.4. This was confirmed by the fact that Avena fatty acid- binding protein also possesses the ability to transfer phosphatidylcholine (Spener and Kader, unpublished). As a reciprocical experiment, lipid transfer protein from spinach leaf was assayed for fatty acid binding. The experiment was successful since a binding of ^{14}C-oleoyl-coenzyme A or ^{14}C-oleic acid was observed

(Rickers et al., 1985). These experiments introduced a novel property of lipid transfer proteins from plants. It is of interest to note that, in animal cells, the properties of transferring and binding lipids are associated to two different types of proteins, the lipid transfer proteins and the fatty acid binding proteins (see Kader, 1985 ; Rickers et al. 1984).

MODE OF ACTION

The mode of action of lipid transfer proteins is not entirely elucidated. However, it is assumed that a complex between phospholipid and protein is formed. This complex interacts with the various membranes ; the transfer of the phospholipid molecule from the complex to the membrane leads to a net transfer of phospholipid. However, an exchange of the phospholipid molecule bound to the complex with the phospholipids of the membranes leads to an overall-processes of exchange of phospholipids ; this explains why these proteins were previously called "phospholipid exchange proteins".

The main problem for such a model is to demonstrate the binding of phospholipids on the transfer protein. Binding of phosphatidylcholine has been found with basic proteins from maize (Douady et al., 1982) and spinach (Kader et al., 1984) while Tanaka and Yamada (1982) observed a binding of phosphatidylcholine and phosphatidylserine on acidic proteins from castor bean. However, this binding, in the case of spinach and maize proteins, is very low. A possible explanation of this observation is that the binding of lipids on the basic protein is reversible and competes with the interaction of lipids with membranes. It is interesting to note that non-specific proteins from animal cells are also unable to bind phospholipids (Kader, 1985). In future experiments, the sites of binding of phospholipids and fatty acids to plant proteins will have to be studied.

STRUCTURE

The major non-specific proteins from plants have been completely analyzed, in the case of spinach and castor bean (Takishima et al., 1986 ; Bouillon et al., 1987). These proteins contains, respectively, 91 and 92 residues ; the spinach protein contain 6 cysteines while the castor bean one has 8 cysteine residues. Among these eight cysteines, six are located in the same place in the molecule, respectively residues 4, 14, 50, 52, 75, 89 of the spinach protein. This could be of physiological significance for the function of these proteins. Some portions of the sequences are characterized by their hydrophobic properties, especially the N-terminal end. This may be related to the interaction of the protein with the membrane.

LIPID TRANSFER PROTEINS AS TOOLS FOR MEMBRANE STUDY

The use of lipid transfer proteins as tools for manipulating the composition of membranes is very promising. This has been done by incubating dipalmitoyl phosphatidylcholine or dioleoylphosphatidylcholine with mitochondria (Guerbette et al., unpublished). The fatty acid composition of mitochondria was modified after incubation since palmitic or oleic acid became more abundant. This is due to an exchange of mitochondrial phosphatidylcholine with the liposomal one. It will be of interest to study the effects of the replacement of lipids on the functions of mitochondrial membranes.

PHYSIOLOGICAL ROLES OF LIPID TRANSFER PROTEINS

All the experiments previously described were performed in vitro. However, it has been assumed that lipid transfer proteins act in vivo by facilitating the intracellular dispatching of phospholipids, which are , in major part, synthesized in the endoplasmic reticulum for phosphatidylcholine, phosphatidylethanolamine and phosphatidylinositol ; see Mazliak et al.; 1982 ; Moore, 1984). Thus, lipid-transfer proteins could play an essential role in the phymology of plant cell by contributing to the biogenesis of membranes. Also, the renewal and repairing of membranes could be facilitated by lipid transfer proteins which are involved in the intracellular lipid dynamics.

In addition, the role of lipid transfer proteins might be important for the movement of galactolipids in the cytosol and within the chloroplast stroma. This movement is necessary to convey galactolipids assembled in envelope membranes to thylakoids (Douce et al., 1984) . It is to be noted that no transfer activity for phosphatidylcholine has been found within the chloroplast stroma (Schwitzeguebel et al. 1984).

However, it cannot be excluded that intracellular transfer of membrane fragments also participate to the biogenesis of membranes, especially plasmalemma (Moreau et al.,1984).

Additional roles for lipid transfer proteins have been assumed : regulation of the acyl-CoA pool within the cytosol and participation in the biosynthesis of linolenic acid via the cooperative pathway involving the plastids and the endoplasmic reticulum (Dubacq et al., 1984). It has been showed that the chloroplast envelope is a efficient membrane acceptor for phosphatidylcholine molecules transported by lipid transfer proteins (Miquel et al., 1987). Also, by transporting linoleoyl-phosphatidylcholine towards chloroplasts by lipid transfer proteins, Ohnishi and Yamada (1982) observed a synthesis of linolenoyl- monogalactosyldiacylglycerol. A phospholipase activity was suggested to be involved in this cooperative process (Oursel et al. 1987) ; the acyl chains could be available for desaturations within galactolipid molecules. The relation between lipid transfer proteins and acyl desaturases remains to be established.

IMMUNOCHEMICAL APPROACHES

A monospecific antibody against maize lipid transfer protein , raised in rabbits, has been used with various immunochemical methods : immunodiffusion , Western blots (Douady et al., 1986), ELISA (Grosbois et al., 1986). Three major results were obtained : i/ lipid transfer proteins are abundant in maize cytosols since they represent 2 to 4 % of total protein extracts; ii/ lipid transfer proteins have been found partly associated to microsomal and mitochondrial membranes. This binding could have a physiological significance ; iii/ the amount of lipid transfer protein increases during the developement of maize seedlings (Grosbois, unpublished).

MOLECULAR BIOLOGY

In order to study the physiological role of lipid transfer proteins , it is interesting to establish a relationship between the biosynthesis of these proteins and the biogenesis of membranes. The approaches of molecular biology could provide the necessary informations for studying the expression of genes involved in the biosynthesis of these proteins. In the first series of experiments, the biosynthesis in vivo (by following the incubation of labeled cysteine in coleoptiles) (Tchang et al., 1987) and in vitro (by using the RNA messengers extracted from coleoptiles) (Tchang et al., 1985) was studied in maize seedlings. These experiments clearly indicated the RNA messenger(s) coding for lipid transfer proteins is (are) present in these seedlings. A synthesis of isoforms of lipid transfer proteins has also been studied in castor bean seedlings (Yamada et al., this volume). Recent progress in the molecular biology of maize lipid transfer protein has been obtained by isolating two cDNA clones from cDNA librairies screened with the antibody against maize lipid transfer protein (Tchang et al., unpublished). One of the two clones contains a full-length cDNA.

With this finding, new perspectives are opened,especially the use of these cDNA as probes to study the variations of the levels of RNA messengers during the developement of maize seedlings or to search genomic clones coding for lipid transfer proteins.

Future experiments will involve the study of the regulation of the expression of genes coding for these proteins as well as the transformation of plants by genetic engineering in order to determine the physiological role(s) in vivo of lipid transfer proteins in plant cells.

REFERENCES

Bouillon P., Drischel C., Vergnolle V., Duranton H. and Kader J.C.- (1987) Complete primary structure of spinach leaf phospholipid transfer protein. Eur. J. Biochem. 166, 387-391

Douady D., Grosbois M., Guerbette F. and Kader J.C. (1982) - Purification of basic phospholipid transfer protein from maize seedlings. Biochim. Biophys. Acta, 710, 143-153.

Douady D., Guerbette F. et Kader J.C. (1985) - Purification of phospholipid transfer protein from maize seeds using a two-step chromatographic procedure. Physiol. Vég., 23, 373-380.

Douady D., Grosbois M. ,Guerbette F. and Kader J.C. (1986) - Phospholipid transfer protein from maize seedlings is partly membrane-bound. Plant Sci. 45, 151-156.

Douce R., Block M.A., Dorne A.J. and Joyard J. , 1984 - Subcell. Biochem. 10, 1-84.

Dubacq J.P., Drapier A., Trémolières A. and Kader J.C. (1984) - Role of phospholipid transfer protein in the exchange of phospholipids between microsomes and chloroplasts. Plant Cell Physiol., 25, 1197-1200.

Grosbois M., Guerbette F., Douady D. and Kader J.C. (1987) - Enzyme immunoassay of a plant phospholipid transfer protein. Biochim. Biophys. Acta. 917, 162-168.

Kader J.C. (1985)- Lipid-binding proteins in plants, Chem. Physics Lipids. 38,51-62.

Kader J.C.,Douady D. et Mazliak P. (1982) -Phospholipid transfer proteins in : Phospholipids, a comprehensive treatise, J.N.Hawthorne et G.B. Ansell éd, Elsevier/North Holland Biomedical Press, Amsterdam, 279- 311.

Kader J.C., Julienne M. and Vergnolle C. (1984) - Purification and characterization of a spinach leaf protein capable of transferring phospholipids from liposomes to mitochondria or chloroplasts. Eur. J. Biochem., 139, 411-416.

Mazliak P., Jolliot A. and Bonnerot C. ,1982 - Biosynthesis and Metabolism of phospholipids in : Biochemistry and Metabolism of Plant Lipids , J.F.G.M. Wintermans and P.J.C. Kuiper eds, Elsevier, pp.89-98.

Miquel M., Block M.A., Joyard J., Dorne A.J., Dubacq J.P. , Kader J.C.and Douce R. (1987) - Protein mediated transfer of phosphatidyl - choline from liposomes to spinach chloroplast envelope membranes. Biochim. Biophys. Acta 937, 219-228.

Moore T.S., 1984 - Biochemistry and biosynthesis of plant acyl lipids in : Structure, Function and Metabolism of Plant Lipids , Siegenthaler P.A.etEichenbergerW.,eds,Elsevier,Amsterdam, p. 93-96.

Moreau P., Lessire R. and Casagne C. (1984) - In vivo membrane transfer of very long chain fatty acids synthesized by etiolated leek seedlings in : Structure Function and Metabolism of plant Lipids , Siegenthaler P.A. et Eichenberger W., eds, Elsevier, Amsterdam, p. 307-310.

Nishida I. and Yamada M., 1986 - Semisynthesis of a spin-labeled monogalactosyldiacylglycerol and its application in the assay for galactolipid-transfer activity in spinach leaves. Biochim. Biophys. Acta 813, 298-306.

Ohnishi J.I. and Yamada M. , 1982 - Glycerolipid synthesis in Avena leaves during greening of etiolated seedlings . Plant Cell Physiol. 23 767-773.

Oursel A., Escoffier A. , Kader J.C., Dubacq J.P. and Trémolières A. , 1987 - Last step in the cooperative pathway for galactolipid synthesis. FEBS Lett. 219, 393-399.

Rickers J., Tober I. and Spener F., 1984 - Purification and binding characteristics of a basic fatty acid binding protein from Avena sativa seedlings. Biochim. Biophys. Acta 794, 313-319.

Rickers J., Spener F. and Kader J.C. (1985) - A phospholipid transfer protein that binds long-chain fatty acids. FEBS Lett., 180, 29-32.

Schwitzguebel J.P., Nguyen T.D., and Siegenthaler P.A. (1984) Are phospholipid exchange proteins present in the stroma from higher plant chloroplasts ? in : Structure Function and Metabolism of plant Lipids , Siegenthaler P.A. et Eichenberger W., eds, Elsevier, Amsterdam, p. 299 - 302.

Takishima, K., Watanabe, S., Yamada, M. and Mamiya, G. , 1986 - The amino acid sequence of the nonspecific lipid transfer protein from germinated castor bean endosperm.Biochim. Biophys. Acta 870, 248-250.

Tanaka T. and Yamada M., 1982 - Properties of phospholipid exchange proteins from germinated castor bean endosperms in : Biochemistry and Metabolism of Plant Lipids, J.F.G.M. and P.J.C.Kuiper Eds, Elsevier, p. 99-106.

Tchang F., Laroche-Raynal M., Vergnolle C., Demandre C., Douady D., Grosbois M., Guerbette F., Delseny M. and Kader J.C., 1985 - In vitro synthesis of a plant phospholipid transfer protein. A study by HPLC. Biochem. Biophys. Res. Commun., 133, 75-81.

Tchang F., Guerbette F., Douady D., Grosbois M., Vergnolle, Jolliot A., Dubacq J.P. and Kader J.C. 1987 - Properties and in vitro synthesis of phospholipid transfer proteins.in : Plant Lipid Metabolism, Plenum Press, P.K. Stumpf, J.B. Mudd, W. Nes ed., 353 - 356.

Watanabe, S. and Yamada, M., 1986 - Purification and characterization of a non-specific lipid transfer protein from germinated castor bean endosperms which transfers phospholipids and galactolipids.Biochim. Biophys. Acta 876, 116 - 123.

Yamada M., Tanaka T., Kader J.C. and Mazliak P. (1978) - Transfer of phospholipids from microsomes to mitochondria in germinating castor bean endosperm. Plant Cell Physiol., 19, 173-176.

BIOLOGICAL ROLE OF PLANT LIPIDS
P.A. BIACS, K. GRUIZ, T. KREMMER (eds)
Akadémiai Kiadó, Budapest and Plenum Publishing
Corporation, New York and London, 1989

GLYCEROL-3-PHOSPHATE ACYLTRANSFERASE AND ITS COMPLEMENTARY DNA

N. Murata, O. Ishizaki and I. Nishida

National Institute for Basic Biology, Myodaiji, Okazaki 444, Japan

CHILLING SENSITIVITY AND LIPID PHASE TRANSITION

Most tropical and sub-tropical plants suffer severe damage when they are exposed to chilling temperature such as 5-10°C (Lyons 1973). This phenomenon is termed chilling injury, and these plants are called chilling-sensitive plants. On the contrary, most temperate plants can survive at 0°C, and they are called chilling-resistant plants. Most crops of high productivity are chilling-sensitive, and the crop productivity in the temperate area is limited by the chilling sensitivity.

In the mechanism proposed by Raison (1973) and Lyons (1973) for the chilling sensitivity of plants, the primary event in chilling injury is the formation of a lipid gel phase in cellular membranes at low temperature. When a membrane goes into the phase separation state in which both gel and liquid crystalline phases co-exist, the membrane becomes leaky to small electrolytes, diminishing ion gradients across the membranes that are essential for the maintenance of physiological activities of the plant cells. We have shown that this mechanism operates in the chilling injury of the blue-green alga, Anacystis nidulans, in which the electrolytes leak out from the cytoplasm to the outer medium when the cytoplasmic membranes enter the phase separation state at low temperature (Ono and Murata 1981a,b, 1982; Murata et al. 1984; Murata and Nishida 1987). However, the lipid phase transition in higher-plant membranes has not been well demonstrated, and the validity of the proposed mechanism for chilling injury is still in question.

Higher-plant cells contain monogalactosyldiacylglycerol (MGDG), digalactosyldiacylglycerol (DGDG), sulfoquinovosyldiacylglycerol (SQDG) and phosphatidylglycerol (PG) in plastid membranes, phosphatidylcholine (PC), phosphatidylethanolamine (PE), diphosphatidylglycerol and PG in mitochondrial membranes, and PC, PE and phosphatidylinositol in endoplasmic reticulum and plasma membranes (Harwood 1980). Temperature for transition between the gel and liquid crystalline phases of glycerolipid molecular species varies markedly with a degree of unsaturation in their fatty acyl chains (Silvius 1982). The molecular species containing only saturated fatty acids such as palmitic acid (16:0) and stearic acid (18:0) reveal phase transition temperatures above 40°C. The molecular species of PC containing a cis-unsaturation bond reveal the phase transition near 0°C and introduction of the second cis-unsaturation bond decreases the phase transition temperature to about -20°C (Phillips et al. 1972). In contrast, the substitution of 16:0 by (3-trans)-hexadecenoic acid (16:1t) at the C-2 position of PG shifted the phase transition temperature only by 10°C (Bishop and Kenrick 1987). These findings suggest that if there are lipid molecular species which induce the membrane phase transition above 0°C, they should be fully saturated or trans-monounsaturated ones. In all glycerolipids from leaf cells, only PG contains high levels of these molecular species (Murata et al. 1982; Murata 1983; Raison and Wright 1983; Murata and Yamaya 1984).

MOLECULAR SPECIES OF PHOSPHATIDYLGLYCEROL AND THEIR BIOSYNTHESIS

Chilling-sensitive plants contain much higher proportions of 16:0/16:0 plus 16:0/16:1t species of PG than chilling-resistant plants (Table 1). In 20 plants examined the sum of the contents of these molecular species ranges from 3 to 19% of the total PG in the chilling-resistant plants, and from 26 to 65% in the chilling-sensitive plants. The phase transition temperatures of 16:0/16:0 and 16:0/16:1t molecular species of PG are 42 and 32°C, respectively (Murata and Yamaya 1984; Bishop and Kenrick 1987). These findings may suggest that these two molecular species are closely associated with the chilling sensitivity of plants. Roughan (1985a) has surveyed the fatty acid composition of PG in 74 plants to confirm the correlation between chilling sensitivity and saturated plus trans-monounsaturated PG molecular species. Li et al. (1987) have found that the correlation also exists among rice varieties having different sensitivities toward chilling. In addition to these two molecular species, 18:0/16:0 and 18:0/16:1t PG species exist at low proportions. Although the

Table 1. Mole percent of 16:0/16:0 plus 16:0/16:1t species to the total of PG from leaves.

Chilling-sensitive plants		Chilling-resistant plants	
Sweet potato	65	Pea	19
Squash	64	Lettuce	17
Taro	62	Cluster amaryllis	11
Rice	44	Red clover	10
Castor bean	42	Japanese radish	8
Tobacco	39	Spinach	6
Sponge cucumber	39	Oat	5
Maize	37	Cabbage	4
Kalanchoe	34	Dandelion	4
Cyclamen	26	Wheat	3

phase transition temperatures of these molecular species have not been measured, they are estimated to be higher than those of 16:0/16:0 and 16:0/16:1t PG species, respectively, since the replacement of 16:0 by 18:0

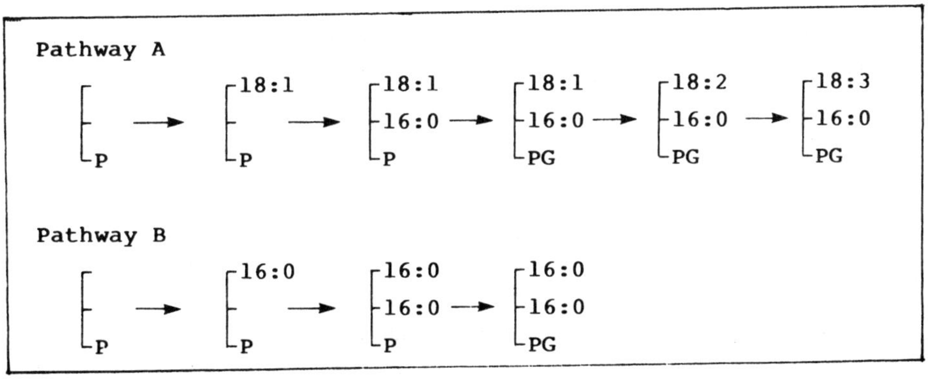

Fig. 1. Two pathways proposed for biosynthesis of the PG molecular species in the chloroplast of chilling-resistant and chilling-sensitive plants. Pathway A is dominant in the chilling-resistant plants, whereas both pathways A and B are of comparable activity in the chilling-sensitive plants. P, phosphate; PG, glycerophosphate.

should increase phase transition temperature. These four molecular species are called, in some occasions, "high-melting point" PG molecular species (Kenrick and Bishop 1986b), or "disaturated" PG molecular species (Roughan 1985a); in the latter case, the _trans_-unsaturated fatty acid is regarded as a saturated one because the _trans_-unsaturation does not decrease the phase transition temperature as much as _cis_-unsaturation (Bishop and Kenrick 1987). Another lipid class which contains saturated molecular species is SQDG (Murata and Hoshi 1984). The content of 16:0/16:0 plus 18:0/16:0 species relative to the total SQDG ranges from 0 to 20%. The phase transition temperatures of 16:0/16:0 species of SQDG is 43°C (Bishop et al. 1986). Although there is little correlation between the saturated SQDG and chilling sensitivity (Murata and Hoshi 1984), the saturated SQDG, if present, may interact with the saturated and _trans_-unsaturated PG molecular species by stimulating the formation of gel phase domains in the chloroplast membranes.

Sparace and Mudd (1982) as well as Roughan (1985b) demonstrated that PG is synthesized in chloroplasts, whereas the other chloroplast lipids, i.e., MGDG, DGDG and SQDG, are synthesized fully or partially under the cooperation between chloroplasts and endoplasmic reticula (Roughan and Slack 1982, 1984). Based on the result of the positional distribution of fatty acids in PG molecular species and the demonstration by Sparace and Mudd (1982) that PG is synthesized from glycerol 3-phosphate and acetate by chloroplasts, we have proposed biosynthetic pathways for the PG molecular species (Murata 1983), as presented in Fig. 1. In pathway A, oleic acid (18:1) is esterified to the C-1 position of glycerol 3-phosphate, and 16:0 to the C-2 position. After the PA thus produced is converted to PG, most of the 18:1 at the C-1 position is desaturated to linoleic acid (18:2) and subsequently to alpha-linolenic acid (18:3). A part of 16:0 at the C-2 position is converted to 16:1t. The combination of the fatty acids thus produced forms a variety of _cis_-unsaturated molecular species which undergo phase transition at about 0°C or below. In pathway B, 16:0 is esterified to both the C-1 and C-2 position of glycerol 3-phosphate. After the conversion of PA into PG, a part of 16:0 at the C-2 position is desaturated to 16:1t, whereas 16:0 at the C-1 position is not desaturated at all, resulting in the formation of only two molecular species, 16:0/16:0 and 16:0/16:1t. In chilling-resistant plants, in which the sum of the 16:0/16:0 plus 16:0/16:1t contents is low, pathway A should be favoured over pathway B. In chilling-sensitive plants, on the other hand, pathway A and pathway B should be comparably active, resulting in production of 16:0/16:0 and 16:0/16:1t PG species in addition to cis-unsaturated molecular species.

If such a scheme is valid, the chilling sensitivity of higher plants, or apparently the variation in the proportion of the 16:0/16:0 plus 16:0/16:1t PG molecular species in higher-plants, should result from the preferential transfer of 16:0 or 18:1 to the C-1 position of glycerol 3-phosphate. Such a preference is possible in one of the following mechanisms: (1) An enzyme which transfers the acyl group to the C-1 position of glycerol 3-phosphate has a different selectivity toward 18:1 and 16:0 between the chilling-sensitive and the chilling-resistant plants; i.e., the enzyme in the chilling-resistant plants has a rather strict selectivity for 18:1 over 16:0, whereas that in the chilling-sensitive plants is unspecific for either of the fatty acids and transfers both 18:1 and 16:0 at comparable rates: (2) Concentrations of the two substrates of this enzyme, 18:1-(acyl-carrier protein) (abbreviated as 18:1-ACP) and 16:0-ACP, differ between the two types of plants; the chilling-resistant plants contain much higher concentrations of 18:1-ACP than 16:0-ACP, whereas the chilling-sensitive plants contain comparable concentrations of the two substrates. However, Roughan (1986) has shown that the concentrations of the two substrates are about the same in both chilling-sensitive and chilling-resistant plants.

ACYL-(ACYL-CARRIER-PROTEIN):GLYCEROL-3-PHOSPHATE ACYLTRANSFERASE

Acyl-(acyl-carrier-protein):glycerol-3-phosphate acyltransferase (EC 2.3.1.15), abbreviated as glycerol-P acyltransferase, in higher-plant chloroplasts transfers the acyl group from acyl-ACP to the C-1 position of glycerol 3-phosphate to synthesize 1-acylglycerol 3-phosphate. This reaction is the first step in the biosynthesis of PG. According to our hypothesis as presented in Fig. 1, the substrate selectivity of this enzyme toward the acyl group in the acyl-ACP is very likely to be a determinant of the chilling sensitivity of higher plants.

This enzyme was purified first by Bertrams and Heinz (1981) in a soluble form from leaves of two chilling-resistant plants, pea and spinach, by ion-exchange chromatography, gel-filtration chromatography and isoelectric focusing. Recently, we have isolated the enzyme from cotyledons of a chilling-sensitive plant, squash, by ion-exchange chromatography, gel-filtration chromatography and affinity chromatography with acyl-carrier protein (Nishida et al, 1987). The squash cotyledons contain three isomeric forms designated as AT1, AT2 and AT3. AT2 and AT3 have been purified to single components after 40,000- and 32,000-fold purification, respectively. We have also found at least two isomeric forms of glycerol-P acyltransferase having different isoelectric points and molecular masses in all the plants examined, such as rice, barley, spinach, sunflower and maize (Dubacq, Douady, Nishida and Murata, unpublished data).

To determine whether the glycerol-P acyltransferase from chilling-sensitive plants exhibits fatty acid selectivities different from those of chilling-resistant plants, activities of enzymes from a chilling-sensitive plant, squash, and from chilling-resistant plants, spinach and pea, are compared (Frentzen et al. 1983, 1987). When about equal amounts of 18:1-ACP and 16:0-ACP are mixed as the substrate of the enzyme reaction, the enzymes from spinach and pea preferentially incorporate 18:1 to the C-1 position of glycerol 3-phosphate. Among the three isomeric forms from squash, AT1 shows a preference to 18:1-ACP as spinach and pea enzymes. AT2 and AT3, on the other hand, hardly discriminate 18:1-ACP and 16:0-ACP. However, the observed selectivity of AT1 preferable to 18:1-ACP is significantly reduced with increase in the pH of the reaction mixture from pH 7.4 to pH 8.0, the stromal pH of illuminated chloroplasts. The fatty acid selectivity of glycerol-P acyltransferase can therefore explain the finding that the content of 16:0/16:0 plus 16:0/16:1t PG molecular species is low in chilling-resistant plants and high in chilling-sensitive plants.

MOLECULAR CLONING OF GLYCEROL-P ACYLTRANSFERASE

A direct verification of the proposed role of glycerol-P acyltransferase in the chilling sensitivity of higher plants would be obtained by studies combined with the molecular cloning of glycerol-P acyltransferase and subsequent transformation of plants by the cloned gene. For this purpose, we have produced polyclonal antibodies against the AT3 by injections into mice. The antisera thus produced reacted with AT2 and AT3.

A lambda-gt11 random-primed cDNA library and a lambda-gt11 oligo(dT)-

primed cDNA library were constructed from poly(A)$^+$RNA from greening squash cotyledons. The random-primed cDNA library was screened with the antisera to obtain one positive clone, designated as lambda-AT01, containing a cDNA insert of 400 bp. The oligo(dT)-primed cDNA library was screened with the cDNA insert of lambda-AT01 to yield one positive clone. This clone, designated as lambda-AT02, contained a cDNA insert of 718 bp which covered the poly(A) tracks. The random-primed cDNA library was rescreened with the cDNA insert of lambda-AT01 to yield a clone, designated as lambda-AT03, containing a cDNA insert of 1,426 bp and an open-reading frame of 1,188 bp. The nucleotide sequence determination indicates that the cDNA inserts of the three clones overlap each other. The amino-acid sequence deduced from the nucleotide sequence of the open-reading frame is compared with the amino-terminal sequence of AT2 and the amino-terminal sequence of a CNBr-fragment of AT2 and AT3, suggesting that the cDNA insert of lambda-AT03 encodes either AT2 or AT3.

The amino-acid sequence deduced from the open-reading frame in lambda-AT03 is composed of 396 amino acids. However, when the deduced amino acid sequence is compared with the amino-terminal sequence determined for AT2, it appears that a precursor protein with a leader sequence of 28 amino acid residues is first synthesized and then processed to become a mature protein of 368 amino-acid residues (Table 2). The presence of the leader sequence is consistent with the suggestion (Feierabend 1982) that the chloroplast glycerol-P acyltransferase is one of the nuclear-coded proteins in which the leader peptide is postulated to be necessary for the transport of protein from the cytoplasm into the chloroplast stroma.

Table 2. Amino acid residues and relative molecular masses of precursor protein, leader peptide and mature protein of AT2 or AT3 deduced from the nucleotide sequence of glycerol-P acyltransferase cDNA.

Type of protein	Number of amino acid	M_r (Da)
Precursor protein	396	43,838
Leader peptide	28	2,927
Mature protein	368	40,929

ACKNOWLEDGEMENTS

This work was supported in part by a Grant-in-Aid for Scientific Research (61440002) from the Ministry of Education, Science and Culture, Japan, and by the NIBB Program for Biomembrane Research.

REFERENCES

Bertrams, M. and Heinz, E. (1981) Positional specificity and fatty acid selectivity of purified sn-glycerol 3-phosphate acyltransferases from chloroplasts. Plant Physiol. 68, 653-657

Bishop, D.G. and Kenrick, J.R. (1987) Thermal properties of 1-hexadecanoyl-2-trans-3-hexadecenoyl phosphatidylglycerol. Phytochemistry 26, 3065-3067

Bishop, D.G., Kenrick, J.R., Kondo, T. and Murata, N. (1986) Thermal properties of membrane lipids from two cyanobacteria, *Anacystis nidulans* and *Synechococcus* sp. Plant Cell Physiol. 27, 1593-1598

Feierabend, J. (1982). Inhibition of chloroplast ribosome formation by heat in higher plants. IN "Methods in Chloroplast Molecular Biology" (Edelman, M., Hallick, R.B. and Chua, N.H., eds.), pp. 671-680, Elsevier Biochemical, Amsterdam

Frentzen, M., Heinz, E., McKeon, T.A. and Stumpf, P.K. (1983) Specificities and selectivities of glycerol-3-phosphate acyltransferase and monoacylglycerol-3-phosphate acyltransferase from pea and spinach chloroplasts. Eur. J. Biochem. 129 , 629-636

Frentzen, M., Nishida, I. and Murata, N. (1987) Properties of the plastidial acyl-(acyl-carrier-protein):glycerol-3-phoshate acyltransferase from the chilling-sensitive plant, squash (*Cucurbita moschata*). Plant Cell Physiol. 28, 1195-1201

Harwood, J.L. (1980) Plant acyl lipids: structure, distribution, and analysis. In "The Biochemistry of Plants", Vol. 4, (Stumpf, P.K., ed.), pp. 1-55, Academic Press, New York

Kenrick, J.R. and Bishop, D.G. (1986) The fatty acid composition of phosphatidylglycerol and sulfoquinovosyldiacylglycerol of higher plants in relation to chilling sensitivity. Plant Physiol. 81, 946-949

Li, T., Lynch, D.V. and Steponkus, P.L. (1987) Molecular species composition of phosphatidylglycerols from rice varieties differing in

chilling sensitivity. Cryo. Lett. 8, 314-321
Lyons, J.M. (1973) Chilling injury in plants. Annu. Rev. Plant Physiol. 24, 445-466.
Murata, N. (1983) Molecular species composition of phosphatidylglycerols from chilling-sensitive and chilling-resistant plants. Plant Cell Physiol. 24, 81-86
Murata, N. and Hoshi, H. (1984) Sulfoquinovosyl diacylglycerols in chilling-sensitive and chilling-resistant plants. Plant Cell Physiol. 25, 1241-1245.
Murata, N. and Nishida, I. (1987) Lipids of blue-green algae (cyanobacteria). In "The Biochemistry of Plants", Vol. 9, (Stumpf, P.K., ed.), pp. 315-347, Academic Press, London
Murata, N. and Yamaya, J. (1984) Temperature-dependent phase behavior of phosphatidylglycerols from chilling-sensitive and chilling-resistant plants. Plant Physiol. 74, 1016-1024
Murata, N., Sato, N., Takahashi, N. and Hamazaki, I. (1982) Compositions and positional distribution of fatty acids in phospholipids from leaves of chilling-sensitive and chilling-resistant plants. Plant Cell Physiol. 23, 1071-1079
Murata, N., Wada, H. and Hirasawa, R. (1984) Reversible and irreversible inactivation of photosynthesis in relation to the lipid phases of membranes in the blue-green algae (cyanobacteria) Anacystis nidulans and Anabaena variabilis. Plant Cell Physiol. 25, 1027-1032
Nishida, I., Frentzen, M., Ishizaki, O. and Murata, N. (1987) Purification of isomeric forms of acyl-(acyl-carrier-protein):glycerol-3-phosphate acyltransferase from greening squash cotyledons. Plant Cell Physiol. 28, 1071-1079
Ono, T. and Murata, N. (1981a) Chilling susceptibility of the blue-green alga Anacystis nidulans. I. Effect of growth temperature. Plant Physiol. 67, 176-181
Ono, T. and Murata, N. (1981b) Chilling susceptibility of the blue-green alga Anacystis nidulans. II. Stimulation of the passive permeability of cytoplasmic membrane at chilling temperatures. Plant Physiol. 67, 182-187
Ono, T. and Murata, N. (1982) Chilling susceptibility of the blue-green alga Anacystis nidulans. III. Lipid phase of cytoplasmic membrane. Plant Physiol. 69, 125-129
Phillips, M.C., Hauser, H. and Paltauf, F. (1972) The inter- and

intra-molecular mixing of hydrocarbon chains in lecithin/water systems. Chem. Phys. Lipids 8, 127-133

Raison, J.K. (1973) The influence of temperature-induced phase changes on kinetics of respiratory and other membrane-associated enzyme. J. Bioenerg. 4, 258-309

Raison, J.K. and Wright, L.C. (1983) Thermal phase transitions in the polar lipids of plant membranes. Their induction by disaturated phospholipids and their possible relation to chilling injury. Biochim. Biophys. Acta 731, 69-78

Roughan, P.G. (1985a) Phosphatidylglycerol and chilling sensitivity in plants. Plant Physiol. 77, 740-746

Roughan, P.G. (1985b) Cytidine triphosphate-dependent, acyl-CoA-independent synthesis of phosphatidylglycerol by chloroplasts isolated from spinach and pea. Biochim. Biophys. Acta 835, 527-532

Roughan, P.G. (1986) Acyl lipid synthesis by chloroplasts isolated from the chilling-sensitive plant Amaranthus lividus L. Biochim. Biophys. Acta 878, 371-379

Roughan, P.G. and Slack, C.R. (1982) Cellular organization of glycerolipid metabolism. Annu. Rev. Plant Physiol. 33, 97-132

Roughan, P.G. and Slack, C.R. (1984) Glycerolipid synthesis in leaves. Trends Biochem. Sci. 9, 383-386

Silvius, J.R. (1982) Thermotropic phase transition of pure lipids in model membranes and their modification by membrane proteins. In "Lipid-Protein Interactions", Vol. 2, (Jost, P.C. and Griffith, O.H., eds.), pp. 239-281, John Wiley and Sons, New York

Sparace, S.A. and Mudd, B. (1982) Phosphatidylglycerol synthesis in spinach chloroplasts: Characterization of the newly synthesized molecule. Plant Physiol. 70, 1260-1264

BIOLOGICAL ROLE OF PLANT LIPIDS
P.A. BIACS, K. GRUIZ, T. KREMMER (eds)
Akadémiai Kiadó, Budapest and Plenum Publishing
Corporation, New York and London, 1989

CLONING AND SEQUENCE DETERMINATION OF A COMPLEMENTARY DNA FOR PLASTID GLYCEROL-3-PHOSPHATE ACYLTRANSFERASE FROM SQUASH

O. Ishizaki, I. Nishida, K. Agata, G. Eguchi and N. Murata

National Institute for Basic Biology, Myodaiji, Okazaki 444, Japan

Acyl-(acyl-carrier protein):glycerol-3-phosphate acyltransferase (EC 2.3.1.15), designated as glycerol-P acyltransferase, in higher-plant chloroplast transfers the acyl group from acyl-(acyl-carrier protein) to the C-1 position of glycerol 3-phosphate to synthesize lysophosphatidic acid. Previously, we purified two isomeric forms of the glycerol-P acyltransferase from squash cotyledons (Nishida et al. 1987). Polyclonal antibodies against one of the isomeric forms were prepared by injection into mice.

Three clones of cDNAs encoding glycerol-P acyltransferase were isolated from λgt11 cDNA libraries made from poly$(A)^+$ RNA of squash cotyledons by immunological selection and cross-hybridization. Fig. 1 shows the sequence of the isolated cDNAs which had 1,781 bp as a whole. One of the resultant clones contained an open-reading frame of 1,188 bp. The amino-acid sequence deduced from the nucleotide sequence matched the partial amino-acid sequence determined for the enzyme. The results suggest that a precursor protein of 396 amino-acid residues is synthesized first, and then processed to a mature enzyme of 368 amino-acid residues.

REFERENCE

Nishida, I., Frentzen, M., Ishizaki, O. and Murata, N. (1987). Purification of isomeric forms of acyl-[acyl-carrier-protein]:glycerol-3-phosphate acyltransferase from greening squash cotyledons. *Plant Cell Physiol.* 28, 1071-1079.

Fig. 1. cDNA nucleotide sequence and deduced amino-acid sequence of the squash glycerol-P acyltransferase. The arrow indicates the predicted processing site. Underlined portions of amino-acid sequence matched the partial amino-acid sequences determined for the glycerol-P acyltansferase.

```
←
   1 GCCTTCTCTAGGGTTTCTCTTCCTCCCTCCTCCTCTTCTCTCAAGCTCCTTCCCCTTTCTCTGCAATTCGGACCTCCCA
  80 AGCTTGCCTCCTCGTGCTCGCTTCGGTTTTCCGCTTCCAGAGCA

 124 ATG GCG GAG CTT ATC CAG GAT AAG GAG TCC GCC CAG AGT GCT GCC ACC GCT GCT GCT GCT
 -28 Met Ala Glu Leu Ile Gln Asp Lys Glu Ser Ala Gln Ser Ala Ala Thr Ala Ala Ala Ala
                                    ↓
 184 AGC TCC GGT TAT GAA AGA CGG AAT GAG CCG GCT CAC TCC CGC AAA TTT CTC GAT GTT CGC
  -8 Ser Ser Gly Tyr Glu Arg Arg Asn Glu Pro Ala His Ser Arg Lys Phe Leu Asp Val Arg

 244 TCT GAA GAA GAG TTG CTC TCC TGC ATC AAG AAG GAA ACA GAA GCT GGA AAG CTG CCT CCA
  13 Ser Glu Glu Glu Leu Leu Ser Cys Ile Lys Lys Glu Thr Glu Ala Gly Lys Leu Pro Pro

 304 AAT GTT GCT GCA GGA ATG GAA GAA TTG TAT CAG AAT TAT AGA AAT GCT GTT ATT GAG AGT
  33 Asn Val Ala Ala Gly Met Glu Glu Leu Tyr Gln Asn Tyr Arg Asn Ala Val Ile Glu Ser

 364 GGA AAT CCA AAG GCA GAT GAA ATT GTT CTG TCT AAC ATG ACT GTT GCA TTA GAT CGC ATA
  53 Gly Asn Pro Lys Ala Asp Glu Ile Val Leu Ser Asn Met Thr Val Ala Leu Asp Arg Ile

 424 TTG TTG GAT GTG GAG GAT CCT TTT GTC TTC TCA TCA CAC CAC AAA GCA ATT CGA GAG CCT
  73 Leu Leu Asp Val Glu Asp Pro Phe Val Phe Ser Ser His His Lys Ala Ile Arg Glu Pro

 484 TTT GAT TAC TAC ATT TTT GGC CAG AAC TAT ATA CGG CCA TTG ATT GAT TTT GGA AAT TCA
  93 Phe Asp Tyr Tyr Ile Phe Gly Gln Asn Tyr Ile Arg Pro Leu Ile Asp Phe Gly Asn Ser

 544 TTC GTT GGT AAC CTT TCT CTT TTC AAG GAT ATA GAA GAG AAG CTT CAG CAG GGT CAC AAT
 113 Phe Val Gly Asn Leu Ser Leu Phe Lys Asp Ile Glu Glu Lys Leu Gln Gln Gly His Asn

 604 GTT GTC TTG ATA TCA AAT CAT CAG ACT GAA GCA GAT CCA GCT ATC ATT TCA TTG TTG CTT
 133 Val Val Leu Ile Ser Asn His Gln Thr Glu Ala Asp Pro Ala Ile Ile Ser Leu Leu Leu

 664 GAA AAG ACA AAC CCA TAT ATT GCA GAA AAC ACG ATC TTT GTG GCA GGG GAT AGA GTT CTT
 153 Glu Lys Thr Asn Pro Tyr Ile Ala Glu Asn Thr Ile Phe Val Ala Gly Asp Arg Val Leu

 724 GCA GAC CCT CTT TGC AAG CCC TTC AGC ATT GGA AGG AAT CTT ATT TGT GTT TAT TCA AAA
 173 Ala Asp Pro Leu Cys Lys Pro Phe Ser Ile Gly Arg Asn Leu Ile Cys Val Tyr Ser Lys

 784 AAG CAC ATG TTC GAT ATT CCT GAG CTC ACA GAA ACA AAA AGG AAA GCA AAC ACA CGA AGT
 193 Lys His Met Phe Asp Ile Pro Glu Leu Thr Glu Thr Lys Arg Lys Ala Asn Thr Arg Ser

 844 CTT AAG GAG ATG GCT TTA CTC TTA AGA GGT GGA TCA CAA CTA ATA TGG ATT GCA CCC AGT
 213 Leu Lys Glu Met Ala Leu Leu Leu Arg Gly Gly Ser Gln Leu Ile Trp Ile Ala Pro Ser

 904 GGT GGT AGG GAC CGG CCG GAT CCC TCG ACT GGA GAA TGG TAC CCA GCA CCC TTT GAT GCT
 233 Gly Gly Arg Asp Arg Pro Asp Pro Ser Thr Gly Glu Trp Tyr Pro Ala Pro Phe Asp Ala

 964 TCT TCA GTG GAC AAC ATG AGA AGG CTT ATT CAA CAT TCG GAT GTT CCT GGG CAT TTG TTT
 253 Ser Ser Val Asp Asn Met Arg Arg Leu Ile Gln His Ser Asp Val Pro Gly His Leu Phe

1024 CCC CTT GCT TTA TTA TGT CAT GAC ATC ATG CCC CCT CCC TCA CAG GTC GAA ATT GAA ATT
 273 Pro Leu Ala Leu Leu Cys His Asp Ile Met Pro Pro Pro Ser Gln Val Glu Ile Glu Ile

1084 GGA GAA AAA AGA GTG ATT GCC TTT AAT GGG GCG GGT TTG TCT GTG GCT CCT GAA ATC AGC
 293 Gly Glu Lys Arg Val Ile Ala Phe Asn Gly Ala Gly Leu Ser Val Ala Pro Glu Ile Ser

1144 TTC GAG GAA ATT GCT GCT ACC CAC AAA AAT CCT GAG GAG GTT AGG GAG GCA TAC TCA AAG
 313 Phe Glu Glu Ile Ala Ala Thr His Lys Asn Pro Glu Glu Val Arg Glu Ala Tyr Ser Lys

1204 GCA CTG TTT GAT TCT GTG GCC ATG CAA TAC AAT GTG CTC AAA ACG GCT ATC TCC GGC AAA
 333 Ala Leu Phe Asp Ser Val Ala Met Gln Tyr Asn Val Leu Lys Thr Ala Ile Ser Gly Lys

1264 CAA GGA CTA GGA GCT TCA ACT GCG GAT GTC TCT TTG TCA CAA CCT TGG TAG
 353 Gln Gly Leu Gly Ala Ser Thr Ala Asp Val Ser Leu Ser Gln Pro Trp AM

1315 TCATTTGCAATCATTTTTCAACATCAATTCATTTGGTAATCAGGTTAGGACATAGTTTTGCATACCACAGGACAACACT
1394 GTCATTCAATTAGTACATGTACTGTAAGATAAAACCGAATTCTTTCTCTCCTCGACGTCTCGATGGATTCTGCTAAATT
1473 ACCAGCCTATATCCCACCCTAAGCACGTGCCATTTCAATCGACATCGAAGCACTTCGAATACTTGGTCTCTCATGGGAA
1552 TGTATAGATCTTTCATATCTCTTACACTAAAGGACTCGCAGAGGTTATTCGTACTTTATTTCTAAAGGTATATGTTCAC
1631 CCATTTTCATGTTTATATTATATGCTTCAAGAATTATTATGTTCAATTTGTATTATTAGTTGATACCTTTTGGCTTAGA
1710 TCATTCAGTCGAAGGTGCTCAAATCTCAACTTTGTTTTTGCTCTTAAGTTGTTGAAGGAACATTTTTAACAC
```

BIOLOGICAL ROLE OF PLANT LIPIDS
P.A. BIACS, K. GRUIZ, T. KREMMER (eds)
Akadémiai Kiadó, Budapest and Plenum Publishing
Corporation, New York and London, 1989

GLYCEROL-3-PHOSPHATE ACYLTRANSFERASE FROM SQUASH CHLOROPLASTS: PROTEIN CHARACTERIZATION BY PURIFICATION AND cDNA CLONING

I. Nishida, O. Ishizaki and N. Murata

National Institute for Basic Biology, Myodaiji, Okazaki 444, Japan

INTRODUCTION

Acyl-(acyl-carrier protein):glycerol-3-phosphate acyltransferase (glycero-P acyltransferase) in higher-plant chloroplasts transfers the acyl group from acyl-(acyl-carrier protein) to the C-1 position of glycerol 3-phosphate to synthesize 1-acylglycerol 3-phosphate. Since this reaction is the first step of glycerolipid synthesis in the chloroplasts, it is of special interest to study this enzyme.

MATERIALS AND METHODS

Isomeric forms of squash glycerol-P acyltransferase were purified from squash cotyledons as described previously (Nishida et al. 1987). Amino-acid sequences were determined with an amino-acid sequence analyzer (Model 470A, Applied Biosystems) equipped with a liquid chromatograph (PTH analyzer, Model 120A, Applied Biosystems). The amino acid composition was analyzed with an amino-acid analyzer (Model 835, Hitachi) after the protein was hydrolyzed with HCl at 110 ^{o}C for 24 h.

RESULTS AND DISCUSSION

Three isomeric forms of glycerol-P acyltransferase, designated as AT1, AT2 and AT3, were found in squash cotyledons (Nishida et al. 1987). AT2 and AT3 were purified to single components after 40,000- and 32,000-fold purification, respectively, on the basis of specific activity. Table 1 summarizes physicochemical properties of the purified AT2 and AT3. AT2 and AT3 were very similar to each other in their relative molecular masses and

Table 1. Molecular mass and isoelectric point of isomeric forms of glycerol-P acyltransferase from squash cotyledons (Nishida et al. 1987)

Isomeric form	Molecular mass, kDa			Isoelectric point
	Gel filtration[a]		SDS-PAGE[b]	
	TSKgel	Superose		
AT2	38	39	39	5.6
AT3	38	39	39	5.5

isoelectric points; about 40 kDa and pI 5.5-5.6, respectively. Immunochemical properties of AT2 and AT3 were also similar; mouse antisera raised against AT3 reacted with AT2. Only one difference so far demonstrated is that AT3 but not AT2 can be adsorbed to a hydroxyapatite column.

Partial amino-acid sequence and amino-acid composition were determined for AT2 and AT3. The amino-acid sequence at the amino terminus of AT2 was X1-Pro2-Ala3-His4-Ser5-Arg6-Lys7-Phe8-Leu9-Asp10-Val11-Arg12-Ser13-Glu14-Glu15-Glu16-Leu17-Leu18; X represents an amino-acid which was present but could not be identified. Both AT2 and AT3 were fragmented with CNBr to yield polypeptides having relative molecular masses of about 18 kDa. The amino-acid sequence at the amino terminus of the 18-kDa fragment from AT2 was X1-X2-Leu3-Tyr4-Gln5-Asn6-Tyr7-Arg8-Asn9-Ala10-Val11-Ile12-Glu13-Ser14-Gly15-Asn16-Pro17-Lys18-Ala19-X20-X21-Ile22-Val23. The amino-terminal sequence of 18-kDa fragment from AT3 determined for 14 amino-acid residues was identical with the corresponding sequence of the 18-kDa fragment from AT2. Table 2 compares the amino-acid compositions of AT2 and AT3. They were almost identical with each other, and the minor difference was within the experimental error.

We have isolated a cDNA clone encoding either AT2 or AT3 and deduced the total amino-acid sequence (Ishizaki et al. 1988). The partial amino-acid sequences determined for AT2 and AT3 were found in the deduced sequence. The amino-acid composition calculated for the deduced sequence was also in good agreement with those determined for AT2 and AT3 (Table 2).

Table 2. Amino-acid composition of AT2 and AT3 and that deduced from nucleotide sequence of cDNA for glycerol-P acyltransferase

Amino acid	AT2	AT3	Deduced from cDNA
Gly	22	24	21
Ala	31	32	30
Val	22	21	22
Leu	37	37	36
Ile	24	25	27
Phe	16	16	16
Ser	25	25	29
Thr	13	12	12
Tyr	11	10	10
Met	7	6	7
Cys	nd*	nd	4
Lys	20	20	20
His	10	10	10
Pro	25	25	25
Arg	18	18	17
Trp	nd	nd	3
Asx	39	39	38
Glx	41	41	41
Total	361	361	368

*nd: not determined.

The hydrophilicity profile of the deduced amino-acid sequence does not show any cluster of hydrophobic regions which may correspond to a trans-membrane structure, but rather a homogeneous distribution of hydrophilic regions (Fig. 1). This is consistent with the finding that the squash glycerol-P acyltransferase is a soluble protein.

Glycerol-P acyltransferase of E. coli is a membrane-bound protein and its DNA has been cloned (Lightner et al. 1983). The homology between the squash and E. coli glycerol-P acyltransferases in their deduced amino-acid

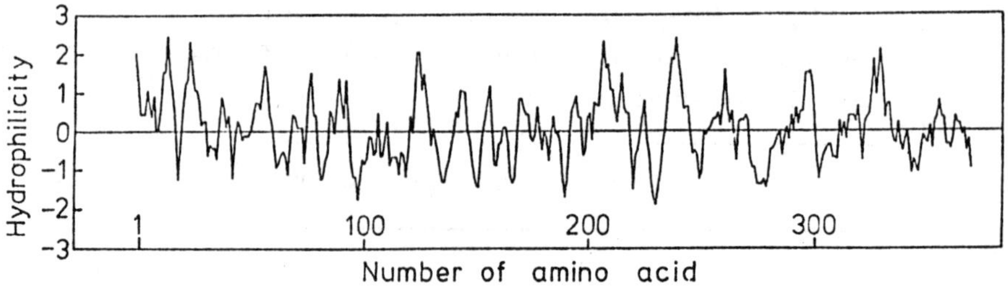

Figure 1. Hydrophilicity profile (calculated according to Hopp and Woods 1981) of the deduced amino-acid sequence of the squash glycerol-P acyltransferase.

sequences was less than 10%, and no high homology was found even in partial sequences.

REFERENCES

Hopp, T. P. and Woods, K. R. (1981). Prediction of protein antigenic determinants from amino acid sequences. Proc. Natl. Acad. Sci. USA 78, 3824-3828.

Ishizaki, O., Nishida, I., Agata, K., Eguchi, G. and Murata, N. (1988). Cloning and sequence determination of a cDNA for plastid glycerol-3-phosphate acyltransferase from squash. In this book.

Lightner, V. A., Bell, R. M. and Modrich, P. (1983). The DNA sequences encoding plsB and dgk loci of Escherichia coli. J. Biol. Chem. 258, 10856-10861.

Nishida, I., Frentzen, M., Ishizaki, O. and Murata, N. (1987). Purification of isomeric forms of acyl-[acyl-carrier-protein]:glycerol-3-phosphate acyltransferase from greening squash cotyledons. Plant Cell Physiol. 28, 1071-1079.

BIOLOGICAL ROLE OF PLANT LIPIDS
P.A. BIACS, K. GRUIZ, T. KREMMER (eds)
Akadémiai Kiadó, Budapest and Plenum Publishing
Corporation, New York and London, 1989

ACYL CARRIER PROTEINS OF BARLEY SEEDLING LEAVES AND CARYOPSES

L. Hansen and Penny von Wettstein-Knowles

Department of Physiology, Carlsberg Laboratory, Gamle Carlsberg Vej 10, 2500 Copenhagen Valby and Genetics Institute, Øster Farimagsgade 2A, 1353 Coppenhagen, Denmark

INTRODUCTION

Transferring dark grown, six day old barley seedlings into the light initiates development of chloroplasts from proplastids. Concomitantly with this biogenesis the amount of fatty acyl chains in ng/g fresh weight approximately doubles. The majority of the biosynthetic steps involved in the latter require that the elongating chains are presented to the enzymes while bound through a thioester linkage to the prosthetic group of a small cofactor-like protein known as acyl carrier protein (ACP). Several years ago primary structure analysis (1) revealed that barley chloroplasts contain two mature ACPs (I and II). As a step toward understanding the small ACP gene family we have prepared cDNAs from mRNAs isolated from greening barley seedlings. This report summarizes our observations on these clones.

RESULTS AND DISCUSSION

Since only two residues from position 38-59 of barley ACP I differ from the analogous spinach ACP I region and a cDNA sequence was available (2) for the latter, a very specific probe was synthesized for screening the library. The resulting 66 base long oligonucleotide was identical to the spinach sequence except that appropriate barley codons were substituted at the two discrepant sites. This probe readily hybridized to poly A^+ RNA from barley seedlings and immature caryopses giving bands at 1020-1050 and 800 bases, respectively (3).

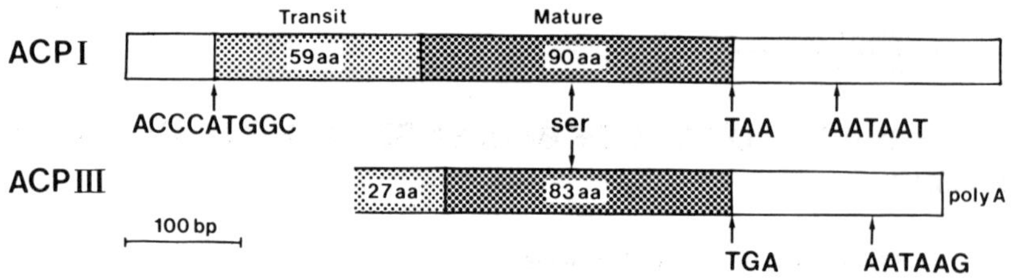

Fig. 1. Diagrams of cDNA sequences of ACP I (pACP 11, 768 bases) and ACP III (pACP 1, 520 bases) genes expressed in greening barley seedlings. Included are an apparent initiation sequence, stop codons, presumed polyadenylation signals, a poly A tail, and the regions coding for a complete or partial transit peptide as well as the mature proteins having the serine residue to which 4' phosphopantetheine is attached.

The lengths of the barley cDNA inserts in the 22 positive clones isolated from the λgt11 library ranged from 300-800 bases. Initially several of the clones differing in length were subcloned into pUC 18 and completely sequenced (3). Fig. 1 is a summary of their salient features. Mature ACP I has 90 residues as deduced from the cDNA sequence of pACP 11, whereas pACP 1 codes for a mature protein of 83 residues designated ACP III. While the overlapping cDNA sequences corresponding to the mature proteins are 71% homologous the 5′ and 3′ regions are not homologous. Also at the primary structure level the deduced 27 amino acid partial transit peptide of ACP III shows no obvious similarities to that of the 59 amino acid transit peptide of ACP I. The deduced 24 amino terminal residues of ACP III differ from those reported for ACP II (1) in nine positions. The above analyses infer that three genes for ACP are expressed in greening barley seedlings. While ACP I and II occur in chloroplasts, the subcellular localization of ACP III is unknown. In addition, a fourth functional ACP gene may exist in barley since the 66 base long probe hybridizes to a significantly smaller poly A^+ RNA fom immature caryopses than from seedlings.

The restriction enzymes Sal I, Pvu II and Ava II can cut ACP I but not ACP III cDNAs, whereas the opposite is true for Pst I and Kpn I (Fig. 2). By this criterium 13 of the clones code for ACP I and three for ACP III. The precise ends of some of these cDNA inserts have been determined by sequencing with the results shown in Fig. 2. Interestingly, nearly all of the ACP I coding inserts end at the same base as pACP 11 does, and none contains a

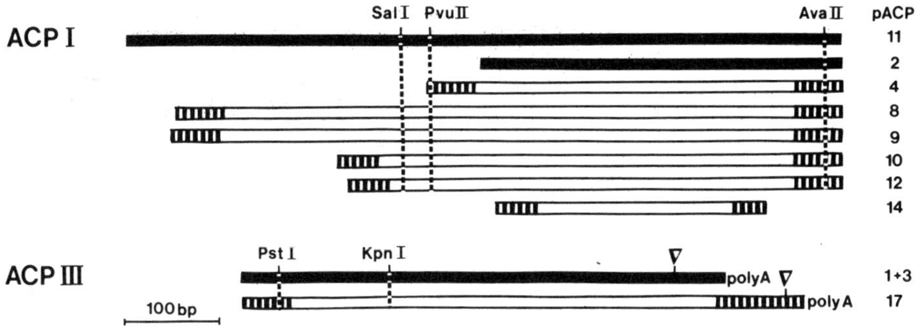

Fig. 2. Diagrams of 11 cDNA clones showing restriction sites which distinguish ACP I from ACP III. Solid bars, completely sequenced (3); broken-open bars, end regions partially sequenced to aid clone identification and precise length determinations. Triangles indicate polyadenylation signals for pACPs 1 and 3 (AATAAG) and pACP 17 (AATATG). The latter is located 100 bases downstream from the former.

poly A tail. Presumably a secondary structure of this mRNA leads to a loss of its 3′ end including the poly A tail during cDNA synthesis. Present information implies that the 13 ACP I inserts represent the same mature mRNA. While the three ACP III cDNAs appear to arise from the same primary transcript, the latter can be processed in two ways by the use of alternate polyadenylation signals (Fig. 2). Initial observations on the other six clones suggest they all code for ACP I.

Fig. 3 shows that ACP I and III inserts hybridize to genomic DNA digested with Bam HI and Xba I. These restriction enzymes were used as neither cuts the cDNAs for ACP I or III. Only a single strong band is seen in each lane suggesting that if more than one copy of either gene exists they are located on the same restriction fragment. Also if introns occur within these genes they do not include Bam HI or Xba I sites. By contrast ACP I cDNA hybridizes to two fragments in genomic DNA digested with Hind III which cuts once in the ACP I insert. The most fascinating observation derived from Fig. 3 is that despite the 71% homology between the 250 overlapping bases coding for the mature ACP I and III proteins, neither insert hybridizes to fragments detected by the other. This infers that the bands seen in Fig. 3 arise primarily from hybridization of the 3′ and 5′ ends of the inserts. Do the faint bands at approximately 3.3 kb result from a weak homo-

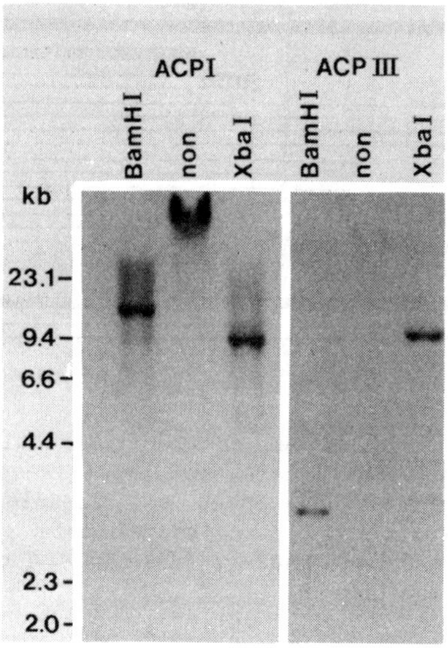

Fig. 3. Hybridization of ACP I and ACP III cDNA inserts to 10 μg non-digested, Bam HI and Xba I digested genomic DNA from Bonus barley. The DNA was electrophoresed in a 0.7% agarose gel. After transfer to a nitrocellulose filter, prehybridization was done in a solution containing 2 x SSC, 1 x Denhardt, 0.1% SDS and 100 μg/ml denatured, sheared carrier DNA at 65°C. Hybridizations were carried out in 50 ml of the same solution at 65°C after addition of the randomly oligo labelled probes (2.5 ng/ml, 10^8 cpm/μg). The filters were washed in 2 x SSC at room temperature and then in 0.2 SSC, 0.1% SDS at 65°C before autoradiography. Hind III digested λ DNA served as size markers. See (3) for abbreviations.

logy between ACP II and III? Certainly the lack of strong bands that can be attributed to ACP II suggests that the 3′ and 5′ ends of its mRNA will not show extensive homology to those of ACPs I and III.

REFERENCES

1. Høj, P.B. and I. Svendsen (1984) Barley chloroplasts contain two acyl carrier proteins coded for by different genes. Carlsberg Res. Commun. 49: 483-492

2. Scherer, D.E. and V.C. Knauf (1987) Isolation of a cDNA clone for the acyl carrier protein-I of spinach. Plant Mol. Biol. 9:127-134

3. Hansen, L. (1987) Three cDNA clones for barley leaf acyl carrier proteins I and III. Carlsberg Res. Commun. 52:381-392

BIOLOGICAL ROLE OF PLANT LIPIDS
P.A. BIACS, K. GRUIZ, T. KREMMER (eds)
Akadémiai Kiadó, Budapest and Plenum Publishing
Corporation, New York and London, 1989

IMMUNOCHEMICAL ASPECTS OF PEA CHLOROPLASTS ACYL TRANSFERASE

J.P. Dubacq[1,2], D. Douady[2] and C. Passaquet[1]

[1]Laboratoire des Biomembranes et Surfaces Cellulaires Végétales 04311, 46 Rue d'Ulm, 75005 Paris,
[2]Laboratoire de Physiologie Cellulaire (UA CNRS 041180), Tour 53/3, 4 Place Jussieu, 75005 Paris, France

INTRODUCTION

The acyltransferases are enzymes catalyzing the first steps of membrane lipid biosynthesis by sucessive esterification of the glycerol-3-phosphate hydroxyl groups with fatty acids. The first acyltransferase which has been extensively studied by Frentzen"et al" (1983),is present in a soluble form in the stroma of chloroplasts where it catalyzes the reaction Acyl-ACP + glycerol-3-phosphate - lysophosphatidic acid + ACP. Depending on plants, several isoforms of this soluble acyltransferase are present. According to Murata (1983),the differences between plants in the selectivity of this acyltransferase for either saturated or monounsaturated fatty acids would be responsible for the occurence of disaturated lipids in chloroplasts and consequently for their chilling sensitivity.Furthermore, the activity of the two successive acyltransferases in chloroplast leads to 1-C_{18} -2-C_{16} glycerolipids which look like the cyanobacterial lipids. This similarity between chloroplast and cyanobacteria for their membrane lipid molecular species can be related to the endosymbiotic theory.Hence investigations concerning the acytransferases are of great interest for both these reasons.

The present work deals with purification and immunological properties of pea chloroplast acyl-ACP:glycerol-3-phosphate acyltransferase.

METHODS

Enzyme activity was assayed according to Bertrams and Heinz (1981). The SDS-PAGE, blotting, immunological detection and enzyme immunoassay were performed as previously described (Douady et al. 1986, Grosbois et al. 1987) The protein concentration was determined according to Bradford (1976).Stroma and thylakoid membranes were isolated from Percoll purified chloroplasts.

RESULTS

The Acyl-CoA or Acyl-ACP: glycerol-3-phosphate acyltransferase was purified (Douady and Dubacq 1987). Homogeneity was achieved by FPLC on a MonoQ column (fig.1A).The activity estimated with Acyl-CoA indicated that the purification factor was about 15000 (Table 1).

	Total activity nmoles x min^{-1}	Yield %	Specific activity nmoles x $min^{-1} mg^{-1}$	Purification factor
Homogenate	8184	100	0.55	1
pH 5.1	9945	121	7.65	14
$(NH_4)_2SO_4$	8060	98	13.48	24
Sephadex	8011	98	87.70	165
D E A E	4514	55	419	761
FPLC monoQ	336	4	8400	15200

Table 1: Purification steps of acyl-CoA:glycerol-3-phosphate acyltransferase from Pea leaves.

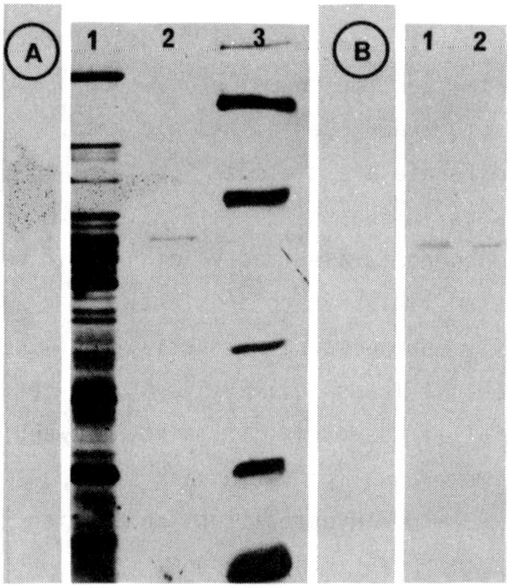

Fig.1: Electrophoresis and immunoblotting of Pea acyltransferase. A-SDS PAGE after silver staining 1- DEAE fraction, 2- FPLC purified acyltransferase, 3-M.W. markers. B- 1 and 2 Western blot of A-1 and 2, respectively, stained with horse radish peroxydase complex.

The acyltransferase isolated from pea (40.5 kDa, pI 6.6) was shown to be selective for oleic acid (ACP or CoA thioesters).This protein differed from the acyltransferases isolated by Nishida "et al." (1987) and Frentzen "et al." (1987), from a chilling sensitive plant (squash) by both pI,molecular mass and enzymatic properties (absence of selectivity for oleic acid) in addition, the pea acyltransferase was sensitive to NaCl and oleic acid (data not shown).

Using this purified protein, antibodies were raised in rabbit.Immunoblotting proved the specificity of antibodies for the pea acyltransferase at different steps of the purification procedure.The figure 1-B shows that, despite the great number of polypeptides present in the DEAE fraction after SDS-PAGE, the antibodies exhibited a very specific reaction with a single bandstained after Western blot and horse radish peroxidase reaction. Antibodies were also effective to inhibit strongly the enzyme activity.

Antibodies were then used to evidence the presence of acyltransferase in the stroma of chloroplast and its absence from thylakoid membranes (fig.2).

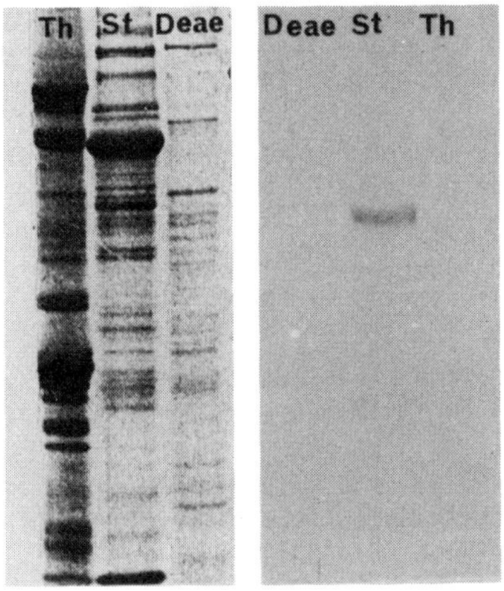

Fig. 2: Electrophoresis and the corresponding immunoblot of thylakoids (Th), stroma (St) and DEAE fraction (Deae).
Samples were obtained from Percoll purified chloroplasts (Th, St) and during acyltransferase purification (Deae).
Electrophoresis was stained by Coomassie blue. Immunoblot was revealed as in Fig. 1.

A titration curve was established by the ELISA technique which shows the proportionnality between absorbance at 405 nm and various quantities of pure acyltranferase in the 1 to 25 ng range. This technique allows a quantification of the acyltransferase in chloroplast fractions (Fig. 3). This very active protein represents less than 0.04% of the stroma proteins.

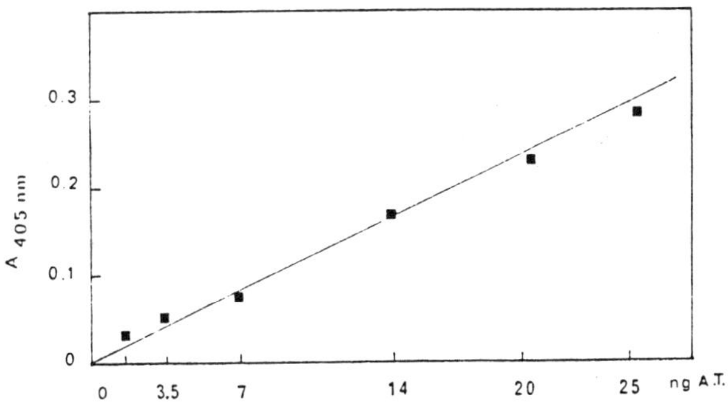

Fig. 3: Enzyme immunoassay standard curve for Pea acyltransferase (A.T.).

The immunological properties described above were also used to study in vitro translation of polyA+ mRNA. Immunoprecipitation of the synthesized proteins indicated that the acyltransferase was probably synthetized with an additional transit peptide of 3.5 kDa (paper in preparation).

Both the purification of the protein and the specificity of the antibodies offer good conditions to further study structural and genetic aspects of acyltransferases.

REFERENCES

Bertrams, M. and Heinz, E. (1976) Experiments on enzymatic acylation of sn-glycerol 3-phosphate with enzyme preparations from pea and spinach leaves. Planta 132, 161-168.

Bertrams, M.and Heinz, E. (1981) Positional specificity and fatty acid selectivity of purified sn-glycerol 3-phosphate acyltransferases from chloroplasts. Plant Physiol. 68, 653-657.

Bradford, M.M. (1976) A rapid and sensitive method for the quantitation of microgram quantities of protein utilizing the principe of protein-dye blinding. Anal. Biochem. 72, 248-254.

Douady, D., Grosbois, M. Guerbette, F. and Kader, J.C. (1986) Phospholipid transfer protein PLTP from maize seedlings is partly membrane bound. Plant Science, 45, 151-156.

Douady, D. and Dubacq, J.P. (1987) Purification of acylCoA : glycerol-3-phosphate acyltransferase from pea leaves. Biochim. Biophys. Acta 921, 615-629.

Frentzen, M. Heinz, E. Mc Keon, T.A. and Stumpf, P.K. (1983) Specificities and selectivities of glycerol-3-phosphate acyltransferase and monoacylglycerol -3-phosphate acyltransferase from pea and spinach chloroplasts. Eur. J. Biochem. 129, 629-636.

Frentzen, M;, Nishida, I. and Murata, N. (1987) Properties of the plastidial acyl-(Acyl-Carrier-Protein): glycerol-3-phosphate acyltransferase from the chilling-sensitive plant squash (Cucurbita moschata). Plant Cell Physiol. 28, 1195-1201.

Grosbois,M., Guerbette, F., Douady, D. and Kader, J.C. (1987 Enzyme immunoassay of a plant phospholipid transfer protein. Biochim. Biophys. Acta, 917, 162-168.

Murata N.(1983) Molecular species composition of phosphatidylglycerols from chilling sensitive and chilling-resistant plants. Plant Cell Physiol. 24, 81-86.

Nishida, I. Frentzen, M., Ishizaki, O. and Murata, N. (1987) Purification of isomeric forms of acyl-(Acyl-Carrier-Protein): glycerol-3-phosphate acyltransferase from greening squash cotyledons. Plant Cell Physiol. 28, 1071-1079.

BIOLOGICAL ROLE OF PLANT LIPIDS
P.A. BIACS, K. GRUIZ, T. KREMMER (eds)
Akadémiai Kiadó, Budapest and Plenum Publishing
Corporation, New York and London, 1989

ISOFORMS OF NON-SPECIFIC LIPID TRANSFER PROTEIN FROM GERMINATED CASTOR BEAN SEEDS

M. Yamada[1], S. Tsuboi[1], Y. Ozeki[1] and K. Takishima[2]

[1]Department of Biology, University of Tokyo, Komaba, Tokyo 153, Japan
[2]National Defense Medical College, Tokorozawa Saitima 359, Japan

A non specific lipid transfer protein (ns-LTP) which enhances the transfer of both phospholipids and galactolipids between biomembranes has been purified from germinated castor bean seeds (Watanabe 1986) and its primary structure has been determined to be 9.3 KD-protein with 92 amino acid residues (Takishima 1986). We now call this nsLTP-9A , because isoforms of nsLTP-9A have been found in the same materials. This paper represents the occurrence of ns-LTP isoforms in germinated castor bean tissues.

Materials and Methods

ns-LTPs were purified from germinated castor bean seeds by Watanabe et al.(1986) and amino acid sequencing of castor bean ns-LTPs by Takishima et al.(1986). Immunoblotting was performed by rutine method using anti-ns-LTP-9A rabbit serum and anti-rabbit IgG conjugated with horseradish peroxidase.

Results and Discussion

Fig. 1 shows the occurrence of the proteins cross-reacted with anti-LTP-9A IgG in 4-day old seedlings. It is evident that ns-LTP-9A was located in the cotyledons and axis (hypocotyls and radicles),not in the endosperms which we expected in the previous paper (Watanabe et al. 1986). Another isoform resposibleto 7 KD-protein was also found in the cotyledons and axis, whereas the isoform responsible to 8 KD-protein was in the endosperm. We call the former ns-LTP-7 and the latter ns-LTP-8. These results lead to the view that ns-LTPs are tissue-specific in higher plants and

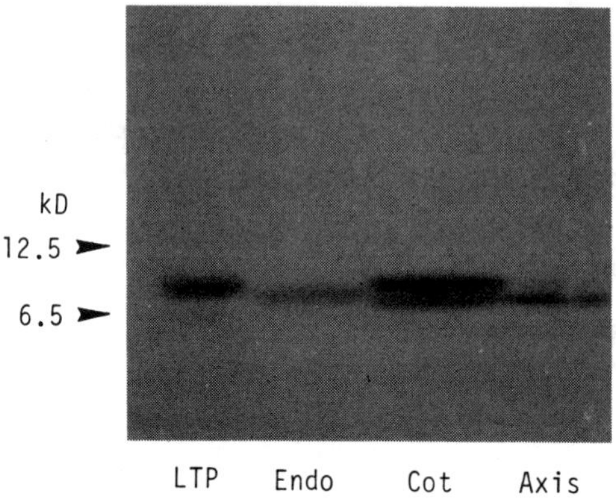

Fig, 1 . Immunoblotting pattern of germinated castor bean tissues by anti-ns-LTP-9A serum.

LTP; Purified ns-LTP-9A; Endo, Endosperm homogenate; Cot, Cotyledon homogenate; Axis, Hypocotyl and radicle homogenate. Arrow, 12.5 kD(cytochrome C); 6.5 kD(aprotinin)

Residue	nsLTP-9A	nsLTP-9B	nsLTP-9C	nsLTP-9S
Asp	5	3	5	1
Asn	4	6	3	6
Thr	6	8	8	5
Ser	10	6	8	7
Glu	1	4	1	0
Gln	4	1	5	0
Pro	6	6	4	6
Gly	5	4	5	15
Ala	15	9	11	14
Cys	8	8	8	6
Val	5	9	7	3
Met	0	1	1	2
Ile	7	5	4	7
Leu	5	6	6	6
Tyr	0	2	0	3
Phe	2	2	3	0
Lys	6	7	11	7
His	0	1	1	2
Arg	3	4	1	1
Trp	0	0	0	0
Total	92	92	92	91
M.W.	9,319	9,847	9,593	8,833

Table 1. Amino acid composition of castor bean 9kD-ns-LTPs and spinach ns-LTP.

M.W.; Molecular weight.

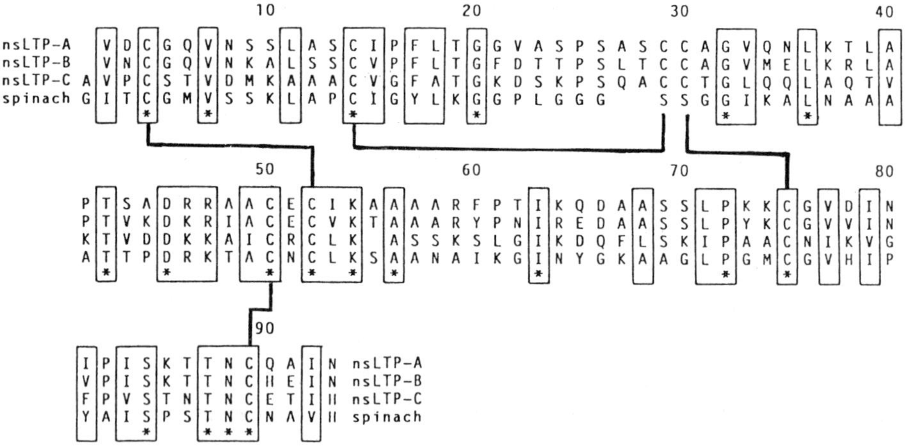

Fig. 2. Primary structure of castor bean 9kD-ns-LTPs and spinach ns-LTP.

Asterisk, conserved residue; Square, Semi-coserved residue; ns-LTP-A, ns-LTP-9A(the same abbreviation for the others). Solid line, disufide bridge.

different functions of the isoforms is expected. ns-LTP-7 and ns-LTP-8 respectively were purified from axes and endosperms from castor bean seedlings, to a homogeniety of the protein on SDS-PAGE. Both purified ns-LTP-7 and ns-LTP-8 enhanced the transfer of phosphatidycholine, phosphatidylethanolamine and monogalactosyldiacylglycerol in the same extent as ns-LTP-9A.

In the final step for the purification of ns-LTP-9A, we used FPLC Momo S. The eluate from the FPLC column gave the main peak of ns-LTP-9A as well as the two minor peaks which are active for lipid transfer. When another lot of castor bean seeds was used as the source for the purification of ns-LTP, either of the minor peaks appeared as the main peak. We call these isoforms ns-LTP-9B and ns-LTP-9C. The purification and amino acid sequencing of ns-LTP-9B and ns-LTP-9C, respectively, were carried out as previously described (Watanabe et al. 1986, Takishima et al. 1986)

Table 1 shows the amino acid composition of ns-LTP-9A and -9B and -9C, in comparison with spinach ns-LTP (Bouillon 1987), as calculated from the data of amino acid sequencing. Characteristic of ns-LTP-9 isoforms was to be rich in non-polar amino acid residues such as alanine, valine, leucine, isoleucine and proline, and absent from tryptophan. Eight cysteines were conserved. Further information is obtained from the amino acid sequence of ns-LTP-9 isoforms (Fig. 2) and the hydropathy pattern of these isoforms (Fig. 3). The amino acid sequence of ns-LTP-9 was varied among three isoforms. Homology of ns-LTP-9B, ns-LTP-9C and spinach ns-LTP with ns-LTP-9A is 68%, 35% and 40%, respectively. Eight cysteines in ns-LTP-9 isoforms were completely conserved , forming four disulfide bridges. Thus, amino-terminus near Cys-4 is close to carboxyl-terminus near Cys-89, together with Cys-50 and Cys-52 in the middle region of the peptide chain, so that the hydrophilic residues-45 to -49 near Cys-50, which are relatively conserved, may be exposed to the surface of ns-LTP-9 molecule, probably forming a hydrophilic patch. This area is expected to be

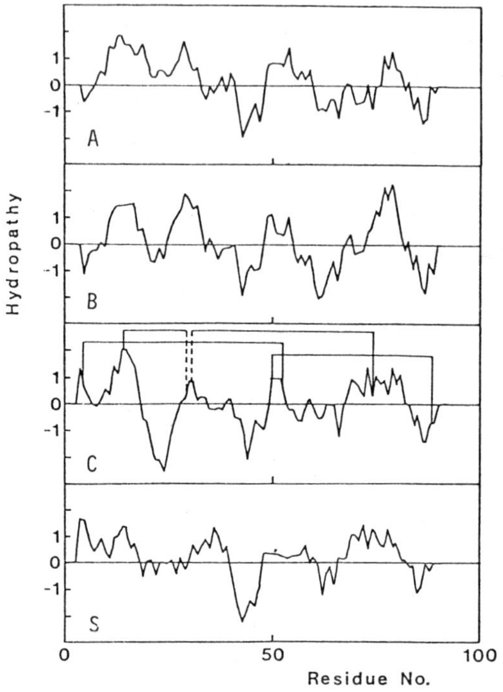

Fig. 3. Hydropathy pattern of castor bean 9 kD-ns-LTPs and spinach ns-LTP.

A, ns-LTP-9A; B, nsLTP-9B C, ns-LTP-9C; S, spinach ns-LTP.

interacted with hydrophilic moieties of the lipid molecule. Hydropathy pattern indicates the presence of four hydrophobic regions around the disulfide bridges of ns-LTP -9. These regions, also being gathered by disulfide bridges, would be interacted with hydrophobic moieties of the lipid molecule.

In summary, there are at least five isoforms of ns-LTPs such as ns-LTP-9A, ns-LTP-9B, ns-LTP-9C, ns-LTP-8 and ns-LTP-7 in castor bean seedlings.

References

Bouillon, P., Drischel, C., Vergnolle, C., Duraton, H. and Kader, J-C. (1987) The primary structure of spinach-leaf phospholipid-transfer protein Eur.J.Biochem.166:387-391.

Takishima, K., Watanabe, S., Yamada, M. and Mamiya, G. (1986) The amino-acid sequence of the non-specific lipid transfer protein from germinated castor bean endosperms Biochem. Biophys. Acta 870,248-255.

Watanabe, S. and Yamada, M. (1986) Purification and characterization of a non-specific lipid transfer protein from germinated castor bean endosperms which trasfers phospholipids and galactolipids Biochim. Biophys. Acta 876,116-123.

BIOLOGICAL ROLE OF PLANT LIPIDS
P.A. BIACS, K. GRUIZ, T. KREMMER (eds)
Akadémiai Kiadó, Budapest and Plenum Publishing
Corporation, New York and London, 1989

PHOSPHOLIPID TRANSFER PROTEINS FROM FILAMENTOUS FUNGI

P. Grondin[1], C. Vergnolle[2], L. Chavant[1] and J.-C. Kader[2]

[1]Laboratoire de Cryptogamie, Université Paul Sabatier, Route de Narbonne, Toulouse, France
[2]Laboratoire de Physiologie Cellulaire (UA 1180 CNRS), Université Pierre et Marie Curie
Place Jussieu, Paris, France

INTRODUCTION

Among the cytosolic non enzymatic proteins, some are able to stimulate phospholipid transfer between two distinct membranes. These proteins are named Phospholipid Transfer Proteins (PLTP).

The phospholipid transfer activity has been found first in animal organisms (Wirtz and Zilversmit, 1969) then in plant organisms (Ben Abdelkader and Mazliak, 1970). In 1974, Bureau and Mazliak detected this same activity in procaryotic extracts from *Bacillus subtilis*. Two years later, Cobon et al. (1976) found it in the yeast *Saccharomyces cerevisiae* then Cohen et al. (1979) in *Rhodopseudomonas sphaeroïdes* extracts.

The first evidence about phospholipid transfer activity in filamentous fungi extracts has been published by Chavant and Kader (1982) on *Mucor mucedo* and *Aspergillus ochraceus*.

De Scheemaeker et al. in 1984 presented the first results about a partial purification of a phospholipid transfer protein from *Mucor mucedo* mycelium. From these results they concluded that, in this organism, an acidic PLTP exhibited a molecular mass higher than 20 kDa. It appears that this protein was quite different from those extracted from plant organisms which were mainly basic and having a small size and can be compared to some animal PLTP. The protocol described used as steps ammonium sulfate precipitation, molecular filtration and anion exchange chromatography.

From these preliminary results, we undertook the elaboration of a new protocol to purify a PLTP from *Mucor mucedo* mycelium. According to other results on other organisms, we try other phases for liquid chromatography. Our aim was to purify and to characterize, for the first time, a PLTP from this type of organisms and to compare it with other PLTPs.

It is to be noted that we adapted the *Mucor mucedo* culture in relatively great volume which give us sufficient biomass to work.

In this work, we use an activity test as described in Douady et al. (1982) with liposomes and mitochondria from potato tuber as acceptor membranes. Liposomes contain [^{3}H]-phosphatidylcholine as transferable molecule and [^{14}C]-cholesteryl oleate as non transferable tracer.

RESULTS

Step 1 : Extraction of cytosoluble proteins :

The mycelium is frozen at least one night at -20°C before grinding with mechanical grinder in buffer (TRIS-HCl 100 mM, pH 8.0; sucrose 400 mM, Na_4EDTA 1 mM, cystein chloride 4 mM, 2-mercaptoethanol 8 mM). From this step all operations are done at +4°C.The extract is filtrated through tissue and centrifuged 20 minutes at 9000xg.

Step 2 : pH 5.1 treatment :

The supernatant is collected, its pH is adjusted to 5.1 by HCl 1M and is stored 15 minutes. After centrifugation 20 minutes at 9000xg, the supernatant, which is cleared of lipid contamination, is collected. From this step, we can determine a phospholipid transfer activity ; this fraction will be considered as a reference for the purification.

Step 3 : Ammonium sulfate precipitation :

Proteins are precipitated by addition of ammonium sulfate to 75 % of saturation and storage during 3 hours. The proteins are collected by centrifugation 30 minutes at 9000xg and are resuspended with a minimum of chromatographic buffer (Sodium phosphate 10 mM, pH 7.5, 2-mercaptoethanol 8 mM, sodium azide 3 mM, glycerol 10 %).

Step 4 : Sephadex G75 chromatography :

The protein extract is loaded on a Sephadex G75 column equilibrated in chromatographic buffer. The elution is performed with the same buffer and an active peak appears around an elution volume corresponding to a molecular mass of 21 kDa (according to former results). All the active fractions are pooled and used rapidly for next step.

Step 5 : Hydroxyapatite chromatography :

The Sephadex G75 fraction is loaded on an hydroxyapatite column equilibrated in chromatographic buffer. The column is washed with the same buffer and the elution of an active fraction is performed by a 100 mM potassium phosphate buffer. The active pool is collected and desalted on a Sephadex G25 column for the next step.

Step 6 : Blue Trisacryl chromatography :

The desalted fraction issued from hydroxyapatite chromatography is loaded on a Blue Trisacryl column equilibrated in chromatographic buffer. The column is washed with the same buffer and elution of an active fraction is performed by a linear gradient of sodium chloride in chromatographic buffer. The sodium chloride gradient is done from 0 to 1 M and the active fractions are detected around 300 mM of this salt.

DISCUSSION

From this protocol, we obtain a purification factor around 45 with a recovery around 5 % even though the ammonium sulfate precipitation and Sephadex G75 chromatography give a recovery of 30 %.

The major problem is the affinity phase, Blue Trisacryl gel, which is very interesting because it retains an active protein fraction. Unfortunately, this step is very destructive for the total activity, only 16 % of the activity loaded is recovered. It is probable that the active protein(s) is(are) denatured or complexed, or agglutinated like the animal PLTP studied by KAMP et al. (1973).

Our results confirm for instance the molecular mass around 21 kDa and permit to purify the active protein(s) 45 times.

Now, we will try to improve the affinity step and to add to this protocol a final step on anion exchanger. It will be later possible to obtain some results about the specificity of the active fraction among the major phospholipids.

REFERENCES :

Ben Abdelkader J.C. and Mazliak P., 1970 - Echanges de lipides entre mitochondries, microsomes et surnageant cytoplasmique de cellules de pommes de terre ou de chou-fleur. Eur. J. Biochem., 15, 250-262.

Bureau G. and Mazliak P., 1974 - Transferts d'acides gras et de protéines entre membranes cytoplasmiques et mésosomes de *Bacillus subtilis* var. Niger. F. E. B. S. Letters, 39, 332-336.

Chavant L. and Kader J.C., 1982 - The presence of phospholipid transfer proteins in filamentous fungi. In "Biochemistry and Metabolism of Plant Lipids". Elsevier Biomedical Press B.V., 125-128.

Cobon G.S., Crowfoot P.D., Murphy M. and Linnane A.W., 1976 - Exchange of phospholipids between mitochondria and microsomes "in vitro" stimulated by yeast cytosol. Biochim. Biophys. Acta, 441, 255-259.

Cohen L.K., Luekin D.R. and Kaplan S., 1979 - Intermembrane phospholipid transfer mediated by cell free extracts of *Rhodopseudomonas sphaeroïdes*. J. Biol. Chem., 254, 721-728.

De Scheemaeker H. Vergnolle C., Tchang F. Chavant L. and Kader J.C., 1984 - Comparison between phospholipid transfer proteins in two filamentous fungi. In "Structure and Metabolism of Plant Lipids". Elsevier Sciences Publishers B.V.

Douady D., Grosbois M., Guerbette F. and Kader J.C., 1982 - Purification of a basic phospholipid transfer protein from maize seedlings. Biochim. Biophys. Acta, 710, 143-153.

Kamp H.H., Wirtz K.W.A. and Van Deenen L.L.M., 1973 - Some properties of phosphatidylcholine exchange protein purified from beef liver. Biochim. Biophys. Acta, 398, 313-325.

Wirtz K.W.A. and Zilversmit D.B., 1969 - Participation of soluble liver proteins in the exchange of membrane phospholipids. Biochim. Biophys. Acta, 193, 105-116.

BIOLOGICAL ROLE OF PLANT LIPIDS
P.A. BIACS, K. GRUIZ, T. KREMMER (eds)
Akadémiai Kiadó, Budapest and Plenum Publishing
Corporation, New York and London, 1989

SOME CHARACTERISTICS OF A PROTEIN A: ACP-I FUSION IN PLANT FATTY ACID SYNTHESIS

D.J. Guerra[1] and P.D. Beremand[2]

[1]Institute of Biological Chemistry, Washington State University, Pullman, WA 99164-6340, USA
[2]USDA/ARS Northern Regional Research Center, Peoria, IL 61604, USA

INTRODUCTION

Much of what we understand of the regulation of plant fatty acid synthesis (FAS) involves acyl carrier protein (ACP). ACP is a soluble, acidic monomeric protein necessary in *de novo* FAS and the accumulation of certain lipids (1). Previous work has revealed that two or more isoforms of ACP exist in higher plants (2,3). These isoforms are tissue-specific in expression and are derived from separate nuclear genes (2). We have shown that ACP isoforms effect the targeting of carbon from the plant plastid (4). Our model for ACP isoform regulation of lipid accumulation suggested that ACP-I was the preferred co-substrate for a specific oleoyl-ACP thioesterase (5). We recently synthesized and cloned a synthetic ACP-I deduced from the authentic purified spinach protein sequence (6). We have since characterized this recombinant ACP-I and have recently fused the coding sequence to the Fc-binding portion of *S. aureus* protein A. We report that our fusion protein is fully functional in plant FAS, but that the rate of *de novo* FAS *in vitro* is distinct from the original recombinant protein. It appears from our findings that the protein A:ACP-I fusion would be most suitable for plant transformation programs designed to alter fatty acid metabolism.

MATERIALS AND METHODS

The 268 bp Bam H1 fragment of pPB 269 was purified and ligated into the Bam H1 site of p^{RiT2T} (7). This construct was introduced into *E. coli* cells carrying a temperature-sensitive repressor to prevent the lethal consequences of over-expression caused by the Pr promoter (7). Restriction analyses were performed to verify appropriate construct design.

The fusion protein was purified to > 90% homogeneity utilizing ion exchange, gel permeation and IgG agarose chromatography (7). This preparation was used for subsequent *in vitro* studies using purified *E. coli* acyl ACP synthetase and spinach leaf FAS (7). Assay conditions and product analyses of enzymatic procedures were as described (7).

RESULTS AND DISCUSSION

The protein A:acyl carrier protein-I gene was partially sequenced and found to be completely homologous to our desired construction (7). The purified protein from *E. coli* N4830 cells was approximately 90% pure as judged from SDS•PAGE using both Coomassie staining and Western immunoblot detection (data not shown). This purified preparation was stable at -20° for several months and was used as source of fusion ACP for *in vitro* enzymatic characterizations.

Table 1. Acyl-ACP synthetase activity of protein A:ACP-I fusion and *E. coli* ACP.

Protein (μg)	Radioactivity Incorporation[1] *E. coli* ACP	Protein A:ACP-I Fusion
	^{14}C DPMX10^{-3}	
0.1	14.4	1.5
0.2	18.7	0.9
0.8	16.3	2.2
1.6	21.3	4.0
2.0	29.1	4.9
4.0	43.7	12.7
8.0	46.2	30.3

[1] Amount of precipitatable ^{14}C-palmitoyl ACP or ^{14}C-palmitoyl protein A:ACP-I formed in 10 min under standard assay conditions.

Table 1 shows that the fusion protein was active in the standard assay for ACP, but that reactivity was not comparable to *E. coli* ACP at lower concentrations (up to 4 µg added to 50 λ reaction; 8.5 µM). We have calculated that the Km app for the fusion ACP was 0.5 mM and the equivalent figure for *E. coli* ACP was 50 µM. This 10-fold increase in Km value suggests that the ACP portion of the fusion requires higher concentration to be recognized by the acyl ACP synthetase. This effect is probably caused by the unrecognized portion (Ca 75% by mass) taken up by protein A. During the linear phase of reactivity (i.e., between 1.6 to 4.0 µg 50 λ^{-1}) it appears that the fusion protein is 25% active as ACP. This result suggests that all of the ACP in the fusion protein is reactive on a M_r basis (i.e., specific activity).

Table 2. Spinach Leaf Fatty Acid Synthesis *in vitro*.

Protein (µM)	Recombinant ACP-I	Protein A:ACP-I Fusion
	fmol product/µl enzyme/min	
4.2	13.3	12.1
12.6	19.4	48.6
21.4	12.7	71.3

Table 2 shows the result of using either recombinant ACP-I alone or the fusion protein in a spinach chloroplast extract for *de novo* FAS (7). In this instance the amount of protein added was based on the acyl-ACP synthetase assay for ACP-equivalent reactivity. Therefore, the amount of protein added was not equivalent (i.e., 4 x more fusion protein at each ACP-equivalent concentration). Our results confirmed previous efforts from our lab (8) that increasing levels of ACP relative to available carbon source caused a decrease in net *de novo* FAS when recombinant ACP-I was used as co-factor/co-substrate. Surprisingly, the fusion protein did not follow this kinetic increasing concentration above 20 μM. This effect is probably not caused by protein concentration since the amount added as ACP source comprises less than 10% of total protein in the reaction vessel even at the highest concentration. Our interpretation of this result is that the fusion ACP represents a mutated form of the protein. Therefore, it is fully functional on a specific activity basis (25% by mass) but is not regulated by or directly regulatory in the role of *de novo* FAS *in vitro*. We are in the process of utilizing the fusion ACP in constructions for oilseed transformation.

REFERENCES

1. Stumpf, P.K. 1984. *In* Numa, S., ed., Fatty Acid Metabolism and its Regulation. Elsevier Science Publ., Amsterdam, pp 155-179.

2. Høj, P.B. and I. Svendsen. 1984. Barley chloroplasts contain two acyl carrier proteins coded for different genes. Carlsberg Res. Commun. 49: 483-492.

3. Ohlrogge, J.B. and T.M. Kuo. 1985. Plants have isoforms for acyl carrier protein that are expressed differently in different tissues. J. Biol. Chem. 260: 8032-8037.

4. Guerra, D.J., J.B. Ohlrogge and M. Frentzen. 1986. Activity of acyl carrier protein isoforms in reactions of plant fatty acid metabolism. Plant Physiol. 82: 448-453.

5. Guerra, D.J., J.B. Ohlrogge and M. Frentzen. 1987. A possible differential role for plant acyl carrier protein isoforms in higher plants. *In* Stumpf, P.K., B. Mudd and D. Nes, eds., The Metabolism, Structure and Function of Plant Lipids. Plenum Publishing Co.

6. Beremand, P.D., D.J. Hannapel, D.J. Guerra, D.N. Kuhn and J.B. Ohlrogge. 1987. Synthesis, cloning and expression in *Escherichia coli* of a spinach acyl carrier protein-I gene. Arch. Biochem. Biophys. 256: 90-100.

7. Beremand, P.D., D.D. Doyle, K.Dziewanowska and D.J. Guerra. 1988. Expression of an active spinach acyl carrier protein-I/protein-A fusion. Plant Molec. Biol. (submitted).

8. Guerra, D.J., K. Dziewanowska, J.B. Ohlrogge and P.D. Beremand. 1988. Purification and characterization of recombinant spinach acyl carrier protein-I expressed in *Escherichia coli*. J. Biol. Chem. 263: 4386-4391.

CHAPTER 5

BIOCIDES, INTERACTION WITH PLANT LIPIDS

BIOLOGICAL ROLE OF PLANT LIPIDS
P.A. BIACS, K. GRUIZ, T. KREMMER (eds)
Akadémiai Kiadó, Budapest and Plenum Publishing
Corporation, New York and London, 1989

PLANT LIPID BIOSYNTHESIS AS TARGET FOR BIOCIDES

H.K. Lichtenthaler

Botanisches Institut II (Plant Physiology and Plant Biochemistry), University of Kalsruhe, Kaiserstr. 12, D-7500 Karlsruhe, FRG

INTRODUCTION

The research on potential inhibitors, herbicides and xenobiotics of plant lipid biosynthesis is not only a matter of general interest for agronomists and those concerned with crop protection but also for the plant biochemist, since active inhibitors represent essential tools for the characterization and better understanding of those metabolic processes and pathways which are involved in the biosynthesis of plant lipids. Theoretically all steps in plant lipid biosynthesis could be a target of inhibitors. However, only a few herbicide groups of plant lipid biosynthesis are known today. This review summarizes our present knowledge of the different types of xenobiotica and their mode of interaction with plant lipid biosynthesis.

FATTY-ACID BIOSYNTHESIS

According to the concept of Stumpf and coworkers, acetyl-CoA is synthesized in the mitochondria by oxidative decarboxylation of pyruvate and becomes hydrolyzed to free acetic acid. The latter can quickly penetrate the envelope membranes of mitochondria and plastids by diffusion. In the plastids, acetic acid is converted to acetyl-CoA by the enzyme acetyl-CoA-synthetase. Activation of acetic acid to acetyl-CoA can be inhibited by 3-nitro-2,5-dichlorbenzoic acid (**dinoben**), which is also used as a herbicide in the soybean (Muslih and Linscott 1977). Dinoben also inhibits the incorporation of malonic acid into the fatty-acid fraction, whereas the incorporation of acetyl-CoA is not affected. This indicates that the thioesterases are the target for the herbicide dinoben.The percentage inhibition of lipid accumulation by dinoben is not equal for the different plant lipid classes and is higher for neutral lipids, MGD, PC and

PE than for DGD, PG and PI. A possible explanation for this matter could be the existence of ACP-isoforms with different Km for acyltransferase and thioesterase reactions, as could be shown for spinach (Guerra et al. 1986).

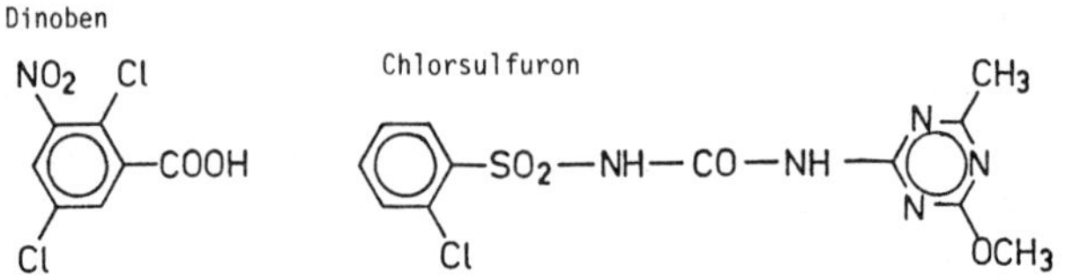

Fig. 1: Chemical structure of the herbicides dinoben and chlorsulfuron.

An alternative pathway for the biosynthesis of acetyl-CoA may be the direct synthesis from intermediates of the Calvin-cycle. Plastids may therefore be autonomous in acetyl-CoA formation. This was shown by several authors in the past (Yamada and Nakamura 1975, Grumbach and Forn 1980, Murphy and Leech 1978). The glycolytic enzymes shown to be present in plastids (Liedvogel and Bäuerle 1986, Reddy and Das 1987, Kurzok and Feierabend 1986) and one critical enzyme, the plastid pyruvate dehydrogenase complex (PDS), are determined and purified from different plants including spinach (Treede and Heise 1986, Camp and Randall 1985). Therefore the question of the origin of acetyl-CoA for fatty acid biosynthesis can only be answered by determining the physiological concentrations of the metabolites and the Km-values of the enzymes. In the past the concentration of free acetic acid in spinach was overestimated (Liedvogel 1985), but nevertheless it was proposed that, in the case of spinach, free acetic acid and not pyruvate is the real substrate for fatty acid biosynthesis (Treede et al. 1986, Treede and Heise 1985). It is proposed that the acetyl-CoA, which is formed by the plastid pyruvate dehydrogenase complex, is channelled into the plastid isoprenoid biosynthesis (Schulze-Siebert and Schultz 1987) or used for the biosynthesis of branched amino acids (Murphy and Stumpf 1981). By blocking the acetolactate synthetase, the key enzyme for the synthesis of branched amino acids, with **chlorsulfuron** (Fig. 1), an increased incorporation of pyruvate into fatty acids is observed (Homeyer et al. 1985). This indicates that fatty acid synthesis can be influenced by xenobiotics which do not interfere with the fatty-acid biosynthesis sequence.
The **chloracetamides** and the **thiocarbamates** (chemical structure see Fig. 2) might also interfere with the metabolism of CoA. Fürst (1987) proposed that the broad spectrum of effects such as inhibition of lipid, isoprenoid and flavonoid biosynthesis is caused by disrupting processes requiring CoA.

This theory is supported by the observation that EPTC, a thiocarbamate, and dichlormid, a chloracetamide , inhibit acetate incorporation into acetyl-CoA as well as conversion of pyruvate to acetyl-CoA (Wilkinson and Oswald 1987).

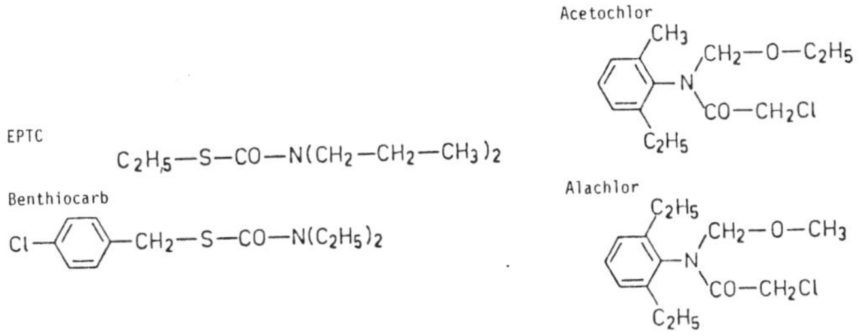

Fig. 2: Chemical structure of two thiocarbamates (left side) and two chloracetamide derivatives (right side).

Acetyl-CoA carboxylase as target

The next enzyme in the fatty acid biosynthesis pathway is the acetyl-CoA carboxylase. The acetyl-CoA carboxylase from grasses (Poaceae) can specifically be blocked by herbicides (see Fig. 3) of the cyclohexane-1,3-dione (**sethoxydim, cycloxydim, clethodim**) and aryloxyphenoxypropionic acid-type (**diclofop, fenoxaprop, haloxyfop, fluazifop**) which was shown by us only recently (Focke and Lichtenthaler 1987, Kobek et al. 1988).

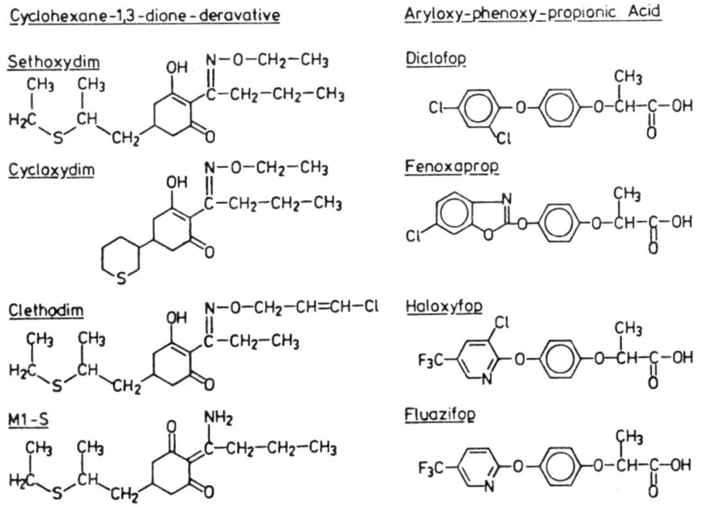

Fig. 3: Chemical structure of **cyclohexane-1,3-diones** (left side) and **aryloxyphenoxypropionic acid-type** herbicides (right side).

The target acetyl-CoA carboxylase was confirmed in independent work for sethoxydim and haloxyfop (Burton et al. 1987), for haloxyfop and tralkoxydim (Secor and Cséke 1988) and for alloxydim and clethodim (Rendina and Felts 1988). For some time these herbicides were thought to be inhibitors of one of the enzymes of the fatty-acid synthetase complex, because in intact plants they also inhibit the incorporation of malonic acid into the fatty-acid fraction (Hoppe and Zacher 1982, Burgstahler 1985, Burgstahler et al. 1986). In an isolated enzyme preparation capable of de novo fatty-acid biosynthesis, no inhibition of malonyl-CoA or malonate incorporation was found (Focke and Lichtenthaler 1987, Burton et al. 1987). In chloroplast preparations the incorporation rate of malonate is very low as compared to acetate, and this very low rate can be blocked by these herbicides (Kobek et al. 1988). The explanation for this is that in the intact plant and also in plastids (there to a lower degree) malonic acid and/or malonyl-CoA are converted by a decarboxylase to acetic acid/acetyl-CoA, the incorporation of which into the fatty acid fraction can then be inhibited at the target acetyl-CoA carboxylase (Kobek et al. 1988).

The **phenoxyphenoxypropionic acid-type herbicides** (chemical structure see Figure 3) have the same enzyme target as the cyclohexane-1,3-diones (Kobek et al. 1988). Diclofop is also regarded as an auxin antagonist (Shimabukuro et al. 1982). This latter function of diclofop may be explained by the hypothesis that diclofop can also act as a protonophore, which shuttles protons across the plasmalemma (Wright and Shimabukuro 1987). But higher herbicide concentrations seem to be required than are needed for fatty-acid synthesis inhibition. In field application the aryloxyphenoxypropionic acid-type herbicides are used in the esterified form. The active form, which inhibits the fatty-acid biosynthesis of isolated chloroplasts, is, however the free acid form (Hoppe and Zacher 1985, Kobek et al. 1988 and Fig.4), generated by a hydrolase, which is extracellular or surface associated with the plasmalemma (Holl et al. 1986). The aryloxyphenoxypropionic acid herbicides occur in two enantiomers, of which the R-enantiomer is the active form (Hoppe and Zacher 1985, Uchiyama et al. 1986 and Fig. 5).
The herbicides of the cyclohexane-1,3-dione-type and phenoxyphenoxypropionic acid-type applied hitherto are interesting new herbicides which selectively control grasses in dicot and certain monocot crop plants with the advantage that the applied amounts per ha are much lower than for the classical herbicides which block photosynthetic electron transport.

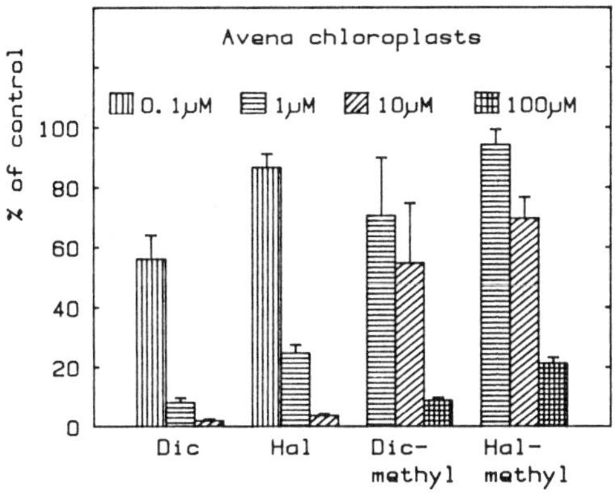

Fig.4: Inhibitory activity of the esterified and free acid-form of diclofop and haloxyfop on de novo fatty acid biosynthesis of oat chloroplasts.

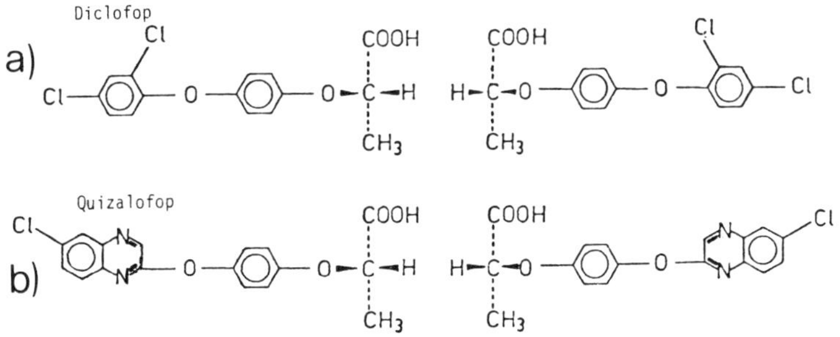

Fig.5: Stereoisomers of the herbicides diclofop (a) and quizalofop (b),the R-enantiomers (right side) are the active herbicides.

Besides the inhibition of fatty acid biosynthesis the cyclohexane-1,3-dione derivates e.g. sethoxydim (Fig.3) also induce several other alterations of the plant composition and metabolism. Due to the herbicide-induced inhibition of fatty acid and biomembrane lipid formation, the development of functional chloroplasts, the formation of thylakoids and the chloroplast replication are also blocked (Lichtenthaler 1984, Lichtenthaler and Meier 1984) as well as the accumulation of chlorophylls and carotenoids (Lichtenthaler 1987, Lichtenthaler et al. 1987) but not their biosynthesis (Lichtenthaler 1987). This resulted in the formation of white leaf parts which are free of photosynthetic pigments. Furthermore, in herbicide-treated maize shoots the accumulation and biosynthesis of plastidic and cytoplasmic phospho- and glykolipids was blocked (Burgstahler and Lichtenthaler 1984 and Burgstahler 1985). The cyclohexane-1,3-dione derivatives block the biosynthesis of certain fatty acids (C16:0, C16:1 and C18:3) to a larger degree than the other fatty acids (Fig. 6).

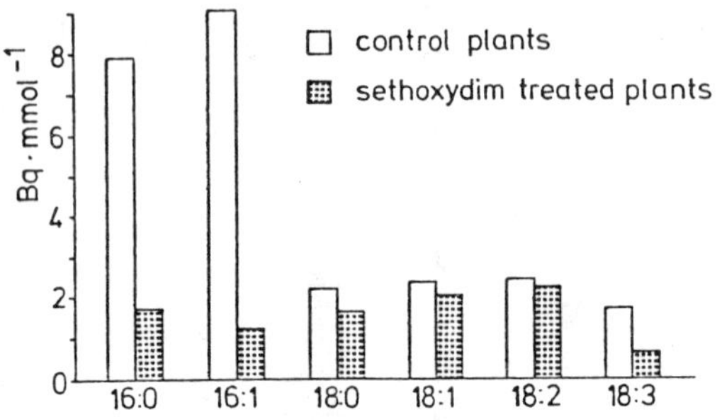

Fig. 6: Specific radioactivities of individual fatty acids labelled from 2-^{14}C acetate in maize leaves treated with (0.125 kg a.i.ha^{-1}) or without sethoxydim (control). The ^{14}C-label was applied via the cut ends of the shoots 1h after herbicide spraying of the intact plant. The incorporation lasted 1h. Mean of 5 replications, S.D 5% (Retzlaff 1986, personal communication).

All the different effects of cyclohexanedione or diphenoxypropionic acid-type herbicides can be explained by their inhibition of de novo fatty acid biosynthesis at the level of the acetyl-CoA carboxylase as main target since neither lipid nor biomembrane formation or growth are possible when fatty-acid biosynthesis is blocked.

The effect of natural antibiotics on fatty-acid biosynthesis

Two antibiotics inhibit enzymes of the fatty-acid synthetase-complex: cerulenin and thiolactomycin.

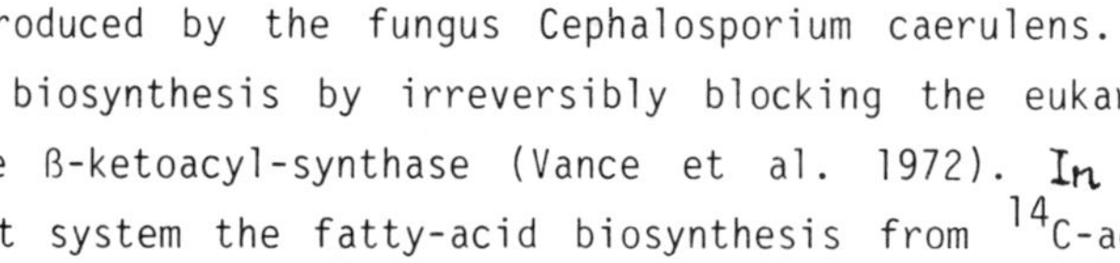

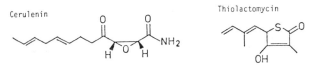

Fig. 7: Chemical structures of the antibiotics cerulenin and thiolactomycin.

Cerulenin is produced by the fungus Cephalosporium caerulens. Cerulenin inhibits lipid biosynthesis by irreversibly blocking the eukaryotic and prokaryotic-type ß-ketoacyl-synthase (Vance et al. 1972). In our Avena chloroplast test system the fatty-acid biosynthesis from ^{14}C-acetate was inhibited to 45 % at 10 µM cerulenin and in an enzyme preparation from barley chloroplasts to 50 % at 10 µM (Lichtenthaler et al., unpublished).

It is of interest in this respect that the synthesis of structurally similar ß-ketoacyl compounds - as catalyzed by the flavanone synthetase and the yeast HMG-CoA synthetase (hydroxy-methyl-glutaryl-CoA synthetase) - is also inhibited by cerulenin (Kreuzaler and Hahlbrock 1975, Ohno et al. 1974). The plant HMG-CoA synthase is, however, not inhibited by cerulenin (Bach 1988). The reduced accumululation of isoprenoid pigments (chlorophylls, carotenoids) in greening barley seedlings (Laskay et al. 1985a) can therefore not be explained by a direct inhibition of the isoprenoid and prenylpigment biosynthesis.

The two ketoacyl-synthetases present in the fatty-acid synthetase complexes of higher plants exhibit differential sensitivities against cerulenin. In plants the ß-ketoacyl-synthetase I (de novo fatty acid synthesis) is affected by cerulenin, whereas the ß-ketoacylsynthetase II, which catalyzes the elongation from 16:0 to 18:0 is relatively insensitive (Jaworski et al. 1974). The ß-ketoacylsynthetase of Cephalosporium caerulens, which produces cerulenin, is much less sensitive than the yeast enzyme (Kawaguchi et al. 1979) and in Escherichia coli two sensitive and one insensitive synthetase were described (Jackowski and Rock 1987).

Treatment of barley with cerulenin resulted in an enhanced formation of medium-chain fatty acids and an increase of linoleic acid (Packter and Stumpf 1975). Laskay et al. (1985b) investigated the effect of cerulenin on the changes in the lipid content during greening and proposed a specific shortage of galactolipids because of a blocked conversion of 18:2 PC to 18:2 MGD, resulting in a higher desaturation rate of 18:2 PC to 18:3 PC.

The second natural antibiotic **thiolactomycin** only inhibit prokarotic- type fatty acid synthetases and the inhibition is reversible. Produced by Nocardia sp, thiolactomycin affects the acetyl-CoA:ACP acyl transferase as well as the ß-ketoacylsynthetase, both of which are regarded as being rate-limiting for fatty acid biosynthesis in plants (Shimakata and Stumpf 1983). Thioloactomycin is more inhibitory to the elongation of 16:0 to 18:0 than to the de novo biosynthesis of 16:0 (Yamada et al. 1987). The level of MGD was strongly decreased in presence of thiolactomycin as well as the linolenic acid content of MGD and the desaturation of PG from the 16:0 to 16:1 and the 18:1 to 18:2 species.

Effect of pyridazinones on fatty acid desaturation

The desaturation of newly synthesized fatty acids apparently proceeds within the plastids or in part also in the cytoplasm. Substituted pyridazinones can have multiple sites of action. Some of them effectively inhibit photosynthetic electron transport and others the carotenoid biosynthesis (Lichtenthaler and Kleudgen 1977).Some of these also block the desaturation of fatty acids. The desaturation of linoleic acid was proposed to be one target (St. John 1976). Treated plants show a higher 18:2/18:3 ratio than untreated plants. Furthermore some pyridazinones inhibit the desaturation of PG 16:0 to PG 16:1 (3-trans) (Khan et al. 1979). Khan proposed that the blocked desaturase may be plastidic, which was proved for Arabidopsis (Norman and St. John, 1987). In general different pyridazinones may act on different desaturases with different I_{50}-values. Another important point is the changed 18:2/18:3 ratio, which may cause a decreased frost-hardening (Fedtke 1982).

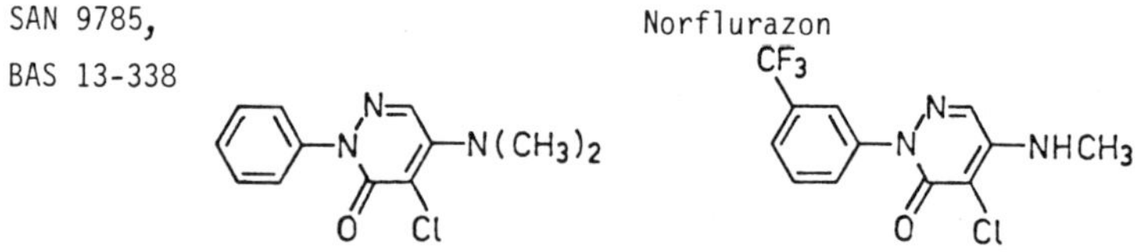

Fig.8: Chemical structure of two pyridazinones.

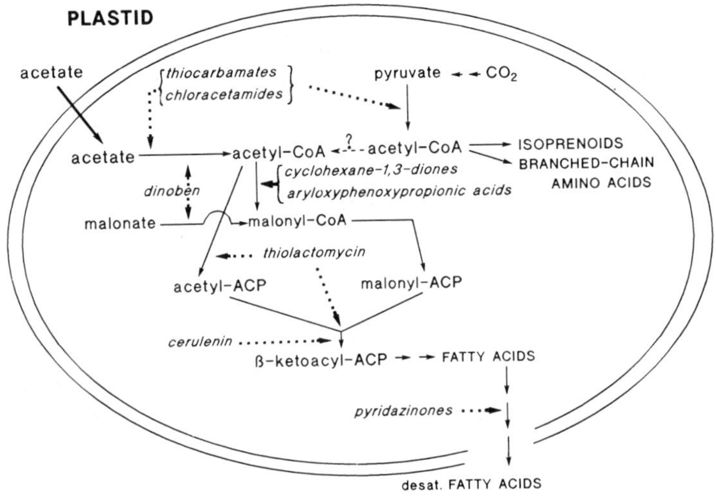

Fig. 9: Scheme of the inhibition by xenobiotics of different steps of de novo fatty-acid biosynthesis.

CONCLUSION

Today several xenobiotics, either natural antibiotics (cerulenin, thiolactomycin) or active herbicides are known to interfere with acetyl-CoA formation, de novo fatty acid biosynthesis, desaturation of fatty acids, biosynthesis of long chain fatty acids and glycerolipid formation. This is summarized in Fig. 9. Certainly other new herbicides which interfere with the same or other particular parts of plant lipid biosynthesis will be developed in the future . From the study of the mode of action of these inhibitors in sensitive and tolerant species, one will not only be able to efficiently control weeds in crop plants and guarantee a better food production but also obtain a new and better understanding of the function of plant lipids and the regulation of plant lipid biosynthesis.

Acknowledgements: I wish to express my sincere thanks to Mr. Manfred Focke for excellent assistance during the preparation of the review.

REFERENCES

Bach, T.J., 1988. Current trends in the development of sterol biosynthesis inhibitors: Early aspects of the pathway. J. Amer Oil Chem. Soc. 65,591-595

Burgstahler R.J., 1985. On the mode of action of sethoxydim (Poast), a cyclohexane-1,3-dione derivative. Karlsruhe Contribution to Plant Physiology 13, 1-111.(ISSN 0173-3133).

Burgstahler R.J. and Lichtenthaler H.K., 1984. Inhibition by sethoxydim of phospho- and galctolipid accumulation in maize seedlings in: Structure, Function and Metabolism of Plant Lipids, P.-A. Siegenthaler and W. Eichenberger ed., Elsevier Science Publishers B.V., Amsterdam, 619-622.

Burgstahler R.J., Retzlaff G. and Lichtenthaler H.K., 1986. Mode of action of sethoxydim: effects on the plant's lipid metabolism Proc. IUPAC Ottawa. 3B-11.

Burton J.D., Gronwald J.W., Somers D.A., Connelly J.A. Gengenbach B.G., and Wyse D.L., 1987. Inhibition of plant acetyl-coenzyme A carboxylase by the herbicides sethoxydim and haloxyfop. Biochem. Biophys. Res., Commun. 148, 1039-1044.

Camp P.J. and Randall D.D., 1985. Purification and charaterization of the pea chloroplast pyruvate dehydrogenase complex. Plant Physiol. 77, 571-577.

Fedtke C., 1982. Pyridazinones In: Biochemistry and Physiology of Herbicide Action, Fedtke C. ed., Springer-Verlag, Heidelberg, pp. 108-110.

Focke M. and Lichtenthaler H.K., 1987. Inhibition of the acetyl-CoA carboxylase of barley chloroplasts by cycloxydim and sethoxydim. Z.Naturforsch. 42c, 1361-1363.

Fürst C.P., 1987. Understanding the mode of action of the chloracetamides and thiocarbamates. Weed Technology 1, 270-277.

Grumbach K. and Forn B. 1980. Chloroplast autonomy in acetyl-coenzyme-A formation and terpenoid biosynthesis. Z. Naturforsch. 35c, 645-648.

Guerra D.J., Ohlrogge J.B., and Frentzen M., 1986. Activity of acyl carrier protein isoforms in reactions of plant fatty acid metabolism. Plant Physiol. 82, 448-453.
Höll G.B., Tritter S.A. and Todd B.G., 1986. Distribution of diclofop-methyl and metabolits in oat (Avena sativa) protoplasts. Weed Res. 26, 421-425.
Homeyer U., Schulze-Siebert D., and Schultz G., 1985. On the specifity of the herbicide chlorsulfuron in intact spinach chloroplasts. Z. Naturforsch. 40c, 917-918.
Hoppe H.H. and Zacher H., 1982. Hemmung der Fettsäurebiosynthese durch Diclofop-methyl in Keimwurzelspitzen von Zea mays. Z. Pflanzenphysiol. 106, 287-298.
Hoppe H.H. and Zacher H., 1985. Inhibition of fatty acid biosynthesis in isolated bean and maize chloroplasts by herbicidal phenoxy-phenoxy propionic acid derivatives and structurally related compounds. Pestic. Biochem. Physiol. 24, 298-305.
Jackowski S. and Rock C.O., 1987. Acetoacetyl-acylcarrier protein synthase, a potential regulator of fatty acid biosynthesis in bacteria. J.Biol. Chem. 262, 7927-7931.
Jaworski, J.G., Goldschmidt, E.E. and Stumpf P.K., 1974. Fat metabolism in higher plants. Properties of the palmityl acyl carrier protein: Stearyl acyl carrier protein elongation in maturing Safflower seed extracts. Arch. Biochem. Biophys. 163, 769-776.
Kawaguchi A., Tomoda H., Okuda S., Awaya J., and Omura S., 1979. Cerulenin resistance in a cerulenin-producing fungus. Isolation of cerulenin insensitive fatty acid synthetase. Arch.Biochem.Biophys. 197, 30-35.
Khan M.-U., Lem N.W., Chandorkar K.R. and Williams J.P., 1979. Effects of substituted pyridazinones (San 6706, San 9774, San 9785) on glycerolipids and their associated fatty acids in the leaves of Vicia faba and Hordeum vulgare. Plant Physiol. 64, 300-305.
Kobek K., Focke M. and Lichtenthaler H.K., 1988. Fatty acid biosynthesis and acetyl-CoA carboxylase as a target of diclofop, fenoxaprop and other aryloxy-phenoxy-propionic acid herbicides. Z. Naturforsch. 43c, 47-54
Kreuzaler F. and Hahlbrock K., 1975. Enzymic Synthesis of an aromatic ring from acetate units. Eur J. Biochem. 56, 205-213.
Kurzok H.-G. and Feierabend, 1986. Comparison of the development and site of synthesis of a cytosolic and a chloroplast isoenzyme of triosephosphate isomerase in rye leaves. J. Plant Physiol. 126, 207-212.
Laskay G., Lehoczki G. and Szalay L., 1985a. Effects of cerulenin on the photosynthetic properties of detached barley leaves during chloroplast development. J. Plant Physiol. 119, 55-64.
Laskay G., Farkas T. and Lehoczki E., 1985b. Cerulenin-induced changes in lipid and fatty acid content of chloroplasts in detached greening barley leaves. J. Plant Physiol. 118, 267-275.
Lichtenthaler H.K., 1984. Chloroplast biogenesis, its inhibition and modification by new herbicides compounds. Z. Naturforsch. 39c, 492-499
Lichtenthaler H.K., 1987. Functional organization of carotenoids and prenylquinones in the photosynthetic membrane in: The metabolism, structure and function of plant lipids. Stumpf P.K., Mudd J.B. and Nes W.D. ed., Plenum Press, New York, 63-75.
Lichtenthaler H.K. and Kleudgen H.K., 1977. Effect of the herbicide San 6706 on biosynthesis of photosynthetic pigments and prenylquinones in Raphanus and in Hordeum seedlings. Z. Naturforsch. 32c, 236-240.
Lichtenthaler H.K. and Meier D., 1984. Inhibion by sethoxydim of chloroplast biogenesis, development and replication in barley seedlings. Z. Naturforsch. 39c, 115-122.

Lichtenthaler H.K., Kobek K. and Ishii K., 1987. Inhibiton by sethoxydim of pigment accumulation and fatty acid biosynthesis in chloroplasts of Avena seedlings. Z. Naturforsch. 42c, 1275-1279.

Liedvogel B., 1985. Acetate concentration and chloroplast pyruvate dehydrogenase complex in Spincacia oleracea leaf cells. Z. Naturforsch. 40c, 182-188.

Liedvogel B. und Bäuerle R., 1986. Fatty acid synthesis in chloroplasts from mustard from mustard (Sinapis alba L.) cotyledons: formation of acetyl-coenzyme A by intraplastid glycolytic enzymes and a pyruvate dehydrogenase complex. Planta 169, 481-489.

Murphy D.J. and Leech R.M., 1978. The pathway of ^{14}C bicarbonate incorporation into lipids in isolated photosynthesizing spinach chloroplasts.FEBS Lett. 88, 192-197.

Murphy D.J. and Stumpf P.K., 1981. The origin of chloroplast acetyl-CoA. Arch Biochem. Biophys. 212: 730-739.

Muslih R.K. and Linscott D.L., 1977. Regulation of lipid biosynthesis in soybeans by two benzoic acid herbicides. Plant Physiol. 60, 730-735.

Norman H.A. and St. John J.B., 1987. Differential effects of a subsituted pyridazinone, BASF 13-338 on pathways of monogalactosyldiacylglycerol synthesis in Arabidopsis. Plant Physiol. 85, 684-688.

Ohno T., Kesado T., Awaya J., and Omura S., 1974. Target of inhibition by the anti-lipogenic antibiotic cerulenin of sterol synthesis in yeast Biochem.Biophys.Res. Commun. 57, 1119-1124.

Packter N.M. and Stumpf P.K. 1975. Fat metabolism in higher plants. The effect of cerulenin on the synthesis of medium- and long-chain acids in leaf tissue. Arch. Biochem. Biophys. 167, 655-667.

Reddy A.R. and Das V.S.R., 1987. Chloroplast autonomy for the biosynthesis of isopentenyldiphosphate in guayule (Parthenium argentatum Gray) New Phytol. 106, 457-464.

Rendina A.R. and Felts J.M., 1988. Cyclohexanedione herbicides are selective and potent inhibitors of acety-CoA carboxylase from grasses. Plant Physiol. 86, 983-986.

Schulze-Siebert D. and Schultz G., 1987. ß Carotene synthesis in isolated spinach chloroplasts. Plant Physiol. 84, 1233-1237.

Secor J. and Csêke C., 1988. Inhibition of acetyl-CoA carboxylase activity by haloxyfop and tralkoxydim. Plant Physiol. 86, 10-12.

Shimabukuro M.A., Shimabukuro R.H. and Walsh W.C., 1982. The antagonism of IAA-induced hydrogen ion extrusion and coleoptile growth by diclofop-methyl. Physiol. Plant. 56, 444-452.

Shimakata T. and Stumpf P.K., 1983. The purification and function of acetyl coenzyme A: acyl carrier protein transacylase. J. Biol. Chem. 258, 3592-3598.

St. John J.B., 1976. Manipulation of galactolipid fatty acid composition with substituted pyridazinones. Plant Physiol., 57, 38-40.

Treede H.J. and Heise K.P., 1985: The regulation of acetyl coenzyme A synthesis in chloroplasts. Z. Naturforsch. 40c, 496-502.

Treede H.J. and Heise K.P., 1986. Purification of the chloroplast pyruvate dehydrogenase complex from spinach and maize mesophyll. Z. Naturforsch. 41c, 1011-1017.

Treede H.J., Riens B. and Heise K.P., 1986. Species-specific differences in acetyl coenzyme A synthesis of chloroplasts. Z. Naturforsch. 41c, 733-740

Uchiyama M., Washio N., Ikar T., Igarashi H. and Suzuki I., 1986. Stereospecific responses to (R)-(+)-and (S)-(-)-quizalofop-ethyl in tissues of several plants. J. Pesticide Sci 11, 459-467.

Vance D., Goldberg I., Mitsuhashi O. and Bloch K., 1972. Inhibition of fatty acid synthetase by the antibiotic cerulenin. Biochem. Biophys. Res.Commun. 48, 649-656.

Wilkinson E.E. and Oswald T.H., 1987. S-ethyl-dipropylthiocarbamate (EPTC) and 2,2-dichloro-N,N-di-2-propenylacetamide (dichlormid) inhibitions of synthesis of acetyl-coenzyme A derivates. Pestic. Biochem. Physiol. 28, 38-43.

Wright J.P. and Shimabukuro R.H., 1987. Effect of diclofop and diclofop-methyl on the membrane potentials of wheat and oat coleoptiles. Plant Physiol. 85, 188-193.

Yamada Y. and Nakamura Y, 1975. Fatty acid synthesis by spinach chloroplasts. II. The path from PGA to fatty acids. Plant Cell Physiol. 16, 151-162.

Yamada M., Kato M., Nishida I., Kawano K., Kawaguchi A. and Ehara T., 1987. Modulation of fatty acid synthesis in plants by thiolactomycin, In: The metabolism, structure and function of plant lipids, Stumpf P.K., Mudd J.B. and Nes W.D. ed., Plenum Press, New York, pp. 447-454.

BIOLOGICAL ROLE OF PLANT LIPIDS
P.A. BIACS, K. GRUIZ, T. KREMMER (eds)
Akadémiai Kiadó, Budapest and Plenum Publishing
Corporation, New York and London, 1989

ACETYL-CoA CARBOXYLASE AS TARGET FOR HERBICIDES

M. Focke and H.K. Lichtenthaler

Botanisches Institut II (Plant Physiology and Plant Biochemistry), University of Kalsruhe, Kaiserstr. 12, D-7500 Karlsruhe, FRG

INTRODUCTION

The cyclohexane-1,3-dione and aryloxyphenoxypropionic acid-type herbicides are used as selective grass herbicides in several economical important crop cultures. They are known to block de novo fatty acid synthesis in the chloroplasts of sensitive grasses (Poaceae). These herbicides have been suggested to be inhibitors of an enzyme of the fatty acid synthetase complex, because they block the incorporation of labelled malonic acid into the fatty acid fraction. The aim of this work was to identify the single target enzyme of this herbicides.

MATERIAL AND METHODS

An active enzyme fraction was isolated according to Høj and Mikkelsen (1982) with some modifications. Details of plant cultivation, enzyme isolation, and enzyme assay were described before (Focke and Lichtenthaler 1987, Kobek et al. 1988). 6-day-old etiolated barley (var. Alexis) were illuminated for six hours. The 40-70% ammonium sulphate saturation fraction of the chloroplast stroma was dialyzed and concentrated against polyethyleneglycol. Enzymic activity was measured by the incorporation of acetate, acetyl-CoA, malonate, or malonyl-CoA into the fatty acid fraction. Herbicides were added in a methanolic solution with a final methanol concentration of 0.2 % in the assay volume.

RESULTS AND DISCUSSION

If 14C-labelled acetic acid in presence of the appropriate cofactors is incubated with the crude enzyme preparation, a part of the label is incorporated into the fatty acid fraction. This indicates that in this preparation all plastidic enzymes from the pathway of acetate to fatty acids, e.g. acetyl-CoA synthase, acetyl-CoA carboxylase, and all enzymes of the fatty acid synthetase complex are present. The incorporation of acetic acid can be inhibited in a dose-dependent manner and to a different degree by the various herbicides. This is shown in Figure 1.

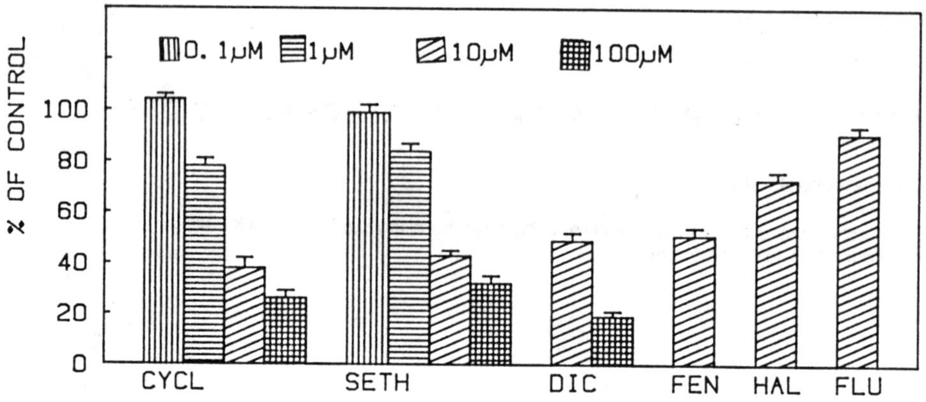

Fig.1: Incorporation of ^{14}C-acetate into the fatty acid fraction. Applied were 0.86µCi (= 14nmol) acetate per condition. Controls incorporated 6nmoles acetate per mg protein and hour. Mean of three determinations. Abbreviation: seth = sethoxydim; cycl = cycloxydim; cleth = clethodim; diclofop; fen = fenoxaprop; hal = haloxyfop; flu = fluazifop.

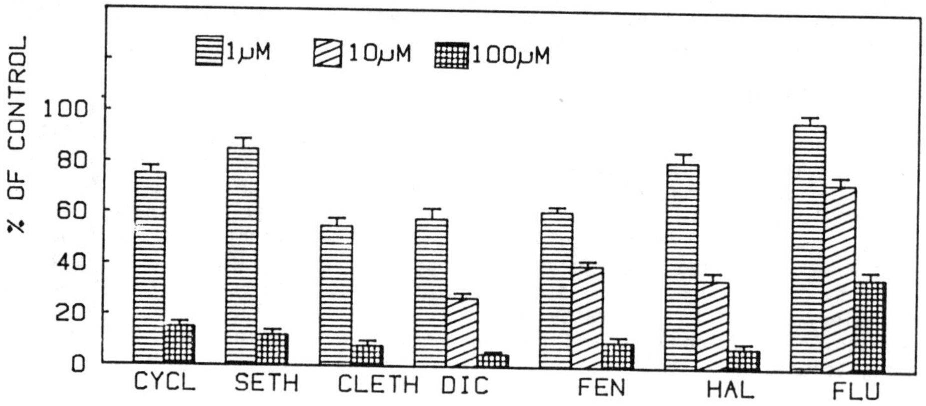

Fig.2: Inhibition of ^{14}C acetyl-CoA incorporation into the fatty acid fraction. Applied were 0.02 µCi (= 8.5 nmoles) acetyl-CoA. The control control incorporated approximately 15nmoles per mg protein and hour. Abbreviations are the same as in Figure 1.

The incorporation ^{14}C-acetyl-CoA is affected in a similar way by these herbicides (Figure 2). Therefore the activation of acetic acid to acetyl-CoA cannot be the target for these herbicides.

The next enzyme in the metabolic pathway leading to the formation of fatty acids, is acetyl-CoA carboxylase. Acetyl-CoA carboxylase catalyzes the ATP-dependent carboxylation of acetyl-CoA to malonyl-CoA, the substrate needed for the elongation steps in fatty acid biosynthesis. Fig. 3 shows that the incorporation of malonate or malonyl-CoA into fatty acids at high concentrations of herbicides resulting in a strong decrease of the incorporation of acetate or of acetyl-CoA is still unaffected. This strongly suggests the acetyl-CoA carboxylase and not an enzyme of the fatty acid synthetase complex to be the real enzymic target for the cyclohexane-1,3-dione and aryloxyphenoxypropionic acid-type herbicides. Recently these results were independently confirmed by three other groups (Burton et al. 1987, Rendina and Felts 1988, Secor and Cséke 1988).

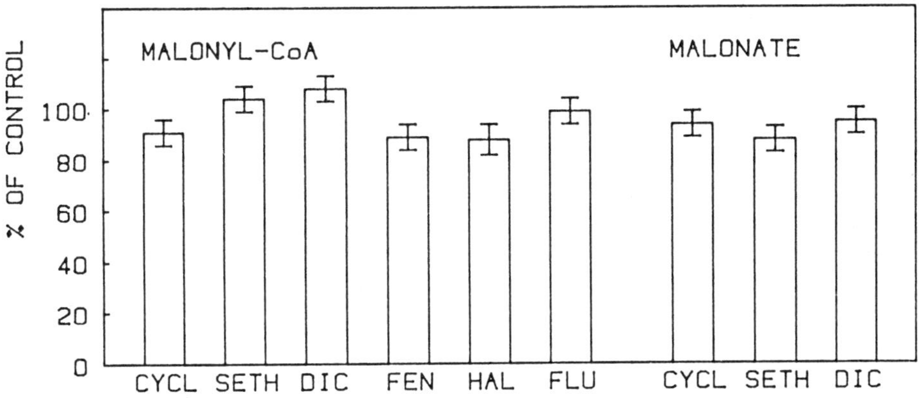

Fig.3: Incorporation of malonate or malonyl-CoA in the presence of herbicides(100 µM). Controls incorporated 14 % of the applied malonyl-CoA (0.05 µCi =12nmoles) or 5 % of the applied malonate (0.3 µCi = 15nmoles), respectively

To explain the inhibition of the incorporation of malonic acid into fatty acid in intact plants by the cyclohexane-1,3-diones (Burgstahler 1985) and aryloxy-phenoxypropionic acid-herbicides (Hoppe and Zacher 1982), we suggest the action of a malonate and/or malonyl-CoA decarboxylase, able to catalyze the formation of acetate/acetyl-CoA, substrates the incorporation of which into fatty acids can then be blocked by these herbicides. Our model is summarized in Figure 4.

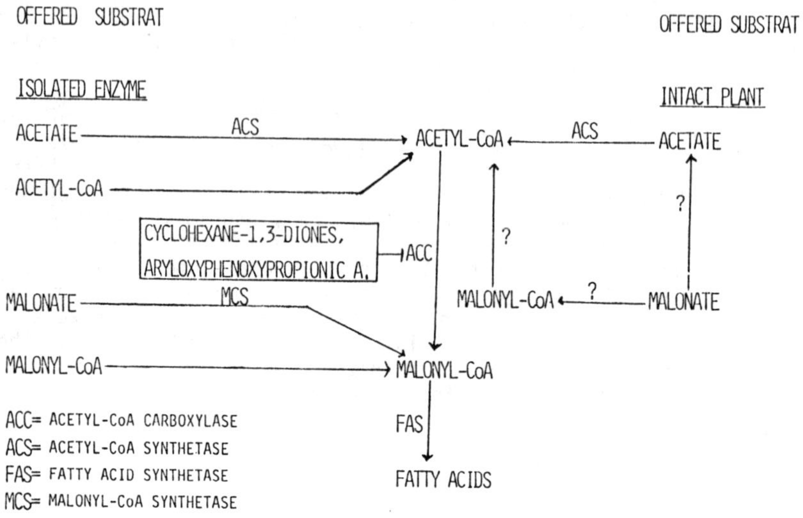

Fig.4: Proposed differences in the incorporation of labelled precursors into fatty acids by intact plants and an isolated enzymic system.

REFERENCES:

- **Burgstahler**, R.J., 1985.On the mode of action of sethoxydim (Poast) ,a cyclohexane-1,3-dione. Karlsruhe Contribution to Plant Physiology. 13, 1-111. ISSN 0173-3133.
- **Burton**, J.D., Gronwald J.W., Somers, D.A., Connelly, B.G., Gengenbach, B.G.and Wyse, D.L., 1987. Inhibition of plant acetyl-coenzyme A carboxylase by the herbicides sethoxydim and haloxyfop. Biochem. Biophys. Res. Commun. 148, 1039-1044.
- **Focke**, M. and Lichtenthaler, H.K., 1987. Inhibition of the acetyl-CoA carboxylase of barley chloroplasts by cycloxydim and sethoxydim. Z. Naturforsch. 42c, 1361-1363.
- **Høj**, P.B. and Mikkelsen, J.D., 1982. Partial separation of individual enzyme activities of an ACP-dependent fatty acid synthetase from barley chloroplast. Carlsberg Res. Commun. 47, 119-141.
- **Hoppe**, H.H. and Zacher, H., 1982. Inhibition of fatty acid biosynthesis in tips of radicles from Zea mays by diclofop-methyl. Z. Pflanzenphysiol. 105, 287-289.
- **Kobek**, K., Focke, M. and Lichtenthaler, H.K., 1988. Fatty acid biosynthesis and acetyl-CoA carboxylase as a target of diclofop, fenoxaprop and other aryloxy-phenoxy-propionic acid herbicides.Z. Naturforsch. 43, 47-54.
- **Rendina**, A.R. and Felts, J.M., 1988. Cyclohexanedione herbicides are selective and potent inhibitors of acetyl-CoA carboxylase from grasses. Plant Physiol. 86, 983-986.
- **Secor**, J. and Cséke, C., 1988. Inhibition of acetyl-CoA carboxylase activity by haloxyfop and tralkoxydim. Plant Physiol. 86, 10-12.

BIOLOGICAL ROLE OF PLANT LIPIDS
P.A. BIACS, K. GRUIZ, T. KREMMER (eds)
Akadémiai Kiadó, Budapest and Plenum Publishing
Corporation, New York and London, 1989

INHIBITION OF THE FATTY ACID BIOSYNTHESIS IN ISOLATED CHLOROPLASTS

K.Kobek and H.K Lichtenthaler

Botanisches Institut II (Plant Physiology and Plant Biochemistry), University of Kalsruhe, Kaiserstr. 12, D-7500 Karlsruhe, FRG

SUMMARY: We studied the effect of the cyclohexane-1,3-dione herbicides sethoxydim and cycloxydim and the phenoxyphenoxypropionic acid herbicide diclofop on the de novo fatty acid biosynthesis of isolated chloroplasts of several sensitive and tolerant plants. The results indicate that the tolerance of different plants against the investigated inhibitors of the de novo fatty acid biosynthesis to be based either on particular properties of the cytoplasm (**Triticum aestivum, Hordeum vulgare, Poa annua**) or on properties of the chloroplasts, including the target enzyme (**Pisum sativum**). The capacity for fatty acid biosynthesis in isolated oat chloroplasts was dependent on the age of the plastids. The highest rates of fatty acid synthesis were found in the youngest chloroplasts isolated from the lower part of oat leaf blades, whereas the lowest rates were measured in the oldest chloroplasts isolated from the upper part of the leaf blades. The percentage inhibiton of fatty acid biosynthesis by cycloxydim (1μM) was clearly higher in younger than in older chloroplasts.

INTRODUCTION

The new herbicides sethoxydim, cycloxydim and diclofop (Fig. 1) are very effective in the postemergence control of grass weeds in dicotyledonous plants and, in the case of diclofop, also in wheat and barley cultures [1-4]. Both herbicide groups inhibit the de novo fatty acid biosynthesis [2-5], which is located in the plastids [6]. The target enzyme of these herbicides is the acetyl-CoA carboxylase [7]. Not only dicotyledonous plants, but also several grasses are, as intact plants, tolerant against these herbicides. Wheat varieties are tolerant against diclofop [1] and Poa annua is resistant against sethoxydim [8]. To elucidate the causes for the tolerance we investigated the capacity of chloroplasts isolated from several sensitive and insensitive plant species to synthesize fatty acids as well as the inhibitory effects of sethoxydim and diclofop on fatty acid biosynthesis.
The fatty acid biosynthesis activity is dependent on the age of the plastids [9]. In order to elucidate whether the percentage inhibition of fatty acid formation by cycloxydim is also dependent on the age and development stage of the chloroplasts, we isolated chloroplasts from the lower, middle and upper part of the primary leaf blade of oat seedlings which represent differential stages of development.

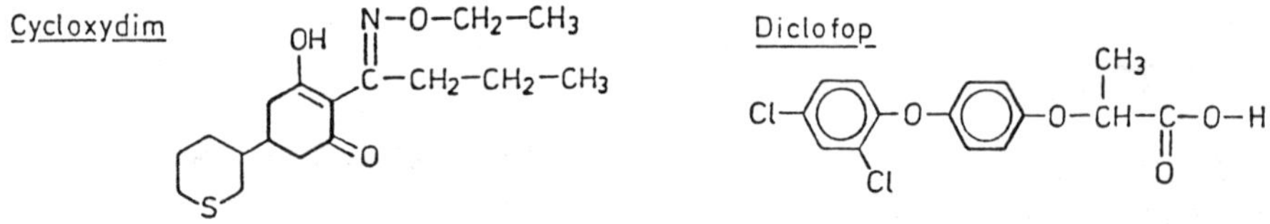

Fig.1: Chemical structure of fatty acid biosynthesis inhibitors.

MATERIALS AND METHODS

Seedlings of **Avena sativa** cv. Flämingnova, **Hordeum vulgare** cv. Alexis, **Triticum aestivum** cv. Max, **Poa annua**, **Poa pratensis** and **Pisum sativum** cv. Kleine Rheinländerin were cultivated and intact chloroplasts were isolated as previously described [3]. The incubation with [2-14C]acetate was carried out for 20 min in a medium containing 300mM sorbitol, 50mM tricine, 50mM phosphate-buffer (pH 7.9), 30 mM $NaHCO_3$, 2.5mM DTE, 2mM ATP, 0.5mM CoA, 0.5mM $MgCl_2$, 70µM [2-14C]acetate (37 kBq per sample) and chloroplasts with a chlorophyll content of ca. 100µg per ml. The light intensity was $1400 \mu mol \cdot m^{-2} \cdot s^{-1}$. Saponification, acidification, extraction of fatty acids and measurement of the incorporated radioactivity was described before [3].

RESULTS AND DISCUSSION

Sethoxydim (1µM) inhibited the biosynthesis of fatty acids in isolated oat chloroplasts (measured via [14C]acetate incorporation into the total fatty acid fraction) to ca. 60 %, whereas in the isolated chloroplasts of the dicotyledonous pea plants their synthesis was not affected by sethoxydim. About 50 % inhibition of fatty acid synthesis by 1µM sethoxydim was detected in experiments with chloroplasts isolated from **Poa pratensis**, a herbicide-sensitive grass. In chloroplasts isolated from **Poa annua**, which as a whole plant is tolerant against sethoxydim [7], fatty acid biosynthesis was also blocked (to ca. 45 %) by 1µM sethoxydim (Fig. 2). This indicates a basic difference in the mechanism of tolerance for **Pisum sativum** and **Poa annua**, respectively. In **Poa annua**, the tolerance is apparently based on

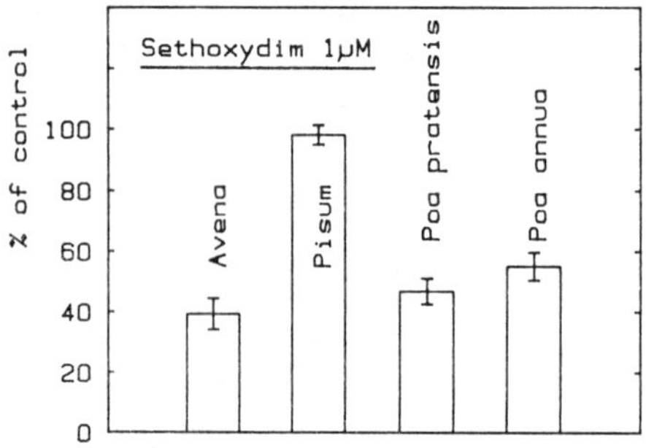

Fig.2: [2-14C]acetate incorporation into the total fatty acid fraction of chloroplasts isolated from different sensitive and tolerant plants in addition of 1µM sethoxydim. The incorporation rates of the controls in nmol acetate $\cdot$ (mg chlorophyll $\cdot$ h)$^{-1}$ were 32 for **Avena sativa**, 38 for **Pisum sativum**, 4 for **Poa pratensis** and 6 for **Poa annua**. Mean values of 4 determinations with SD.

as yet unidentified properties of the cytoplasm, whereas in pea the causes for the tolerance must be localized within the chloroplasts. One could think on a modification of the target enzyme acetyl-CoA carboxylase. The diphenoxypropionic acid derivative diclofop (1µM) also caused a strong inhibition (ca. 90%) of fatty acid biosynthesis in isolated oat chloroplasts. Their biosynthesis in isolated pea chloroplasts, in turn, was not affected by 1µM diclofop. In isolated chloroplasts of the two essential cereals **Hordeum vulgare** and **Triticum aestivum** the fatty acid biosynthesis was blocked by 1µM diclofop to 75 % and 80 %, respectively (Fig. 3). As whole plants, however, barley and wheat are tolerant against diclofop, which permits the use of this herbicide in barley and wheat cultures for control of grass weeds. The tolerance of **Triticum aestivum** cv. Waldron as a whole plant is apparently due to the metabolisation of diclofop to a non phytotoxic decomposition product [1].

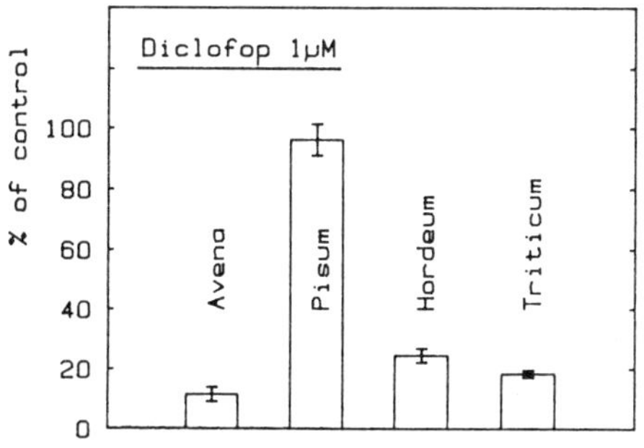

Fig.3: [14C]acetate incorporation into the total fatty acid fraction of chloroplasts, isolated from several sensitive and tolerant plants with addition of 1µM diclofop. The incorporation rates of the untreated control - chloroplasts were 26nmol acetate·(mg chl.·h)$^{-1}$ for barley and wheat. The rates of oat and pea are given in Fig. 2. Mean values of 4 determinations with SD.

The de novo fatty acid biosynthesis of isolated oat chloroplasts as well as its inhibition by cycloxydim was dependent on the age of the chloroplasts. The youngest oat chloroplasts isolated from the lower part of the primary leaf blades had the highest capacity for fatty acid synthesis which was inhibited by 1µM cycloxydim to a larger degree than in older chloroplasts. In those chloroplasts, isolated from the middle and upper part of the

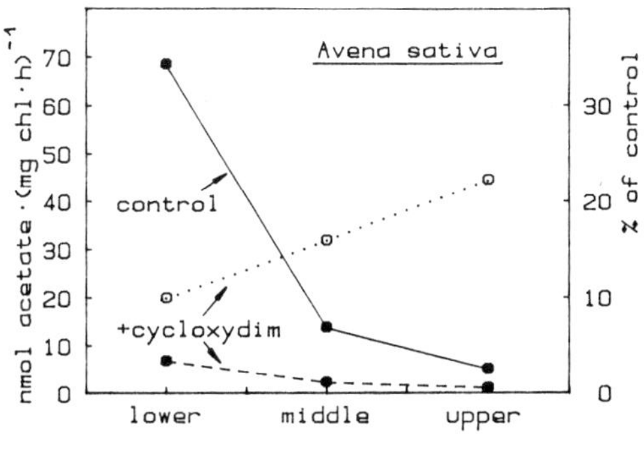

Fig.4: Incorporation of [14C]acetate into the total fatty acid fraction of chloroplasts isolated from the lower, middle and upper part of the primary leaf blades of 8d old oat seedlings. •———•: control rate, •- - - - -•: rate of [14C] acetate incorporation in presence of 1µM cycloxydim and o·········o :remaining percentage of the incorporation activity in presence of 1µM cycloxydim (% of control). Mean values of 4 determinations, SD 5% or less.

leaf blades, the capacity for fatty acid biosynthesis was stepwise decreased. In parallel the efficacy of 1μM cycloxydim was also decreased and the remaining capacity for fatty acid formation increased with the age of the chloroplasts. It reached ca. 22 % of the untreated controls in the oldest chloroplasts whereas in the youngest chloroplasts only 10% of the fatty acid biosynthesis activity was retained (Fig. 4).

Similar results, concerning the incorporation rates of [14C] acetate into the fatty acids of isolated chloroplasts, were found by Hawke et al. [9] , who measured the highest rates of acetate incorporation into the fatty acids in the youngest chloroplasts from leaf blades of maize seedlings.

ACKNOWLEDGEMENTS: We wish to thank the BASF, Limburgerhof and the Hoechst AG, Frankfurt for providing herbicides and for their interest in this research.

REFERENCES:

[1] **Jacobson, A.,** Shimabukuro, R.H. and McMichael, C., 1985. Response of wheat and oat seedlings to root applied diclofop-methyl and 2,4-dichlorphenoxyacetic acid. Pestic. Biochem. Physiol. 24, 61-67.

[2] **Lichtenthaler, H.K.,** Kobek, K. and Ishii, K., 1987. Inhibition by sethoxydim of pigment accumulation and fatty acid biosynthesis in chloroplasts of Avena seedlings. Z. Naturforsch 42c, 1275-1279.

[3] **Kobek, K.,** Focke, M., Lichtenthaler, H.K., Retzlaff, G. and Würzer, B., 1988. Inhibition of fatty acid biosynthesis in isolated chloroplasts by cycloxydim and other cyclohexane-1,3-diones. Physiol. Plant. 72, 492-498.

[4] **Kobek, K.,** Focke, M. and Lichtenthaler, H.K., 1988. Fatty acid biosynthesis and acetyl-CoA carboxylase as a target of diclofop, fenoxaprop and other aryloxy-phenoxy-propionic acid herbicides. Z. Naturforsch. 43c, 47-54.

[5] **Hoppe, H.H.** and Zacher, H., 1985. Inhibition of fatty acid biosynthesis in isolated bean and maize chloroplasts by herbicidal phenoxy-phenoxypropionic acid derivatives and structurally related compounds. Pestic. Biochem. Physiol. 24, 298-305.

[6] **Ohlrogge, J.B.,** Kuhn, D.N. and Stumpf, P.K. Subcellular localization of acyl carrier protein in leaf protoplasts of Spinacia oleracea, Proc. Natl. Acad. Sci. USA 76, 1194-1198.

[7] **Focke, M.** and Lichtenthaler, H.K., 1987. Inhibition of the acetyl-CoA carboxylase of barley chloroplasts by cycloxydim and sethoxydim. Z. Naturforsch. 42c, 1361-1363.

[8] **Struve, I.,** Golle, B. and Lüttge, U., 1987. Sethoxydim uptake by leaf slices of sethoxydim resistant and sensitive grasses. Z. Naturforsch. 42c, 279-282.

[9] **Hawke, J.C.,** Rumbsby, M.G. and Leech, R.M., 1974. Lipid biosynthesis by chloroplasts isolated from developing Zea mays. Phytochem. 13, 403-413.

BIOLOGICAL ROLE OF PLANT LIPIDS
P.A. BIACS, K. GRUIZ, T. KREMMER (eds)
Akadémiai Kiadó, Budapest and Plenum Publishing
Corporation, New York and London, 1989

PERTURBATION OF FUNGAL GROWTH AND LIPID COMPOSITION BY CYCLOPROPENOID FATTY ACIDS

Katharine M. Schmid[1] and G.W. Patterson[2]

[1]Department of Botany and Plant Pathology, Michigan State University, East Lansing, MI 48824, USA
[2]Department of Botany, University of Maryland, College Park, MD 20742, USA

Cyclopropenoid fatty acids (CPE) are found in at least two phylogenetically distant groups of plants [1,2,10]. Although they are best known as components of seed oils, CPE do occur in vegetative plant parts [5,10,12]. We have found that, in representatives of seven Malvaceous genera, CPE are more concentrated in roots than in shoot tissues [9]. In Malva neglecta, for example, malvalate and sterculate together comprise 53% of root neutral lipids, 14% in the stem, 4% in petioles, and <2% in leaf blades. Phospholipids contain far smaller proportions of CPE, but the same distribution pattern holds. In M. neglecta, CPE make up 7.5% of root phospholipids, but only 0.1% of leaf blade phospholipids. Since root CPE also exceed shoot CPE on a dry weight basis, it seems unlikely that the low proportions in leaves are due merely to masking by chloroplast lipids. Nor does CPE content strictly parallel triacylglycerol (TAG) content of the organs. In cotyledons from cotton seeds, which are quite rich in TAG, CPE make up less than 0.1% of neutral lipid acyl groups. Radicles from the same seeds contain 26% CPE, while hypocotyls, like stems, have intermediate levels.

These localization data raised two major questions. First, what determines the uneven distribution of CPE between plant organs and lipid fractions? Second, what are the implications of organ specific expression of CPE? We wondered whether the distribution of CPE might be saying something about their function. Is there a need for expensive, root specific storage lipids?

When fed CPE-containing oils, hogs and cattle develop hard lard, cows produce milk yielding sticky butter, and chickens lay eggs with pink egg whites and pasty yolks. All of these disorders have been linked to a block of the animal stearate desaturation system by CPE [2]. On this basis we hypothesized that CPE would have antifungal activity due to an inhibition of fatty acid desaturation [8]. Early work by Reiser et al. [6] had indicated inhibition of yeast desaturation, and during the course of our studies, 18:0 accumulation was reported in several other yeast genera treated with CPE [4,7]. Here we discuss effects of CPE on two fungi of interest to plant pathologists.

METHODS

Methyl sterculate and methyl malvalate were prepared from trans-esterified Sterculia foetida oil by a combination of urea clathration and reverse phase HPLC [8]. The methyl esters were added to cultures of Ustilago maydis (DC) Cda. and Fusarium oxysporum Schlecht in small quantities of methanol. Control cultures received methanol alone.

Chloroform/methanol (2:1) extracts were fractionated by silicic acid chromatography and TLC. Gas chromatographic analyses of fatty acids were performed following transesterification with sodium methoxide [8,9].

RESULTS AND DISCUSSION

Ustilago maydis, the causal agent of corn smut, is a popular organism for fungicide testing; sporidia grown in liquid culture are quite uniform in size and shape, permitting easy visualization of morphological perturbations. During both sterculate and malvalate treatments, we found that some sporidia remained unicellular and cigar-shaped, while others became branched (Fig. 1) and multicellular. During 30uM CPE treatments, the number of U. maydis sporidia remained virtually constant for about 24 hours. Dry weight lagged behind control levels, but continued to accumulate during this period, presumably due to the growth of branched sporidia. In addition to branching, treated sporidia showed irregular wall deposition as judged by Calcofluor staining. These symptoms are very

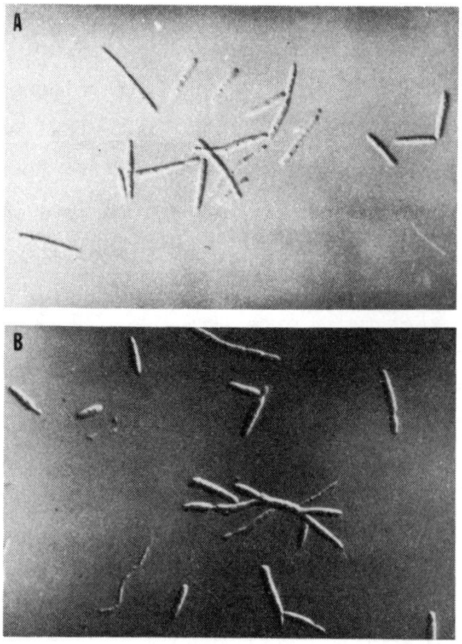

Fig. 1. Ustilago maydis sporidia. A. Control. B. 30uM sterculate treatment.

similar to those of U. maydis treated with sterol biosythesis inhibitors and have been attributed to perturbation of membrane function [3]. CPE treated sporidia have normal ergosterol content, but would be expected to suffer membrane irregularities if sterol desaturation is inhibited.

The second target organism, Fusarium oxysporum, grew normally at concentrations of sterculate that inhibited Ustilago maydis dry weight accumulation at least 50%. Nevertheless, TAG compositions of F. oxysporum and U. maydis implied inhibition of stearate desaturation by CPE in both organisms. The level of 18:0 rose from 9% to 44% in U. maydis given 30uM sterculate, and from 7% to 32% in F.oxysporum. U. maydis, unlike F. oxysporum, incorporated a sizable amount of sterculate or malvalate into its TAG. However, TAG are a normal deposition site for unusual fatty acids, and this would not be expected to harm the fungus.

Differences between phospholipid compositions of the two organisms were more striking. In F. oxysporum, 30uM sterculate treatment resulted in a slight rise in 18:0 and slight incorporation of sterculate. In U. maydis, on the other hand, 18:0 jumped from 2 to 22%. Sterculate and malvalate also made up about 20% of phospholipids from U. maydis after 30uM treatments. Such gross changes in membrane lipids could well account for the morphological effects of CPE on U maydis sporidia.

The comparative resistance of F oxysporum to sterculate may reflect differences either in desaturation of 18:0 or in uptake or metabolism of CPE. It may be significant that we recovered less than 10% of CPE when F. oxysporum was treated with 30uM sterculate, compared to 50% for comparable Ustilago cultures. In fact, the lipid profiles of F.oxysporum given 30uM sterculate resembled those of U. maydis given 1uM sterculate. At the lower dose of CPE, U.maydis sporidia showed a sizeable increase in TAG 18:0, but little incorporation of CPE or elevation of phospholipid 18:0.

Finally, we attempted to link the morphological effects of CPE on U. maydis with a block in 18:0 desaturation by supplying 18:1, the desaturation product. 18:1 alone had little effect on branching, dry weight accumulation or sporidial number. 30uM sterculate increased the proportion of sporidia showing branching, decreased dry weight accumulation, and drastically reduced sporidial numbers. When sterculate treatments were carried out in the presence of equimolar 18:1, all sterculate induced symptoms were alleviated.

In conclusion, CPE does appear to inhibit stearate desaturation in some plant pathogenic fungi, and the effects of CPE on growth and morphology of Ustilago maydis are consistent with an antifungal role for sterculate and malvalate. The levels of CPE inhibiting fungal growth were probably within the range available in Malvaceous roots. It is difficult to calculate meaningful concentrations of such hydrophobic compounds. However, root levels in the species we have examined range from 0.1 to 13umol/g fresh weight. If one gram of fresh weight were equivalent to one ml of water, these would correspond to 100uM and 1.3mM respectively, figures in excess of the 30uM used in these studies.

REFERENCES

1. M.B. Bohannon, R. Kleiman. 1978. Cyclopropene fatty acids of selected seed oils from Bombacaceae, Malvaceae and Sterculiaceae. Lipids 13: 270-273.
2. W.W. Christie. 1970. Cyclopropane and cyclopropane fatty acids. In F.D. Gunstone, ed., Topics in Lipid Biochemistry. London, Logos, Vol. 1, pp. 1-49.
3. T. Kato. 1986. Sterol biosynthesis in fungi, a target for broad spectrum fungicides. In Chemistry of Plant Protection. 1. Sterol Biosynthesis Inhibitors and Anti-Feeding Compounds. Berlin, Springer-Verlag, pp. 1-24.
4. R.S. Moreton. 1985. Modification of fatty acid composition of lipid accumulating yeasts with cyclopropene fatty acid desaturase inhibitors. Appl. Microbiol. Biotechnol. 22: 41-45.
5. H.E. Nordby, G. Yelenosky. 1987. Influence of cold temperatures on leaf lipids by Hibiscus rosa-sinensis. Phytochemistry 26: 3151-3157.
6. R. Reiser, C.K. Parekh, W.W. Meinke. 1963. The metabolism of cyclopropenyl fatty acids. In A.C. Frazer, ed., Biochemical Problems of Lipids. Amsterdam, Elsevier, pp. 251-256.
7. C.E. Rolph, R.S. Moreton, I.S. Small, J.L. Harwood. 1987. Acyl lipid metabolism in Rhodotorula gracilis (CBS 3043) and the effects of methyl sterculate on fatty acid desaturation. In P.K. Stumpf, J.B. Mudd, W.D. Nes, eds., The Metabolism, Structure, and Function of Plant Lipids. New York, Plenum Press, pp. 437-440.
8. K.M. Schmid, G.W. Patterson. 1988. Effects of cyclopropenoid fatty acids on fungal growth and lipid composition. Lipids 23: 248-252.
9. K.M. Schmid, G.W. Patterson. 1988. Distribution of cyclopropenoid fatty acids in Malvaceous plant parts. Phytochemistry, in press.
10. J.R. Vickery, F.B. Whitfield, G.L. Ford, B.H. Kennett. 1984. The fatty acid composition of Gymnospermae seed and leaf oils. J. Am. Oil Chem. Soc. 61: 573-575.
11. J.D. Weete. 1980. Lipid Biochemistry of Fungi and Other Organisms. New York, Plenum Press.
12. I. Yano, B.W. Nichols, L.J. Morris, A.T. James. 1982. The distribution of cyclopropane and cyclopropene fatty acids in higher plants (Malvaceae). Lipids 7: 30-34.

BIOLOGICAL ROLE OF PLANT LIPIDS
P.A. BIACS, K. GRUIZ, T. KREMMER (eds)
Akadémiai Kiadó, Budapest and Plenum Publishing
Corporation, New York and London, 1989

STEROLS AND FATTY ACIDS IN THE PLASMA MEMBRANES OF TAPHRINA DEFORMANS CULTURED AT LOW TEMPERATURE AND WITH PROPICONAZOLE

M. Sancholle[1], J.D. Weete[2] and A. Rushing[2]

[1]Lab. de Cryptogamie, Univ. Paul Sabatier, 118 Route de Narbonne, 31062 Toulouse Cedex, France
[2]Dept. of Botany and Microbiology, Auburn University, Alabama 36849, USA

Propiconazole is a triazole that blocks the C-14 demethylation of lanosterol, preventing its conversion to ergosterol or functionally equivalent sterols (Weete, 1987). Taphrina deformans, a foliar pathogen of peach that proliferates by budding in laboratory culture, is highly sensitive to this inhibitor in that only 0.073 µg/ml are required for 50% suppression of growth (Sancholle et al., 1984). Brassicasterol, as the principal sterol of this fungus (Weete et al., 1983; 1985a), is decreased in favor of 24-methylene-24, 25-dihydrolanosterol and obtusifoliol in the presence of sublethal doses of this inhibitor (Weete et al., 1983), a response that also occurs in the plasma membrane (Weete et al., 1985b). Alterations in phospholipid-fatty acids also accompany the changes in sterol content, i.e. increase in unsaturation due to decreased oleic acid and increased linoleic and linolenic acids (Weete et al., 1983; 1985b). Since T. deformans infects peach leaves prior to bud break when the temperature is relatively cold, we conducted the study reported here for the purpose of assessing whether the sensitivity of this fungus differs at the infection temperature (8-13°C) compared to that of the laboratory cultivation temperature (18°C), and whether the above changes in plasma membrane composition occur.

MATERIALS AND METHODS

Culture conditions: Taphrina deformans (D1 strain) was grown in yeast extract/dextrose (1%/2%, w/v) medium at 18°C on an incubator-shaker at 140 rpm. Seven 500 ml Erlenmeyer flasks each containing 100 ml medium were inoculated with a 36 h culture. After 40 h growth, the cultures are collected by centrifugation at 3500 rpm for 10 min in sterile tubes. The cells were pooled and resuspended in 100 ml fresh medium. Four 500 ml flasks, each containing 80 ml medium, were inoculated with 20 ml of the above cell suspension, bringing the final cell concentration to 5.98 $\times 10^8$ ml. Propiconazole in acetone was added to two of the flasks designated as "treated" to a final concentration of 0.075 µl/ml. The two "control" cultures contained the same amount of acetone. One culture of each treatment was incubated at 18°C, and the two remaining flasks at 8°C, for 48 h. The cells in the remaining 20 ml (see above) were analyzed as a "reference". Cell numbers were determined using a hemactymeter.

Protoplast production: After the 48 h incubation, the cells were counted, collected by centrifugation, washed with magnesium sulfate solution, and then weighed. They were resuspended in osmotic medium containing Novozyme (30 mg/g cells) and BSA (20 mg/g cells), and maintained overnight for cell wall digestion at the original incubation temperature (8°C or 18°C). Protoplasts were collected by centrifugation in "trapping buffer". They were resuspended in 0.1M tris Buffer containing EDTA, cycloheximide and chloramphenicol (200 mg/L each) and 0.4M sucrose, and then broken by mild sonication (35 pulses/s at intensity 3).

Isolation of plasma membranes: The plasma membrane-enriched fractions from the respective treatments were collected on sucrose gradients as described before (Sancholle 1984; Fonvieille 1985). The sucrose concentrations from the bottom of the centrifuge tube were: 50% (4.8 ml), 40% (8 ml), 35% (11.2 ml), 25% (4.8 ml). The plasma membrane fraction migrated to the 25/35% interface. Purity of the membrane fractions was checked by electron microscopy according to the procedures of Weete et al. (1985b), except that phosphate buffer (pH 7.2) was used.

Lipid analyses: The total fatty acid and sterol fractions were isolated by direct saponification of the membranes as described by Weete et al. (1983) for total lipid.

RESULTS AND DISCUSSION

The plasma membrane-enriched fractions of the *T. deformans* reference cells contained membrane-bound vesicles typical of such fractions (Weete et al., 1985b). The major fatty acids of these membranes were those expected for this fungus (Sancholle, 1984; Weete et al., 1983), i.e. palmitic, stearic, oleic, linoleic and linolenic acids (Table 1). The Δ/mole was relatively low at 0.84 which indicated that the lipids of these membranes had a relatively high amount of saturated fatty acids. Brassicasterol and an unidentified C28 diene were detected as previously reported for this species at 84.5% and 15.5%, respectively (Weete et al., 1985a).

There was only a 33% increase in cell number in the cultures after incubation for 48 h at 18°C at the relatively high initial cell density (5.98×10^8 cells/ml), and the plasma membrane-enriched fraction contained vesicles as before. Although the relative amounts of brassicasterol and the C28 diene were essentially the same as in the reference cells, the Δ/mole increased to 1.24 indicating an increase in the degree of fatty acid unsaturation which involved changes in the relative amounts of palmitic, stearic, oleic and linoleic but not linolenic acids (Table 1).

There was only a 14% increase in cell number over the reference cells, and only 7.6% inhibition of growth by the propiconazole. There was a slight measurable growth effect of the inhibitor, and only very modest but reproducible changes in the relative proportions of fatty acids, and only a slight change in the sterol composition (Table 1). There was a 4% decrease in brassicasterol and corresponding increase in 24-methylene-24, 25-dihydrolanosterol plus obtusifoliol. The explanation for the modest effect on growth and plasma membrane composition is that the cell/inhibitor ratio was much higher than typically used when 50% growth inhibition is achieved with 0.075 μg/ml (Sancholle et al., 1984).

There was a 60% increase in cell number when the control cultures were incubated at 8°C for 48 h, indicating a relatively rapid adaptation from 18°C to 8°C. This is consistent with the fact that this organism infects peach buds in early spring at relatively low temperatures.

These cultures contained 30% more cells than corresponding control cultures at 18°C. The plasma membrane-enriched fraction from these cells contained vesicles as before. The relative amount of brassicasterol in this membrane fraction was 6% higher than in the corresponding controls incubated at 18°C. The Δ/mole was 87% and 27% higher than those of the reference and corresponding control cells (18°C), which was due mainly to an increase in linoleic acid and decrease in oleic acid (Table 1).

Table 1. Fatty acids of plasma membrane-enriched fractions from Taphrina deformans cells cultured with or without propiconazole and at 18°C and 8°C.

Fatty acid	Reference	18°C		8°C	
		C[b]	T	C[b]	T
C16:0	31.3	12.5	11.8	13.5	11.0
C18:0	19.2	15.2	14.5	9.4	12.5
C18:1	25.2	29.3	29.4	12.2	13.0
C18:2	14.3	34.5	35.6	49.5	48.8
C18:3	9.9	8.5	8.6	15.4	14.6
Δ/mole[c]	0.84	1.24	1.26	1.57	1.54

[a] Values are reported as the percent of fatty acids obtained by saponification of the membrane fraction.

[b] C = Control, T = treated.

[c] $$\Delta/\text{mole} = \frac{(1 \times \%\ C18{:}1) + (2 \times \%\ C18{:}2) + (3 \times \%\ C18{:}3)}{100}$$

There was only a 28% increase in cell number over the reference culture in the cultures containing propiconazole and incubated at 8°C, which is 20.4% inhibition compared to the controls incubated at the same temperature. Again, the plasma membrane-enriched fraction contained vesicles as before. Propiconazole had a substantial effect on the sterol composition of membranes in this fraction, with a decrease of brassicasterol by 22% and increase in 24-methylene-24, 25-dihydrolanosterol from essentially 0 to 26% of the total sterol fraction. The Δ/mole for fatty acids in this fraction was essentially the same as that for the control cells incubated at 8°C.

In summary, T. deformans readily adapted to 8°C incubation temperature, and had a higher growth rate than cells incubated at 18°C. Chemical changes due to reduced temperature were manifested both in the fatty acids and sterols of the plasma membrane-enriched fraction, i.e. a 25% increase in the degree of unsaturation whereby linoleic acid changed from 24% to 49% of the fatty acids, as might be expected in cold temperature adapted poikilothermic organisms (Thompson, 1985), and 64% increase in the sterol/fatty acid ratio (data not given). The cells were 13% more responsive to propiconazole when cultured at 8°C compared to 18°C, suggesting that effectiveness of the inhibitor may be related to growth rate of the organism. These results also suggest that growth inhibition is due to changes in sterol content resulting from blocking the C-14 demethylation of lanosterol, rather than to the increase in fatty acid unsaturation that typically follows inhibition of C-14 demethylase (Weete, 1987), since the same changes in fatty acids occur in control cultures at 8°C that have the highest growth rate.

Acknowldegements: The work was conducted while the first author was on sabbatical leave at Auburn University. The research was supported by the Alabama Agricultural Experiment Station project ALA-5-887. We are grateful for the assistance of Mr. Shailesh Gandhi.

LITERATURE CITED

Fonvieille, J. L. 1985. Physiologie Composition lipidique du mycélium, isolement et etude biochimique des systèmes membranaires de Scopulariopsis brevicaulis. Thèse Doct. Sc. Nat. Toulouse, No. 1189.

Sancholle, M. 1984. Etude des lipids de quelques champignons: effets du propiconazole inhibiteur de synthèse stérols. Thèse Doct. Sc. Nat. Toulouse, No. 1165.

Sancholle, M., J. D. Weete, and C. Montant. 1984. Effects of triazoles on fungi: I. Growth and physiological responses. Pestic. Biochem. Physiol. 21, 31-41.

Thompson, G. A. 1985. Mechanisms of membrane response to enviromental stress. pp. 347-357 in Frontiers of Membrane Research (J. St. John, E. Berlin, P. C. Jackson, eds.) Rowman and Allan, Totowa, N. J.

Weete, J. D. 1987. Mechanism of fungal growth suppression by inhibitors of ergosterol biosynthesis, pp. 268-285 in Ecology and Metabolism of Plant Lipids (G. Fuller and W. D. Nes, eds.) ACS Symposium Series #325, Washington, D. C.

Weete, J. D., M. Kulifaj, C. Montant, and M. Sancholle. 1985a. Distribution of sterols in fungi. II. Brassicasterol in Tuber and Terfezia species. Can. J. Microbiol. 31, 1127-1130.

Weete, J. D., M. Sancholle, and C. Montant. 1983. Effects of triazoles on fungi. II. Lipid composition of Taphrina deformans. Biochim. Biophys. Acta 752, 19-29.

Weete, J. D., M. Sancholle, J. M. Touze-Soulet, J. Bradley, and R. Dargent. 1985b. Effects of triazoles on fungi. III. Composition of a plasma membrane-enriched fraction of Taphrina deformans. Biochim. Biophys. Acta 812, 633-642.

BIOLOGICAL ROLE OF PLANT LIPIDS
P.A. BIACS, K. GRUIZ, T. KREMMER (eds)
Akadémiai Kiadó, Budapest and Plenum Publishing
Corporation, New York and London, 1989

MEMBRANE LIPID COMPOSITION OF TRICHODERMA STRAINS AND THEIR SENSITIVITY TO SAPONIN AND POLYENE ANTIBIOTICS

Katalin Gruiz and P.A. Biacs

Technical University, Budapest, 1111 Gellért tér 4. Hungary
Central Food Research Institute, 1022 Herman Ottó 15. Hungary

INTRODUCTION

The principle of the fungistatic activity of both the saponins and the polyene antibiotics is an interaction with fungal membrane sterols, and probably with proteins and phospholipids [1,6]. Organisms which have no sterols in their membranes are not sensitive to saponins or polyene antibiotics [5,7]. Electron micrographs of the membranes of erythrocytes [2,6] or yeasts [8,9,10] treated with saponin or polyene antibiotics show convincingly the destruction of the membrane, but the sterol-saponin or sterol-polyene complex has not yet been isolated and determined.

In the course of our previous investigations it has become clear, that there is an interaction between membrane sterols and saponins: an increased ergosterol content was measured after saponin-treatment, but there was no difference between sensitive and non-sensitive strains [3]. Fatty acid measurements showed, that resistance to saponin is accompanied with higher unsaturation and with the increase of the relative amount of certain fatty acids [4].

MATERIALS AND METHODS

Four taxologically similar, but in sensitivity different *Trichoderma* strains (Fungi Imperfecti) were investigated: *Trichoderma viride* G, *Tr. koningii* UAMH 1206, *Tr. viride* BKM F-1117, and *Tr. reesei* QM 9414. The effects of aescin-saponin and the polyene antibiotics; candicidin, amphotericin B, and nystatin were measured after 24h or 48h growth of the cells. The moulds were grown in a potato-dextrose nutrient medium. Saponin, amphotericin B and candicidin were added immediately at the beginning, after inoculation. Nystatin was added in the intensive growth phase.

Sterols were determined by the UV-spectra of the hexane-extract of the saponified wet cells. Fatty acid composition was measured by capillary gas-chromatography on PEG-1000 stationary phase. The ion permeability of the fungal cell membrane was measured by a conductometric method with water-washed, ion-free mycelia.

RESULTS

Present comparative study is made on the purpose to find connection between sensitivity to saponin or polyene antibiotics and cell membrane composition respectively to characterize the changes in the membrane on the effect of saponin and polyene antibiotics.

On the effect of saponin the increase of the ergosterol content (mg ergosterol/g dry cell) is characteristic, but independent of sensitivity (Table 1.). On the effect of amphotericin B and candicidin ergosterol content is increased too; a slight assosiation can be found with the scale of sensitivity.

Table 1. CHANGES IN THE ERGOSTEROL CONTENT OF SENSITIVE AND NON-SENSITIVE *TRICHODERMA* STRAINS ON THE EFFECT OF SAPONIN AND ANTIBIOTICS

Mould strains	control mg/g	%	saponin %	amphotericin %	candicidin %
Trichoderma viride G	2,4	100	108 +	108 +	129 ++
Trichoderma koningii	1,9	100	110 +	163 +	110 +
Trichoderma vir. BKM	1,3	100	238 –	276 ++	215 ++
Trichoderma reesei	1,5	100	120 –	160 +/–	266 ++

++ very sensitive, + sensitive, +/– slightly sensitive, – non-sensitive

Table 2. showes the results treating the cells with nystatin, after 24h. All of the investigated strains were very sensitive to nystatin. The increase of the ergosterol content is considerable.

Table 2. CHANGES IN THE ERGOSTEROL CONTENT OF SENSITIVE *TRICHODERMA* STRAINS ON THE EFFECT OF NYSTATIN

Mould strains	control mg/g	%	nystatin mg/g	%
Trichoderma viride G	3,6	100	7,9	219 ++
Trichoderma koningii	2,7	100	7,3	270 ++
Trichoderma vir. BKM	3,6	100	9,6	266 ++
Trichoderma reesei	2,8	100	6,7	231 ++

The conclusion can be drawn that the increased ergosterol content cannot influence the inhibition or that the ergosterol content is not responsible for resistance.

The complex membrane structure and function can be characterized by the ion-permeability of the cell membrane. Ion-permeability was measured with mycelia without growth and metabolism. Ion-permeability (I) was calculated according to the formula on Fig.1.

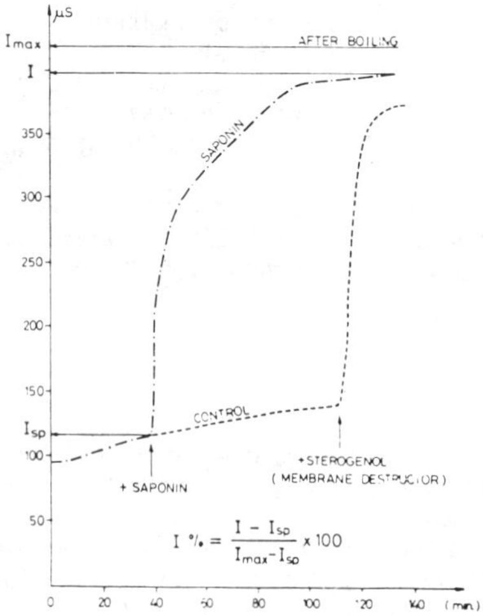

Isp: conduction before adding saponin
I: final-value on the effect of saponin
Imax: conduction after boiling the cells, when all the ions are out of the destroyed cells

Figure 1. Conductometric ion-permeability

Table 3. ION PERMEABILITY OF *TRICHODERMA* MOULD STRAINS ON THE EFFECT OF SAPONIN AND POLYENE ANTIBIOTICS

Mould strain	treatment	sensitivity	ion-permeability %
Tr. vir. G	control		17,9
	saponin	+	80,1
	amphotericin	+	50,7
	andicidin	++	64,0
	nystatin	++	68,2
	pimaricin	+++	37,2
Tr. koningii	control		2,3
	saponin	+	79,7
	amphotericin	+	14,5
	candicidin	+	7,3
	nystatin	++	118,5
	pimaricin	+++	41,1
Tr. vir. BKM	control		16,8
	saponin	−	83,9
	amphotericin	+	26,6
	candicidin	++	44,2
	nystatin	++	62,3
	pimarcin	+++	26,4
Tr. reesei	control		3,1
	saponin	−	73,4
	amphotericin	+/−	27,3
	candicidin	++	15,3
	nystatin	++	35,4
	pimaricin	+++	25,8

Ion-permeability increased both in the case of sensitive and non-sensitive cells on the effect of both saponin and polyene antibiotics. The scale of increase does not agree with the scale of sensitivity. This means, that the membrane construction of the cells without active metabolism is not able to compensate the increased ion-permeability, not even in the case of the resistant strains.

The question was arised, how can resistant cells eliminate the inhibition and membrane destruction effect of saponin and polyenes? For this fatty acids had to be investigated (Table 4 and 5).

Table 4. CHANGES IN THE FATTY ACID COMPOSITION ON THE EFFECT OF SAPONIN AND POLYENE ANTIBIOTICS

Mould strain			14:0	16:0	18:0	18:1	18:2
Tr. vir. G	control		1,6	30,6	21,2	10,2	35,7
	saponin	+	1,4	38,4	6,2	28,6	23,2
	amphoter	+	1,0	38,3	5,7	29,4	23,2
	candicidin	++	0,0	69,5	10,0	20,5	0,0
Tr. kon.	control		1,4	34,2	23,4	9,3	31,3
	saponin	+	1,6	45,3	13,4	28,5	11,4
	amphoter	+	3,0	38,3	9,0	28,5	21,2
	candicidin	+	3,7	36,2	9,3	25,9	21,4
Tr. vir. BKM	control		3,2	38,7	9,7	29,5	18,1
	saponin	−	1,9	33,5	7,9	28,5	28,2
	amphoter	++	0,0	74,7	13,1	12,2	0,0
	candicidin	++	3,4	62,6	8,3	20,0	5,7
Tr. reesei	control		3,4	37,7	8,4	19,2	29,5
	saponin	−	4,3	38,0	9,3	13,2	35,2
	amphoter.	+/−	3,6	38,5	9,7	10,6	37,6
	candicidin	++	4,4	79,4	11,6	4,6	0,0

On the effect of saponin and polyenes the relative amount of linoleic acid increased in the case of resistant strains, but decreased in the case of the sensitives; the scale of decrease corresponds with the scale of sensitivity. With the sensitivity increased paralelly the relative amount of palmitic acid too.

When nystatin was added to the growing mycelium in the intensive growing phase, the effect of the antibiotic on the lipid metabolism can better be seen. Sensitivity is followed by the relative decrease of linoleic and stearic acids and the relative increase of palmitic and linolenic acids. These changes and the increased amount of the shorter fatty acids suggest that the inhibition in the fatty acid biosynthesis is responsible for the sensitivity of the moulds to saponin or polyene antibiotics.

Table 5. CHANGES IN THE FATTY ACID COMPOSITION OF SENSITIVE *TRICHODERMA* STRAINS ON THE EFFECT OF NYSTATIN

Mould strain		12:0	14:0	16:0	18:0	18:1	18:2	18:3
Tr.vir. G	control	0,2	0,5	27,6	14,6	26,1	38,2	2,7
	nystatin	3,3	0,9	38,7	4,2	20,0	26,6	6,2
Tr.kon.	control	0,4	1,1	37,1	17,2	10,9	31,4	1,9
	nystatin	11,5	0,3	40,2	6,2	17,9	17,2	6,7
*Tr.v.*BKM	control	2,1	2,1	26,1	8,7	30,0	26,6	2,4
	nystatin	4,8	14,2	29,0	3,4	30,9	14,1	3,6
Tr.reesei	control	0,3	2,3	28,7	16,2	24,7	24,8	3,0
	nystatin	14,7	2,7	35,5	6,8	21,0	14,2	5,0

When desaturation and elongation is not inhibited, the cells are able to compensate the changes in the membrane structure, e.g. the increased rigidity of the membrane structure on the effect of saponin and polyenes, increasing the membrane fluidity by a higher linoleic acid content.

REFERENCES

1. ASSA,Y.; CHET,I.; GESTETNER,B.; GOVRIN,R.; BIRK,Y.and BONDI,A. (1975) Arch.Microbiol. 103,77.
2. BAGHAM,A.D.and HORNE,R.W.(1962) Nature 196, 952.
3. BIACS,P.A.and GRUIZ,K.(1982) In:Wintermans,J.F.G.M.and KUIPER,P.J.C.(eds) Biochemistry and Metabolism of Plant Lipids, Elsevier,Amsterdam, pp 573-576.
4. BIACS,P.A.and GRUIZ,K.(1984) In:Siegenthaler,P-A.and Eichenberger,W.(eds) Structure, Function and Metabolism of Plant Lipids, Elsevier,Amsterdam, pp 353-356.
5. DEFAGO,G.(1978) Ann. Phytopathol. 10,157.
6. GLAUERT,A.M.;DINGLE,J.and LUCK,J.A.(1962) Nature 196, 953.
7. KINSKY, S.C.(1967) In:Gottlieb,D.and Shaw,P.D.(eds) Antibiotics Vol I. Springer, Berlin pp 122.
8. KITAJAMA,Y.; SEKIYA,T.and NOZAWA,Y.(1976) BBA 455, 452.
9. DE KRUYFF,B.and DEMEL,R.A.(1974) BBA 339, 57.
10. PESTI,M.;NOVAK,E.K.;FERENCZY,L.and SVOBODA,A.(1981) Sabouraudia 19, 17.

BIOLOGICAL ROLE OF PLANT LIPIDS
P.A. BIACS, K. GRUIZ, T. KREMMER (eds)
Akadémiai Kiadó, Budapest and Plenum Publishing
Corporation, New York and London, 1989

INFLUENCE OF HALOXYFOP-ETHOXYETHYL (GRASS HERBICIDE) ON POLAR LIPIDS IN TOLERANT (PEA) AND SENSITIVE (OAT, WHEAT) PLANTS

A. Banas[1], Ingemar Johansson[2], G. Stenlid[2] and S. Stymne[2]

[1]Department of Plant Physiology, Agricultural and Teachers University in Siedlce, Poland
[2]Department of Plant Physiology, Swedish University of Agricultural Sciences, Uppsala, Sweden

INTRODUCTION.

The ethoxyethyl ester of haloxyfop 2-(4-((3-chloro-5-(trifluoromethyl)-2-pyridinyl)oxy)phenoxy)propanoic acid belongs to the p-oxyphenoxypropanoate herbicide family. The mode of action of this herbicide is unsufficiently known. In this study the influence of haloxyfop-ethoxyethyl on lipid metabolism was investigated.

MATERIAL AND METHODS.

For lipid analysis the basal part /0 - end of coleoptile sheath/ of oat and wheat plants and the youngest leaves around the apical meristem of pea plants treated /oat and wheat 10 nmol/plant 19 hours before, pea 20 nmol/plant 48 hours before/ and not treated by haloxyfop-ethoxyethyl was used. Lipid extraction, separation and analysis of fatty acids was done as described by Griffiths et al. /Planta 173: 309-316 1988/. The basal part of wheat plants was also used for incubation with [1-^{14}C] labelled acetate /0,5 uCi/ml, 25°C, light/. The radioactivity incorporated to lipids was determined by liquid scintilation counting.

RESULTS AND DISCUSSION.

In sensitive plants, the level of unsaturation of the fatty acids in polar lipids was strongly changed by the herbicide, whereas no effect was found in tolerant plants /Fig. 1/. In treated plants the relative amount of linolenic acid /18:3/ increased, while that of oleic /18:1/ and linoleic /18:2/ acids decreased. The largest

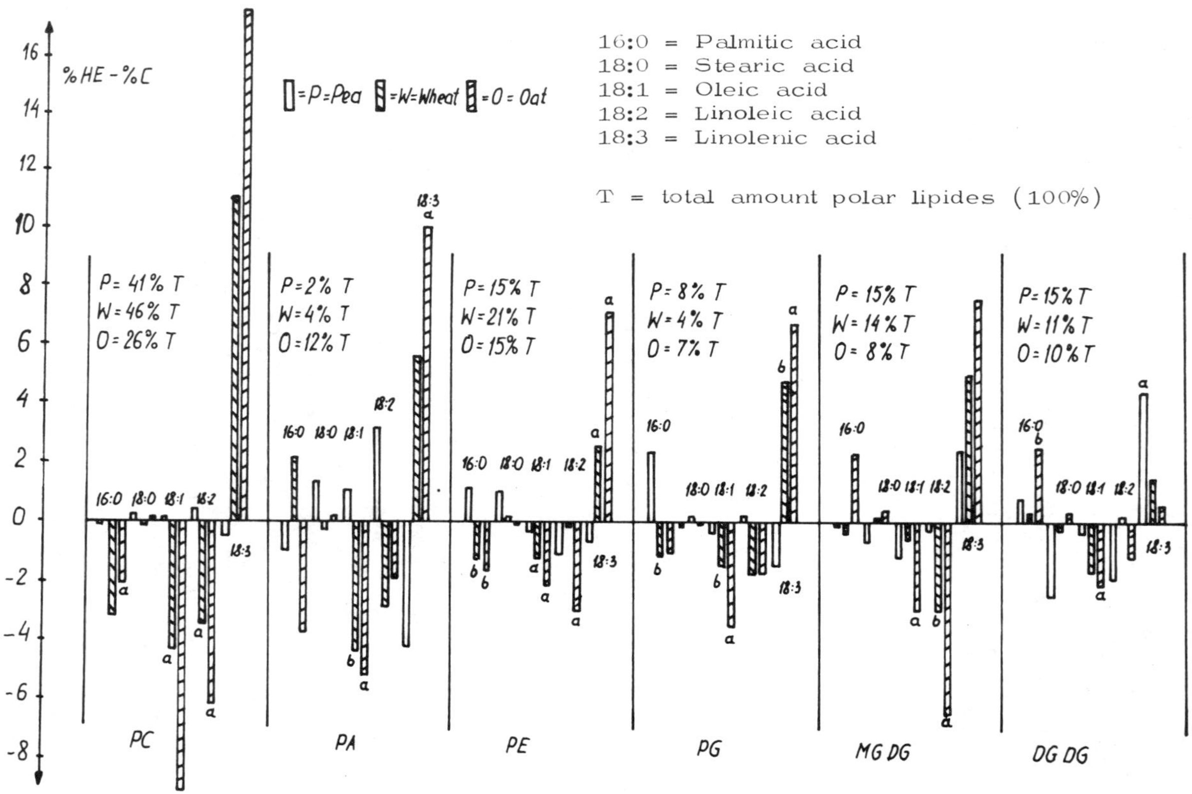

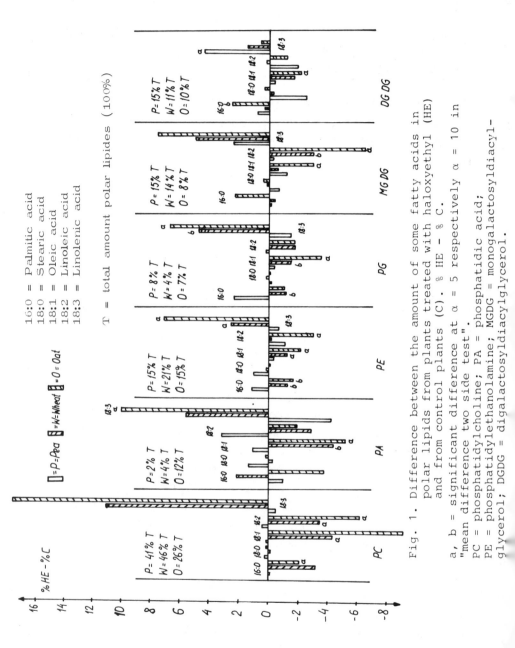

Fig. 1. Difference between the amount of some fatty acids in polar lipids from plants treated with haloxyethyl (HE) and from control plants (C). % HE - % C.

a, b = significant difference at α = 5 respectively α = 10 in "mean difference two side test".
PC = phosphatidylcholine; PA = phosphatidic acid;
PE = phosphatidylethanolamine; MGDG = monogalactosyldiacylglycerol; DGDG = digalactosyldiacylglycerol.

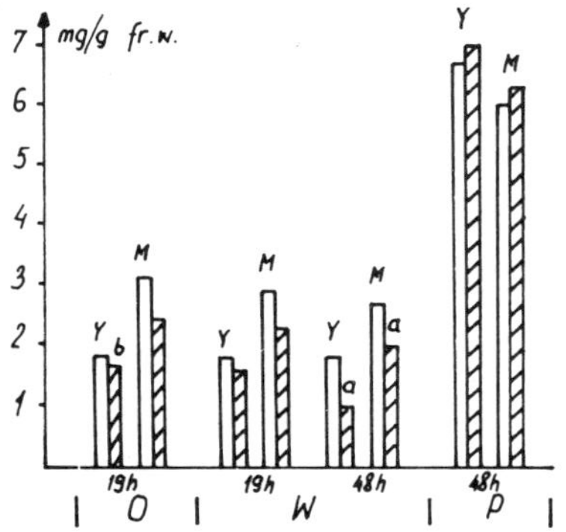

Fig. 2.

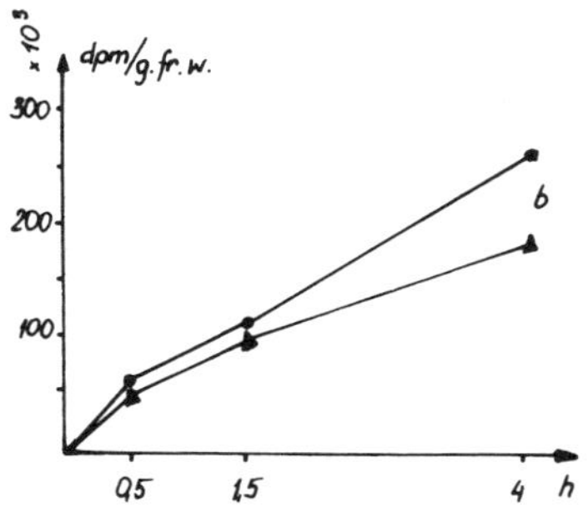

Fig. 3.

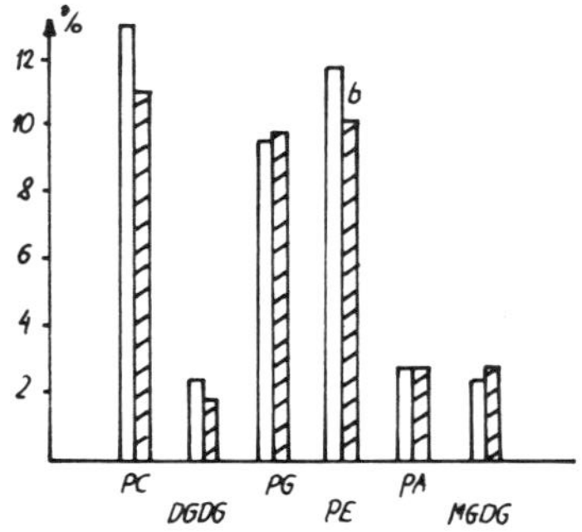

Fig. 4.

Fig. 2. Total amount of fatty acids in young (Y) and mature (M) parts of seedlings of oat (O), wheat (W) and pea (P) plants treated (hatched) and not treated (white) with haloxyfop-ethoxyethyl 19 or 48 hours after treatment.

Fig. 3. Incorporation of radioactive acetate into lipid fraction by excised pieces from wheat plants treated (▲——▲) and not treated (●——●) with haloxyfop-ethoxyethyl, 19 hours before extraction.

Fig. 4. Incorporation of [1-^{14}C] acetate label into polar lipids by excised pieces from wheat plants treated (hatched) and not treated (white) with haloxyfop-ethoxyethyl 19 hours before axtraction.

change was observed in phosphatidylcholine - the most abundant polar lipid. The total amount of lipids decreased in the sensitive plants /Fig. 2/. No effect was observed on the activity of fatty acid synthetase. Pieces from treated and not treated plants incorporated similar amount of radioactive acetate to the lipids /Fig. 3/ and distributed it to polar lipids in similar way /Fig. 4/.

It is not possible to decide if the effects on lipids are direct or indirect, but the lack of effects on tolerant plants suggests that there is some connection between herbicidal effect and lipid metabolism.

BIOLOGICAL ROLE OF PLANT LIPIDS
P.A. BIACS, K. GRUIZ, T. KREMMER (eds)
Akadémiai Kiadó, Budapest and Plenum Publishing
Corporation, New York and London, 1989

EFFECT OF PARAQUAT ON LIPID METABOLISM IN PARAQUAT RESISTANT AND SUSCEPTIBLE CONYZA CANADENSIS

E. Lehoczki[1] and T. Farkas[2]

[1]Research Team of Department of Botany, Hungarian Academy of Sciences, Szeged, Hungary
[2]Institute of Biochemistry, Biological Research Center, Szeged, Hungary

INTRODUCTION

Paraquat (1,1'dimethyl-4,4'dipiridinium dichloride) is used as a broad-spectrum total-kill herbicide. Paraquat enters the chloroplast and interferes with electron transport by efficiently accepting electrons from photosystem I. As activated paraquat it reacts with oxygen forming toxic superoxide radical (Summers, 1980). The superoxide radical (O_2^-) have been shown to do a great deal of damage to biological material, either by itself or its protonated form, but more often via the formation of highly reactive hydroxyl-radical. Hydroxyl-radical reacts rapidly with almost every molecule found in living cells, however one of its most damaging reactions is to distract a hydrogen atom from the hydrocarbon side chain of membrane lipids, initiating the chain reaction of lipid peroxidation. This is a possible reason for the loss of membrane integrity. Although the membrane--disorganizing action of paraquat is well documented, no deta-iled analysis on the membrane lipid content following paraquat treatment has been reported. In order to address this particular question, we carried out a comparative study with the paraquat susceptible biotype of *Conyza canadensis* and a recently appeared one wich exhibited an extreme insensitivity to paraquat (Pölös et al., 1988).

OBSERVATIONS

A comparison of the total amount and composition of fatty acids and lipids from leaves of paraquat resistant (R) and sensitive (S) biotypes of Conyza canadensis in the rosetta stage showed: 1) Leaves of the R biotype contain a higher amount of fatty acids (FA), than the S biotype. There were no significant differences in the total FA composition between the two biotypes. 2) Paraquat treatment of leaves for 24 and 48 h resulted in a decrease (20 %) of total FA in S and in an increase (10 %) of FA in R biotype. A marked (60-90 %) reduction in galactolipids especially in MGDG, and a smaller (25-60 %) reduction in PG content was found in S biotype, while only a minor 10-15 % decrease was observed for all lipids in R biotype. 3) Paraquat treatment resulted also in a decrease of linolenic acid content and in an increase in the level of palmitic and linolenic acids in total FA, MGDG and DGDG of S biotype. 4) The amount and fatty acid composition of PG did not differ significantly in the two biotypes. However, paraquat treatment brought about a pronounced reduction of linolenic and especially Δ3-trans-hexadeceonic acid content in PG of S biotype.

CONCLUSION

In contrast to the S plants, practically no lipid degradation and FA peroxidation was observed in the R biotype treated with paraquat, which may indicate that the toxic effects of oxygen radicals did not develop following paraquat treatment in this biotype. The data on the lipid and FA composition also indicate that the high constitutive level of superoxide dismutase, ascorbat peroxidase and glutation reductase proposed by Shaaltiel and Gressel (1986), can protect the R plant from damage of paraquat.

REFERENCES

Pölös E., Mikulás J., Szigeti Z., Matkovics B., DoQuy H., Párducz Á. and Lehoczki E. (1988) Pesticide Biochem, Physiol. 30: 142-154.

Shaaltiel Y. and Gressel J. (1986) Pesticide Biochem. Physiol. 26: 22-28.

Summers L.A. (1980) The Bipyridinium Herbicides. Academic Press London.

BIOLOGICAL ROLE OF PLANT LIPIDS
P.A. BIACS, K. GRUIZ, T. KREMMER (eds)
Akadémiai Kiadó, Budapest and Plenum Publishing
Corporation, New York and London, 1989

HERBICIDE ACTION ON THE CAROTENOGENIC PATHWAY

M.P. Mayer, D.L. Bartlett, P. Beyer and H. Kleinig

Albert-Ludwigs-Universität, Institut für Biologie II, Zellbiologie, Schänzlerstrasse 1, 7800, Freiburg, FRG

Daffodil chromoplasts (Narcissus pseudonarcissus L.) convert $|^{14}C|$-phytoene in more desaturated and cyclic carotenes. Recently, we succeeded in the dissection in vitro of these membrane bound reactions into three independent segments, the desaturation of phytoene to ζ-carotene, the desaturation of ζ-carotene to lycopene, and the cyclization of lycopene to α/β-carotene. Due to the high enzymatic activities $|^{14}C|$ labelled or nonlabelled intermediates may be used as substrates.

In the course of our investigations the inhibiting properties of several herbicides and related compounds (Fig. 1) on the two desaturation segments were studied.

Norflurazone inhibits the desaturation of phytoene in a non-competive and reversible manner as shown in a Lineweaver-Burk plot in Fig. 2. The Apparent K_m is 0.59 μmol l^{-1} and the K_i for Norflurazone determined from the secondary Lineweaver-Burk plot is 0.39 μmol l^{-1}.

From the I_{50} values for Norflurazone and derivatives, as well as for Fluridone and Racer (Table 1) it becomes evident that the trifluoromethyl function is relevant for the inhibitory effect.

The desaturation of ζ-carotene, on the other hand, was not influenced by these compounds, even when administered in very high concentrations (100 to 1000 times the I_{50} concentration for phytoene desaturation).

The J-compounds (J_{334}, J_{793}, J_{892}) and Dichlormate are known to cause the accumulation of ζ-carotene (S. M. Ridly, Carotenoids and herbicide action, in "Carotenoid Chemistry and Bio-

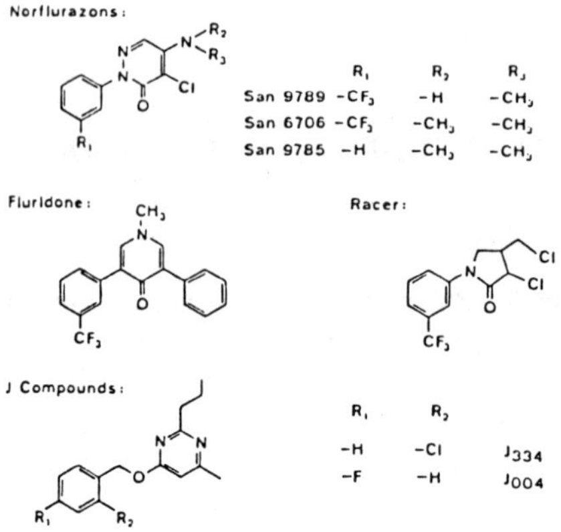

Fig. 1 Structure formulas of the compounds tested.

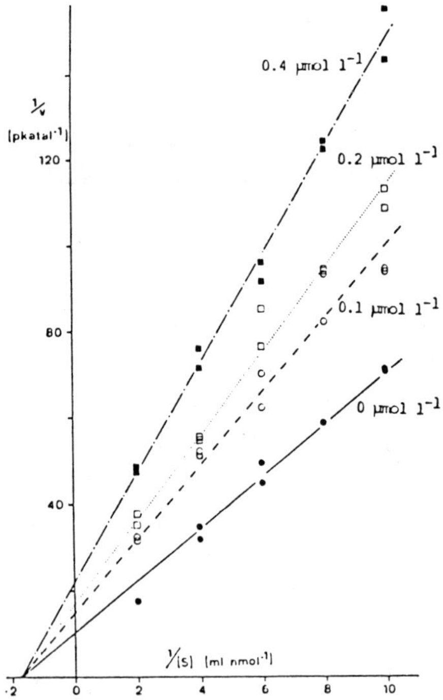

Fig. 2 Lineweaver-Burk plot of the inhibition of phytoene desaturation with Norflurazone.

Table 1 I_{50} values for the inhibition of phytoene desaturation (phytofluene and ζ-carotene formation) in the order of the effectivity of the compounds tested.

Herbicide	I_{50} (μmol 1^{-1})
Fluridone	0.07
Norflurazone (SAN 9789)	0.40
Racer	4.5
Metflurazone (SAN 6706)	8.6
Des(trifluormethyl)flurazone (SAN 9785)	900

chemistry", G. Britton and T. W. Goodwin, Eds, pp. 353-369, Pergamon Press, Oxford, 1982). We show that J_{334} and J_{004} inhibit the desaturation of phytoene and more effectively the desaturation of ζ-carotene (Table 2).

Table 2 Inhibition of phytoene- and ζ-carotene desaturation by J_{334} and J_{004} using radiolabelled phytoene and ζ-carotene as substrates.

	J_{334}	J_{004}
	I_{50} μmol 1^{-1}	
Desaturation of phytoene	16.07	38.5
Desaturation of ζ-carotene	7.3	5.6

In the sequence of carotene desaturation are two targets for herbicide interaction: the step from phytoene to ζ-carotene and the step from ζ-carotene to lycopene. It is assumed that two individual desaturase systems work in sequence.

BIOLOGICAL ROLE OF PLANT LIPIDS
P.A. BIACS, K. GRUIZ, T. KREMMER (eds)
Akadémiai Kiadó, Budapest and Plenum Publishing
Corporation, New York and London, 1989

INVOLVEMENT OF THYLAKOID MEMBRANE LIPIDS IN THE BINDING AND INHIBITORY PROPERTIES OF ATRAZINE AND DIURON

J.-P. Mayor, A. Rawyler and P.-A. Siegenthaler

Laboratoire de Physiologie Végétale, Université de Neuchâtel, 20 Chemin de Chantemerle
2000 Neuchâtel, Switzerland

In thylakoid membranes (TM) diuron-type herbicides block electron flow between the primary Q_A and the secondary Q_B plastoquinone |1|. Since, the binding site is located on the 32 Kd herbicide-binding protein (32 Kd H-B protein), which is embedded in both thylakoid monolayers |2|, one can ask whether acyl lipids play a particular role in the herbicide-protein interaction, as suggested elsewhere |3|.

Phospholipase A_2 (PLA_2) from *Vipera russelli* and lipase from *Rhizopus arrhizus* (LRa) were used to obtain almost complete phospholipid (PL) and galactolipid (GL) depletion in the outer monolayer of thylakoid membranes from atrazine-susceptible (S) and -resistant (R) biotypes of *Solanum nigrum*. PL- and GL-depletion inhibited uncoupled H_2O/ferricyanide (FeCy) electron flow by 7 and 27%, respectively (not shown). Binding and inhibitory properties of DCMU and atrazine were then studied (Table I).

Atrazine (up to 0.2 µM) did not bind to R-TM. In all other cases, herbicide incorporation increased linearly up to 0.1 µM, then decreased progressively (saturable binding). DCMU incorporation was slightly higher in S- than in R-TM. In S-TM, atrazine binding displayed a biphasic pattern, which was conserved after both lipase treatments. However, PLA_2 and LR_a treatments decreased atrazine binding (Kd_1) to the "low" affinity sites whereas they increased it (Kd_2) to the "high" affinity sites. The number of sites (n_1 and n_2) were not greatly affected (Tab. I).

Table I Equilibrium "binding" constants, Kd_1 and Kd_2 (nM), total number of binding sites n_1 and n_2 (nmol/mg Chl) and |herbicide| needed to inhibit 50% (I_{50}, µM) of the uncoupled H_2O/FeCy activity.

		SUSCEPTIBLE			RESISTANT		
		control	PLA_2	LRa	control	PLA_2	LR_a
ATRAZINE	Kd_1	39.00	58.30	44.40	----	----	----
	n_1	2.12	2.47	2.45	----	----	----
	Kd_2	11.20	6.30	6.10	----	----	----
	n_2	0.88	0.49	0.64	----	----	----
	I_{50}	0.140	0.095	0.075	40.0	120.0	50.0
DCMU	Kd_1	9.20	12.10	4.00	21.30	30.00	16.80
	n_1	2.44	2.24	1.88	2.61	1.80	1.53
	Kd_2	----	----	0.20	----	----	----
	n_2	----	----	0.38	----	----	----
	I_{50}	0.027	0.020	0.050	0.034	0.060	0.040

In both S- and R-TM, PLA_2 decreased the affinity of DCMU (Kd_1), whereas LRa increased it. LRa reduced the number (n_1) of DCMU binding sites more efficiently than did PLA_2. The intrinsic susceptibility of S-TM to both herbicides was increased by PLA_2. The intrinsic resistance of R-TM to herbicides was increased more by PL than by GL depletion. On the other hand, LRa reduced the DCMU susceptibility whilst it increased the atrazine susceptibility in S-TM (Tab. I).

Therefore, PL would favour the binding of the two herbicides to the 32 Kd H-B protein in both S- and R-TM. In contrast, GL would facilitate atrazine binding but hinder that of DCMU. Thus GL and PL would cooperate for atrazine binding whereas they would behave as antagonists for DCMU binding.

Contrarily to expectations, an increase (resp. decrease) in herbicide affinity was associated to a decrease (resp. increase) in the herbicide sensitivity, except in the case of DCMU sensitivity of PL-depleted R-TM (Tab. I). This suggests that at least in the S-biotype, (some) acyl lipids in the outer monolayer interact with the 32 Kd H-B protein so as to maintain it in a conformation that allows an optimal compromise between binding affinity and inhibitory power of herbicides.

Technical assistance of Ms N. Guinnard and financial support of the Swiss National Science Foundation (3.346-0.86 to PAS) are gratefully acknowledged.

REFERENCES

1. A. Trebst (1987) The Three-Dimensional Structure of the Herbicide Binding Niche on the Reaction Center Polypeptide of Photosystem II. Z. Naturforsch. *42c*, 272-750
2. D.J. Murphy (1986) The molecular organisation of the photosynthetic membranes of higher plants. Biochim. Biophys. Acta *864*, 33-94
3. N. Adir, J. Hirschberg I. Ohad (1987) Interaction of the Q_B protein of various species with cross-linking reagents and their use for its isolation. In J. Biggens (ed.), *Progress in Photosynthesis Research*, Vol. III, M. Nijhoff Publ., Dordrecht, pp 791-794

BIOLOGICAL ROLE OF PLANT LIPIDS
P.A. BIACS, K. GRUIZ, T. KREMMER (eds)
Akadémiai Kiadó, Budapest and Plenum Publishing
Corporation, New York and London, 1989

SOME EFFECTS OF PYRENOCINE A ON ONION ROOT AND SPINACH CHLOROPLAST LIPID METABOLISM

S.A Sparace[1], R. Menassa[1] and J.B. Mudd[2]

[1]Plant Science Department, Macdonald College of McGill University Ste Anne de Bellevue Quebec H9X 1C0, Canada
[2]Plant Cell Research Institute, 6560 Trinity Court, Dublin, California 94568, USA

INTRODUCTION

Pyrenochaeta terrestris, the causal agent of the onion pink root disease, induces a proliferation of host membrane prior to cellular disruption several cell layers in advance of the infection front (1). The pathogen produces a phytotoxin known as pyrenocine A (5-crotonyl-4-methoxy-6-methyl-2-pyrone) (2). We believe that pyrenocine A may be involved in this response and report some observations on the effects of pyrenocine A on limited aspects of plant lipid metabolism.

METHODS

Seven day old pregerminated onion seeds were preincubated with pyrenocine A at various concentrations or for various durations. Pretreated seedlings were rinsed with water, 1 cm root tips excised and then incubated in the presence of 14C-acetate for 1 h. Chloroplasts were incubated in the simultaneous presence of acetate and pyrenocine for 1 h. All other methods are described elsewhere (3,4).

RESULTS AND DISCUSSION

Pyrenocine A has bimodal effects on onion root and spinach chloroplast lipid metabolism. Low concentrations and shorter preincubation periods promote lipid synthesis from 14C-acetate while high concentrations and longer preincubation periods inhibit lipid synthesis. Maximum stimulation of lipid synthesis (approximately 75%) was observed when onion roots were preincubated for .5 h with 25 μg/ml pyrenocine A while maximum inhibition (approximately 80%) was observed when onion roots were pretreated with 100 ug/ml pyrenocine A for 24 h (Tables 1 and II). Over the range of onion treatments with pyrenocine A, there was a marked increase in the proportions of glycerolipid synthesized at the expense of sterols (50%

TABLE I. EFFECTS OF 24 HOURS PREINCUBATION WITH INCREASING CONCENTRATIONS OF PYRENOCINE A ON ONION ROOT LIPID METABOLISM FROM 14C-ACETATE.

	PYRENOCINE A, µg/ml					
	0	5	1	25	50	100
PMOL/ H/MG	83	70	64	68	32	13
			%			
STER.	63	66	66	56	52	53
G.LIP.	36	32	32	42	46	44
	DISTR. AMONG G.LIP., %					
PC	40	35	34	34	33	25
PE	22	21	20	22	19	16
PA	14	14	17	17	22	25
TG	10	14	13	10	10	18
PI	7	8	9	8	9	9
PG	7	7	8	9	8	8
	TOTAL FATTY ACIDS, %					
16:0	26	35	32	34	39	36
18:0	10	12	10	9	9	0
18:1	52	45	45	47	51	64
20:0	13	8	1	10	0	0

TABLE III. EFFECTS OF COINCUBATION WITH INCREASING CONCENTRATIONS OF PYRENOCINE ON CHLOROPLAST LIPID METABOLISM.

	PYRENOCINE A, µg/ml				
	0	1	10	50	100
NMOL/H/ MG CHL.	105	119	124	90	80
	DISTRIBUTION, %				
FFA	45	43	44	42	40
DG	29	31	34	39	41
PA+LPA	8	9	7	4	4
PG	6	6	6	5	6
MG	5	6	6	5	5
SQDG	3	2	1	1	1
PC	2	2	2	2	1

TABLE II. EFFECTS OF INCREASING PREINCUBATION TIMES WITH 25 µg/ml PYRENOCINE A ON ONION ROOT LIPID METABOLISM FROM 14C-ACETATE.

	HOURS PREINCUB. WITH PYRENOCINE A								
	0	.17	.5	1	2	4	6	8	16
PMOL/ H/MG	9	15	16	9	8	5	7	4	5
					%				
STER.	63	59	53	48	35	30	33	37	46
G.LIP.	33	34	43	48	61	65	62	58	49
	DISTR. AMONG GLYCEROLIPIDS, %								
PC	29	23	25	25	24	24	24	24	26
PE	23	23	23	24	24	25	23	23	17
TG	15	20	17	13	13	15	14	17	14
SQDG	12	10	9	11	11	9	9	9	9
PA	11	11	13	13	12	13	15	15	19
PG	6	8	8	9	9	9	8	6	8
PI	4	5	6	6	6	6	6	6	7
	TOTAL FATTY ACIDS, %								
16:0	28	35	39	33	46	47	35	44	36
18:0	5	6	4	7	5	10	5	7	5
18:1	39	27	32	32	23	29	38	30	44
20:0	17	11	7	12	9	8	11	9	6
22:0	11	20	19	16	17	7	11	11	8

change, Table II), however, relatively small or inconsistent effects were observed on the proportions of individual glycerolipids and fatty acids synthesized in roots. Chloroplasts were intermendiate in their response to pyrenocine A (Table III). 10 µg/ml pyrenocine A stimulated lipid synthesis by 20% and resulted in a 10% increase in the proportion of DG synthesized.

These results indicate that pyrenocine A can affect plant lipid metabolism in such a way that might contribute to increased lipid synthetic activities that must accompany the observed membrane proliferation (1). We propose that pyrenocine A may have this role in the onion pink root disease and that the additional host membrane lipids are subsequently utilized by the advancing mycelium during pathogenesis. The precise mode of action of pyrenocine A in plant lipid metabolism remains to be determined.

REFERENCES

1. Hess,W.M.1969.Amer.J.Bot.56:832-845.
2. Sato,H.,K.Konoma,S.Sakamura,A.Furusaki,T.Matsumoto,T.Matsuzaki.1981. Agric.Biol.Chem.45:795-797.
3. Sparace,S.A.,J.B.Mudd,B.A.Burke,A.J.Aasen.1984.Phytochem.23:2693-2694.
4. Mudd,J.B.,R.DeZacks.1981.Arch.Biochem.Biophys.209:584-591.

BIOLOGICAL ROLE OF PLANT LIPIDS
P.A. BIACS, K. GRUIZ, T. KREMMER (eds)
Akadémiai Kiadó, Budapest and Plenum Publishing
Corporation, New York and London, 1989

FLUAZIFOP, A GRASS-SPECIFIC HERBICIDE, ACTS BY INHIBITING ACETYL-CoA CARBOXYLASE

K.A. Walker[1], S.M. Ridley[2] and J.L. Harwood[1]

[1]Dept. of Biochemistry, University College, Cardiff CF1 1XL, UK
[2]I.C.I. Agrochemicals, Jealott's Hill, Bracknell, Berks RG12 6CY, UK

Fluazifop (2-[4-(5-trifluoromethyl-2-pyridyloxy)phenoxy]propionate) is a grass-specific herbicide. It is an aryloxyphenoxypropionate compound and these herbicides have been shown to kill sensitive species by causing necrosis of meristematic tissues. A number of studies on the mode of action of the aryloxyphenoxypropionates have only been able to demonstrate an inhibition of lipid metabolism. No detectable effects on CO_2 fixation or carbohydrate, amino acid or nucleic acid metabolism have been shown (Walker _et al_., 1988_a_). Cyclohexanedione herbicides, such as sethoxydim, also appear to act in a similar manner (Harwood _et al_., 1988).

Since the absorption, translocation and metabolism of fluazifop seems to be the same in sensitive and resistant plants (Walker _et al_., 1988_a_), we have looked for a target enzyme involved in lipid biosynthesis. The mode of action of fluazifop (Harwood _et al_., 1988) appears to be through an inhibition of fatty acid and, hence, acyl lipid formation. Of the two enzyme systems involved, the following results show that acetyl-CoA carboxylase is the sensitive protein.

When either [^{14}C]acetate of [^{14}]malonate were used to label the fatty acids of leaf pieces of barley (sensitive) or pea (resistant) plants, fluazifop inhibited the total incorporation of radioactivity only in barley. The inhibition only affected fatty acids made _de novo_ and the labelling of very long chain fatty acids was virtually unaffected (Walker _et al_., 1988_a_). Because a higher percentage of label from [^{14}C]malonate was incorporated into very long chain fatty acids, then fluazifop causes less inhibition of fatty acid labelling from this precursor. Although the inhibition by fluazifop of labelling from both precursors could be interpreted as showing fatty acid synthetase to be the target, the evidence is not unequivocal because decarboxylation can lead to the generation of

[^{14}C]acetate from [^{14}C]malonate in leaves. Indeed, we noted that, at low concentrations of fluazifop causing only a partial inhibition of labelling, a shift in the ratio of products towards shorter chains was seen. Such a result would be expected if acetyl-CoA carboxylase was inhibited, thus leading to an increase in the acetyl/malonyl ratio.

When fatty acid synthetase was measured directly in chloroplast stromal fractions with [^{14}C]malonyl-CoA, fluazifop caused no inhibition. However, if unlabelled acetyl-CoA was added simultaneously to ensure that primer concentrations were not limiting then fluazifop addition caused less reduction in labelling than for the addition of acetyl-CoA alone. Such a result would be expected since inhibition of acetyl-CoA carboxylase would prevent large-scale dilution of [^{14}C]malonyl-CoA with unlabelled malonyl-CoA.

Although [^{14}C]acetate is a relatively poor precursor for fatty acid synthesis in stromal fractions, fluazifop did inhibit labelling from this compound. By blocking labelling via acetyl-CoA:ACP transacylase (with unlabelled acetyl-ACP) or via acetyl-CoA carboxylase (with avidin and unlabelled malonyl-CoA) we were able to show that the fluazifop action was at the level of acetyl-CoA carboxylase.

Finally, direct experiments on partly purified acetyl-CoA carboxylases have shown that the enzyme from barley leaves but not from peas is sensitive to fluazifop. The inhibition is stereospecific with only the R-isomer being active (Walker et al., 1988b). This agrees with the herbicidal specificity of the compound.

These results raise two intriguing questions. First, what is different in grasses which makes acetyl-CoA carboxylase susceptible to inhibition? Second, the relative insensitivity of very long chain fatty acid synthesis to fluazifop inhibition poses a problem as to the source of malonyl-CoA for the elongation reactions.

References

Harwood, J.L., Ridley, S.M. and Walker, K.A. (1988) in Herbicides and Plant Metabolism (Dodge, A.D., ed.) Cambridge Univ. Press, in press.

Walker, K.A., Ridley, S.M. and Harwood, J.L. (1988a) Biochem J. **255**, in press.

Walker, K.A., Ridley, S.M., Lewis, T. and Harwood, J.L. (1988b). Biochem. J., in press.

BIOLOGICAL ROLE OF PLANT LIPIDS
P.A. BIACS, K. GRUIZ, T. KREMMER (eds)
Akadémiai Kiadó, Budapest and Plenum Publishing
Corporation, New York and London, 1989

CHLOROACETAMIDE INHIBITION OF FATTY ACID SYNTHESIS IN THE GREEN ALGA SCENEDESMUS ACUTUS

H. Weisshaar and P. Böger

Lehrstuhl Physiologie und Biochemie der Pflanzen, Universität Konstanz, 7750 Konstanz, FRG

INTRODUCTION

The chloroacetamides alachlor, 2-chloro-N-(2,6-diethyl-phenyl)-N-(methoxymethyl) acetamide and metazachlor, 2-chloro-N-(2,6-dimethylphenyl)-N-(1H-pyrazol-1-ylmethyl) acetamide are selective pre-emergent or early post-emergent herbicides used to control annual grasses and many broad-leaved weeds in cotton, brassicas, maize, oilseed rape, peanuts, radish, soya-beans and sugarcane.

This herbcides are absorbed by the germinating shoots and translocated to the vegetative parts of the plants. Although many sites of action are described in the literature the primary mode of action is unknown.

In this contribution we show some indications that fatty acid synthesis in the green alga *Scenedesmus acutus* is inhibited by this herbcide group.

RESULTS and DISCUSSION

In the green alga *Scenedesmus acutus* the uptake of 2-^{14}C) acetate, a precursor of fatty acid synthesis, is inhibited with 100 μM alachlor and 100 μM metazachlor within an incubation time of 4 h (1). There is also shown that the incorporation of the labeled acetate in acyl lipids is inhibited. After 16 h and 40 h incubation with a herbicide concentration of 5 μM the fatty acid content of the green alga changed, as shown in the table.

Table Influence of metazachlor and alachlor (5 μM) on the fatty acid content of *Scenedesmus acutus*

No.	Fatty acid	Fatty acid content %					
		after 16h			after 40h		
		Control	Metazachlor	Alachlor	Control	Metazachlor	Alachlor
1	18:3	37.9	33.1	32.6	32.0	5.9	9.6
2	18:1	9.1	21.0	18.4	2.7	37.3	36.2
3	16.0	20.5	24.7	27.9	22.5	37.9	36.7

The fatty acid extraction was done after a method described by Roughan et al. (1979) Biochem. J. 184, 193-202.

The linolenic acid (18:3) content decreased and the palmitic acid (16:0) and the oleic acid (18:1) content increased. The incorporation of (2-^{14}C) acetate in the hexane extracted fatty acid fraction could also be inhibited through the chloroacetamide herbicides tested, data not shown.

Our data suggest an inhibition of fatty acid biosynthesis between the elongation step of palmitic acid and the desaturation of oleic acid yielding linolenic acid.

REFERENCE

Hans Weisshaar and Peter Böger (1987) Primary Effects of Chloroacetamides Pestic. Biochem. Physiol. 28, 286-293

CHAPTER 6

BIOTECHNOLOGY OF LIPIDS, NUTRITIONAL ASPECTS

BIOLOGICAL ROLE OF PLANT LIPIDS
P.A. BIACS, K. GRUIZ, T. KREMMER (eds)
Akadémiai Kiadó, Budapest and Plenum Publishing
Corporation, New York and London, 1989

PLANT LIPIDS AS RENEWABLE SOURCES OF INDUSTRIAL SURFACTANTS

P.J. Quinn

Department of Biochemistry, King's College London, Campden Hill, London W8 7AH, UK

INTRODUCTION

The annual worldwide production of vegetable oil is currently about 60 m metric tons of which nearly one quarter is derived from soybeans. The majority of oil produced is used as edible oil in the nutrition of man and domestic animals. With a few notable exceptions the composition of plant oils has apparently remained little changed over the thousands of years of domestication and selective breeding where the main emphasis has been to produce hardy and productive varieties.

The diversity of plant lipids is remarkable. More than 500 different fatty acids and other lipid structures have been described by lipid chemists most of which originate from the 10,000 or so species of plant that have so far been subjected to any form of analysis. This represents only about 5% of the estimated number of plant species so we may expect that a very large number of molecular species of lipid have yet to be described.

Some of these unusual fatty acids, such as those of seed oils of castor bean, rape seed, coconut and palm kernel or tung nut have been exploited as raw materials for the chemical industry for some considerable time. They represent a stable and variable, albeit relatively minor fraction, of the world vegetable oil market. The quest for new oil seed crops which possess unusual fatty acids and are able to yield sufficient quantities of them on an economic scale have been prompted by diversification of industrial applications, changes in agricultural practices and production and, more recently, unreliable supplies of mineral oils which serve as the raw materials of the petrochemical industry.

Feedstocks

A large number of plant species have been screened over the years to determine whether they have lipids of use in industrial processes. Some of the recent trends have been landmarked by the discovery that the wax of Jojoba is a close substitute for whale sperm oil [1] and the successful breeding of rape to produce erucic-acid-free seed oil which has created a renewed demand for the supply of long-chain fatty acids. Furthermore, concerted efforts have been made to breed plants which are able to supply useful alternatives for substituted hydrocarbons derived from mineral oil refining.

Chemical Modification of Vegetable Oils

The economics of extraction and purification of vegetable oils is a major factor that determines the utility of plant sources as feedstocks of oleochemical manufacture. Once purified, however, the vegetable oil can be processed on an industrial scale using well-known conventional chemical reactions.

The main chemical derivatives of plant oils and lipids are shown in Table 1. It can be seen that a remarkably wide range and diversity of

Table 1. Chemical derivation of basic oleochemicals.

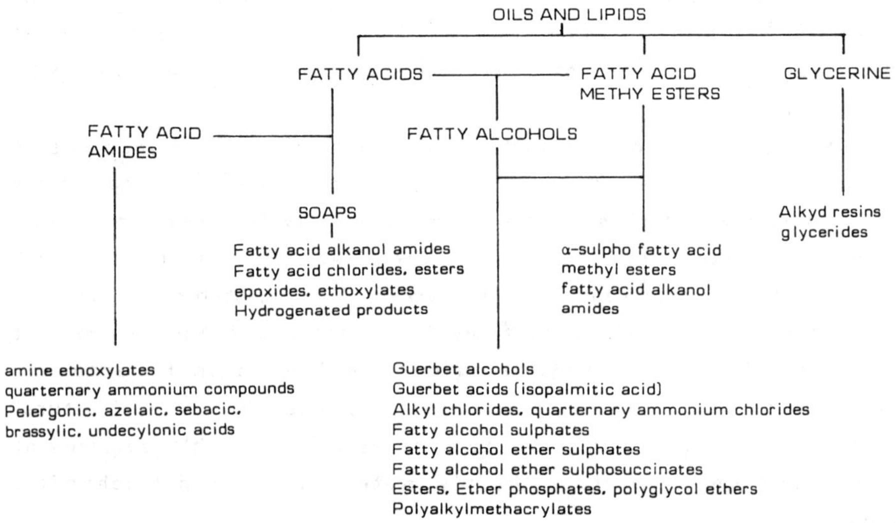

products can be derived by relatively simple chemical reactions. A number of these chemical reactions, such as oxidation, formation and reaction products of epoxides, hydrogenation and various other reactions of acetylinic and olefinic systems has been the subject of a recent review [2]. The review also documents recent discoveries of new fatty acids reported in a variety of plants and other organisms that could be potentially exploited, after suitable chemical modification, for industrial uses.

Sources of Surfactant Precursors

Conventional plant breeding and agronomic methods have been used in the successful production of plants with high yields of fatty acids of use in specialty surfactant applications. These include trienoic fatty acids with conjugated double bonds from *Calendula*, hydroxy fatty acids from *Lasquerella* and long-chain fatty acids from species of *Crambe*. The short chain fatty acids commonly used in surfactant manufacture, however, are generally products of the petrochemical industry. The primary reason for this is that short chain fatty acids are not formed abundantly, especially in storage organs of plants. Nevertheless, some biochemical processes do result in formation of short chain fatty acids.

Acetic acid occurs as alcohol acetates in many plants and in some plant triacyl glycerols. It is widespread in the form of a thio ester in compounds like acetyl CoA where it participates as an energy substrate intermediate and as a precursor utilised in lipid biosynthesis. Fatty acids with chain lengths of six and eight carbons are only rarely found in biological systems such as the oils in seeds of *Leguminosae* and *Borginaceae* and although carbon-10 is often found and is also widespread, it constitutes only a minor fraction of total fatty acids. By contrast, laurate and myristate are widely distributed fatty acids and are major components in many oil seeds of the family *Palmae* which includes coconut. However, while lauric acid is confined to plants, myristic acid is found as a component of green algae, brown algae, bacteria, cyanobacteria, fungi as well as animals. In Gram-negative bacteria, the lipopolysaccharide of the envelope contains an abundance of laurate and myristate. Myristate is also a prominent constituent of the marine algae where it represents 5-15% of the total fatty acids rising to 30% in some diatoms.

Recently, attention has been focussed on species of the genus *Cuphea* as a source of fatty acids in the chain length region of 8-14 carbon atoms. The pathway of biosynthesis of these short chain fatty acids appears to be different from the enzymes involved in acyl-ACP chain termination and also those of ACP-dependent fatty acid synthesis. Moreover, the coconut endosperm ACP systems do not synthesise these fatty acids.

Different species of Cuphea produce fatty acids of a predominant and characteristic chain length. The distribution of fatty acids and the rate of synthesis from a radioactive precursor of [^{14}C]-acetate in developing seeds of C. lutea is shown in Table 2. It can be seen that carbon-10 and carbon-12 fatty acids predominate in this species. The following species

Table 2. Fatty acid composition of Cuphea lutea seeds.

Fatty acid	8:0	10:0	12:0	14:0	16:0	18:0, 1&2
Mole %	2	33	39	11	3	12
Distribution of [^{14}C] acetate	18	40	13	4	6	18

of Cuphea have seeds in which the fatty acids shown in parenthesis are the major component; C. hookerianna (8:0); C. inflata (10:0); C. plustris (8:0 + 14:0); C. laminuligera (12:0). The rate of synthesis of these fatty acids is also very rapid compared to that in other oil seeds, being twice that of safflower, three-fold that of rape and more than an order of magnitude greater than soybean when grown under identical conditions. During the rapid accumulation of triacylglycerol with these short chain fatty acids in the developing seeds of C. lutea the proportion of the different fatty acids remains fairly constant.

An interesting observation has been reported in Cuphea plaustris which has a seed oil rich in caprylic (40 mol %) and myristic (50 mol %) acids. Analysis of the triacylglycerols showed the major species to be carbon-33 (15%) and carbon-39 (59%) which correspond to (carbon-8+8+14) and (carbon-8+14+14) molecular species respectively. This bimodal acyl specificity is the result of a simultaneous biosynthesis of the two types of fatty acid in the same cotyledon cell either by different mechanisms within the cell or at different compartmentalised sites within the cell.

PRODUCTION OF NEW VARIETIES

There are several approaches to the development of new plant species and varieties that possess the required attributes for production of surfactant feedstocks. The principles underlying these methods include conventional selection and breeding and techniques of gene manipulation and transfer.

Breeding Strategies

Plant breeding continues to provide cultivars that can adapt to different environments and sustain high yields, have improved quality by modification of fatty acid composition and that resist diseases, insects and other pests. The steps involved in the plant breeding process include the collection and evaluation of the germ plasm of suitable plant species, characterisation of their lipid composition and methods of extraction and purification, evaluation of their agronomic performance and improvement by breeding and selection methods culminating in large-scale production and commercialisation. The success of this process in terms of cost and time depend among other things on the level of domestication of the particular species, its ecological adaptation and potential for its yield. Obviously there is a minimum requirement for a multiplicity of traits and a deficiency in just one character may not be compensated by a high value performance in the remaining indicators.

Techniques of Molecular Biology

It is true to say that despite an effective armoury able to perform genetic manipulations, our present knowledge of the molecular genetics of plant lipid metabolism is fairly rudimentary. It is natural, nevertheless, for there to be expectations that the tools developed by the molecular biologist will eventually be exploited to achieve even greater versatility and productivity of fatty acids and complex lipids capable of serving as useful industrial materials. A recent example of this skill is the development of a vector system consisting of the organism responsible for crown gall disease, Agrobacterium tumefaciens, at the Montanso Company [3] which has been used to produce transgenic Brassica napus plants. The method should be applicable to the many plant species that can be infected by the vector and where plants can be regenerated from tissue culture. The possibility of being able to alter genes and subsequently reintroduce them into plants offers the first step for promising future in this direction.

As far as the manipulation of plant lipids is concerned, there remains much to be done on the biochemistry of lipid synthesis in oil seed plants. There have, nevertheless, been considerable advances in our knowledge of the biochemical processes of fatty acid synthesis but more yet is to be learnt about the enzymology and points of regulation of the process as well as mechanisms of specificity of fatty acid desaturases and elongation reactions.

In photosynthetic tissues, or example, *de novo* fatty acid synthesis is localised in the chloroplast where the central cofactor/cosubstrate for carbon-8 to carbon-18 fatty acid synthesis is acyl carrier protein. There is some evidence that fatty acid chain length and fatty acid distribution within the different cellular organalles may be regulated by isoform structure of acyl carrier protein. In the developing oil seed, on the other hand, the most probable site of acyl carrier protein-mediated fatty acid synthesis is the plastid. The complex interrelationship between *de novo* fatty acid synthesis and the Kennedy pathway will need to be understood before genetic manipulation of the acyl carrier protein gene can be contemplated on a rational basis.

The rich diversity of fatty acids found in the triglycerides of the seed oil contrasts markedly with the relatiely simple fatty acid spectrum observed in the membrane polar lipid fraction. This implies that fatty acids synthesized *de novo* in the chloroplast and plastids are subsequently modified in other enzyme catalysed reactions. The steps associated with many of these reactions are unknown as are the mechanisms whereby they end up in the oil rather than the various membranes.

APPLICATIONS OF NATURAL SURFACTANTS

Plant lipids and soaps produced by their saponification have been used by man since antiquity as cleansing and disinfecting agents. Besides their general cleansing characteristics, soaps have been shown to be useful antiseptic agents and their antifungal and bacteracidal properties have been extensively investigated [4].

One of the characteristics of long-chain fatty acids and particularly the complex polar lipids is their relatively low critical micelle concentration which renders them weakly surface active. They are therefore useful in food processing applications and other situations where toxicity needs to be avoided. The relationship between the critical micellar concentration and detergent strength is illustrated in Figure 1. Complex lipids, such as phospholipids, have critical micellar concentrations in the nM region while typical synthetic detergents have critical micellar concentrations of about mM. The solubility of complex lipids in aqueous solvents, as reflected in the critical micellar concentration, is low because of the predominance of the non-polar hydrocarbon substituents over the polar head group of the molecules. The consequence, however, is that although difficult to disperse in water,

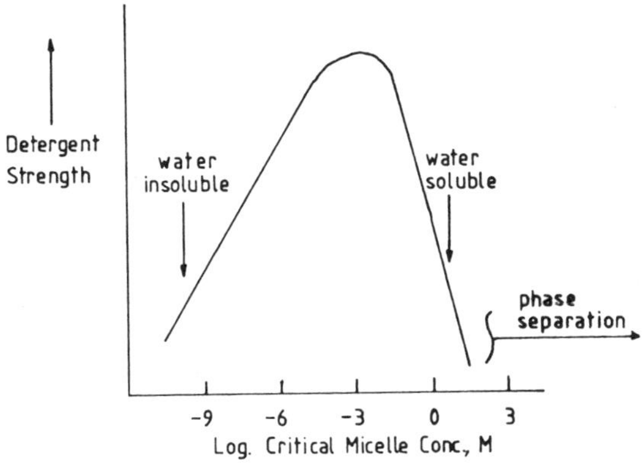

Fig.1 Relationship between detergency and critical micelle concentration.

complex lipids form extremely stable emulsions which often consist of multibilayer liposome structures.

Complex Lipids as Formulation Aids

Liposomes have particular application in the formulation of medicaments and cosmetics. One of the more recent developments has been their application as formultion aids for pesticides where recent patents have been granted [5] and this will serve as a useful example of how natural surfactants can be utilised in an industrial application.

The universal solvent for spray applications of pesticides to arable land or crops is water. Many agrichemical agents, however, are not readily soluble in water or they are unstable in aqueous solutions and degrade to inactive products. In order to form stable dispersions or emulsions a variety of formulation aids are employed including synthetic surfactants, solvents and particulates that are sufficiently fine to form colloidal or particulate suspensions. In addition to these base formulation aids other ingredients of wettable powders or emulsifiable concentrates are designed to improve the spray characteristics of the formulation such as antifoaming agents or to facilitate the interaction of the spray with the target surface. Examples of such additives include wetting and sticking agents that assist in the spread of the spray droplets over the waxy cuticular layer of plant foliage and in retention at the target under weathering conditions respectively.

One particular problem encountered with many formulations is that solvents and detergents are often phytotoxic and concentrations in the spray need to be low to avoid damage to plant tissues. Since dilute tank mixes are required this can hamper spray operations in wet conditions where heavy machinery must be used on soft ground and in situations where low volume applications are desired such as aerial spraying. Another difficulty with conventional formulations is that incompatibilities are frequent in tank mixes of different pesticides or pesticides with other agrichemical agents like fertilizers, plant growth regulators, desiccants, etc.

In recent years the use of various vegetable oils have been explored as alternative carriers to water for pesticide applications. Studies in North America have been undertaken using cotton seed oil alone or as an oil-in-water emulsion with insecticides and herbicides. Other vegetable oils under the trade name of Codicide oil have also been used in European trials for the application of pre-emergent herbicides. Many advantages are claimed by the use of oils alone for application of a variety of pesticides including protection of the agrichemical agent against degradation, leaching from soil and weathering on exposed targets, potentiation of uptake into pests and plants etc. The major disadvantage of these practices, however, is that conventional spraying equipment is not able to handle viscous oils and special machinery is required. Controlled droplet application machines are available on the market which show performances giving desirable drop characteritics but in general heavy capital investment is needed to employ these methods. The use of oil-in-water emulsions avoids these problems but substantial amounts of synthetic detergent agents are required to create stable emulsions with attendant increase in costs and phytotoxicity.

Soya lecithin has been found to be a useful substitute as a pesticide formulation aid. It is a plentiful byproduct of soya crushing and enjoys a competitive price with synthetic detergents. Methods have been devised to produce stable dispersions of soya lecithin and the capacity of the hydrocarbon domain to solubilise pesticides of low water solubility is considerably greater on a molar basis than are surfactants of lower critical micellar concentration. Furthermore, because of the marked order imposed on the phospholipid molecules by their arrangement in a bimolecular leaflet there is also a tendency for the pesticide molecule to be oriented within the structure. This has important implications with respect to the packaging of pesticides with agents that protect them from degradation or enhance their stability under various environmental conditions. The capacity of soya lecithin dispersions to emulsify various pesticides depends on the chemical structure of the agent. Larger amounts of pesticides which have partial solubility in water can be emulsified by soya lecithin compared to those that are highly insoluble in water.

Effect on Spray Characteristics

There have been a number of reports that surfactants and emusifying agents affect the spray characteristics particularly the size of droplets formed and the distribution of droplet sizes. A particularly important droplet index is the proportion of droplets with a diameter less than 100 um since such droplets represent the upper limit of sizes subject to respiratory inhalation. When the proportion of such droplets is high hazards arise because breathing in spray mist causes intoxication. Furthermore, sprays with these characteristics are prone to drift and action on non-target areas is more difficult to control.

A study has been peformed to examine the effect of soya lecithin on droplet spray characteristics [6] and its performance in terms of the spectrum of droplet sizes and proportion of driftable drops was superior to that of conventional synthetic surfactants. Contact angles of droplets targetted to leaf surfaces were in the range of 140-150° which, although not as effective as synthetic wetting agents, showed considerable improvement over water (contact angle < 30°). The rationale for this wetting action is believed to be the formation of a lipid bilayer covering the droplet immediately after atomisation. Thus the surface layer is hydrophobic in character and comptable with the waxy leaf surface. The layer also reduces evaporation of water from the drop and is probably the major factor in preventing development of drops less than 100 um diameter.

Protective Packaging of Biologically Active Agents

Residue trials performed with the pyrythroid insecticide, Permethrin, applied to field crops indicates that the insecticide is maintained in an active form at the target site considerably longer than conventional formulation of pyrethroid [7]. One possible reason for this is that leaching from the target site is prevented and the active agent is protected by hydrolysis by partitioning into the hydrocarbon domain of the liposome structure. Thus one of the main advantages of formulatin of pesticides in liposomal preparations is that the active ingredient can be stabilised within the lipid structure. Other more active methods can also be employed in such systems to prevent chemical degradation of the pesticide at the target site. Such degradation can arise from oxidation or from irradiation by ultraviolet light. In the case of oxidation, formulation of pesticides susceptible to oxidative degradation together with lipophilic antioxidants of natural (eg tocopherols, etc.) or synthetic (butalated hydroxytoluene, etc.) origin can afford considerable protection against such degradation. The physico-chemical properties of the pesticide and antioxidant can be matched to ensure that they partition into the phospholipid structure so that they reside in close proximity. A similar strategy can be employed to protect pesticides under ultraviolet light, in this case by the use of ultraviolet screening agents.

To confirm these effects experiments have been conducted with Permethrin which, although considerably more stable than many other synthetic and natural pyrethroid insecticides, is nevertheless degraded at a relatively rapid rate on exposure to ultraviolet light. The inclusion of ultraviolet absorbing agents into liposomal dispesions of permethrin in soya lecithin causes a considerable reduction in the rate of degradation of pyrethroid under ultraviolet irradiation. The general strategy of packaging ultraviolet screening agents was found to be a very effective method of increasing the biologically-effective half-lives of pesticides. This means that a range of agrichemical pesticides that are effective chemical agents but otherwise unsuitable because of their unstable character can be protected and stabilised at the target site by packaging them together with protective agents in phospholipid liposomes thereby increasing the range of pesticides available. Furthermore, the possibility of reducing rates of application of pesticides can be contemplated if more effective use is made of these agents. The use of comparatively unstable agents also means that when they are leached from the target site and exposed to the elements of the environment they will tend to degrade more readily and the build-up of residues will be prevented.

CONCLUSION

It can be seen that renewable sources of surfactants derived from vegetable oils and complex lipids offer considerable opportunities for replacing surfactants traditionally supplied from the petrochemical industry. Our knowledge of the biochemistry of lipid synthesis in oil seeds is expanding and the next decade will witness the application of molecular biological techniques combined with plant breeding programmes to produce high yielding varieties of a range of plants producing surfactant feedstocks to serve the needs of future industry.

REFERENCES

[1] Miwa, T.K. (1984). Structual determination and uses of Jojoba oil. J. Am. Oil. Chem. Soc., 61, 407-410.

[2] Gunstone, F.D. (1987). Fatty acids and glycerides. Nat. Prod. Rep., 4, 95-112.

[3] Horsch, R.B., Ry, J.E., Barnason, A., Metz, S., Rogers, S.G. and Fraley, R.T. (1987). Genetic transformation for oil crop improvement. J. Am. Oil Chem. Soc., 64, 1246.

[4] Kabara, J.J. (1984). Antimicrobial agents derived from fatty acids. J. Am. Oil Chem. Soc., 61, 397-403.

[5] Spray method and formulation for use therein. United States Patent 4,666,747 (1987). Acacia Chemicals Limited.

[6] Quinn, P.J., Perrett, S.F. and Arnold, A.C. (1986). An evaluation of soya lecithin in crop spray performance. Atomisation. Spray Tech. 2, 235-246.

[7] Quinn, P.J. (1985). The chemico-physical properties of membrane lipids and their relevance to plant growth and protection. In: Frontiers of Membrane Research in Agriculture, (St. John, J.B., Berlin, E. and Jackson, P.C., eds.), Rowman and Allanheld, Towota, N.J. pp. 55-75

BIOLOGICAL ROLE OF PLANT LIPIDS
P.A. BIACS, K. GRUIZ, T. KREMMER (eds)
Akadémiai Kiadó, Budapest and Plenum Publishing
Corporation, New York and London, 1989

ENZYMIC MODIFICATION OF SUNFLOWER LECITHIN

Anna P. Erdélyi

Department of Agricultural Chemical Technology, Technical University of Budapest
1521 Budapest, Hungary

Lecithin could be produced from any crude vegetable oil, but because of the huge quantities of soybean grown and processed, and because of the relatively high percentage of phosphatides in soybean oil, practically in the world soybean oil is the principal commercial source of natural and modified lecithins as well. The world's consumption of lecithin is estimated at 100,000 tons per year. In western Europe 30,000 tons are produced and more than the half of it is applied as modified lecithins /4/. In Hungary sunflower is the major oilseed crop, on this account sunflower lecithin has been the subject of this study.

Although emulsifying properties of sunflower lecithin are not considered to be worse than those of soybean lecithin, users are reluctant to apply it because its consistency is even more paste-like than that of natural soybean lecithin, so it is difficult to handle. Reviewing the technical literature, it was obvious that some modification can solve this problem.

Modification of lecithin, on industrial scale, is performed in three main ways: modification by physical means
modification by chemicals
modification by enzymes

Considering the industrial implementation and the properties of products obtained by modification, from the numerous modifying methods the enzymic way was chosen. Selective cleavage of phospholipid molecules can be carried out by a number of specific hydrolase enzymes, but on industrial scale only phospholipase A_2 is applied. Partial selective hydrolysis of the 2-position fatty acids of lecithin results in a product with inhanced o/w emulsifying properties, in addition, the procedure demands no special machinery and is easy to be attached to the manufacture of natural lecithin.

Enzymic hydrolysis of lecithin has not been exhaustively published in papers but rather patented. Actually, no exact data on the reaction conditions have been available, e.g., enzyme level has not been specified, as well as reaction time and temperature have varied in a wide range. On the other hand, some parameters prescribed cannot be carried out industrially, e.g., to achieve the optima of minimum 6 mM Ca^{2+} per litre and the pH to be kept at 8, calcium salt and alkali should be added. These substances remaining in the product should increase the impurity content. Because of high thermal stability of the enzyme, there seemed some difficulty in inactivation too.

Considering the facts mentioned above, the aim of our experiments was as follows:

- to study the effect of reaction parameters, i.e., reaction time and temperature and enzyme level, without adding Ca^{2+} ions and adjusting pH during the reaction,
- to block the reaction,
- to produce a product with decreased viscosity,
- to improve the original emulsifying properties mainly for o/w emulsions.

In the experiments phospholipase A_2 was a NOVO product of porcine pancrease origin /Lecitase 10 L/ having the activity of 10,000 IU/ml. Enzymic modifications were carried out on sunflower lecithin gums of different water content in a wide range of phospholipase A_2 level, i.e. 60,000 IU/100 g of phosphatide, at 40-90°C increasing the temperature stepwise by 5°C. Reaction time varied between 1-5 hrs. To follow the hydrolysis, two dimensional TLC and acid value determination were used.

Summarising our results we have found that

- In the range of enzyme level and temperature studied, phospholise A_2 catalyzes the hydrolysis of 2-position fatty acids even without adding Ca^{2+} ions and adjusting pH.
- High enzyme concentration, i.e. 60.000-6,000 IU/100 gramme of phosphatide, yields to too high a degree of hydrolysis of most valuable components and generation of unwanted highly decomposed substances.
- Applying high enzyme concentration, temperature dependence of the reaction cannot be observed.
- Inactivating methods recommended by the technical literature proved not to be sufficient at the suggested level, but according to our experiments, min. 2.5 mM Zn^{2+}/litre is needed to block the reaction, which is three times as much as suggested by NOVO /5/.
- In contrast to certain papers /2,3,5/ recommending 40-50°C for operation, temperature optimum of the enzyme was found to be at 70-75°C.
- The effect of water content of gums on the degree of hydrolysis was found to be significant. As is shown in Figure 1, the higher the water content is, the more advantageous the specific enzyme requirement is.

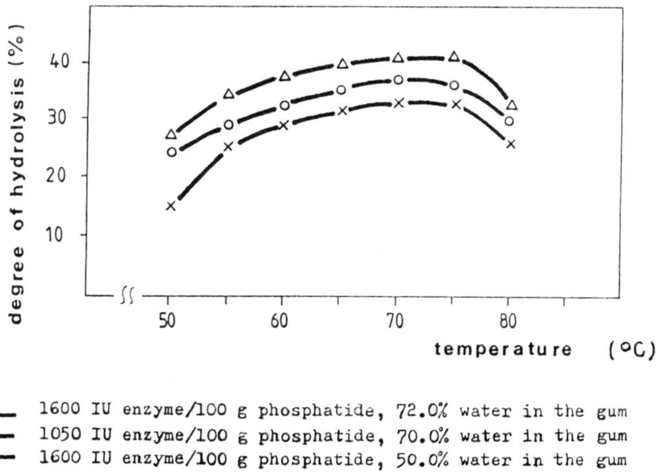

Fig.1. Effect of water content of gum on the degree of hydrolysis

- Under the conditions applied, a product with improved emulsifying properties and viscosity can be obtained by reaching or exceeding the degree of hydrolysis of 35.

As a conclusion of the laboratory experiments, the reaction parameters /reaction temperature and time as well as the approximate enzyme level and water content of gum/ could be determined, but there remained further two factors influencing the quality of the end product, namely acid value and acetone insoluble content of the starting material. Knowing the acid value and the AI content of the starting material, they can be calculated for the new product obtained as a function of hydrolysis degree.

$$AV = \frac{DH \times 56.1 \times 100}{750/100 - DH/ + DH/469 + 282/} \qquad \text{Eq. /1/}$$

AV = increase of acid value in the acetone insoluble part

DH = degree of hydrolysis achieved

750 = presumed molecular weight of phosphatides

469 = presumed molecular weight of lysophosphatides, if the 2-position fatty acid hydrolyzed is oleic acid

282 = molecular weight of oleic acid

Obviously, the starting acid value has to be considered because if it is too high, the acid value of the product can exceed the limit of 45.

Calculation of decrease of AI content is shown in Eq. /2/.

$$AI\ \% = \frac{750/100 - DH/ + 469\ DH}{750 \times 100} \qquad \text{Eq. /2/}$$

AI% = percentage of decrease in the original AI content

DH = degree of hydrolysis achieved

750 = presumed molecular weight of phosphatides

469 = presumed molecular weight of lysophosphatides, if the 2-position fatty acid hydrolyzed is oleic acid

Evidently, to obtain a hydrolyzed product with the desired AI content, AI content of the gum should be adjusted in advance. At this point we have to deal with the manufacture of natural lecithin. There are two factors influencing mainly the acetone insoluble content of the gum: degumming temperature and discharging pressure of the separator /1/.

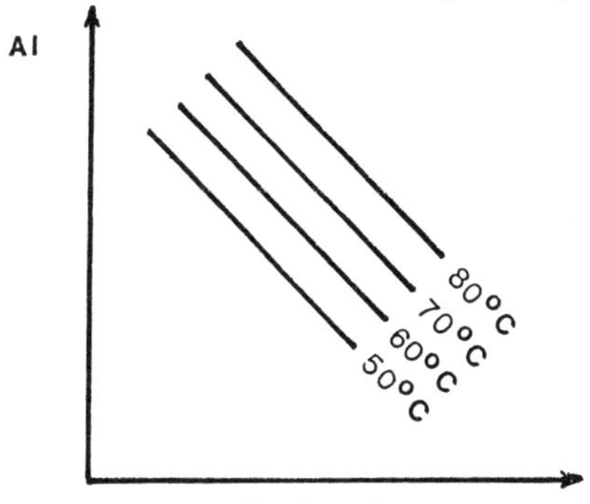

Fig.2. Relationship between AI and discharging pressure at given temperature /1/

The higher the degumming temperature at a given discharging temperature is, the higher the AI content is. Raising the discharging pressure results in lower AI. The relationship varies with design of separators. Recently these parameters have been determined in one processing plant so that industrial experiments could be carried out.

× This study was supported by the Hungarian Ministry of Food and Agriculture.

REFERENCES

1. Flider,F.J. /1985/ Manufacture of soybean lecithins. In: AOCS Mongr. 12, Lecithins, pp 21-37
2. de Haas,G.H.,Postema,N.M., Nieuwenhuizen,W. and van Deenen, L.L.M. /1968/ Purification and properties of phospholipase A from porcine pancrease. Biochim.Biophys.Acta 159, 103-117
3. de Haas,G.H.,Postema,N.M., Nieuwenhuizen,W. and van Deenen, L.L.M. /1968/ Purification and properties of an anionic zymogen of phospholipase A from porcine pancrease. Biochim. Biophys. Acta 159, 118-129
4. van Nieuwenhuyzen,W. /1981/ The industrial uses of special lecithins. JAOCS, 58, 886-888
5. Novo Enzymes: Lecitase™ /product sheet/

BIOLOGICAL ROLE OF PLANT LIPIDS
P.A. BIACS, K. GRUIZ, T. KREMMER (eds)
Akadémiai Kiadó, Budapest and Plenum Publishing
Corporation, New York and London, 1989

NON-CALORIC FAT SUBSTITUTE AND NEW TYPE EMULSIFIERS ON THE BASIS OF SUCROSE ESTERS OF CARBOXYLIC ACIDS

W. Engst, A. Elsner and G. Mieth

Central Institute of Nutrition, Academy of Science of the GDR, 1505 Potsdam-Rehbrücke, GDR

Successful reduction of fat consumption requires both qualified fat substitutes and effective emulsifiers. Esters of sucrose and fatty acids can act either as non-caloric fats / 1 / or as surfactants / 2 / depending on the number of bound fatty acids. The objectives of this investigation were to synthesize a non- or low-caloric fat substitute as well as emulsifiers on the basis of sucrose esters of carboxylic acids in a solvent-free synthesis procedure suitable for human nutrition.

1. Materials and Methods

Materials and methods used for synthesis and product characterization are described in detail by Mieth et al. / 3, 4 / and Elsner et al. / 5 /. The reaction conditions of the alkali metal catalyzed solvent-free interesterification procedures are summarized in Table 1.

Table 1 Reaction conditions for the synthesis of sucrose fatty acid esters

	Synthesis procedure A	Synthesis procedure B	Synthesis procedure C
Reactants	Sucrose octaacetate	Sucrose octaacetate	Sucrose octaacetate
	Fatty acid methylester	Triglyceride	Monoglyceride
Molecular ratio	1 : 10	1 : 3,5	1 : 3
Time (h)	3 - 6	4	2,5
Temperature (K)	383 - 393	393 - 413	393 - 413
Pressure (kPa)	2,0 - 3,0	0,7	101,3
Catalyst	Sodium 1,5 %	Sodium 1,5 %	Sodium 1,0 %

2. Results

As the result of the reaction of sucrose octaacetate and methylpalmitate under described conditions (Synthesis A) yields of 80 - 90 % of sucrose carboxylic acid totalester (STE) are obtained with an average content of bound long-chain fatty acids of 7 molecules. Different from this a mixture of triglycerides, acetoglycerides and STE results from the interesterification of sucrose octaacetate with triglycerides (so-called "Sucro-acetoglycerides" - SAG). The yield of STE with an average number of bound fatty acids of 5 amounts to 50 % in SAG. The chemical, physico-chemical and physiological characterization of products proves excellent properties of STE and SAG as low-caloric fat substitute. They are miscible in triglycerides and insoluble in water, and the lipolysis by pancreatic enzymes (in vitro) is drastically reduced (Fig. 1). Consequently, the resorption is limited and the caloric intake is lowered.

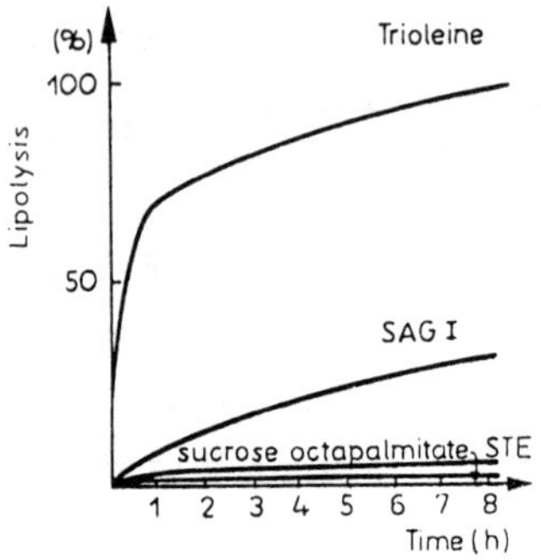

Fig. 1

In vitro lipolysis of STE and SAG in comparision with trioleine

STE - Synthesis procedure A
SAGI - Synthesis procedure B
Sucrose octaacetate - isolated from A

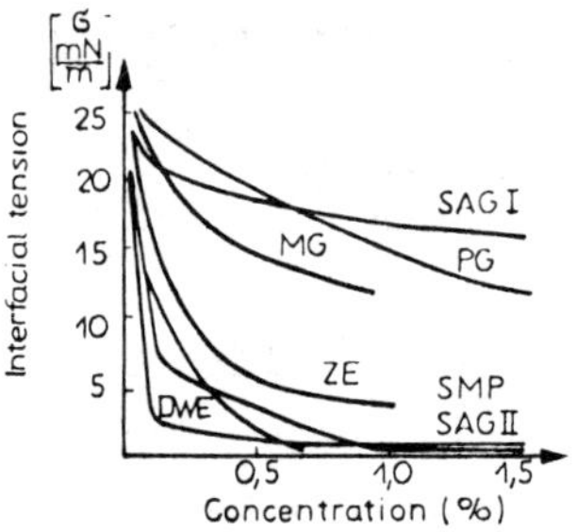

Fig. 2

Interfacial tension of SAG in comparison with commercial emulsifiers

MG - Monoglyceride
PG - Mono- and diglyceride
ZE - Citric acid-fatty acid-glycerol esters
DWE - Diacetyl tataric acid-fatty acid-glycerol esters
SMP - Sucrose monopalmitate

In vivo results by Aust et al. / 6 / confirm these data and reveal a specific metabolic behaviour (higher cholesterol excretion etc.) of these sucrose fatty acid esters. On the other hand SAG containing only small amounts of non-caloric components shows emulsifying behaviour in both model systems and foods / 7 /. Modifying the SAG-synthesis (Reaction C) it is possible to enhance this interfacial activity (see Fig. 2).

Literature

/ 1 / Boggs, R.W. (1986): Sucrose Polyester (SPE)-A Non-Caloric Fat. Fette, Seifen, Anstrichm. 88, 154-158

/ 2 / Adams, W.F. & Schuster, G. (1985): Zuckerester in Emulgatoren für Lebensmittel, Schuster, G., Akademieverlag, Berlin

/ 3 / Mieth, G., Elsner, A. & Weiss, A. (1983): Zur Synthese und Charakterisierung von Saccharosefettsäurepolyestern, 1. Mitt. Über ein neues Syntheseverfahren. Nahrung 27, 748-751

/ 4 / Mieth, G. Elsner, A. & Engst, W. (1988): Zur Synthese und Charakterisierung von Saccharosefettsäurepolyestern, 2. Mitt. Modifizierung des Syntheseverfahrens von Saccharosefettsäuretotalestern zur Gewinnung eines grenzflächenaktiven Pseudofett-Typs. Nahrung, in press.

/ 5 / Elsner, A., Engst, W., Mieth, G. & Schliemann, H. (1988): Zur Synthese und Charakterisierung von Saccharosefettsäurepolyestern, 3. Mitt. Verfahren zur Herstellung grenzflächenaktiver Verbindungen. Nahrung, in press.

/ 6 / Aust, L., Mieth, G., Proll, J., Elsner, A., Behrens, H., Gerhard, W., Brückner, J. & Noack, J. (1988): Orientierende Untersuchungen zum Stoffwechsel verschiedenartiger akalorischer Verbindungen mit fettähnlichen Eigenschaften bei der Ratte. Nahrung 32, 49-57

/ 7 / Engst, W., Elsner, A., Mieth, G. & Schliemann, H. (1988): Zur Synthese und Charakterisierung von Saccharosefettsäurepolyestern, 4. Mitt. Applikation von Sucroacetoglyceriden in lebensmitteladäquaten Systemen. Nahrung, in press.

BIOLOGICAL ROLE OF PLANT LIPIDS
P.A. BIACS, K. GRUIZ, T. KREMMER (eds)
Akadémiai Kiadó, Budapest and Plenum Publishing
Corporation, New York and London, 1989

CONTAMINANTS AND MINOR COMPONENTS OF VEGETABLE OILS

Mária Jeránek and Zsuzsa Weinbrenner

Research Institute for Vegetable Oils and Detergents Industry, 1106 Budapest, Maglódi út 6, Hungary

Besides triglycerides, the main amount of components of vegetable oils, there are several other natural components present in small quantity in oils, and such materials, too, which are forming or getting into oils in course of storage, transport or processing.

Due to development of the analytical methods, these natural components could have been better and better recognized. To that group belong sterols, tocopherols, colour materials, triterpene alcohols, certain hydrocarbons and wax esters. The proportion of the single groups and the characteristic components is different according to the kind of oil. F.e., sterol content varies between 0,2 and 1%. A value of about 1% is characteristic of the corn oil. β-sitosterol is the main component in majority of vegetable oils. The proportion and quality of the other sterol components present in these oils are different. F.e., beside β-sitosterol, sunflower oil contains 5-avenasterol, campesterol and stigmasterol;corn oil contains campesterol; soy oil contains campesterol and stigmasterol, and rapse oil contains brassicasterol and campesterol.

For determination of oil kind, the examination of sterol composition was proposed by Codex Alimentarius' Fats and Oil Committee for that cases, where the type of oil couldn't be identified on the basis of its fatty acid composition. By effect of processing, the quantity of sterols decreased by 20-40%, depending on the parameters of the technology.

The high biological effect tocopherols are present in an amount of 0,02-0,1%. Arachis oil and corn oil are rich in tocopherols. These components as natural antioxidants influence the oxidation stability of polyunsaturated fatty acids. In course of edible oil production, the tocopherol content also decreases, it can be 30-50%, depending on the parameters of the deodorization. The more colourless and odourless product is produced, the less tocopherol remains in the oil. But it has to be mentioned that, at the same time, the distillate is an important raw material for natural tocopherol preparations.

Most of the natural constituents of the edible oils produced in large quantity in the world, does not cause any hazard to human health. But there are exceptions, f.e., the gossypol in cottonseed oil. By processing, these components have to be removed.

Table: Effect of processing

	Altered natural fat components	Removed of reduced components and contaminants
pressing and solvent extraction	-	-
degumming	-	- hydratable non-oil materials (proteins, carbohydrates) removed - hydratable phospholipids partially removed - chlorophyll partially removed
alkali refining	-	- free fatty acids removed - residual phospholipids removed - heavy metal - colouring material reduced
bleaching	- peroxides destroyed - conjugated fatty acids formed	- carotenoids, chlorophyll and its decompositon products removed - gossypol removed (in case of cottonseed oil) - polycyclic aromatic hydrocarbons removed (if activated carbon is used)
deodorization	- geometrical isomers and dimers, polymers formed	- free fatty acids, peroxide decomposition products removed - tocopherols and sterol reduced - pesticide residues removed - mycotoxins removed - solvent residue removed
winterization	-	- wax esters and higher-melting triglycerides removed

For healthy nutrition, year-by-year more attention is payed by researchers and producers to the contaminants getting into oils from the environment. To this group belong herbicides, pesticides, processing aids and contaminations occuring in course of storage and transport. The majority of these materials don't expose the health to dangers, however, there are also toxic compounds among them.

More hundred herbicides are used worldwide. The majority involves such volatile compounds which decompose in some day or week and can't be detected in the oilseeds, raw materials. The risk of their use is avoided by strict control made by the authorities before giving permission for use. A further ensurance for removal of their residues being occasionally present, is the oil production itself, namely, the repeated heat treatment and vacuum distillation (deodorization). Before processing of oilseed cultivated on several culture plots worldwide, also today a great attention is payed to the control of the so-called chlorinated hydrocarbons (DDT, eldrine, dieldrine etc.). These pesticides have been suppressed in our country many years ago but there was not the same situation everywhere in the world. They, namely, accumulate in soil and cause a long-term contamination. In oilseeds cultivated in Hungary, these compounds can't be detected at a level of 10 ppb.

The seed to be processed may be contaminated by very dangerous carcinogenic polyaromatic hydrocarbons during drying of the seeds or fruit with combustion gases. Recognizing this risk, the majority of plants has changed over to indirect drying with heated air. In coconut oil and fish oil samples, Sagredos and his co-workers (F.S. 1988) found more hundred ppb of polyaromatic hydrocarbons, while in crude oils produced in Europe, less than 50 ppb were measured; and within that, the amount of the dangerous heavy components (containing 5-7 rings) was found below 10 ppb. In course of processing, polyaromatic hydrocarbons can be removed from oils by bleaching with active carbon.

Occasionally, impurities of aflatoxin can be found in peanuts. However, that isn't any problem from the viewpoint of the oil. It dissolves in oil poorly, and can totally be removed by neutralization and bleaching.

For processing aids necessary for the production of edible oils, a proposal was elaborated by Codex Alimentarius' Fats and Oil Committee. The materials proposed in this list (inorganic acids, alkalis, solvents for extraction, bleaching materials) are not dangerous for health.

In case of appropriate regulation of technological processes, the residues of these materials can be removed, f.e., the amount of solvent residues, remaining in oil, may have a level of about 1 ppm only.

The metal traces present in oils cause generally no problems concerning the health, but as prooxidative components they influence the oil quality unfavourably. Alkali refining and bleaching reduce the metal content.

Contamination of mineral oil may occure in course of transport. In order to avoid it, a proposal for transport and storage of the oils was

was elaborated by Codex Alimentarius' Fats and Oil Committee. Besides purity of the tanks, transport apparats, that proposal summarizes the requirements on the structural materials and the rules of oil handling, too.

The third group of components present in small amounts in edible oils contains compounds forming during storage and processing. To that group belong the autoxidative products - ketones, aldehides, peroxides - as well as dimer and polymer triglycerides, fatty acid isomers forming under thermic influence. The autoxidation products are removed in course of oil refining - bleaching and deodorization. The formation of thermal alteration compounds depends on the deodorization circumstances; at a temperature of about 240°C a polymer content of 0,2-0,4% and a transisomer content of 0,3-0,5% were found. That is unfavourable from the respect that the amount of natural fatty acid components, mainly essential polyunsaturated fatty acids, decreases. To the autoxidation as well as to the polymerization should be payed great attention firstly in course of use of edible oils. As a result of a well-regulated technology, far less compounds are forming than, f.e., in case of cooking when oil is longer used at presence of air.

I hope, I managed to give a short survey on natural minor components and contaminants of the vegetable oils, and on the influence of the technological process on these components.

At last, I would like to pay attention to the native cold-pressed oils. These products have become popular during the last years. These oils contain all contaminants come from environment by forming, storage and handling of raw materials. Because of the lack of refining, it needs very careful oilseed production and handling.

In order to obtain more accurate knowledge on minor compounds of oils, the IUPAC Fats and Oil Committee is as well just dealing with the elaboration of the proposal for newer up-to-date methods (for detection of metals, polyaromatic hydrocarbons, polymers, mineral oils).

BIOLOGICAL ROLE OF PLANT LIPIDS
P.A. BIACS, K. GRUIZ, T. KREMMER (eds)
Akadémiai Kiadó, Budapest and Plenum Publishing
Corporation, New York and London, 1989

METAL IMPURITIES OF CRUDE AND EDIBLE VEGETABLE OILS

Anna Fábics Ruzics

Research Institute for Vegetable Oil and Detergent Industry, 1106 Budapest, Maglódi út 6, Hungary

It is getting on for 10 years since a research work has been begun at the Research Institute for Vegetable Oil and Detergent Industry that aimed the quantitative and qualitative recovery of those components in vegetable oils and fats which are undesirable from the viewpoint of food hygyene.

The results of that research work have already been reported in several lectures and reports that were dealing with the examination results of monocyclic and polyaromatic carbohydrates and of prooxidant and toxic metals to be found in vegetable oils.

This lecture reports on examination results of metal content in sunflower oil - which is the kind of oil produced in highest quantity in Hungary -, in its raw material and products used for feeding and food industrial (human) purposes.

The determination of metal content in products of the vegetable oil industry is supported by foreign as well as domestic prescriptions.

These prescriptions also urged for an evaluation of vegetable oils and by-products arising during their processing.

In course of sunflower seed processing, oil, meal, hull and phosphatide will be produced from the raw material. The amount of the single products arising during production is shown in percentage distribution as follows:

s u n f l o w e r   s e e d

hull	oil	phosphatide	meal
(for feeding and other purposes)	(for human purposes)	(for feeding and human purposes)	(for feeding purposes)
15%	42%	0,4%	40%

Our intention was to establish, how the metal contents of the above products developed. The examinations were carried out using flame atomic absorption spectrophotometer made by Zeiss.

The preparation of samples for measurement was made by different methods depending on the material of the sample.

For determination, we used sunflower seed samples taken from large-scale production and delivered to the factories as well as that originated from factories producing crude oil.

Minimum 6 samples were taken from each of the products and were examined in 5-5 simultaneous measurements.

Prior to measurements, we tested the standard deviation of the methods, and determined the per cent of the recoveries,

In course of preparation of oil and phosphatide samples, we used the ruling standard methods; one one hand the so-called acidic extraction method, and on the other hand, the dry ashing method following a pre-destruction with sulphuric acid.

Applying these methods, that were a little modified, we made the measurements of iron, copper and zinc from the acidic solutions. For determination of lead and cadmium, we carried out an extraction with MIBK, while using APDC, too. The preparation of sunflower seed, meal and hull was made by means of wet destruction according to the prescriptions of the Hungarian Standard MSZ 3612/1-76.

By using these preparation methods, a suitable amount of more than 90% of the mentioned metal contents should be recovered in every cases, depending on the kind of metal, the amount of the metal content, and the way of the preparation.

According to our experiences, the degree of the standard deviation of the methods depends on the described parameters, and it may be acceptable. We got the highest value for standard deviation in case of lead content determination, it was 28% when very low, 0,06 ppm was found. In the most measurements, the standard deviation didn't exceed 10%, that is generally acceptable in atomic spectrophotometric measurements.

The preparation methods we used are not the most up-to-date methods, there are better and most reliable methods and instruments, too, by use of which quicker and more accurate measurements can be made.

Our examination results are summarized in the following tables:

Iron

Sunflower seed
24,0-75,0 mg/kg

Hull	Oil	Phosphatide	Meal
5,0-28,0 mg/kg	2,0-7,0 mg/kg	30,0-300,0 mg/kg	56,0-80,0 mg/kg

Copper

Sunflower seed
14,0-70,0 mg/kg

Hull	Oil	Phosphatide	Meal
98,0-105,0 mg/kg	0,2-1,0 mg/kg	4,7-9,5 mg/kg	34,0-35,0 mg/kg

Zinc

Sunflower seed
35,6-63,3 mg/kg

Hull	Oil	Phosphatide	Meal
12,5-39,5 mg/kg	0,7-0,8 mg/kg	8,0-9,0 mg/kg	88,7-94,0 mg/kg

Lead

Sunflower seed
1,0-1,6 mg/kg

Hull	Oil	Phosphatide	Meal
below 0,05 mg/kg	0,05-0,4 mg/kg	0,05-0,07 mg/kg	below 0,05 mg/kg

Cadmium

Sunflower seed
0,05-1,0 mg/kg

Hull	Oil	Phosphatide	Meal
below 0,05 mg/kg	0,2-0,23 mg/kg	0,05-0,1 mg/kg	below 0,05 mg/kg

On basis of the described results, it can be established that during processing the metal content of raw material distributes among the single products in the following way:

In general, it can be established that the metal content, that is originally present in seed, doesn't develop according to the percentage distribution of the certain products. Concerning the iron content, we can find an enrichment in meal and in phosphatide products.

The proportion of oil, as main component in seed, is favourable, as its iron content is much lower than it would have been believed on base of its proportion. From the seed, ca. 42% oil is obtained in the plant, and the amount of iron present in that oil is ca. 6%.

For copper content, its enrichment in hull as well as in meal can be observed.

The metal content of oil is as well here favourable.

With exception of meal, the zinc content is lower in every products than it would have been thought, considering its proportion.

The lead and cadmium contents of the products are low when comparing to that of the raw material. That may be due to the fact that a part of them will eliminate during the processing technology.

Owing to our technical possibilities, we couldn't detect the degree of such metal impurities in sunflower hull and meal. The degree of impurities in oil and lecithin is acceptable.

Summarizing the results of our evaluation, we can establish the following:

In course of sunflower oil production, the metal content of crude sunflower oil is favourably influenced by the seed processing technology.

The element content of sunflower hull, meal and lecithin as by-products for feeding purposes, examined by us, agrees to the Hungarian prescriptions.

BIOLOGICAL ROLE OF PLANT LIPIDS
P.A. BIACS, K. GRUIZ, T. KREMMER (eds)
Akadémiai Kiadó, Budapest and Plenum Publishing
Corporation, New York and London, 1989

INTERACTIONS OF CHLOROPHYLL PIGMENTS WITH LIPIDS OF RAPESEED OIL

M. Holasova[1], H. Parizková[1], J. Pokorny[2] and J. Davidek[2]

[1]Research Institute of Food Industry, Na bélidle 21, Prague 5, Czechoslovakia
[2]Dept. Food Chem., Prague Institute of Chemical Technology, Suchbátarova 5, Prague 6, Czechoslovakia

Chlorophyll pigments represent natural components of oilseeds and are partially extracted into crude oils during industrial oilseed processing by pressing and extraction. They act as pro-oxidants in oils in interaction with light i.e.as sensitizers producing singlet oxygen. Carotenoids present in oils are effective as singlet oxygen quenchers.

Czechoslovak zero -erucic rapeseed oils were examined for the levels of chlorophyll pigments in dependence to glucosinolate content of the rapeseed and its technological processing. Chlorophyll pigment interactions with some other components of the oil were also followed.

About 70 samples of oils were thus analyzed for the contents of chlorophylls a and b and pheophytins a and b. The samples analyzed included crude (pressed and extracted) oils with normal and reduced glucosinolate contents, samples of these oils after the individual industrial processing operations, and stored oils.

The analytical procedures applied in the study were: fluorometric determination of chlorophyll pigments /1/, absorption spectrophotometry for the analysis of carotenoids /2/. Hydroperoxides were determined by the iodometric titrational or spectrophotometric procedures /3,4/.

The investigation led to the following findings:
Crude Czechoslovak zero-erucic rapeseed oils contain 18 - 40 mg of chlorophyll pigments/kg of oil with pheophytin a predominating (up to 90% of total content).

During the manufacture of edible oils bleaching reduces the chlorophyll pigment content to the highest extent, other stages of refining have a small effect.
During the manufacture of shortenings,prerefining removes only a part of pigments.After hydrogenation only traces of them remain,probably due to adsorption on the catalyst.
No significant difference was found between pigment contents in oils from rapeseed with normal and reduced glucosinolate levels
The degradation rate of total chlorophyll pigments at 80^{0}C,in the dark, under reduced oxygen concentration is strongly influenced by hydroperoxides (butyl oleate hydroperoxide added to oil). Secondary oxidation products (1-hexanal added as model component) and tocopherol degradation products (tocopheryl quinone) have a lower though not insignificant effect.
Under the conditions of Schaal oven test (60^{0}C, in the dark, free access of oxygen, 10 mm layer) carotenoids decompose first, thus protect chlorophyll pigments against destruction, and prolong the initiation period of autoxidation. After carotenoid decomposition chlorophyll pigments are rapidly destroyed by hydroperoxides.
In the dark or in oils packed in amber bottles, chlorophyll pigments have no prooxidative effect on edible oils, while on light they are active prooxidants because of their photosensibilizing activities.

REFERENCES

/1/USUKI,R., SUZUKI,T., ENDO,Y., KANEDA,T.: Residual amounts of chlorophylls and pheophytins in refined edible oils, " J.Amer Oil Chem.Soc." 61, (4),785-788 (1984)
/2/ Czechoslovak Standard,:Determination of vitamin A and its provitamins,"CSN 560053-1987"
/3/ Czechoslovak Standard: Methods for testing of fats and oils. Peroxide value determination, "CSN 580130, part 4 - 1986"
/4/ TAKAGI,T., MITSUNO,Y.,MASUMURA,M.: Determination of peroxide value by the colorimetric iodine method with protection of iodide as cadmium complex, "Lipids", 13, (2),147-152 (1978)

BIOLOGICAL ROLE OF PLANT LIPIDS
P.A. BIACS, K. GRUIZ, T. KREMMER (eds)
Akadémiai Kiadó, Budapest and Plenum Publishing
Corporation, New York and London, 1989

THE BIOLOGICAL ROLE OF SOY LIPIDS IN THE NUTRITION

K. Lindner

College of Commerce and Catering, Budapest, Hungary

Nearly during 3000 years the soybean had been consumed, utilised in forms of food, of which the most important were soybean sprouts, steamed green beans, soy milk, soy sauce and paste, soy flour and soy bean curd. All of this food products contains the genuine, unaltered soy lipids.

After the Second World War for the increased meat production were inadequate the traditional sources of feed-protein (fish meal and scarps from meat-processing plants) to meet these increases in demand. Therefore in this time the USA has started growing soybeans on large scale, primarily for their oil and for the toasted, defatted meal, used as animal feed.

It is well known that the production of "refined edible soybean oil" (vegetable oil), during the different steps of thechnology (toasting, degumming, neutralisation, bleaching, hydrogenation, winterising, deodoration) has wery drastic effects on vitamins, phospholipids, sterols and unsaturated ligands of fatty acids too.

The major effects of the processing are given in Table 1.

Table 1. Effects of processing on soybean oil

Process	Fat components altered/changed	Nutrients removed/reduced
Toasting	Modification of the fatty acids if the heating is excessive	Carotenoids and tocopherels partially reduced

Continuation of the Table 1.

Process	Fat components altered/changed	Nutrients removed/reduces
Solvent extraction	Residual solvents in small amounts in the oil	
Degumming		Hydratable non glyceridic lipids, such as phospholipids, partially removed
Alkali refining		Residual phospholipids removed
Bleaching	Conjugated acids formed and peroxids destroyed	Carotenoids removed
Deodorization	Formation of geometrical isomers in sensitive fatty acids. Formation of linear and cyclic dimers-polymers	Sterols and sterolesters reduced in quantity; Tocopherols reduced in quantity
Hydrogenation	Partial saturation Formation of positional and geometric isomers Formation of linear and cyclic dimers/polymers	Essential fatty acids reduced in quantity
Winterisation	Enhancement of saturated triglycerides	Higher-melting triglycerides removed

Therefore it is a very important recommendation of the FAO Expert Consultation (1980) on dietary fats and oils in human nutrition, that

- loss of essential fatty acids during hydrogenisation and other processes should be minimized;

- specific nutrients such as tocopherols and carotenes, if removed in nutritionally significant amounts during refining, should be repleaced whenever feasible;
- Prolonged (repeated) exposure of polyunsaturated fats to high temperatura during refining should be discouraged.

On the basis of this very important recommendations, the research team of the REKORG Marketing, Puplicity and Supplying Firm have developed a new processing, to produce cold press, unaltered, nature "virgin soybean oil" and fatty or meagre soybean meal for human nutrition. The main point of the manufacturing is the low temperature (below 90 °C) handling and the semi-autofermentation.

This soy-products has neutral taste and their storage time, without deterioration is over two years, based on the antioxidant effects of genuine tocopherols and carotenes.

The unaltered lipid constituens of this new soybean oil are given in the Table 2.

Table 2. The biological important nutrients in the new soybean oil

Nutrient	New soybean oil	Data from literature
Linoleic acid	50,6 %	51,0 - 54,7 %[a]
Linolenic acid	9,1 %	5,8 - 9,0 %[a]
Tocopherols	382 - 1400 mg/kg	913[c] ; 1400 mg/kg[a]
Total phospholip.	2,1 %	
Lecithin	0,61 %	
Kephalin	0,65 %	
Lypositol	0,84 %	
Total steroids	3170 mg/kg	3000 mg/kg[b]
Campestrol	27,5 %	26,8 %[d] ; 21,3 %[b]
Beta-sitosterol	47,1 %	48,8 %[d] ; 53,5 %[b]
Stygmasterol	25,5 %	17,2 %[d] ; 19,1 %[b]

a Kurnik (1962); b Scher and Vogel (1976);
c Perédi and Balogh (1981); d Perédi and Balogh (1983)

In addition to the data of the nutrients it is necessary to mention, that the so called antinutritive factors totally disappeared during the plant cell dystroing treatment, so thus the urease activity value is diminished to zero and the tripsin inhibitor activity below 1 TIU/mg.

From point of view of the nutrition, the nutritive value of the "virgin soybean oil" is given in the Table 3.

Table 3. The nutritive importance of the main components of the genuine soybean oil

Component	Function	Importance
Essential fatty acids		
Linoleic acid (18:2, n-6); alfa-Linolenic acid (18:3, n-3)	-normal growth and structure/ funktion of all tissues	-curative effect for the various skin disorders of children Conteract effect on: -increasing of serum cholesterol -the thrombotic activity of blood platellets
Vitamins		
Carotene	-interconverting to retinol/rhodopsin -stimulating effects on the metabolism and function of cells	-dark adaptation of eye -inhibition effect on corneal destruction
Tocopherole	-defense against oxigen-containing radicals in the cell	-adjunct effect on the therapy of different diseases
Phosphatide	-major lipid components of most membranes in animal cell	-as polar lipid promote the incorporation of the biofactors throught the cell membrane

Continuation of the Table 3.

Component	Function	Importance
Phytosterine	-antirachitic and specific hormonal effect	-increased steroid metbolism

In addition to these specific biological roles is very important the combinative influence of the oil components on the different biological function of the whole organism. This matter needs profound research work.

Recognizing that the main thing is put into practice the new procedure and products, the REKORG Firm has a research and developing program to applicate in the suitable section of the Food Industry (baking-, meat-, sweets-industra etc.). The bread with 20 % whole soymeal is now in the commerce.

It seems the possibility to enlarge the application of the genuine "virgin soybean oil" to pruduce skin protective cosmetic creams and skin regenerative (burn) pharmaceutical products too.

REFERENCES

1. FAO Food and Nutrition Series No. 20. (1980): Dietary Fats and Oils in Human Nutrition, Rome
2. Kurnik, E. (1962): A szója. Akadémiai Kiadó, Budapest
3. Perédi, J. and Balogh A. (1981): Vizsgálati adatok a hazai növényolajok tokoferol tartalmáról. Olaj Szappan Kozmetika 30. 1-5.
4. Perédi , J. and Balogh A. (1983): Növényi olajok gőzölésekor felfogható párlat összetétele és értékesitésének lehetősége. Olaj Szappan Kozmetika 32. 100-105.
5. Scher, H. and Vogel, H. (1976): Sterine von Soje in verschiedenen pflanzlichen Oelen. Fette Seifen, Anstrm. 78. 301-304.

BIOLOGICAL ROLE OF PLANT LIPIDS
P.A. BIACS, K. GRUIZ, T. KREMMER (eds)
Akadémiai Kiadó, Budapest and Plenum Publishing
Corporation, New York and London, 1989

THE ROLE OF LIPIDS IN THE ABSORPTION OF VITAMIN A

A. Blaskovits, L. Gampe and J. Borvendég

National Institute of Pharmacy, Budapest, Hungary

The absorption of several fat soluble biologically active substances may take place in the presence of lipids.
The chemical structure, the digestibility and the fatty acid composition of lipids show considerable variance. There are great differences in their absorption. Vitamin A is one of the fat soluble vitamins thus its absorption and bioavailability are influenced considerably by the fats.

Our work was aimed to study:

- how do the different fatty acid composition of oils and fats influence the plasma level and liver accumulation of retinyl-palmitate.

Sunflower-, rapeseed-, and coconut oils, besides fat, butterfat and Liga margarine were used in the trial. Their fatty acid composition show considerable differences. The coconut oil contains mainly lauric acid, sunflower oil comprises the essential linoleic acid, butterfat comprehends palmitic- and oleic acid, in the Liga margarine there is the oleic acid in the highest quantity.

A fluorometric method was used for the determination of vitamin A in the blood.
After the samples were saponified the vitamin A content of the liver was measured by high-performance liquid chromatography.

170-200 g body-weight LATI CFYmale rats were used in the study, kept in standard conditions. The animals were divided into six groups, each group contained six animals.
The animals were given 3000-3000 IU retinyl-palmitate dissolved in sunflower oil, in rapeseed oil, in coconut oil, in fat, in butterfat and in Liga margarine in 0,5 ml volumes by p.o. intubation.
For the study of vitamin A absorption 100-100 μl blood was taken from the caudal vein directly before the administration of the oils and fats and subsequently in the 1st-2nd-3rd-4th-6th-8th hours.
Vitamin A accumulation in the liver was studied in a separate trial. Animals were kept, prepared, grouped and sampled by p.o. intubation similarly to the first trial. We used a control group, too. 16 hours after the administration of vitamin A dissolved in the different oils and fats, animals were bled to death and vitamin A contents of the livers were determined.
On the basis of our work we can say that the intestinal absorption of retinyl-palmitate and its accumulation in the liver - compared to retinol - were not influenced by the oil and fat in which it had been dissolved.
The highest plasma level of vitamin A was found in the 4th hour after its administration in every case and plasma level was not influenced by the different fatty acid composition and absorption mechanisms of the oils and fats.
The vitamin A content of the liver increased significantly and in the same amount /about 40-50%/ in each treated groups as compared to the control animals. The vitamin A content of the liver of the control animals was found 46,3 $\pm$ 3,7 μg vitamin A/g liver.

On the base of this observation it can be said that retinyl-palmitate might be used instead of retinol in pharmaceutical preparations, because the bioavailability of retinyl-palmitate as compared to that of retinol is independent of the oil and the fat in which it has been dissolved.

BIOLOGICAL ROLE OF PLANT LIPIDS
P.A. BIACS, K. GRUIZ, T. KREMMER (eds)
Akadémiai Kiadó, Budapest and Plenum Publishing
Corporation, New York and London, 1989

INVESTIGATION OF COFFEE LIPIDS

F. Örsi and B. Dicházi

Technical University of Budapest, P.O. Box 92, 1521 Budapest, Hungary

1. INTRODUCTION

The lipids of coffee play an important part in the storeability of roasted coffee. During storage the lipids can be oxidized if oxigen is present and it changes the aroma of coffee, too.

The aims of our investigation were to separete the most important classes of coffee lipids and to analyse their change during processing (roasting, storage and making coffee).

2. MATERIALS AND METHODS

2.1. Investigated coffee samples

The investigated coffee beans imported by Hungarian Confectionary Industry are summarised in the Table 1.

Table 1. The investigated coffee beans

Name	species
Uganda Angola	Coffea canephora var. Robusta
Burundi Ecuador New-Guinea Peru Santos Viktoria	Coffea arabica

The coffee beans were roasted by INFRAMAX infra-red roast apparatus in laboratory scale. 50 g sample was roasted for 8 min.

One part of beans was stored in climat box at 40 ^{o}C and 100% relative humidity for 10 days to promote the degradation of coffee lipids.

The beans were grinded and in AUTOPRESS apparatus coffee drink was made from 36 g freshly grinded coffee with 330 cm^3 water.

2.2. Isolation, separation of lipids

The lipids of grinded beans were extracted by chloroform methanol = 7:3 mixture in Besson apparatus. The lipids of coffee drink were extracted by chloroform in shaker funnel. The extracted lipids were separated by silica gel coloumn chromatography in three fraction: nonpolar lipids eluated by chloroform, glycolipids eluated by aceton, phospholipids eluated by methanol.

The nonpolar lipids (120 mg solved in 5 cm^3 hexane) were separated by silica gel coloumn chromatography in hydrocarbons eluated by hexane, glycerides eluated by hexane- ether, free fatty acids eluated by ether-hexane, monoglycerides and sterols eluated by ether.

The identification of lipids and control of separation were made by TLC and detected by iodine vapour.

The triglycerid fraction was separated on $AgNO_3$ impregnated silica gel layer and detected by sulfuric acid. The fatty acids of fractions was analysed by GLC. The apparatus CHROM41 column: 10% polyethyleneglycoladipate on Chromosorb WHP, at 190 ^{o}C.

The sterols were separated by HPLC on μ Bondapack 10 C18 with 95% methanol eluent and detected at 280 nm in UV detector.

3. RESULTS

There is characteristic difference between the two species of the coffee in the lipid content in harmony with the literature.

Table 1. Characterization of beans

Sample	lipids		water		lost of weight at roasting	
	%	sd	%	sd	%	sd
C. canephore	8,9	2,4	10,2	1,3	13,35	0,5
C. arabica	14,2	0,8	8,9	1,3	13,82	2,3

sd = standard deviation

There is significant difference between the two species in the quantity of lipids can be extracted. The effect of roasting and storage on the lipid contant, and the lipid quantity may be extracted from the coffee drink showed in the Table 2. Effect of roasting, storage and making coffee on the lipid quantity.

Table 2.

Sample	raw		roasted		stored		drink	
	%	sd	%	sd	%	sd	%	sd
C. canephore	8,9	2,4	14,2	1,6	11,4	3,0	2,7	1,5
C. arabica	14,2	0,8	20,0	1,7	17,3	4,0	1,8	1,0

sd = standard deviation

The roast increases the extractable lipid and at the end of storage decreases again. The coffee drink have 10% of original roasted beans.

The lipid classes in the caffee separated by chromatographic methodes are presented in the Table 3.

Table 3. The lipid classes in coffee

Sample	lipid class	raw %	roasted %	stored %	drink %
C. cane.	hydrocarbon	2,1	3,3	3,3	1,2
	glycerid	61,4	57,7	72,8	61,5
	free fatty acid	4,6	7,7	1,8	4,1
	sterols	24,0	31,8	22,8	33,2
	glycolipid	3,8	0	0	0
	phospholipid	4,0	0	0	0
C. arab.	hydrocarbon	2,5	3,0	3,5	1,2
	glycerid	65,6	58,7	65,8	62,2
	free fatty acid	2,4	3,5	3,4	2,9
	sterols	25,6	34,7	28,1	33,7
	glycolipid	0	0	0	0
	phospholipid	4,1	0	0	0

var. coef. = 15%

There was no drastic change of lipid classes only the polar glyco-, and phospholipids degraded during the roasting. There is significant change in the quantity of free fatty acids.

The change of triglycerids was characteristic on $AgNO_3$ impregnated silica gel layer. The results are summarized in the Table 4.

The fatty acids of trigycerides fraktion was summarized in the Table 5.

Only the linolenic acid content changed drasticaly.

The sterol fraction of nonpolar lipids and raw coffee extract was investigated by HPLC and in the C. Arabica species some carecteristic sterols was found which are absent in the C. Canephore species.

Table 4. The triglycerid fractions on the number of unsaturated bonds

Sample	number of unsaturation	raw %	roasted %	stored %	drink %
C. cane-phora	0-1	40,2	20,0	43,1	36,1
	2	8,5	25,5	4,6	5,2
	3	6,6	2,1	12,6	17,2
	3-4	17,1	15,1	6,5	10,5
	4-6	10,4	22,5	15,1	15,1
	6-9	27,2	14,8	18,1	25,5
C. ara-bica	0,1	39,5	23,5	20,6	19,5
	2	5,3	24,0	12,8	5,7
	3	4,1	3,0	16,4	31,0
	3-4	9,8	9,1	9,6	5,4
	4-6	11,4	30,6	14,1	21,4
	6-9	29,9	9,8	26,5	28,3

var. coef. = 15%

Table 5. Fatty acids of triglycerides

Samplex	fatty acid	raw %	roasted %	stored %	drink %
C. caneph.	14:0	3,3	1,4	1,6	2,6
	16:0	36,0	37,1	35,4	34,3
	18:0	7,2	5,7	5,8	8,4
	18:1	12,5	13,5	12,2	10,9
	18:2	37,2	41,1	43,9	43,7
	18:3	3,8	1,2	1,0	0
C. arabica	14:0	0,6	1,1	1,6	2,6
	16:0	38,4	42,2	34,1	31,6
	18:0	5,5	4,4	7,6	10,7
	18:1	7,4	6,2	9,6	9,7
	18:2	46,3	44,3	44,4	45,7
	18:3	1,6	1,8	1,2	0

var. coef. = 5,5%

4. SUMMARY

The most important lipid classes were analysed in the two major botanical species Coffea arabica and Coffea canephora var. Robusta by HPLC, GC and TLC.

There is difference between the two species in the quantity of total lipid, the contant of sterol fraction and glycolipid. The in chloroform soluble lipid contant increases during the roasting and decreases during storage.

The unsaturated triglycerides change during the storage and change the saturated/unsaturated fatty acid proportion, too. The coffee drink has 10-15% of total lipid and the quantity of sterol fraction higher.

BIOLOGICAL ROLE OF PLANT LIPIDS
P.A. BIACS, K. GRUIZ, T. KREMMER (eds)
Akadémiai Kiadó, Budapest and Plenum Publishing
Corporation, New York and London, 1989

STUDIES ON CEREAL LIPASES

Julie O'Connor, H.J Perry and L. Harwood

Department of Biochemistry, University College, Cardiff CF1 1XL, UK

Recent changes in nutritional habits have resulted in an increased demand for wholemeal cereal and bran-enriched products. However, wholemeal cereal products are unstable on storage, resulting in oxidative rancidity, loss of nutritional value and reduced baking performance. This deterioration is initiated by an endogenous lipase(s), thought to be concentrated primarily in the bran, which causes triacylglycerol hydrolysis. The resultant unesterified polyunsaturated fatty acids can then be attacked by lipoxygenase. The peroxidative reactions of the latter are enhanced by the manipulation of flour during baking.

In this study, we have been examining the lipases responsible for the initial lipid hydrolysis in wheat and oat flours. Properties of the enzymes have been investigated, solubilisation and preliminary purification procedures evaluated, and the results reported.

Materials and Methods

Finely-milled (0.5mm) wholemeals were obtained by coarse grinding followed by fine milling (Retsch mill), and defatted by three successive hexane extractions (10ml/g) [1]. Extracted lipid classes were separated by thin layer chromatography and identified using specific stain reagents and co-chromatography against known standards. Lipase activity was solubilised by shaking defatted flour overnight at 4°C in the presence of 50mM phosphate buffer and detergent. Lipase was measured either by a modification of the copper-soap assay of Sahasrabudhe _et al_. [2], or by using $[^{14}C]$triacylglycerol as substrate. Esterase activity was measured fluorometrically [3].

Results

In a survey of different cereals, the lipase activity of oats was found to be much greater than that from wheat or barley varieties. However, incubation of flour preparations with [^{14}C]triacylglycerols revealed that transacylation reactions were quite active relative to lipase action in wheat, whereas the lipase activity dominated in oat flour preparations. Comparable activities were measured with triolein and tricaprylin substrates. Moreover, analysis of the removal of endogenous and exogenous triacylglycerols by lipase activity did not reveal a preference for any specific molecular species.

In order to study the lipases further we have begun purification. Since the enzymes are membrane-localised, our initial experiments have been directed at solubilisation and the maintenance of stable solubilised activity. A variety of ionic and non-ionic detergents have been tested and the wheat lipase found to solubilise best with Tween 20 while for that from oat flour we use Triton X-100. In the latter case, the concentration of Triton X-100 in the assay system affects total activity.

Solubilisation of the lipases from both oat and wheat flours also leads to the release of esterase activity. Some partial separation of activity could be achieved in the case of wheat flour by differential solubilisation with varying concentrations of Tween 20. This technique also leads to some enrichment of lipase activity with respect to protein.

The solubilised oat and wheat lipases have been fractionated by ion-exchange chromatography in the presence of detergent. Two distinct lipase peaks were separated in the case of wheat extracts, both of which also contained esterase activity. Further work to characterise and purify these peaks is being undertaken.

*The financial support of MAFF is gratefully acknowledged. We would also like to thank Dr. T. Galliard (RHM Research) for helpful discussions and advice.

References

1. Galliard, T. (1986). J. Cereal Science 4, 33-50.
2. Sahasrabudhe, M.R. (1982) J. Am. Oil Chem. Soc. 59, 345-355.
3. Saunders, R.M. and Heltved, F. (1985) J. Cereal Science 4, 79-86.

BIOLOGICAL ROLE OF PLANT LIPIDS
P.A. BIACS, K. GRUIZ, T. KREMMER (eds)
Akadémiai Kiadó, Budapest and Plenum Publishing
Corporation, New York and London, 1989

LIPID AND ANTIOXIDANT CONTENT OF RED PEPPER

H.G. Daood, P.A. Biacs, N. Kiss-Kutz, F. Hajdú and B. Czinkotai

Central Food Research Institute, 1022 Budapest, Herman Ottó út 15, Hungary

INTRODUCTION

Lipid oxidation is a major cause of quality changes in food. It can occur in plant products all the way from the growing of the plant, through the ripeness, processing, storage and distribution to the final treatment for consumption. Lipid oxidation in food systems is affected by a number of promoting and inhibiting factors. Among inhibitors one finds those that are naturally occurring e.g. effective antioxidants such as tocopherols and ascorbic acid.[1] Lipid oxidation reactions often lead also to secondary reactions, which can influence food attributes. One of the most important negative changes caused by the secondary reactions of lipid oxidation is the decolorization of food pigments[2]. The purpose of this work was to investigate the content and composition of lipids, carotenoids and effective antioxidants in paprika fruit.

Materials and Methods

Paprika /Capsicum annuum L. Sz-20 cv./ fruits were obtained from the experimental fields of the szegedi füszerpaprika enterprise. The fruits were harvested at the various stages of ripening. Fatty acid methyl esters and tocopherols were from Sigma. Organic acids were from Serva.

Extraction and preparation of samples

The fruit was separated to pell, pulp and seeds by a plastic knife. Carotenoid pigments were extracted by chloroform-acetone-isopropyl alcohol /2:1:1/ after acetone extraction as described earlier[3]. Lipids of different parts were extracted and their fatty acid composition were analysed by GLC[4]. Tocopherols of different parts of the fruits were extracted, saponified and prepared for HPLC analysis according to Speek and co-workers[5]. Organic acids were prepared by a method described previously[6]. Following preparation the samples were redissolved in a minimal volume of the HPLC eluent.

HPLC separation and identification of components

A Beckman series of liquid chromatograph equipped with a Model 165 variably wavelength detector, a Modell 114 solvent delivery pump, a Model 420 controller and a Model C-R3A Shimadzu integrator was used to perform the chromatographic separation of carotenoids, tocopherols and organic acid under different conditions which are shown in the figures or tables.
The individual pigments of paprika were identified according to their absorption spectra and comparing them with those reported in the literature[7]. Fatty acids, organic acids and tocopherol were identified by using internal standards.

Results and Discussion

GLC analysis of fatty acid in different parts of paprika fruit showed that unsaturated fatty acids such as oleic, linoleic and linolenic are predominant /Fig. 1./. Lipid content and fatty acid composition of peel and pulp were similar but lipid of seeds comprised the highest percentage /45%/ of the total lipid.

Carotenoid composition of paprika fruit is shown in fig.2. The fruit distributed mono- and di-fatty acid esters of capsanthin and capsorubin in addition to free pigments. GLC analysis of fatty acids esterified with the carotenoid pigment indicated that capsanthin and capsorubin esterified mainly with saturated fatty acid to form diesters. Although unsaturated fatty acids are predominant in paprika, they were detected only with the fraction of capsanthin monoesters.

As to antioxidants, paprika fruit contains ascorbic acid and several isomers of tocopherol. The pulp distributed mainly α-tocopherol while the seeds contained γ -tocopherol as the abundant isomer /Fig.3/. Despits the high biological activity of γ -tocopherol, its content in paprika seeds is not of high technological importance especially when the seeds are ground together with the paprika powders /Table 1/. The most effective amount of tocopherol was found in the pulp of the fruit.

Table 1: Change in tocopherol content /mg/100 g.d.wt./ of paprika fruit and products

Product state	Dry matter %	α-Tocopherol /Mesocarp/	γ -Tocopherol /Seed/
GREEN FRUIT	13.16	14.2 ∓ 0.64	3.7 ∓ 0.07
CHLOROPH.DISCOL. Stage-1	15.03	44.6 ∓ 1.77	4.6 ∓ 0.13
CHLOROPH.DISCOL. Stage-2	15.1	86.0 ∓ 4.3	7.01 ∓ 0.24
RED FRUIT	14.4	236.0 ∓ 8.26	9.35 ∓ 0.23
3 WEEKS AFTER HARVESTING	94.02	78.0 ∓ 4.2	12.7 ∓ 0.41
GROUND PAPRIKA	92.1	11.6 ∓ 0.29	-
GROUND + SEEDS /10-15%/	92.0	20.0 ∓ 0.4 1.4 ∓ 0.035 γ -tec.	

At the advanced ripeness there was an increase in lipid content of paprika /Fig. 4/ in accordance with chlorophyll degradation and appearance of red colour on the skin of the fruit. The main carotenoid, capsorubin and capsanthin, esterified simultaneously with fatty acid to form relatively stable red pigments. Manner of red igments, ascorbic acid and tocopherol during ripening of paprika is illustrated in fig. 5. It could be concluded that ascorbic acid was very effective antioxidant when water content of the fruit was high while the considerable effect of tocopherol could be observed at low water activity /postharvest stages/. Concentration of ascorbic acid and tocopherol should be kept not less than 100 and 200 mg/g dry weight respectively in order to get paprika products with high colour stability.

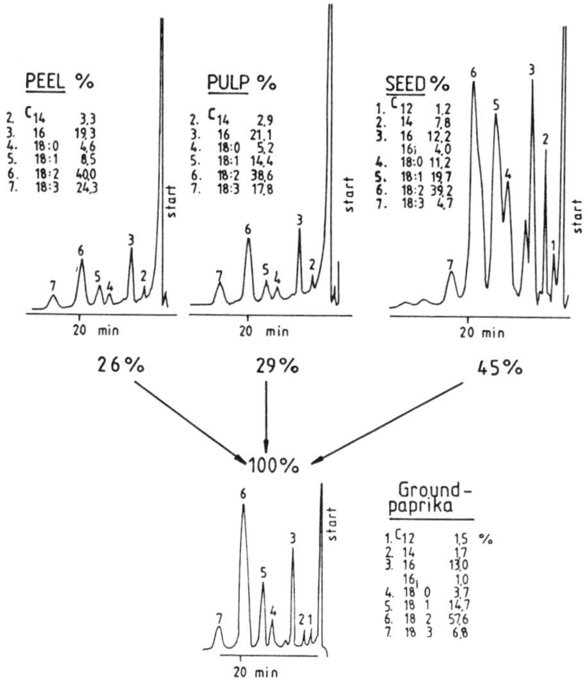

Fig. 1.: Fatty acid composition of the different parts of paprika fruit

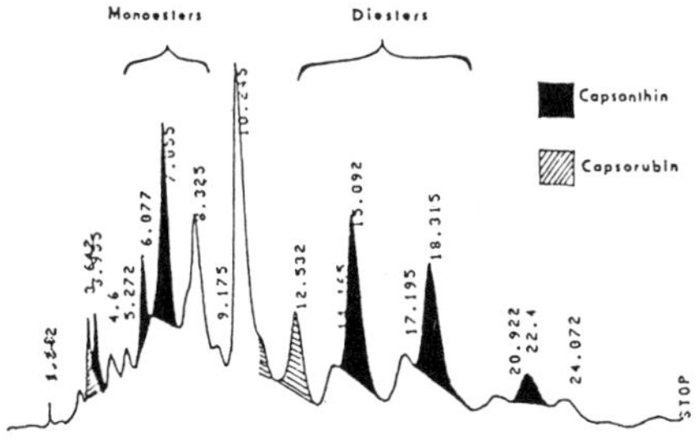

Fig.2.: HPLC chromatogram of carotenoids from paprika

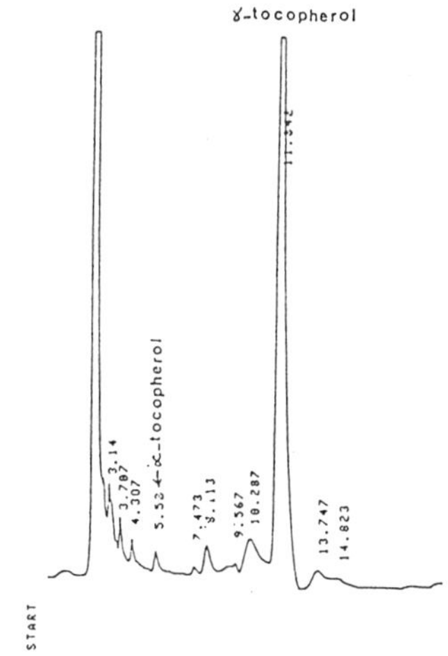

Fig.3.: HPLC chromatogram of tocopherols from paprika SEEDS

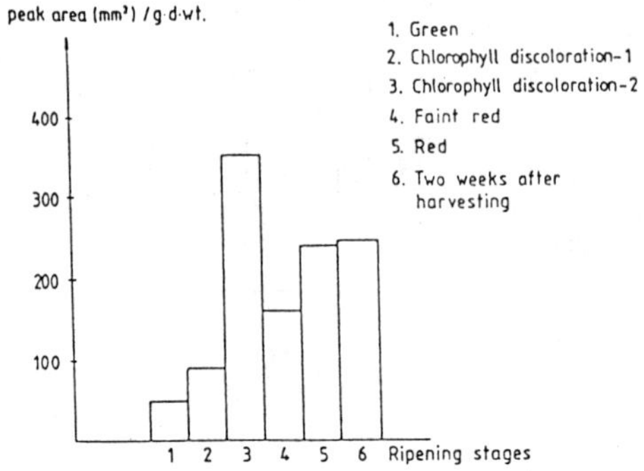

Fig.4.: Lipid content of paprika fruit /GLC analysis/

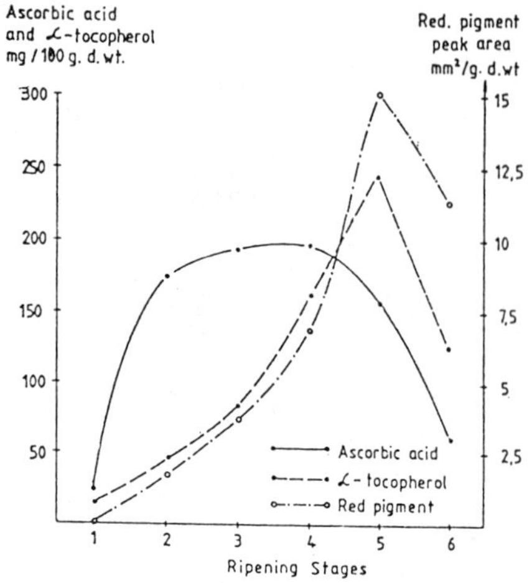

Fig.5.: Changes in antioxidant and red pigment content of paprika fruit

The biotechnologist when aiming at increasing color stability of paprika, either by breeding experiments or by addition of food preservatives, have to take those metabolites into consideration. Special attention should also be paid to the conditions of processing, packaging, storage, and others that need to be investigated.

REFERENCES

1 ERIKSSON, E.C. /1982/: Food Chem. 9, 9-13

2 ESKIN, N.A.M., GROSSMAN, S. & PINSKY, A./1977/: CRC-Critical Rev Fd Sci. Nutr. 9, 1-40

3 PAVISA, Cs.A., HAJDU, F., HOSCHKE, Á., BODNÁR, J. & BIACS, P. /1987/: Acta Aliment. 16, 129-142

4 BIACS, P., GRUIZ, K. & SZEKERES, E. /1980/: Acta Aliment. 9, 86-87

5 SPEEK, A.J., SCHRIJVER, J. & SCHREURS, W.H.P. /1985/: J.Food Sci. 50, 121-124

6 BIACS, P., DAOOD, H., PAIS, I., FEHÉR, M. & HAJDU, F. /1987/: in Chromatography-87 /KALÁSZ, H. and ETTRE, L. eds./ pp.39-60

7 BUCKLE, K.A. & RAHMAN, M.M. /1979/: J.Chromatog. 171, 385-391

CHAPTER 7

DEVELOPMENT, ENVIRONMENT, STRESS

BIOLOGICAL ROLE OF PLANT LIPIDS
P.A. BIACS, K. GRUIZ, T. KREMMER (eds)
Akadémiai Kiadó, Budapest and Plenum Publishing
Corporation, New York and London, 1989

METABOLIC RESPONSES OF PLANT CELLS TO STRESS

G.A. Thompson, Jr., K.J. Einspahr, S. Ho Cho, T.C. Peeler and Martha B. Stephenson

Department of Botany, University of Texas, Austin, Texas 78713, USA

INTRODUCTION

Plants are routinely exposed to a bewildering variety of environmental stresses. Considering the extremes of temperature, salinity, moisture level, etc. that must be overcome, it is perhaps surprising that plant life is distributed as widely as it is. The key to survival lies in the remarkable capacity of many plants to acclimate to changing conditions. Such an acclimation is necessary because the patterns of cellular composition and metabolism favoring productive growth under one set of conditions usually do not suffice under different conditions.

If we understood more clearly just how these acclimation responses operate, we might be able to modify plant growth to our great benefit. In this discussion I shall review certain recent observations which illustrate the current state of research in this field. My remarks will emphasize the role of lipids, particularly as they are involved in the response of plants to temperature and salinity change.

The most thoroughly studied stress response is that triggered by low temperature. It is now generally accepted that major physiological dysfunctions associated with low temperature stress are caused by the decreased fluidity of membrane lipid bilayers (Lyons, et al, 1979). The ability of plants (as well as animals) to reestablish a functional membrane fluidity is a prerequisite to survival. This is true for plants chilled to near freezing temperatures as well as those subjected to below freezing conditions. In the latter case, osmotic changes resulting from the extracellular formation of ice crystals compound the problems that must be overcome in restoring adequate membrane function (Steponkus, 1984).

As an increasing number of reports appear on the subject of low temperature acclimation in plants, it becomes more and more apparent that a variety of lipid compositional changes are effected. The principal ones include 1) altered levels of fatty acid unsaturation, 2) altered molecular species composition within lipid classes, and 3) altered proportions of the different lipid classes. These three categories of change will be discussed separately.

TEMPERATURE-INDUCED CHANGES IN FATTY ACID UNSATURATION

Plant fatty acids, primarily palmitic acid (16:0) and oleic acid (18:1), are synthesized in the chloroplast and distributed to various parts of the cell for incorporation into membrane lipids. The insertion of from 1 to 4 double bonds into the fatty acids, catalyzed by a family of fatty acid desaturates, is the classical mechanism for refluidizing membranes made rigid by exposure to low temperature. Desaturation may occur either in the endoplasmic reticulum, as it does in animal cells, or in the chloroplast (Fig. 1).

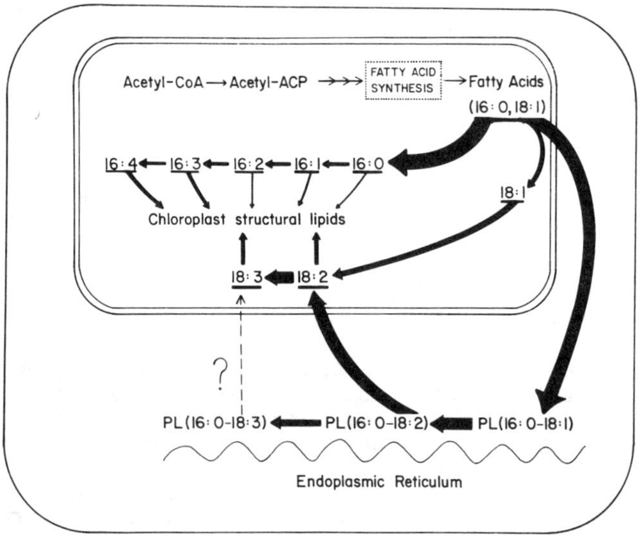

Fig. 1. Fatty acid desaturation in plants

Extensive desaturation in the chloroplast is found only in the so-called 16:3 plants, which synthesize a sizeable portion of their galactolipids in this cell compartment (Roughan and Slack, 1982).

A number of studies, including several done in my laboratory with the ciliate *Tetrahymena pyriformis* and the green alga *Dunaliella salina*, have implicated microsomal desaturases as the site controlling the degree of cellular fatty acid unsaturation. The more extensive *Tetrahymena* studies indicate 2 distinct and sometimes simultaneous processes regulating desaturase activity in chilled cells (Thompson and Nozawa, 1984). The mechanism most readily available is the unusually high retention of fatty acid desaturase activity in chilled cells, permitting a continuation of desaturation at a time when the net synthesis of saturated fatty acids has been almost totally inhibited. A complementary mechanism in *Tetrahymena* features the low temperature-induced synthesis of additional fatty acid synthetase molecules.

While the precise mechanism underlying the rise in unsaturation of microsomal fatty acids in Dunaliella is not as well understood, the increase in double bonds is substantial and relatively quick in chilled cells. As with animals, most of the desaturation appears to take place on fatty acyl chains covalently attached to phospholipids (Stymne and Appleqvist, 1978). In Dunaliella cells rapidly chilled from 30° to 12°C, by far the greatest rise in fatty acid unsaturation was seen in microsomal membranes, with the increase in chloroplast unsaturation being slow in developing and quantitatively less extensive (Lynch and Thompson, 1984a). On the other hand, Williams et al (1988) report enhanced desaturation of Brassica napus monogalactosyldiacylglycerol (MGDG) by the chloroplast desaturase system when plants were grown at 5° and 10°C., as compared with the normal 20° or 30°C-grown plants. There is a real need for more studies comparing the relative contributions of microsomal and chloroplast desaturases under conditions of stress.

LIPID MOLECULAR SPECIES RETAILORING UNDER TEMPERATURE STRESS

It has recently become evident that phospholipid molecular species retailoring, that is, the enzymatic deacylation and subsequent reacylation of existing phospholipids to yield different fatty acid pairings, occurs as the earliest known response following the chilling of Tetrahymena (Lynch and Thompson, 1984b). In metabolically remote compartments of the Tetrahymena cells, such as the cilia, molecular species retailoring is the only mechanism available for short term lipid compositional change since the cilia (and indeed nearly all of the cell's membranes) contain phospholipase A, acyl CoA synthetase, and acyltransferase, the three enzymes required for the retailoring process (Fig. 2), but not those catalyzing most other reactions of lipid metabolism (Lynch and Thompson, 1988).

Molecular species retailoring is also seen as an early microsomal response of Dunaliella to chilling, coinciding with the initial rise in fatty acid unsaturation. While these microsomal changes were detectable within 12 hr following a shift from 30° to 12°C (Lynch and Thompson, 1984c), equivalent but smaller alterations were not detectable in chloroplasts until more than 30 hr had elapsed (Lynch and Thompson, 1984a). Since membrane fluidity can be significantly altered merely by redistributing existing fatty acids into new molecular species combinations through deacylation and reacylation (Ramesha and Thompson, 1983), this series of reactions must be considered as an important mechanism for effecting early membrane fluidization under stress.

Just why a different pattern of molecular species is synthesized at low temperature remains uncertain. At the present time attention is directed to the Dunaliella phospholipase A enzymes because of their elevated activity at low temperature (Norman and Thompson, 1986). Since these hydrolases display a specificity towards certain phospholipid molecular species (H. Norman and G. Thompson, unpublished

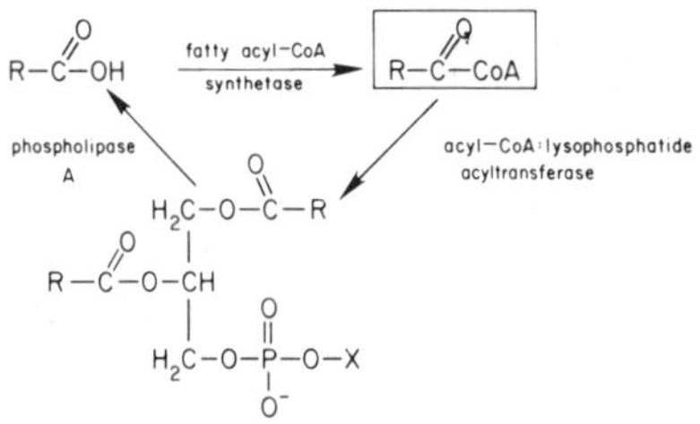

Fig. 2. Retailoring of phospholipid molecular species.

observations), increased deacylation could alter the preexisting composition. Little information is available regarding the alternative option, namely, a chilling-induced modification of acyltransferase activity.

CHANGES IN LIPID CLASS PROPORTIONS

Many cases are known where the percentages of particular lipid classes differ significantly in plants grown at different temperatures. Examples of these are shown in Table I. Based on periodic lipid analyses during the time course of low temperature acclimation in <u>Tetrahymena</u> and in <u>Dunaliella</u> it has been concluded that lipid class alterations in these organisms come in a later stage of the low temperature acclimation process (Thompson, 1988). They may be required mainly to restore an average cylindrical shape to bilayer lipids following a temporary unbalancing of their average conformation towards a conical shape by extensive desaturation of the hydrophobic moiety (Israelachvili, et al, 1980). Additional examples are needed to determine more definitely the role of lipid class changes at low temperatures. In a study of acclimation to high instead of low temperature, Süss and Yordanov (1986) detected a significant increase in <u>Phaseolus</u> DGDG with respect to MGDG (Table I). The pattern of $^{14}CO_2$ incorporation into DGDG stressed at 50°C disclosed a preferential synthesis of more saturated molecular species, e.g. 18:1/16:0 DGDG, as compared with the pattern of synthesis in control 25°C-grown plants.

Another interesting instance of proportional changes in lipid classes was reported in plasma membrane isolated from rye seedlings (Lynch and Steponkus, 1987). If the seedlings had been acclimated to 2°C, their plasma membranes contained higher amounts of phospholipids and free sterols than did plasma membranes from 20°C-grown control seedlings. Conversely, the plasma membranes of non-acclimated cells were enriched in steryl glucosides and glucocerebrosides. It appears that lipid compositional changes are responsible for the marked differences in physical properties that have been observed between protoplasts of acclimated and non-acclimated cells upon freezing. Thus both the plasma membrane of acclimated protoplasts and that of non-acclimated protoplasts which had been altered by fusion with

Table I. Examples of temperature-induced changes in lipid class composition.

Tissue (ref.)	Temperature	Lipid Composition
Phaseolus chloroplast (Süss & Yordanov, 1986)	35°C 50°C (5hr.)	MGDG/DGDG=1.3 MGDG/DGDG=0.9
Dunaliella chloroplast (Lynch & Thompson, 1982)	30°C 12°C	MGDG/DGDG=3.2 MGDG/DGDG=2.0
Secale plasma membrane (Lynch & Steponkus, 1987)	20°C 2°C	Free sterol=33%, cerebroside=16% Free sterol=44% cerebroside= 7%

unsaturated phosphatidylcholine liposomes formed exocytotic extrusions during the type of osmotic stress caused by freezing and thawing, while the equivalent preparations from non-acclimated protoplasts formed endocytotic vesicles (Balsamo, et al, 1988). It is thought that following thawing the rye cell is able to recover full viability if it has undergone exocytotic extrusions but not if inward vesiculation has occurred.

A SPECIAL ROLE FOR PHOSPHATIDYLGLYCEROL?

I now come to a low temperature phenomenon unique to chloroplasts. Considerable excitement was aroused when Murata (1983) recognized that chilling sensitivity of many plants is closely correlated with the content of high-melting phosphatidylglycerol molecular species. These lipids, namely 16:0/16:0 phosphatidylglycerol and 16:0/*trans*-3-16:1 phosphatidylglycerol, are present only in chloroplasts and constitute the sole natural phosphoglycerides undergoing a phase transition above 0°C. For this reason their presence in high concentrations was postulated to trigger a damaging lipid phase separation in the temperature range where chilling injury occurs.

More recent surveys (Roughan, 1985; Bishop, 1986) have identified exceptions to Murata's generalization; nevertheless, many workers remain convinced that PG has special functions in the low temperature-stressed thylakoid. One example of a somewhat different sort is well illustrated by the findings of Krupa et al (1987) with cold-hardened and non-hardened rye. The light harvesting complex isolated from hardened thlakoids contained a 54% lower level of *trans*-Δ3-16:1-containing PG than did non-hardened controls and, when resolved by electrophoresis on non-denaturing polyacrylamide gels, also had a 60% lower ratio of oligomeric LHC/monomeric LHC. Because a variety of complementary findings (reviewed by Dubacq and Trémolières, 1983) supports the concept that *trans*-Δ^3-16:1-containing PG stabilizes LHC in its active, oligomeric form, it seems logical to suspect that this lipid affects the flow of electrons through the photosystems. Perhaps by altering the content of these particular PG molecular species the plant can compensate for the effects of changing temperature *per se* on LHC oligomer stability.

EFFECTS OF SALINITY STRESS ON LIPID METABOLISM

Changes in the lipid composition of plants grown under different levels of salinity have been known for some time (e.g., Harzallah-Skhiri et al, 1980). Here I would like to discuss a different involvement of lipids in acclimation to salinity stress. Our recent studies of the halotolerant alga *D. salina* have been focused on the role of lipids in signal transduction during the initial phase of osmotic shock. When *D. salina* cells grown in a medium containing 1.7 M NaCl are rapidly diluted to 0.85 M NaCl, the concentration of phosphatidylinositol 4,5-bisphosphate (PIP_2) in the cells' plasma membrane drops by 30% within 2 min. (Einspahr et al, 1988a). Within the same time period there is a sharp rise in the level of phosphatidic acid, suggesting a marked increase in PIP_2 turnover in response to the hypoosmotic shock. On the other hand, when *D. salina* is exposed to hyperosmotic shock, i.e., transfer from 1.7 M NaCl to 3.4 M NaCl, an opposite reaction can be measured, namely, PIP_2 rises in content while phosphatidic acid drops (Einspahr et al, 1988b). Because the reactions to these two osmotic shocks are so distinctive, we feel that we are seeing more than just a non-specific stress response.

Current work has confirmed the presence in the *D. salina* plasma membrane of a PIP_2-specific phospholipase C (K. Einspahr, unpublished observations). There is also strong evidence for a protein kinase C in the alga (R. Wayne, L. Tung, S. Roux, and G. Thompson, unpublished observations). Although there has as yet been no systematic analysis of this signal transduction system's involvement in other stress responses, the time is ripe to pursue these ends. The increasing number of literature reports on elements of this signal system in higher plants as well as algae argues for its universal presence in the plant world.

ACKNOWLEDGEMENTS

Work in the author's laboratory was supported in part by the National Science Foundation, the Robert A. Welch Foundation, and the Texas Advanced Technology Research Program.

REFERENCES

Balsamo, R.A., M. Uemura, and P.L. Steponkus, Plant Physiol. 86:53 (1988). abstract. Transformation of the cryobehavior and ultrastructure of the plasma membrane of rye protoplasts by modification of the lipid composition.

Bishop, D.G., Plant, Cell and Environment 9:613-616 (1986). Chilling sensitivity in higher plants: the role of phosphatidylglycerol.

Dubacq, J.-P., and A. Trémolières, Physiol. Veg. 21:293-312 (1983). Occurrence and function of phosphatidylglycerol containing △3-trans-hexadecenoic acid in photosynthetic lamellae.

Einspahr, K.J., T.C. Peeler, and G.A. Thompson, Jr., J. Biol. Chem 263:5775-5779 (1988a). Rapid changes in polyphoinositide metabolism associated with the response of Dunaliella salina to hypoosmotic shock.

Einspahr, K.J., M. Maeda, and G.A. Thompson, Jr., J. Cell Biol. in press (1988b). Concurrent changes in Dunaliella salina ultrastructure and membrane phospholipid metabolism after hyperosmotic shock.

Harzallah-Skhiri, F., T. Guillot-Salomon, and M. Signol, in Biogenesis and Function of Plant Lipids, Mazliak, P. P. Benveniste, C. Costes, and R. Douce, eds. Elsevier, Amsterdam (1980) pp. 99-102. Lipid changes in plastids isolated from alfalfa seedlings grown under salt stress.

Israelachvili, J.N., S. Marčelja, and R.G. Horn, Quart. Rev. Biophysics 13:121-200 (1980). Physical principles of membrane organization.

Krupa, Z. N.P.A. Huner, J.P. Williams, E. Maissan, and D.R. James, Plant Physiol. 84:19-24 (1987). Development at cold-hardening temperatures. The structure and composition of purified rye light harvesting complex II.

Lynch, D.V., and G.A. Thompson, Jr., in Advances in Membrane Fluidity, Vol 3, Aloia, R., C.C. Curtain, and L.M. Gordon, eds. Alan R. Liss, New York (1988) pp. 323-360.

Lynch, D.V., and G.A. Thompson, Jr. Plant Physiology 74:198-203 (1984a). Chloroplast phospholipid molecular species alterations during low temperature acclimation in Dunaliella.

Lynch, D.V., and G.A. Thompson, Jr., Trends Biochem. Sci. 9:442-445 (1984b). Retailored lipid molecular species: a tactical method for modulating membrane properties.

Lynch, D.V., and G.A. Thompson, Jr., Plant Physiol. 74:193-197 (1984c). Microsomal phospholipid molecular species alterations during low temperature acclimation in Dunaliella.

Lynch, D.V., and P.L. Steponkus, Plant Physiol. 83:761-767(1987). Plasma membrane lipid alterations associated with cold acclimation of winter rye seedlings (Secale cereale L cv. Puma).

Lyons, J.M., D. Graham, and J.K. Raison, Low Temperature Stress in Crop Plants. The Role of the Membrane. Academic Press, New York (1979) pp. 565.

Murata, N., Plant and Cell Physiol. 24:81-86 (1983). Molecular species composition of phosphatidylglycerols from chilling-sensitive and chilling resistant plants.

Norman, H.A., and Thompson, G.A., Jr., Biochim. Biophys. Acta 875:262-269 (1986). Activation of a specific phospholipid fatty acid hydrolase in *Dunaliella salina* microsomes during acclimation to low growth temperature.

Ramesha, C.S., and G.A. Thompson, Jr., Biochim. Biophys. Acta 731:251-260 (1983). Cold stress induces *in situ* phospholipid molecular species changes in cell surface membranes.

Roughan, P.G., Plant Physiol. 77:740-746 (1985). Phosphatidylglycerol and chilling sensitivity in plants.

Roughan, P.G., and C.R. Slack, Annu. Rev. Plant Physiol. 33:97-132 (1982). Cellular organization of glycerolipid metabolism.

Steponkus, P.L., Annu. Rev. Plant Physiol. 35:543-584 (1984). Role of the plasma membrane in freezing injury and cold acclimation.

Stymne, S., and L.Å. Appelqvist, Eur. J. Biochem. 90:223-229 (1978). The biosynthesis of linoleate from oleoyl-CoA via oleoyl-phosphatidylcholine in microsomes of developing safflower seeds.

Süss, K.-H., and I.T. Yordanov, Plant Physiol. 81:192-199 (1986). Biosynthetic cause of *in vivo* acquired thermotolerance of photosynthetic light reactions and metabolic responses of chloroplasts to heat stress.

Thompson, G.A., Jr., J. Bioenergetics and Biomembranes, in press, 1988. Membrane acclimation by unicellular organisms in response to temperature change.

Thompson, G.A., Jr., and Y. Nozawa, in Membrane Fluidity, Kates, M., and L.A. Manson, eds., Plenum, New York (1984) pp. 397-432. The regulation of membrane fluidity in *Tetrahymena*.

Williams, J.P., G. Johnson, M.U. Khan, and K. Mitchell, these proceedings. Biosynthesis of Galactosyldiacylglycerol molecular species from *Brassica napus* grown at low temperatures.

BIOLOGICAL ROLE OF PLANT LIPIDS
P.A. BIACS, K. GRUIZ, T. KREMMER (eds)
Akadémiai Kiadó, Budapest and Plenum Publishing
Corporation, New York and London, 1989

MEMBRANE RESPONSE TO ENVIRONMENTAL STRESSES: THE LIPID VIEWPOINT - INTRODUCTORY OVERVIEW

P. Mazliak

Université P. et M. Curie, UA 1180, Laboratoire de Physiologie Cellulaire, Tour 53/3
4 Place Jussieu, 75252 Paris Cedex 05, France

INTRODUCTION : MEMBRANE HOMEOSTASIS.

Most plants respond to environmental stresses by changes in the lipid composition of their cell-membranes. Main transformations, observed in a long term period, are concerned with the phospholipid composition of membranes and the degree of unsaturation of component fatty acids (1). In the short term, some "retailoring" of lipid molecular species (intermolecular rearrangements of lipid acyl chains) can be observed (2). All these chemical changes would result in modifications of the physical state of membrane lipids.

It has been often postulated that a proper functionning of plant cell metabolism (enzymes, metabolite transport, ion permeation, etc...) would require the physical properties of membranes (fluidity for instance) to be maintained within an optimal range (3). A direct effect of environmental stresses would be to perturb the physical properties of membranes by changing the physical organization of membrane lipids. Consequently, the response of plant cells to these perturbations would be a series of chemical modifications in the lipid composition of membranes in order to recover the initial organization (and initial physical properties) of cell-membranes. This central hypothesis is called the "membrane homeostasis" hypothesis (4). The ability of an organism to maintain the physical porperties of its membranes within an optimal range in response to any environmental stress would thus be a clue for the stress-resistance of this organism (5).

I LIPID DOMAINS IN FLUID MEMBRANES.

Recent progresses in the analysis of membrane lipids have shown the great variety of molecular species componing the central lipid bilayer of membranes. As an example, table 1 shows some twenty different molecular species of phospholipids found in potato tuber membranes.

Table 1 - Main phospholipid molecular species found in potato tuber microsomes. Analyses were carried out by HPLC, as described in (6).

	mol %		
Molecular species	Phosphatidyl-choline	Phosphatidyl-ethanolamine	Phosphatidyl-inositol
18:3/18:3	tr	tr	-
18:3/18:2	13.4	11.1	-
18:2/18:2	31.8	34.4	4.1
16:0/18:3	10.1	9.1	22.3
16:0/18:2	35.5	38.2	73.6
18:0/18:3	2.1	tr	-
18:0/18:2	6.9	7.1	-

Besides these different phospholipids, different molecular species of di- and tri-acylglycerols, sterols and sterol-esters, etc... are present in the lipid matrix of membranes. As a consequence of the chemical heterogeneity, a structural heterogeneity would be created in the lipid bilayer which, for physical reasons, would be fragmented in multiple adjacent lipid domains where molecular species with similar physical properties would be assembled. Lipid domains have been shown to exist in crystalline membranes, at low temperatures. Experimental evidences for lipid domains in fluid membranes have been brought (7). Membrane domains would not be expected to be static structures but to be time-averaged.

II CHILLING SENSITIVE AND CHILLING RESISTANT PLANTS.

It has been shown by Raison and Lyons (8) that the chilling sensitivity of plants can be correlated with the presence of relatively large rigid (crystalline) domains in the plant membranes of chilling-sensitive plants, below 10°C. In contrast, chilling resistant plants are able to maintain the liquid-crystalline (fluid) organization in the lipid bilayer at low temperatures (0-12°C). Further progress was accomplished by Raison and Wright, in 1983 (9), when these authors could show that thermal phases transitions in the polar lipids of plant membranes was induced, below 10-12°C, by disaturated molecular species of phospholipids, representing less than 1% of total phospholipid mass.

A decisive step in the understanding of the lipid contribution to chilling-sensitivity of plants was later realized by Murata (10) when this author showed (by fluorescence polarization measurements) that the phosphatidylglycerols from chilling-sensitive plants, but no other lipids, contained large proportions of "saturated" molecular species which undergo phase transition at room temperature or above. Since a major part of phosphatidylglycerol (particularly those molecular species containing palmitic, stearic and trans-hexadecenoic acids) are located in chloroplast, phase-transitions related to chilling-injury could probably occur in chloroplast membranes.

III THERMAL ACCLIMATION OF CHILLING RESISTANT PLANTS (COLD HARDENING)

L. Vigh and collaborators (11) have clearly shown that cold herdening of certain wheat varieties was accompanied by a clear maintainance of plasmalemma initial fluidity (Non-hardening varieties were unable to maintain initial fluidity, as measured by ESR spectroscopy). Many investigators have measured changes in fatty acid unsaturation in the attempt to find a connection with chilling-resistance or frost hardiness. In some reports, an increase in unsaturation parallels hardiness (12) ; in others, any relationship fails (13). Accumulation of phosphatidylcholine (14) in cold-hardened tissues could well play a much more decisive role in the maintainance of the high fluidity of plant-cell membranes submitted to cold-stress.

IV RESPONSE TO DROUGHT OR SALINITY.

Water stress provokes modifications in lipid and fatty acid composition of plant cell membranes. Pham Thi et al. (15), using ^{14}C-oleate and ^{14}C-linoleate as precursors, demonstrated directly that the two successive desaturation steps (oleic to linoleic and linoleic to linolenic) were affected by water stress, particularly the transformation of linoleic into linolenic acid, occurring in the galactolipids of green leaves. Increased salinity levels in soil-waters have been shown, too, to reduce plant lipid unsaturation (16).

CONCLUSIONS

Part of the mechanism of plant resistance to environmental stress surely depends on lipid rearrangements in membranes. However, by no means, all adaptative mechanisms could be restricted to these lipid rearrangements.

Proteins are also essential membrane components. New proteins (heat-shock proteins, cold acclimation proteins (17)) are synthesized in response to thermal stresses. Important changes in plant cell energy-charges, metabolic pathways, ion transports, etc... are also occurring. Lipid biochemists just bring some pieces of information to the general puzzle of plant acclimation to environmental stresses.

REFERENCES

(1) A. Sakai, W. Larcher - Frost survival of plants - Responses and adaptation to freezing stress. Springer-Verlag, Heidelberg, 321 pp. (1987).

(2) D.V. Lynch and G.A. Thompson - Retailored lipid molecular species : a tactical mechanism for modulating membrane properties. Trends in Bioch. Sc., 9, 442-445 (1984).

(3) P.W. Hochachka and G.N. Somero - Strategies of Biochemical adaptation - W.B. Saunders Comp., Philadelphia, 358 pp. (1973).

(4) M. Sinensky - Homeoviscous adaptation - A homeostatic process that regulates the viscosity of membrane lipids in Escherichia coli. Proc. Nat. Acad. Sci. USA, 71, 522-525 (1974).

(5) N.F. Hadley - The adaptive role of lipids in Biological Systems. Wiley and Sons, New York, 319 pp. (1985).
(6) C. Demandre, A. Trémolières, A.M. Justin et P. Mazliak - Analysis of molecular species of plant polar lipids by high-performance and gas-liquid chromatography. Phytochemistry, 24, 481-485 (1985).
(7) D.L. Melchior - Lipid domains in fluid membranes : a quick-freeze differential scanning calorimetry study. Science, 234, 1577-1580 (1986).
(8) J.M. Lyons, D. Graham and J.K. Raison - Low temperature stress in crop plants, Academic Press, Naw York, 565 pp. (1979).
(9) J.K. Raison and L.C. Wright - Thermal phase transitions in the polar lipids of plant membranes. Their induction by disaturated phospholipids and their possible relation to chilling injury. Biochim. Biophys. Acta, 731, 69-78 (1983).
(10) N. Murata and J. Yamaha - Temperature-dependent phase behavior of phosphatidylglycerols from chilling-sensitive and chilling-resistant plants. Plant Physiol., 74, 1016-1024 (1984).
(11) L. Vigh, I. Horvath, L.I. Horvath, D. Dudits and T. Farkas - Protoplast plasmalemma fluidity of hardened wheats correlates with frost resistance. FEBS Letters, 107, 291-294 (1979).
(12) C. Willemot - Simultaneous inhibition of linolenic acid synthesis in winter wheat roots and frost hardening by BASF 13-338, a derivative of pyridazinone. Plant Physiol., 60, 1-4 (1977).
(13) G. Smolenska and P.J.C. Kuiper - Effect of low temperature upon lipid and fatty acid composition of roots and leaves of winter rape plants. Physiologia plantarum, 41, 29-35 (1977).
(14) I. Horvath, L. Vigh and T. Farkas - The manipulation of polar head group composition of phospholipids in the wheat Miranovskaja 808 affects frost tolerance. Planta, 151, 103-108 (1981).
(15) A.T. Pham Thi, C. Borrel-Flood, J. Veira da Silva, A.M. Justin and P. Mazliak - Effects of drought on 1-^{14}C-oleic and 1-^{14}C-linoleic acid desaturation in cotton leaves. Physiologia plantarum, 69, 147-150 (1987).
(16) M. Ellouze, N. Gharsalli and A. Cherif - Action du chlorure de sodium sur la composition lipidique des feuilles de Tournesol (Helianthus annuus, L.) et de "Lime Rangpur" (Citrus limonia Osbeck). Physiologie végétale, 18, 1-10 (1980).
(17) L. Meza-Basso, M. Alberdi, M. Raynal, M.L. Ferrero-Cardinanos and M. Delseny - Changes in protein synthesis in rape seed (Brassica napus) seedlings during a low temperature treatment. Plant Physiol., 82, 733-738 (1986).

BIOLOGICAL ROLE OF PLANT LIPIDS
P.A. BIACS, K. GRUIZ, T. KREMMER (eds)
Akadémiai Kiadó, Budapest and Plenum Publishing
Corporation, New York and London, 1989

MEMBRANE STABILITY UNDER THERMAL STRESS

P.J. Quinn

Department of Biochemistry, King's College London, Campden Hill, London W8 7AH, UK

INTRODUCTION

Plant membranes, like that of membranes of animal cells, are composed of lipids and proteins that are often glycosylated. There are no components found universally in membranes and the composition from one membrane type to another is highly heterogeneous and presumably reflects functional specialities. There is some evidence to suggest that the composition, particularly of the lipid component, may change in response to environmental conditions such as temperature, salt concentration etc. These alterations occur in addition to changes associated with growth, development and ultimately senescence of the cell. It is believed that these changes are required to adjust the physical characteristics of the membranes so that they may perform their necessary physiological tasks when environmental factors change. If the environmental conditions are altered beyond the normal limits within which the plant survives, the cell membranes are often found to undergo gross structural changes. These structural perturbations include phase separation of the membrane constituents and are associated with characteristic disturbances of function such as loss of selective permeability and transport processes.

The observed phase separations during temperature stress appear to be driven by phase changes in the membrane lipids. Some lipids extracted from algae and plant membranes are known to exist in a bilayer gel phase when dispersed in aqueous systems at the growth temperature while other lipid fractions are in a liquid-crystalline state. Nearly all membranes contain varying proportions of their lipid complement that do not form bilayer structures under such conditions and most commonly they adopt an hexagonal-II arrangement. The importance of maintaining the balance of lipid phase structure is believed to be the dominant factor in membrane stability under conditions of thermal and other environmental stresses.

EFFECT OF TEMPERATURE

The response of organisms to environmental temperature, as reflected in membrane structure and stability, depends on the rate of exposure or temperature gradient and the degree to which the temperature change extends beyond the optimum growth conditions. Adaptive changes in membrane composition are often observed and involve alteration in the ratio of lipid:protein changes in proportion of the various lipid classes present in the membrane and/or the nature of their associated fatty acyl chains. These changes may involve alteration in the length, degree of unsaturation and the position at which the fatty acids are acylated to the membrane lipids as well as introduction of branched chain hydrocarbons. The physical significance of such changes has been the subject of considerable debate [1]. It is commonly believed, however, that the changes in membrane lipids are required to maintain the fluidity of plant membranes within limits required for the efficient function of the particular membrane. Such changes, therefore, can be regarded as an adaptive change which results in the preservation of a constant membrane fluidity under conditions of varying environmental temperature. Membrane instability and irreversible changes in structure associated with exposure of plant membranes to temperatures beyond the normal growth temperatures is probably the major limiting factor in plant growth conditions.

Based on knowledge of the phase behaviour of membrane lipids it is possible to predict the events that accompany exposure of biological membranes to extremes of temperature. Because the destabilizing events are fundamentally different the effect of low temperature and high temperature on membrane structure and stability will be considered separately.

LOW TEMPERATURE

Upon cooling organisms from the growth temperature, the first membrane lipid phase transition that is likely to occur is that from hexagonal-II or other non-lamellar phase to liquid-crystalline bilayer. There is no evidence that this transition results in irreversible damage and indeed membranes and organelles are generally isolated for biochemical characterisation under temperature conditions that most probably are below this transition temperature. One manifestation, however, could be that the membrane proteins may not be packaged into efficient functional units below a critical temperature and this may explain the characteristic discontinuities in Arrhenius plots of many membrane-bound enzyme activities.

Further decreases in temperature result in transition of lipids from liquid-crystalline to gel phase bilayer configuration. The creation of discrete gel phase regions can be detected by freeze-fracture electron microscopy and appear as smooth, particle-free domains in longitudinal fracture planes within the membrane. The creation of such domains has been associated with permeability changes in membranes but the precise lesion has yet to be identified. The type of lipid that initially forms a gel phase are those which have the highest gel to liquid-crystalline phase transition temperature. These are likely to be the non-bilayer forming lipids as the bilayer-forming lipids with equivalent acyl chain composition generally have lower T_c values.

Cooling to temperatures below ice crystallisation results in dehydration of membranes as the bulk water freezes and solutes are zone-refined to the regions of unfreezable water at the membrane aqueous interface. Saturated solutions of electrolytes and solutes affect membrane phase behaviour by screening charges on acidic lipids which are known to be important for the overall phase behaviour of the membrane.

Damage to the membrane may also result when the system is thawed. Under these conditions if the phase changes take place during the cooling process are not reversed quickly then permeability barrier properties will be drastically impaired by the creation of domains of non-bilayer lipid arrangements. Inverted micelles sandwiched between the bilayer leaflet would be particularly damaging as the creation and disappearance of these structures could result in passage of macromolecules across the membrane.

HIGH TEMPERATURES

The upper limit of temperature at which most plant species can survive appears to be governed by the stability of the chloroplast membranes [2]. Changes in intrinsic fluorescence yield associated with temperature stress signify a disruption of the chlorophyll-a/b light-harvesting protein complexes and the photochemical reaction centres of photosystem II. Changes in the arrangement of thylakoid membranes resulting from brief exposure to high temperatures have been examined by freeze-fracture electron microscopy. Thus, a 5 minute exposure to temperatures between 35° and 45°C causes membrane destacking. This has been interpreted as dissociation of the photosynthetic apparatus in the stacked grana regions brought about by the phase separation of the monogalactosyldiacyglycerol into non-bilayer structures. Brief exposure to higher temperatures (>45°C) provides evidence of gross phase separations of lipid into tubular inverted micellar arrangements of lipid if the thermal stress is applied to chloroplasts that are initially stacked. Extensive membrane fusion and vesiculation occurs when the membranes are unstacked during thermal stress.

All of these events can be ascribed to the release of constraints imposed on non-bilayer lipids to maintain them in a bilayer phase required to achieve a stable membrane structure. Exposure to high temperatures serves to phase separate the non-bilayer lipids into discrete domains that appear to be devoid of membrane proteins. It is likely also that the permeability barrier properties are altered by this type of phase separation. In general, where gross non-bilayer phase separations have taken place it is difficult to imagine how the normal distribution of membrane components can be achieved and especially at a rate that would be consistent with the continued survival of the cell.

CONCLUSIONS

Membrane lipids exhibit complex polymorphism as a function of temperature. A balance of lipid phase structure is believed to result from interaction of lipids with other membrane components and solutes in the aqueous phase. In general, this balance results in a formation of a fluid lipid bilayer matrix. Phase separations of lipid from other membrane constituents can be driven by exposure of membranes to temperatures outside the normal growth temperature. These can be the creation of gel phase domains at low temperature or the formation of non-bilayer structures at high temperature. Both types of lipid phase separation are associated with functional changes in the membrane including loss of selective permeability barrier properties.

REFERENCES

[1] Quinn, P.J. (1981). The fluidity of cell membranes and its regulation. Progr. Biophys. Mol. Biol., 38, 1-104.

[2] Quinn, P.J. and Williams, W.P. (1985). Environmentally induced changes in chloroplast membranes and their effects on photosynthetic function. In: Photosynthetic Mechanisms and the Environment (Barber, J. and Baker, N.R., eds) pp 1-47, Elsevier, Amsterdam.

BIOLOGICAL ROLE OF PLANT LIPIDS
P.A. BIACS, K. GRUIZ, T. KREMMER (eds)
Akadémiai Kiadó, Budapest and Plenum Publishing
Corporation, New York and London, 1989

ARE EICOSAPOLYUNSATURATED FATTY ACIDS INVOLVED IN THE HYPERSENSITIVE RESPONSE IN POTATO?

M.N. Merzlyak, O.B. Chivkunova, I.V. Reshetnikova, N.I. Maximova and M.V. Gusev

Department of Cell Physiology and Immunology, Faculty of Biology, Moscow State University
119899 GSP Moscow W-234, USSR

The hypersensitive cell death providing resistance of plants in race-cultivar systems are based on a specific recognition of certain metabolites of pathogen microorganisms |1|. It was found that main part of elicitor activity of *Phytophthora infestans* is attributed to lipids and due to eicosatetra- and eicosapentaenoic fatty acids (EPUFA) |2-4|. However, these experiments were carried out with mycelial extracts and it is not clear whether the EPUFA are responsible for the events occuring *in vivo*. In this article we report the results of fatty acid analysis in the zoospores of *P. infestans* and the effect of modification their fatty acid composition on the compatibility of the fungus with potato plants.

Using g.l.c. (Silar-10 C, SP-2330) the fatty acid composition was determined in zoospores of several *P. infestans* races. The main fatty acids were $C_{16:0}$, $C_{18:1}$, $C_{18:2}$ and $C_{20:5}$ (Ref. |5,6|, Table 1) as in the mycelium but in somewhat different proportion. The relative contents of EPUFA in zoospores of different races varied from 10 to 30%. The total fatty acid concentration was in the range of 50-100 fmoles/zoospore depending on the race used and growth conditions. Thus the total quantity of EPUFA in the zoospore suspension required to induce hypersensetive reaction in potato is of the order of 50 to 100 pmoles. These values are 100-500 times lower than those of exogenic $C_{20:4}$ |cf. 3,4| inducing browning and phytoalexin accumulation in potato tuber discs. The ratio seems to be even lower on the assumption that only "surface" EPUFA pool are involved in recognition at early stages of interaction. Indeed t.l.c. analysis showed that a gross part of acyl-containing lipids is neutral lipids serving probably as energy source during germination and development of zoospore. According to microscopic observation, the main fluorescence of hydrophobic probe pyren was concentrated inside the cell, possibly in lipid bodies. These considerations suggest that if EPUFA directly or *via* an unidentified intermediate(s) induce plant cell death the substances have to operate at very low concentrations. However, recent data indicate that the elicitor activity of the exogenic EPUFA is enhanced by glucans obtained from *P. infestans* mycelia |3|.

The fatty acid composition was determined in three isolates of *P. infestans* having high aggressivity to potato leaves but losing the ability to colonize the plants after several passages on artificial medium. No correlation between total EPUFA content in zoospores and aggressivity of the fungus was found in these experiments |5|.

To obtain additional information about involvement of EPUFA in immune response of potato plants the hydrogenation of lipids in *P. infestans* zoospores was carried out using $Pd(QS)_2$ as a catalyst |7|. The procedure (Table 1) caused the decrease in the content of eicosapentaenoic acid to 8.4% and the appearance of almost equivalent quantity of arachidic acid. Simultaneously the conversion of $C_{18:2}$ into $C_{18:1}$ and further into $C_{18:0}$ took place. The incubation of zoospores under nitrogen induced some loss of unsaturated fatty acids. The germination of zoospores subjected to the treatment by the inert gases in the presence of $Pd(QS)_2$ was lowered approximately to 50%. However, the ability to colonize the aged potato tuber discs from compatible potato cultivars was retained. The application of zoospore suspensions on the discs obtained from the tubers of incompatible potato cultivars caused the necrotization and browning of the tissues independently on hydrogenation of zoospore lipids. Thus considerable reduction in the EPUFA and linoleic acid contents in zoospores with concomitant appearance of arachidic and production of increased quantities of saturated and monoenic acids of C_{18}-series that are not active as elicitors of the hypersensitive reaction |2| does not change

Table 1. Hydrogenation of fatty acids in P. infestans zoospores. Zoospore suspensions (3 mln/ml) were incubated with $Pd(QS)_2$ (200 ug/ml) for 1.5 h in air (Control) or at a pressure (1.2 MPa) of N_2 (Nitrogen) or H_2 (Hydrogen).

	Fatty acid content, mol%						
	$C_{16:0}$	$C_{16:1}$	$C_{18:0}$	$C_{18:1}$	$C_{18:2}$	$C_{20:0}$	$C_{20:5}$
Control	24.9	4.0	1.2	11.4	34.2	-	24.3
	28.9		46.8			24.3	
Nitrogen	39.7	6.1	1.3	6.0	27.4	-	19.5
	45.8		34.7			19.5	
Hydrogen	31.0	1.7	26.0	12.5	9.5	10.9	8.4
	32.7		48.0			19.5	

the immune response in potato plants to the infection. Nevertheless it should be stressed that EPUFA hydrogenation attained in our experiments was not complete (less than 65%). On the other hand it is possible to assume that with water-soluble catalyst used the hydrogenation occurs primary within surface lipids.

The data thus obtained indicate that at least an absolute level of total EPUFA in zoospores is not directly responsible for the induction of plant cell death and that the compatibility in *Phytophthora* - potato is under control of other factor(s). According to Doke et al. |1|, certain fungal metabolites in incompatible, but not in compatible, race - cultivar systems elicit production of superoxide anion on the surface of plant cell possibly related to hypersensitive reaction. Recently we found that incubation of *P. infestans* zoospores in the systems generating $O_2^{\bar{\cdot}}$ and other activated oxygen species causes the selective destruction of EPUFA |6|. Thus it is possible (in addition to the circumstances discussed in |1|) to propose that certain oxidized EPUFA products formed during incompatible interaction of the partners take part in the induction of cell death during hypersensitive response of potato plants.

REFERENCES

1. Doke N., Chai H.B., Kawaguchi A. 1987. Biochemical basis of triggering and suppression of hypersensitive cell response. In: *Molecular Determinants of Plant Diseases* (Ed. by S. Nishimura et al.) Japan Sci. Soc. Press, Tokyo/Springer-Verlag, Berlin, 235-251.
2. Bostock R.M., Kuć J.A., Laine R.A. 1981. Eicosapentaenoic and arachidonic acids from *Phytophthora infestans* elicit fungitoxic sesquiterpenes in the potato.- Science, *212*, 67-69.
3. Preisig C.L., Kuć J.A. 1985. Arachidonic acid related elicitors of the hypersensitive response in potato and enhancement of their activities by glucans from *Phytophthora infestans* (Mont.) deBary.- Arch. Biochem. Biophys., *236*, 379-389.
4. Ozeretskovskaya O.L., Avdjushko S.A., Chalova L.I., Chalenko G.I., Yzrganova L.A. Arachidonic and eicosapentaenoic acids as an active principle of the inductor from late blight of potato agent.- Dokl. AN SSSR (Rus.), 1987, *292*, 738-741.
5. Golovkin K.A., Chivkunova O.B., Rybakova I.N., Merzlyak M.N., Djakov Yu.T. 1989. The changes of cytomorphological, physiological and biochemical properties of *Phytophthora infestans* during growth on artificial media.- Mycol. Phytopathol. (Rus.), in press.
6. Reshetnikova I.V., Chivkunova O.B., Maximova N.I., Merzlyak M.N. 1988. Superoxide-mediated oxidative destruction of unsaturated fatty acids in zoospores of *Phytophthora infestans*. In: *Oxygen Radicals in Chemistry, Biology and Medicine* (Ed by I.B. Afanas'ev, N.I. Goldstein and M.N. Merzlyak), Riga Med. Inst., Riga, 266-271 (Rus.).
7. Horváth G., Droppa M., Szito T., Mustardy K., Horváth L.I., Vigh L. Homogeneous catalytic hydrogenation in the photosynthetic membrane: effects on membrane structure and photosynthetic activity.- Biochim. Biophys. Acta, 1986, *349*, 325-336.

BIOLOGICAL ROLE OF PLANT LIPIDS
P.A. BIACS, K. GRUIZ, T. KREMMER (eds)
Akadémiai Kiadó, Budapest and Plenum Publishing
Corporation, New York and London, 1989

EFFECTS OF WATER STRESS ON THE LIPID METABOLISM OF DURUM WHEAT

A. Kameli and Dorothy M. Lösel

Department of Plant Sciences, University of Sheffield, S10 2TN, UK

INTRODUCTION

Changes in lipid content and composition have frequently been observed in tissues of plants subjected to environmental stress (Harwood, 1974; Kuiper, 1985) but the physiological significance of such responses remains controversial. In the present study, the effects of water stress on lipid metabolism were compared in two varieties of *Triticum durum*, Capdur (from the National Institute of Agricultural Research, Cambridge) which, in preliminary screening of a range of wheat varieties, showed low drought tolerance and Mohamed Ben Bachir (from the Institut Nationale des Grand Cultures, Algiers), which was more resistant to water stress.

METHODS

Plants grown at 20°C with 16 h light period were exposed to water stress by a rapid method which affects growth in a comparable manner to conventional drought induction (Raynal *et al.*, 1985). Seedlings supported on lids of culture containers were grown for three weeks with their roots suspended in mineral nutrient solution which was replaced every three days. Stress was then applied to half of the plants by transferring lids bearing sets of 5 plants to similar, empty containers for increasing lengths of time (2, 4, 6, 8 h) on successive days then returning them to the nutrient solution for the remainder of each 24 h cycle. Water potentials of leaves were measured at the end of the final stress cycle, using a pressure chamber.

DG, TG = di-, tri-acylglycerols; PC, PE, PG, PI = phosphatidyl-choline, -ethanolamine, -glycerol, -inositol; NL, PL = neutral, polar lipids; S = sterol; SE = sterol esters; MGDG, DGDG = mono-, di-galactosyl diacylglycerols; SL = sulpholipid.

remainder of each 24 h cycle. Water potentials of leaves from each set of plants were measured at the end of the final stress cycle, using a pressure chamber.

Samples of leaves of corresponding age, harvested from replicate sets of stressed and control plants, were weighed, extracted with boiling isopropanol, followed by chloroform:methanol (2:1 then 1:2, v/v) and washed by the procedure of Folch et al. (1957) to separate the lipid fraction. Thin layer chromatography and estimations of lipids were carried out as described previously (Lösel and Lewis, 1974). In studies of incorporation of ^{14}C-labelled photosynthate into lipids, 0.5 g samples of leaves were laid on damp filter paper in a transparent plastic box, allowed to photosynthesize in $^{14}CO_2$ for 30 minutes and for a chase period of 1h in normal air before extraction, TLC separation of lipid classes and scintillation counting.

RESULTS

Table 1. Effects of water-stress on lipid content of wheat leaves (means of 4 replicates ± standard error).

	MBB			Capdur		
	Stress		Control	Stress		Control
Water potential (MPa)	-2.2		-1.0	-2.3		-0.8
Total lipid (mg.g^{-1})	14.7±0.67	**	10.1±0.38	18.6±0.52	*	10.2±2.4
Neutral lipid (mg.g^{-1})						
TG	0.6±0.04	**	0.2±0.04	0.4±0.03	*	0.2±0.02
DG	0.4±0.04	*	0.2±0.04	0.3±0.02	**	0.2±0.03
SE	0.3±0.04	*	0.2±0.03	0.2±0.04	*	0.1±0.02
S	0.2±0.04	*	0.3±0.03	0.1±0.03		0.2±0.05
Phospholipid (μg P.g^{-1})						
PC	41.3±1.99	**	29.0±1.31	33.1±2.26		36.8±0.84
PI	6.5±0.47		4.7±0.42	6.6±0.37	**	4.5±0.52
PE + PG	13.9±0.98	*	11.0±0.82	13.1±0.72	*	16.7±1.02
Total	76.4±0.98	**	57.9±4.59	60.8±8.25	*	72.0±7.52

*, ** indicate significance of differences at 0.05 and 0.01 levels

The effects of water stress on the amounts of individual lipid classes in leaves of stressed and unstressed plants are indicated in Table 1. In both varieties, stress increased the total lipid content. Among the neutral lipids, greater amounts of DG and SE were noted but the major increase was in TG in MBB, in DG in Capdur. Sterols increased in MBB but not in Capdur. The total phospholipid content increased with stress in both but more in MBB, where amounts of PC were much higher.

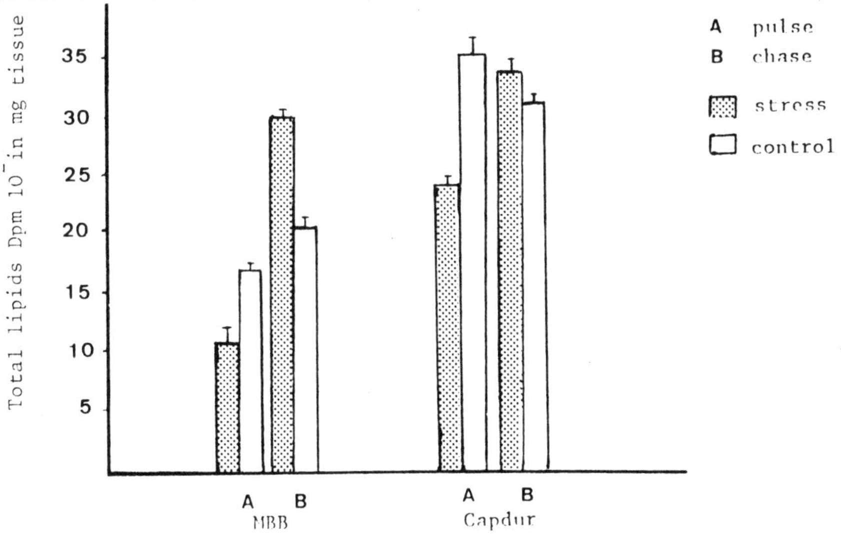

Fig. 1. Radioactivity of total lipids from independent sets of plants: A, after 30 min photosynthesis in $^{14}CO_2$; B, after chase period of 1h.

The incorporation of ^{14}C-labelled photosynthate into lipid fractions was investigated in stressed and unstressed plants of both varieties (Fig. 1). Both immediately after photosynthesis in labelled CO_2 and following a chase period of 1 hour, the radioactivity of lipid fractions was significantly lower in leaves from stressed plants of MBB than in control leaves but higher in stressed leaves of Capdur. During the chase period, the total lipid radioactivity doubled in both stressed and control leaves of MBB, while smaller but significant increases occurred in leaves of Capdur from both treatments.

Investigation of the distribution of radioactivity among individual lipid classes showed that in MBB, a significantly higher percentage of the total lipid label was incorporated in neutral lipids of stressed than of control leaves, whereas no significant effect of water stress on incorporation of radioactivity by neutral lipids was detected in Capdur. Within the neutral lipid classes of MBB, the percentage of lipid-^{14}C incorporated into DG, TG and hydrocarbons was significantly increased by water stress, which did not significantly affect the proportions of radioactivity in these lipids in the variety Capdur. The percentage incorporation of radioactivity by polar lipids was significantly decreased by water stress in MBB but no significant change was detected in the proportion of labelled photosynthate present in polar lipids of Capdur leaves following stress treatment.

Table 2. Incorporation of ^{14}C-labelled photosynthate by neutral and polar lipids of wheat leaves as percentages of total lipid dpm (means of 4 replicates, standard errors given in brackets).

	Percentage of total lipid radioactivity				
	DG	TG	Hydrocarbon	NL	PL
MBB					
Stress	6.0	2.1*	3.5*	11.6	88.4*
	(0.75)	(0.20)	(0.98)	(1.73)	(1.73)
Control	3.8	1.3	2.1	7.2	92.8
	(0.44)	(0.23)	(0.76)	(1.05)	(0.91)
Capdur					
Stress	3.6	1.5	2.0	7.1	93.0
	(0.26)	(0.08)	(0.54)	(0.72)	(0.72)
	3.3	0.8	1.0	5.1	95.0
	(0.95)	(0.24)	(0.32)	(1.44)	(1.40)

Within the polar lipid fraction, no consistent stress-induced changes in the ^{14}C content of individual phospholipid classes of either variety of wheat were detected in this experiment. Amongglycolipids, however, the radioactivity of MGDG increased significantly in both varieties following water stress (Table 3).

Table 3. Distribution of label among glycolipids after photosynthesis in $^{14}CO_2$. Means of 4 replicates, standard errors given in brackets.

		Percentage of total polar lipid dpm MGDG	DGDG	SL
MBB	Stress	21.7 ± 1.82**	17.7 ± 1.33	6.8 ± 0.63
	Control	14.5 ± 0.99	20.6 ± 1.30	6.5 ± 1.12
Capdur	Stress	19.6 ± 1.13	18.2 ± 2.08	5.0 ± 0.36
	Control	13.7 ± 0.85	16.2 ± 3.42	6.4 ± 0.19

DISCUSSION

Although lipid accumulates in both wheat varieties during water stress these observations, unlike those of Pham Thi *et al.* (1985) on cotton varieties of differing drought-resistance, support the hypothesis that resistant plants respond to water stress by more marked changes in lipid synthesis than those found in drought-sensitive varieties.

Financial support of the Ministry of Higher Education of Algeria for A. Kameli is gratefully acknowledged.

REFERENCES

Folch, J., Lees, M. and Sloane-Stanley, G.H. (1957) J. biol. Chem. 226, 497-509.

Harwood, J.L. (1984) Effects of the environment on the acyl lipids of algae and higher plants. *In* Structure, function and metabolism of plant lipids (Eds. P.-A. Siegenthaler and W. Eichenberger), pp. 543-550. Elsevier, Amsterdam.

Kuiper, P.J.C. (1984) Environmental changes and lipid metabolism of higher plants. Physiol. Plant. 64, 118-122.

Lösel, D.M. and Lewis, D.H. (1974) Lipid metabolism in leaves of *Tussilago farfara* during infection by *Puccinia poarum*. New Phytol. 73 1157-1169.

Pham Thi, A.T., Borrel-Flood, C., Da Silva, J.V., Justin, A.M. and Mazliak, P. (1985) Effects of water stress on lipid metabolism in cotton leaves. Phytochemistry 24, 723-727.

Raynal, D.J., Grime, J.P. and Boot, R. (1985) A new method for the experimental droughting of plants. Ann. Bot. 55, 893-897.

BIOLOGICAL ROLE OF PLANT LIPIDS
P.A. BIACS, K. GRUIZ, T. KREMMER (eds)
Akadémiai Kiadó, Budapest and Plenum Publishing
Corporation, New York and London, 1989

ENZYMATIC BREAKDOWN OF POLAR LIPIDS IN COTTON LEAVES UNDER WATER STRESS

A.T. Pham Thi, L. El-Hafid, Y. Zuily-Fodil and J. Vieira da Silva

Laboratoire d'Ecologie Générale et Appliquée, Université Paris VII
2 Place Jussieu, 75251 Paris Cedex 05, France

Water stress provokes a decrease in leaf polar lipid content (Pham Thi et al.1982).This decrease is due to an inhibition of the biosynthesis from ^{14}C-acetate (Pham Thi et al.1985), but certainly also to an acceleration of degradative phenomena. The enzymes responsible for polar lipid catabolism in plants have received little attention (Galliard et al.1974). This paper describes the enzymatic systems acting on MGDG and PC breakdown and their evolution under water stress .

Materials and methods

Plant material : Cotton, Gossypium hirsutum L. cv.Mocosinho are grown in controlled conditions. Plants are subjected to water stress by withholding irrigation .

Measurement of lipolytic activities : Leaves are homogenized in Tris buffer containing protein protectors. After filtration through Miracloth, the leaf homogenate is centrifuged at 100.000g for 1h. The supernatant contains the soluble enzymes. The membranous enzymes of the pellet are rendered soluble in Tris buffer containing 1% Triton X-100. Radioactive products formed by the degradation of ^{14}C- MGDG and ^{14}C-PC by enzyme extracts are analysed by TLC on silica gel plates. After radioautography, the bands are scraped off and their radioactivity is counted .

Results

Degradation of MGDG - Autoradiography and enzyme activities calculated from radioactivity countings (Fig.1) show that the degradation of MGDG leads to the formation of free fatty acids (FFA), monoacylglycerols (MG), and to a

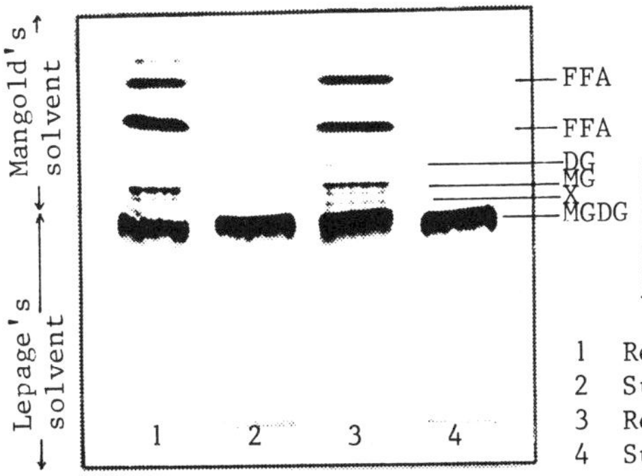

Fig.1- Products of MGDG degradation ($mg.gdw^{-1}.h^{-1}$)

	X(unknown)	MG	DG	FFA
S	0.256	0.933	0.164	4.355
M	0.355	0.618	0.171	7.648

1 Reaction with membranous enzymes (M)
2 Substrate + boiled membranous enzymes
3 Reaction with soluble enzymes (S)
4 Substrate + boiled soluble enzymes

lesser extent, to diacylglycerols (DG). These results indicate the presence of at least 2 types of MGDG-lipolytic enzymes : Those which remove the acyl groups from the polar lipid and those which attack the molecule at the bond between glycerol and galactose. The former are probably the lipolytic-acyl-hydrolases described by Galliard et al.(1974), the latter is a galactosyl hydrolase. Results also show that these two types of enzymes occur in the soluble as well as in the membranous fractions.
The presumed reactions are as follows :

(1) MGDG $\xrightarrow{LAH}$ MGMG + FFA $\xrightarrow{LAH}$ MGG + FFA

(2) MGDG $\xrightarrow{galactosyl\ hydrolase}$ DG + Galactose $\longrightarrow$ MG + FFA

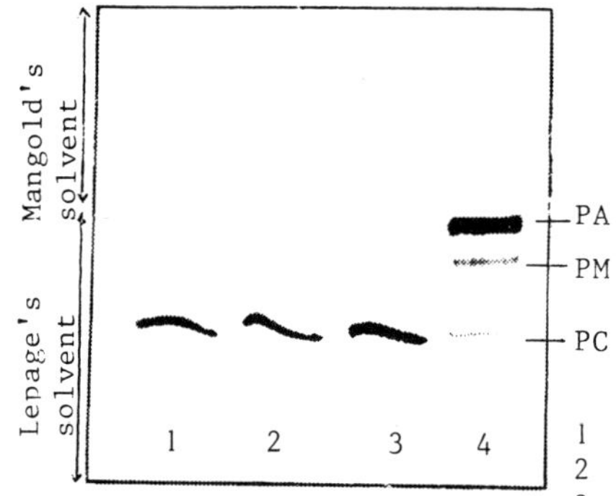

Fig.2 Products of PC degradation ($mg.gdw^{-1}.h^{-1}$)

	PM *	PA
S	–	–
M	2.385	21.615

* Phosphatidylmethanol is an artefact due to the reactivity, in vitro, between methanol and PA

1 Substrate + boiled soluble enzyme
2 Reaction with soluble enzyme (S)
3 Substrate + boiled membranous enzyme
4 Reaction with membranous enzyme (M)

Degradation of PC - Autoradiography and enzyme activities calculated from radioactivity countings (Fig.2) indicate the presence in Cotton leaves , of a unique PC-hydrolysing system , the phospholipase D, which is attached to the membranes and leads to the liberation of phosphatidic acid (PA) .

Effects of water stress

Fig.3 shows that under water stress, the enzymatic degradation of MGDG and PC increases, specially when water potentials fall down beyond -1.3 to -1.5 MPa. This increase is undoubtly responsible of the decrease in Cotton

Fig.3 Effect of water stress on MGDG and PC degradation

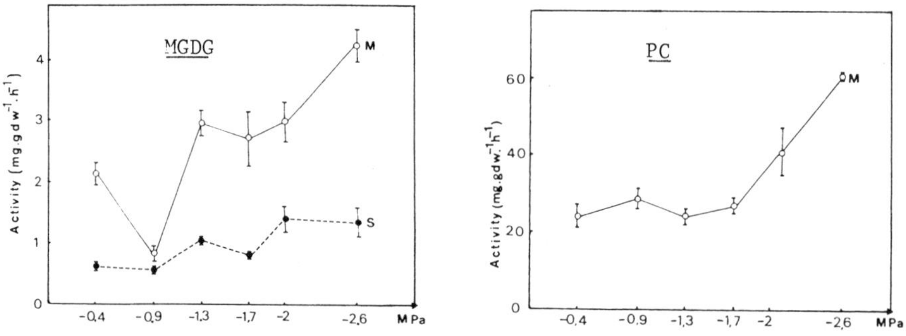

leaf polar lipid content (Pham Thi et al.1982) . It indicates an accelerated degradation of the cell membranes .

References

Galliard T., Dennis S.(1974)- Phytochem.13 : 1731-1735

Pham Thi A.T., Flood C., Vieira da Silva J.(1982)- in "Biochemistry and Metabolism of Plant Lipids" (Wintermans JFGM and Kuiper PJC eds) pp 451-454 , Elsevier, Amsterdam

Pham Thi A.T., Flood C.,Vieira da Silva J.,Justin A.M., Mazliak P. (1985)- Phytochem. 24 : 723-727 .

BIOLOGICAL ROLE OF PLANT LIPIDS
P.A. BIACS, K. GRUIZ, T. KREMMER (eds)
Akadémiai Kiadó, Budapest and Plenum Publishing
Corporation, New York and London, 1989

EFFECTS OF WATER STRESS ON MOLECULAR SPECIES COMPOSITION OF POLAR LIPIDS FROM VIGNA UNGUICULATA LEAVES

A.T. Pham Thi[1], F. Monteiro de Paula[1], G. Herbert[1], A.M. Justin[2], C. Demandre[2] and P. Mazliak[2]

[1]Laboratoire d'Ecologie Générale et Appliquée, Université Paris VII, 2 Place Jussieu, 75251 Paris
[2]Laboratoire de Physiologie Cullulaire, Université P. et M. Curie, 4 Place Jussieu, 75230 Paris Cedex 05, France

Under water stress, polar lipid composition of leaves undergoes noticeable changes (Pham Thi et al.1982). On the other hand, recent works (Murata 1983) have demonstrated the existence of a relationship between molecular species composition of plant lipids, particularly of PG, and chilling sensitivity. In this paper, we examine the influence of water deficits on the molecular species composition of the main polar lipids from two varieties of Cowpea .

Materials and methods

Vigna unguiculata L., a drought-resistant (cv.Epace), and a drought-sensitive (cv.IT 83D), were grown in controlled conditions. When plants were 3 weeks old, they were subjected to water stress by withholding irrigation until the leaf potentials reached -2.0 MPa. Leaf lipids were extracted in chloroform/methanol. The main polar lipids were purified by HPLC on a silicic column, and molecular species were separated by HPLC using a C_{18}column (Demandre et al. 1985).

Results (Tab.1)

In the two Vigna varieties, water stress provokes a decrease in the percentage of 18:3/18:3 MGDG and a drastic decrease in all 16:1trans-containing molecular species of PG . These variations could be related to changes in the properties of the thylakoid membranes and therefore in the photosynthetic activity of the leaves. Molecular species composition of DGDG is affected by water stress in a different manner in the two varieties : in the drought-resistant variety, the unsaturation level increases, due to a greater proportion of 18:3/18:3, while in the drought-susceptible variety, it decreases . In PC, the percentage of 18:3/18:3 increases, in a greater proportin in Epace than in IT 83D .

Tab.1 - Effects of water stress on the molecular species composition of polar lipids from Vigna leaves. C=well-hydrated plants ; S=water-stressed plants

A. MGDG

		18:3 18:3	18:2 18:3	18:2 18:2	18:0 18:3	16:0 18:3
Epace	C	95.4	2.0	0.3	0.2	2.1
	S	90.3	6.7	0.4	-	2.6
IT 83D	C	90.9	3.0	0.6	1.3	4.2
	S	85.1	2.8	1.0	3.6	7.5

B. DGDG

		18:3 18:3	18:2 18:3	18:2 18:2	18:0 18:3	16:0 18:3	16:0 18:2
Epace	C	71.9	1.2	0.8	1.3	24.7	-
	S	76.4	1.1	0.4	1.7	17.0	3.4
IT 83D	C	66.1	3.6	0.6	3.8	25.3	0.5
	S	62.4	7.7	1.5	6.1	16.7	5.6

C. PG

		18:2 18:2	18:3 16:1t	18:2 16:1t	18:3 16:0	18:0 16:1t	18:1 16:0	16:0 16:1t	16:0 16:0
Epace	C	1.4	19.4	4.6	7.7	10.2	5.2	39.4	12.2
	S	-	5.8	6.3	12.1	3.0	18.9	12.1	41.8
IT 83D	C	2.3	13.4		10.8		6.7	39.0	27.8
	S	4.5	8.6		12.5		18.7	22.6	33.1

D. PC

		18:3 18:3	18:2 18:3	18:2 18:2	18:0 18:3	18:1 18:2	16:0 18:3	16:0 18:2
Epace	C	6.1	17.7	14.9	3.1	6.7	24.5	27.0
	S	18.8	11.7	12.3	4.0	-	35.8	17.4
IT 83D	C	13.4	12.4	3.3	16.1	5.1	40.7	9.0
	S	18.4	15.8	2.7	10.9	1.6	40.4	10.1

References

Demandre C., Trémolières A., Justin A.M., Mazliak P. (1985)- Analysis of molecular species of plant polar lipids by high performance and gas liquid chromatography . Phytochem. 24 : 481-485

Murata N. (1983)- Molecular species composition of phosphatidylglycerols from chilling-sensitive and chilling-resistant plants . Plant Cell. Physiol. 24 : 81-86

Pham Thi A.T., Flood C., Vieira da Silva J. (1982)- Effects of water stress on lipid and fatty acid composition of Cotton leaves . in Biochemistry and Metabolism of Plant Lipids (Wintermans JFGM and Kuiper PJC eds.) , pp. 451-454 , Elsevier , Amsterdam .

BIOLOGICAL ROLE OF PLANT LIPIDS
P.A. BIACS, K. GRUIZ, T. KREMMER (eds)
Akadémiai Kiadó, Budapest and Plenum Publishing
Corporation, New York and London, 1989

STRATEGIES OF MODIFICATION OF MEMBRANE FLUIDITY BY CATALYTIC HYDROGENATION

F. Joó[1], L. Vígh[2]

[1]Institute of Physical Chemistry, Kossuth Lajos University, Debrecen 4010, P.O. Box 7, Hungary
[2]Institute of Biochemistry, Biological Research Center of the Hungarian Academy of Sciences, Szeged 6701, P.O. Box 521, Hungary

INTRODUCTION

Catalytic hydrogenation of unsaturated fatty acids esterified in lipids of the most diverse membrane systems is a well known technique of lipid chemistry. However, the usual heterogeneous catalysts of hydrogenation cannot be applied for efficient modification of membranes, since the possibility of reaching the catalyst surface by the acyl moieties, buried into the membrane is rather limited. On the other hand, there is a vast number of soluble transition metal complexes which are capable of activation of molecular hydrogen and catalyze the reduction of not only of olefinic but -C≡C- , =C=O , =C=N-, etc., functions as well. However, there are strict requirements to be met by a homogeneous catalyst intended for use in biomembrane hydrogenations, concerning biocompatibility, high specific activity under physiological conditions, selectivity, stability, and the possibility of removal after completion of the reaction.

It is well established by now that *in vivo* or *in vitro* hydrogenation of unsaturated fatty acid residues of lipids can be carried out using complexes of platinum metal ions (mainly rhodium, ruthenium and palladium) containing as ligands various water soluble phosphines [1], or Alizarin Red (sodium 1,2-dihydroxyanthraquinone-3-sulfonate, QS), and this method proved to be a fruitful way for investigating the connection between membrane fluidity and the various properties and functions of the membranes, or the intact cells[2-4]. It has been shown, as well, that hydrogenation of pea chloroplast resulted exclusively in reduction of unsaturated fatty acids, leaving chlorophylls, carotenoids and plastoquinone uneffected[5]. However, the question of selectivity should be considered in a sense broader that chemoselectivity (including the different activity of the catalyst towards _cis_-, and _trans_-isomers of olefinic substrates). We have already demonstrated [6], that with a careful choice of reaction time cytoplasmic membranes can be extensively hydrogenated without the saturation of the C=C double bonds in the thylakoid membranes of the blue-green alga, _Anacystis nidulans_. Site-selectivity, however, remains one of the most intriguing problems of wholecell hydrogenations, one of the important aspects being the preparation of new catalysts, the presence of which can be simply detected inside the different parts of the cell, _e.g._ by fluorescence microscopy. A different approach involves binding of active hydrogenation catalysts to insoluble or soluble macromolecules, hindering this way their penetration into the inside areas of the cell. Functionalized macromolecules can themselves be ligands of transition metal complexes acting as hydrogenation catalysts [7]. In addition to the questions of selectivity, other parameters should also be considered. Though the most widely used catalyst, $Pd(QS)_2$ is outstandingly active even at 3 ^{o}C and under atmospheric pressure of H_2, the use of H-donors, other than molecular hydrogen is clearly desirable.

MEMBRANE HYDROGENATIONS BY PHOTOCHEMICALLY-DRIVEN, METAL COMPLEX CATALYZED HYDROGEN TRANSFER

It has been reported that a solution of $RuCl_2(bipy)_3$ and $RhCl(mSPØ_2)_3$ (bipy: 2,2'-bipyridyl, $mSPØ_2$: mono-sulfonated triphenylphosphine) in aqueous ascorbate buffer (pH= 3-7), generates molecular hydrogen upon irradiation by visible light [8]. We have found that the same system is capable of hydrogenating lipid dispersions under ambient conditions. For example, when a sonicated dispersion of dioleoylphosphatidylcholine (DOPC, 0.2 mg/ml) in aqueous ascorbate buffer (pH = 5,

0.1 M) was irradiated in the presence of $RuCl_2(bipy)_3$ and $RhCl(mSPØ_2)_3$ (0.2 mM, each) under nitrogen at 35 °C, 45 % conversion of oleate (18:1) to stearate (18:0) was observed in 30 min. Under comparable conditions, egg yolk lecithin reacted much faster than DOPC (80 % conversion in 15 min) while the lipids in cod liver oil dispersions (prepared by co-sonication with dipalmitoylphosphatidylcholine) could be hydrogenated only to 30 % conversion in 120 min. Attempts to replace ascorbate by other H-donor compounds (reducing sugars, EDTA, amines, and propan-2-ol) failed. A most important result of these studies is in that the membranes of <u>Tetrahymena pyriformis</u> could also be hydrogenated this way; the amount of unsaturated fatty acids in total lipids decreased by 20 % upon 60 min irradiation.

In contrast to the Ru + Rh bimetallic system described above, $K_4[Pt_2(H_2P_2O_5)_4]$, potassium tetrakis (μ-pyrophosphito)-diplatinate(II) is itself capable for playing the role of both the photosensitizer and the hydrogen transferring species. However, in this case only propan-2-ol could be used as H-donor. Oleate in DOPC, dispersed in phosphate buffer of pH = 5, containing 1 % propan-2-ol, was hydrogenated to stearate with 13 % conversion upon 30 min irradiation.

Both $RuCl_2(bipy)_3$, and $K_4[Pt_2(H_2P_2O_5)_4]$ are strongly fluorescent allowing detection by fluorescence microscopy.

MEMBRANE HYDROGENATIONS BY MACROMOLECULAR CATALYSTS

Preliminary experiments were done with high molecular weight complexes of palladium(II) with polyvinylpyrrolidinone, Pd-PVP. It was found that after a careful preptreatment of a solution of Pd-PVP, prepared from PVP of M.W. 24000, oleate in DOPC was hydrogenated to stearate with 14 % conversion, at room temperature and atmospheric hydrogen pressure (solvent: 0.1 M phosphate buffer of pH = 7.5). The catalyst system is rather a stabilized colloid of palladium metal or of high nuclearity Pd-clusters, than a homogeneous solution. Experiments to check the permeability of the catalyst are underway.

Acknowledgement: This work was supported by the Hungarian National Scientific Research Foundation (OTKA 543/86 to L.V.).

References

1. F. Joó, and Z. Tóth: Catalysis by water-soluble phosphine complexes of transition metal ions in aqueous and two-phase media, J. Molecular Catalysis 8 (1980) 369-383
2. L. Vígh, I. Horváth, F. Joó, and G.A. Thompson, Jr.: The hydrogenation of phospholipid-bound unsaturated fatty acids by a homogeneous, water-soluble, palladium catalyst, Biochim.Biophys.Acta 921 (1987) 167-174
3. I. Horváth, A.R. Mansourian, L. Vígh, P.G. Thomas, F. Joó, and P.J. Quinn: Homogeneous catalytic hydrogenation of the polar lipids of pea chloroplasts in situ and the effects on lipid polymorphism, Chem.Phys.Lipids 39 (1986) 251-264
4. Z. Gombos, K. Barabás, F. Joó and L. Vígh: Lipid saturation induced microviscosity increase has no effect on the reducibility of flash-oxidized Cytochrome f in pea thylakoids, Plant.Physiol. 86 (1988) 335-337
5. B. Szalontai, M. Droppa, L. Vígh, F. Joó, and G. Horváth: Selectivity of homogeneous catalytic hydrogenation in saturation of double bonds of lipids in chloroplast lamellae, Photobiochem.Photobiophys. 10 (1986) 233-240
6. L. Vígh, Z. Gombos, and F. Joó: Selective modification of cytoplasmic membrane fluidity by catalytic hydrogenation provides evidence on its primary role in chilling susceptibility of Anacystis nidulans, FEBS Lett. 191 (1985) 200-204
7. E. Bayer, and W. Schumann: Liquid phase polymer-based catalysis for stereo- and regio-selective hydrogenation, J.Chem.Soc.Chem.Commun. (1986) 949-952
8. S. Oishi: A water-soluble Wilkinson's complex as homogeneous catalyst for the photochemical reduction of water, J.Molecular Catalysis 39 (1987) 225-232

BIOLOGICAL ROLE OF PLANT LIPIDS
P.A. BIACS, K. GRUIZ, T. KREMMER (eds)
Akadémiai Kiadó, Budapest and Plenum Publishing
Corporation, New York and London, 1989

SODIUM CHLORIDE EFFECT ON GLYCEROLIPIDS, CHANGES DURING RIPENING OF OLIVE

B. Marzouk, M. Zarrouk and A. Cherif

Centre de Biologie et de Ressources Génétiques. I.N.R.S.T. Borj-Cedria 2050 Hamman Lif, Tunisie

INTRODUCTION

Several works have been made about salt stress effects on Higher Plants and particularly on their biochemical and physiological functions. On the other hand, very few studies concerned with the salinity effect on the olive tree physiology. Indeed, El Amami (1975) remarked that oil production didn't affect when olive trees (var. Chetoui) were irrigated with water containing 3 grams per liter of dry residue.
Bouaziz (1976) obtained the same result after irrigation of olive trees (var. Chemlali) using brackish water ($4gl^{-1}$ of dry residue).
Excepted these essays, we know of no other studies concerning the sodium chloride action on lipid metabolism in olive fruit. In the present paper, we are investigating the influence of sodium chloride contained in irrigation water on oil content, fatty acid composition and amounts during the olive fruit ripening because the more interesting problem is the valorization of the important brackish water reserves for irrigation in olive tree cultivation.

MATERIALS AND METHODS

One year-old-plants were separated in four homogeneous groups and cultivated on sand in tanks. Each group was irrigated weekly using nutrient solution of various sodium chloride concentrations: 0 , 50 , 100 and 150mM. Olive plants fruited after four years and olive fruits were harvested at different periods during the ripening cycle. Lipid extraction and separation and methyl esters analysis have been described in a previous paper (Marzouk et Cherif, 1981).

RESULTS

Sodium chloride effect on glycerolipid composition in fully ripe olive fruits.

In fully ripe olive fruits, we observed that sodium chloride modified oil fatty acids composition. It provoked an oleic acid percentage dropping when NaCl concentration increased in culture medium first in favour of linoleic acid percentage and accessorily to benefit of palmitic and stearic acids.

Table 1 : Sodium chloride effect on fatty acid composition in fully olive fruits.

mM NaCl	% of total fatty acids					
	$C_{16:0}$	$C_{16:1}$	$C_{18:0}$	$C_{18:1}$	$C_{18:2}$	$C_{18:3}$
0	12,2	1,1	3,3	65,6	15,4	2,2
50	13,2	1,8	3,6	57,9	21,7	1,7
100	12,4	0,8	6,9	54,1	23,9	1,6
150	15,3	1,2	5,8	49,8	25,6	2,0

Then, the saturated fatty acid percentage accused an increase at150mM NaCl concentration. Similar changes have been found in triacylgycerols fatty acid composition (Table 2 because triacyglgycerols (TAG) in olive fruits constitute about 96% of total glycerolipids (Marzouk et Cherif, 1981).

Table 2 : Sodium chloride effect on triacylglycerols fatty acid composition in fully ripe olive fruits :

mM NaCl	% of triacylglycerols total fatty acids					
	$C_{16:0}$	$C_{16:1}$	$C_{18:0}$	$C_{18:1}$	$C_{18:2}$	$C_{18:3}$
0	13,4	0,5	2,6	66,9	16,5	Tr
50	15,3	0,6	2,5	58,8	22,8	Tr
100	14,2	0,6	5,2	54,0	25,8	Tr
150	16,6	0,5	5,2	51,1	26,5	Tr

The diacyglycerols fatty acid composition changed in the same way as total fatty acids and TAG fatty acids because they are considered as metobolic intermediate in TAG biosynthesis pathway (Slack and Roughan, 1978). Concerning polar lipids, sodium chloride affected linolenic acid composition only leading to a percentage increase.

Effect of sodium chloride on lipid contents in fully ripe olive fruits.

The lipid quantitative changes expressed in milligrams per gram of fresh matter showed a regular decrease content of total fatty acids (TEA), triacylglycerols (TAG) and oleic acid whereas linoleic and palmitic acids contents remaimed almost unchanged under sodium chloride effect (Fig 1). So, NaCl provoked a partial inhibition of oleic acid biosynthesis.

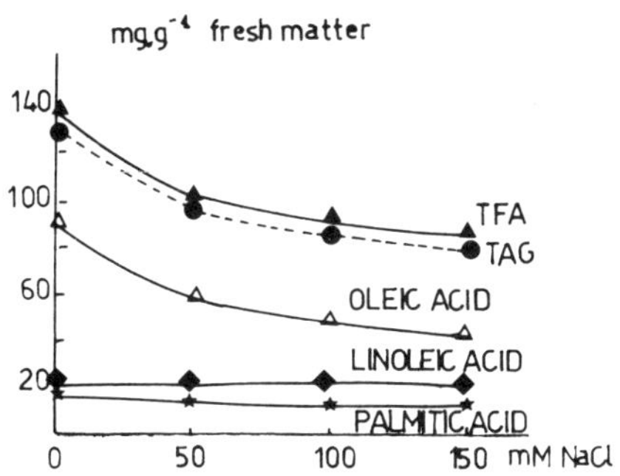

Fig 1 : Sodium chloride effect on TFA,TAG and main fatty acid contents in fully ripe olive fruit

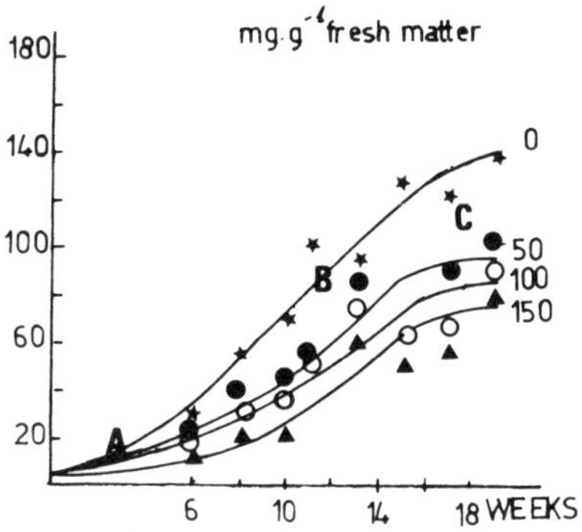

Fig 2 : Sodium chloride on TFA con - tents during olive fruit ripening.

Sodium chloride effect on the total fatty acid contents during olive fruit ripening.

Glycerolipids fatty acids analysis in olive fruit during ripening stages permitted to notice that lipidic accumulation curves have the same aspect as well for the control as for the treated samples (Fig 2) and lipids accumulated in three phases :

a slow lipid biosynthesis phase in newly formed olive fruits in region A of curves.

a rapid biosynthesis phase in region B.

a mature phase in region C without appreciable further lipid accumulation.

All the curves slope up decreased regularly with sodium chloride concentra tions increase ; in addition, at the ripening term, these curves reached a landing more rapidily for the treated samples. This indicated that the salt caused a fatty acid biosynthesis decrease and a lower accumulation of fats in the treated olive fruits.

Changes in fatty acid composition in olive fruit under saline conditions.

The evolution of principal fatty acids during olive fruit ripening showed a rapid increase of the oleic acid percentage at the beginning of maturation to reach a landing at the 8th week. Its percentage remainded stable for the control whereas it accused a regular and pronounced decrease when the NaCl concentration became important. Concerning palmitic and linoleic acids, their percentage decreased until the 8th week and after that, the palmitic one continued to decrease when the linoleic one increase until the end of maturation. (Fig 3)

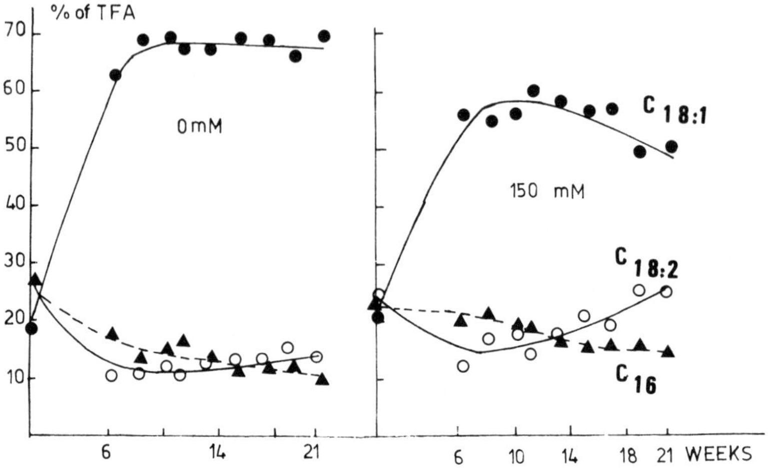

Fig 3 : Sodium chloride effect on fatty acid composition during olive fruit ripening.

CONCLUSION

The obtained results showed that total fatty acids, triacylglycerols and oleic acid accumulation during olive fruit ripening decreased when NaCl concentrations augmented in irrigation water. This oleic acid synthesis decrease could be explained by a photosynthetic acitivity inhibition because the chloroplast is an important site of oleic acid production in plant cells (Stumpf and al. , 1980).

REFERENCES

Bouaziz E. (1976) Comportement le l'Olivier var. Chemlali irrigué à l'eau saumâtre dans les conditions arides de la Tunisie Centrale. Les Cahiers du C.R.G.R,Tunis , 20 , 21p.

El Amami S.(1975) Comportement de l'Olivier irrigué à l'eau douce et à l'eau salée. Les Cahiers du C.R.G.R, Tunis, 11, 5p

Marzouk B. et Cherif A. (1981) La lipogenèse dans l'olive : I. Formation des lipides neutres. Oleagineux, 36, N° 2, 77-82.

Slack C.R. and Roughan P.G. (1978)- Rapid temperature-induced changes in the fatty acid composition of certains lipids in developing linseed and soybean cotyledons. Biochem.J., 170 , 437-439.

Stumpf P.K., D.N.Kuhn, D.J. Murphy, M.R. Pollard,T.Mckeon and J. McCarty. Oleic acid, the central substrate, p-3 in Biogenesiss and function of plant lipids. Mazliak P., Benveniste P., Costes C., Ed. Elsevier N. Holland (1980).

BIOLOGICAL ROLE OF PLANT LIPIDS
P.A. BIACS, K. GRUIZ, T. KREMMER (eds)
Akadémiai Kiadó, Budapest and Plenum Publishing
Corporation, New York and London, 1989

EFFECT OF SALT ON LIPID RESERVES OF COTTON SEEDS

A. Smaoui and A. Cherif

Centre de Biologie et de Ressources Génétiques, I.N.R.S.T., B.P. 95, 2050 Hamman-Lif, Tunisie

INTRODUCTION

In our country, Cotton is irrigated in some areas with relatively brackish water. Seeds, secondary crop, have high fat and protein content, and have to be worth valorizing. At this viewpoint, we have studied sodium chloride action on total fatty acids during different seed ripening stages.

MATERIAL AND METHODS

Four sets of <u>Gossypium hirsutum</u> L. (Cv S_4) plants were grown in pots at controlled conditions. Plants were irrigated with tap water containing added sodium chloride (3, 6 and 9g/l), the control set with tap water only. Seeds were harvested at different stages after anthesis. Seeds total lipids were extracted and analyzed by gas chromatography.

RESULTS

<u>Total lipid composition of control mature seed</u>

Composition of seed fatty acids showed that linoleic acid (C18:2) is the most important one (56%) while oleic acid (C18:1) represented only 16% of total fatty acids. The palmitic acid (C16:0) was the principal saturated one (25%).

<u>Sodium chloride action on total fatty acids during seed maturation</u>

At earlier stages, 1 and 2 weeks after anthesis, the total lipids level remained lower and relatively stable, but raised from the 3 to the 5th(Fig.1). Beyond 5 weeks, there was not important lipogenesis until bolls dehiscence. Fatty acids variation of salt treated plants ressembled to control one with slight oil synthesis stimulation at 3g/l. However, total lipids level decreased for salt concentration more than 6g/l.

<u>Sodium chloride action on the oil content of mature seeds</u>

Table 1 showed that NaCl stimulated lipid synthesis at 3g/l. For 9g/l NaCl set, oil content decreased slightly whereas plant growth was reduced to 29%.

NaCl Concentration g/l	0	3	6	9
Total fatty acids mg/g dry matter	109	116	113	103

Table 1 : NaCl effect on oil content mature seeds.

Sodium chloride action on the fatty acids contents

Palmitic acid increased slightly at 3g/l of NaCl and decreased slowly for the higher concentrations (Fig.2). So at 9g/l, its contend decreased only of 7% compared with control. Oleic acid content was practically invariable for salt treatments . while linoleic acid decreased from 6g/l of NaCl.

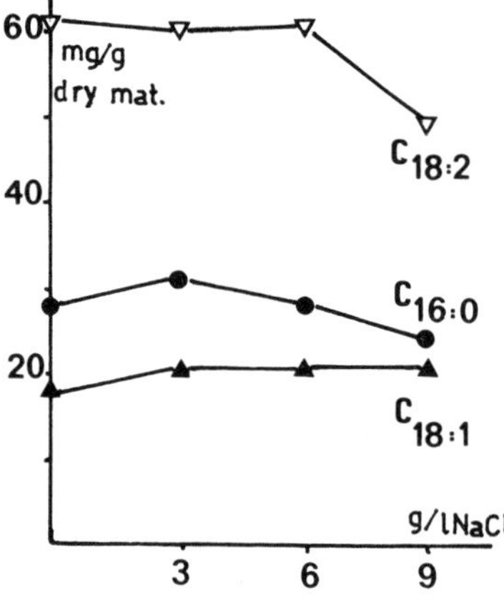

Fig.1: NaCl action on total fatty acids during seed maturation.

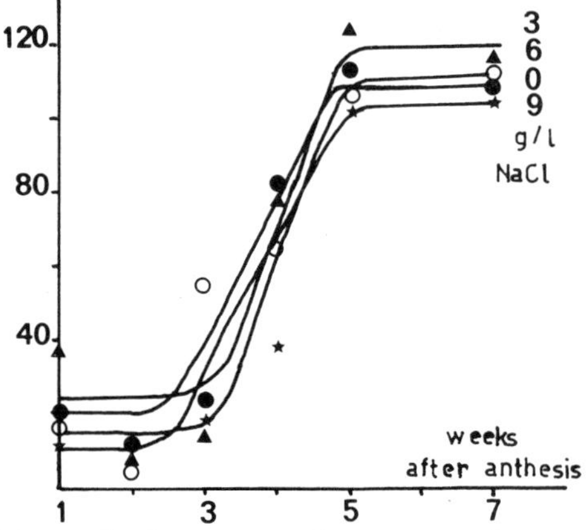

Fig.2: NaCl action on the fatty acids contents.

CONCLUSIONS

High salt concentration did not delay lipogenesis during ripening seeds but delay anthesis date and did not seem to decrease drastically total lipids content. This finding could be related to the salt tolerance of Cotton plant. However low sodium chloride concentration (3g/l) stimulated oil synthesis. Sodium chloride reduced the content of the major fatty acid (linoleic acid) of Cotton seeds as it did for the principal olive fatty acid (oleic acid) (Marzouk et al. 1986).

REFERENCE

Marzouk B., Zarrouk M., Cherif A., 1986 - Lipids changes in olive fruit under saline conditions. Proceedings : world conference on emerging technologies in fat and oil industry. A.R. Baldwin, Cannes, France, 408-409.

BIOLOGICAL ROLE OF PLANT LIPIDS
P.A. BIACS, K. GRUIZ, T. KREMMER (eds)
Akadémiai Kiadó, Budapest and Plenum Publishing
Corporation, New York and London, 1989

METABOLISM OF INOSITOL PHOSPHOLIPIDS IN RESPONSE TO OSMOTIC STRESS IN DUNALIELLA SALINE

K.J. Einspahr, T.C. Peeler and G.A. Thompson, Jr.

Department of Botany, University of Texas, Austin, Texas 78713, USA

One of the most active areas of animal cell biochemistry deals with the transduction of hormonal and environmental signals across the plasma membrane by a mechanism involving polyphoinositide degradation. Surprisingly little attention has been directed towards identifying an equivalent system in plants. Isolated elements of the transduction pathway have been reported (e.g. Drobak and Ferguson, 1985; McMurray and Irvine, 1988; Morse, et al., 1988), but only preliminary evidence has been established concerning a physiological role for signal transduction by this mechanism. We have characterized such a system which may mediate the acclimation of *Dunaliella salina*, a halotolerant green alga, to osmotic stress.

D. salina cells grown in our standard 1.71 M NaCl medium contained phosphatidylinositol 4,5-bisphosphate (PIP_2) and phosphatidylinositol 4-phosphate (PIP) as well as phosphatidylinositol. PIP and PIP_2 are highly localized within the plasma membrane, where they account for 7.4 and 2.1 mol% of the total phospholipids (Einspahr et al., 1988a).

When cells grown in 1.71 M NaCl were suddenly diluted to 0.85 M NaCl, they swelled within 2-4 min to a volume 1.76 times larger and a surface area 1.53 times larger than found in control cells. The extensive structural alterations triggered by this expansion have been described by Maeda and Thompson (1986). During the first 2 min. following dilution the cellular content of PIP and PIP_2 fell to 79% and 65%, respectively, of controls. Over the same time period the level of phosphatidic acid (PA) rose to 141% of control values.

The plasma membrane, but not other membranes, contains a phospholipase C specific for hydrolyzing PIP_2 to inositol phosphates, mainly inositol 1, 4, 5-trisphosphate, *in vitro*. *D. salina* also contains an active, Ca^{2+}-requiring protein kinase C (Tung, H.Y.L., R. Wayne, S. Roux, and G.A. Thompson, Jr., unpublished observations).

If the cells were subjected to a hyperosmotic shock by quickly raising the medium salt concentration from 1.71 M NaCl to 3.42 M NaCl, the PIP_2 mass rose to 130% of control values and PA dropped to 56% of control values (Einspahr et al., 1988b). Thus hypoosmotic and hyperosmotic shock produce opposite effects. Efforts are currently underway to confirm the involvement of a GTP-binding protein in the signal transduction pathway.

ACKNOWLEDGEMENTS

This research was supported in part by the National Science Foundation, the Robert A. Welch Foundation, and the Texas Advanced Technology Research Program.

REFERENCES

Drobak, B.K., and I.B. Ferguson, Biochem. Biophys. Res. Commun. *130*: 1241-1246 (1985). Release of Ca^{2+} from plant hypocotyl microsomes by inositol-1,4,5-trisphosphate.

Einspahr, K.J., T.C. Peeler, and G.A. Thompson, Jr., J. Biol. Chem. *263*: 5775-5779 (1988a). Rapid responses in polyphosphoinositide metabolism associated with the response of *Dunaliella salina* to hypoosmotic shock.

Einspahr, K.J., M. Maeda, and G.A. Thompson, Jr., J. Cell Biol. in press (1988b). Concurrent changes in *Dunaliella salina* ultrastructure and membrane phospholipid metabolism after hyperosmotic shock.

Maeda, M., and G.A. Thompson, Jr., J. Cell Biol. *102*: 289-297 (1986). On the mechanism of rapid plasma membrane and chloroplast expansion in *Dunaliella salina* exposed to hypoosmotic shock.

McMurray, W.C., and R.F. Irvine, Biochem. J. *249*: 877-881 (1988). Phosphatidylinositol 4,5-bisphosphate phosphodiesterase in higher plants.

Morse, M.J., G.G. Cote, R.C. Crain, and R.L. Satter, Plant Physiol. *86*: 93 (1988). Light-modulated phosphatidlyinositol turnover in *Samanea saman*.

BIOLOGICAL ROLE OF PLANT LIPIDS
P.A. BIACS, K. GRUIZ, T. KREMMER (eds)
Akadémiai Kiadó, Budapest and Plenum Publishing
Corporation, New York and London, 1989

HOMEOVISCOUS REGULATION OF MEMBRANE PHYSICAL STATE IN THE BLUE-GREEN ALGA, ANACYSTIS NIDULANS

Z. Gombos and L. Vígh

Institute of Plant Physiology and Biochemistry, Biological Research Center of the Hungarian Academy of Sciences, 6701 Szeged, P.O. Box 521, Hungary

INTRODUCTION

The blue-green alga, Anacystis nidulans has proven to be a useful organism to study the mechanism of the self-regulation of membrane fluidity. On the base of the self-regulating fluidity-restoration mechanism one can predict that any stimulus inducing alterations in the membrane components which tend to alter fluidity should be counteracted by a change with an opposite effect on overall fluidity. The growth-temperature shift induced regulatory changes both in the composition and thermal behaviour of membrane lipids is well understood. Recently we have been particularly interested in those alternative perturbatory means which affords to examine the restoration mechanism in the absence of temperature change. Due to the fall of the ratio of protein to lipid (p/l), nitrate starvation resulted in a fluidizing effect in Anacystis cells, but it was simultaneously compensated by chain lenghtening and saturation of fatty acids (1). In this study, we present further evidences on the existence of previously proposed compensatory changes by using ESR technique on the envelope polar lipids of nitrate starved cells. Since the chloramphenicol-induced block in protein synthesis triggered lipid saturation analogous to response observed after nitrate-starvation, the p/l decline is apparently counteracted by the same mechanism in this organism.

METHODS

Nitrate starvation of Anacystis nidulans cells has been accomplished as described before (2). The procedure for isolation and characterization of the cell envelope was as reported in (3). Protein and lipid analysis was performed as in (4).

Determination of lipid phase transition within the cytoplasmic membrane in vivo by detecting the temperature-dependent absorption change at 390 nm has been described in (5).

The temperature profiles of the empirical motion parameter of 16-NS motion, h_0/h_{30}, was measured on multilamellar liposomes formed from envelope polar lipids. (For details of similar measurements see ref.6).

RESULTS

As we reported in our previous studies, the nitrate-deprivation-induced changes in fatty acid composition, ie. elevated level of longer chain and more saturated fatty acids, have also been reflected in altered physical properties of hydrated envelope polar lipids (1). Over the temperature range evaluated, sets of fluorescence polarization values, obtained from nitrate-starved cells, were markedly higher than those found in corresponding controls. On the other hand, lipid phase transition could not be demonstrated with any of fluorescence parameters. This fact stimulated us to apply the ESR technique. To examine the molecular motion within bilayers of hydrated envelope lipids, the relative central peak amplitude (h_0 at actual temperature versus $h_{0/30}$ at 30^{0}C) was calculated and has been plotted against the temperature (Fig.1). Microviscosity of that sample originated from 24 h nitrate-starved cells was apparently higher than in control samples.

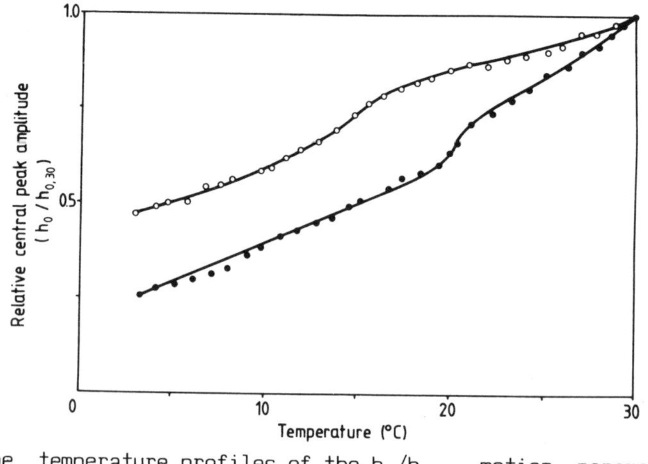

Fig.1. The temperature profiles of the $h_0/h_{0/30}$ motion parameter of 16-NS, measured on liposomes formed from envelope polar lipids of control (o) and nitrate-starved (24 h) cells (●).

Moreover, the liquid cristalline to gel transition could also be detected by the fatty acid spin probe, 16-NS. The begining of fluid to gel transition was at around 15-17^{0}C in control and at about 21-23^{0}C in nitrate-deprived cells.

To gain further evidences on the existence of homeoviscous regulatory responses triggered by p/l decline, next the cells were treated with chloramphenicol (CP) (100 µg/ml) for different time intervals. Parallel with the gradual decrease of cell protein content (data not shown) increasing lipid saturation, highlighted by the ratio of 16:0/16:1, could also be evidenced (Table 1).

Table 1. Effect of CP treatment (0-24 h) on lipid saturation (16:0/16:1) tested on whole cell level.

Duration of CP treatment (h)	0	4	8	12	24
Ratio of 16:0/16:1	1.26	1.47	1.73	1.70	1.65

As we proposed elsewhere (2), low-temperature induced shifts in the light absorption spectrum of the intact *Anacystis* cells can be used to detect the phase-transition of the cytoplasmic membrane *in vivo*. The temperature-dependent absorption change ($A_{390\ nm}$) in intact, CP-treated cells is shown on Fig.2.

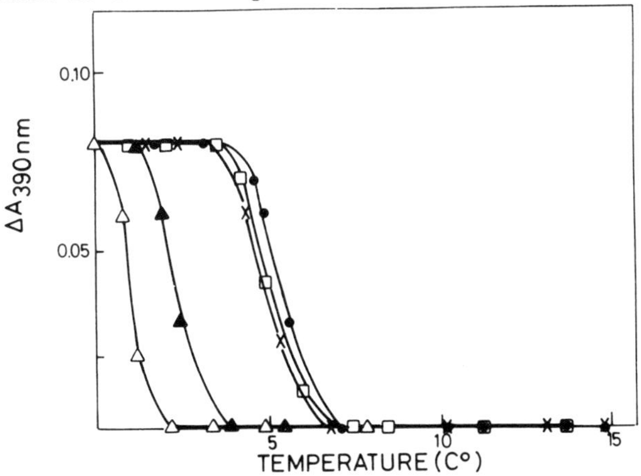

Fig.2. Temperature-dependent absorption changes ($A_{390\ nm}$) in intact *Anacystis* cells, grown at 32^{0}C, and CP-treated for 0 h (●), 4 h (□), 8 h (✖), 12 h (▲), 24 h (Δ), respectively. Cells were treated at chilling temperature for 20 min. Spectra were recorded at 25^{0}C.

According to Fig.2 thermal characteristics of the cytoplasmic membranes have not been affected up to 8 hour long treatment by CP. If the cells kept for 12 and 24 hours in the presence of CP, however, a downward shift in the phase separation of surface membranes became detectable.

CONCLUSION

Anacystis cells grown in the absence of nitrogen source are capable to maintain their membrane physical state at a fairly constant level: the fluidizing effect of the declined protein to lipid ratio is counteracted by the microviscosity increase due to alkyl chain saturation. If the fluidity perturbation induced by p/l fall is attained by an alternative way (CP-treatment), *Anacystis* cells respond by the apparently same strategy.

Acknowledgement: This work was supported by the Hungarian National Scientific Research Foundation (OTKA 543/86 to L.V.).

REFERENCES

1. Z. Gombos, M. Kis, T. Páli, L. Vigh Eur.J.Biochem. 165, 461-465 (1987)
2. Z. Gombos, L. Vigh Plant Physiol. 80, 415-419 (1986)
3. N. Murata, N. Sato, T. Omata, T. Kuwabara Plant Cell Physiol. 22, 855-866 (1981)
4. L. Vigh, I. Horváth, L.I. Horváth, D. Dudits, T. Farkas FEBS lett. 107, 291-294 (1979)
5. T.A. Ono, N. Murata Plant Physiol. 67, 176-181 (1981)
6. L. Vigh, H. Huitema, J. Woltjes, P.R. van Hasselt Physiol. Plant. 67, 92-96 (1986)

BIOLOGICAL ROLE OF PLANT LIPIDS
P.A. BIACS, K. GRUIZ, T. KREMMER (eds)
Akadémiai Kiadó, Budapest and Plenum Publishing
Corporation, New York and London, 1989

EFFECTS OF CO_2 CONCENTRATION DURING GROWTH ON FATTY ACID COMPOSITION IN CHLORELLA VULGARIS 11H

M. Tsuzuki[1], E. Kenjo[2], T. Takaku[1], S. Miyachi[1] and A. Kawaguchi[2]

[1] Institute of Applied Microbiology, University of Tokyo, Yayoi, Bunkyo-ku, Tokyo 113, Japan
[2] Department of Biology, University of Tokyo, Komaba, Meguro-ku, Tokyo 153, Japan

INTRODUCTION

The affinity for inorganic carbon in photosynthesis of microalgae as well as submersed angiosperm is reduced when CO_2 concentration is elevated to 1-5% [1]. It is gradually assumed that this is due to decreases in carbonic anhydrase and in the capacity of accumulating inorganic carbon in high-CO_2 cells. The electron microscopy revealed that the chloroplast envelope was electronically denser in low-CO_2 cells than in high-CO_2 cells, while the opposite effect of CO_2 was observed for the plasma membrane of Dunaliella tertiolecta [2]. From these results, we assumed that CO_2 concentration during growth gives some effects on membrane lipids in algae. In the present study, we investigated the fatty acid composition in low-CO_2 and high-CO_2 cells as well as the variations in fatty acid composition after the shift of CO_2 concentration.

MATERIALS AND METHODS

Chlorella vulgaris 11h was grown at 30°C under constant illumination with fluorescent lamps (7-10 klux). Cell suspension was bubbled with ordinary air to obtain low-CO_2 cells or air enriched with 2% CO_2 to obtain high-CO_2 cells. Fatty acid methyl esters were analyzed by gas-liquid chromatography using docosanoic acid as an internal standard.

RESULTS

Effect of CO_2 concentration on fatty acid composition --- Major fatty acids in C. vulgaris 11h were palmitic (16:0), linoleic (18:2) and α-linol-

Table 1 Fatty acid composition of Chlorella vulgaris 11h

Fatty acid	Fatty acid composition (mol%)	
	Low-CO_2 cells	High-CO_2 cells
14:0	2.6	1.9
16:0	22.2	30.3
16:1	4.8	0
16:2	6.1	6.0
16:3	4.1	1.9
18:1	2.6	7.1
18:2	23.4	34.7
18:3	34.3	18.1

enic (18:3) acids (Table 1). The remarkable changes associated with CO_2 concentration were observed in these fatty acids. In high-CO_2 cells, the contents of 16:0 and 18:2 were higher while that of 18:3 was lower than in low-CO_2 cells. As a result, the desaturation expressed as the average number of double bonds in a fatty acid molecule decreased from 1.82 to 1.49 on increasing CO_2 concentration from 0.04% to 2%. In spite of these changes in unsaturation, the relative contents of C_{16} and C_{18} acids remained almost unchanged. Since the fatty acid composition was fairly constant in the range from 0.5 to 6 ml packed cell volume/l, fatty acid alterations shown in Table 1 were dependent on the CO_2 concentration during growth.

Changes in fatty acid composition provoked CO_2-concentration shift from 2 to 0.04% --- When high-CO_2 cells were transferred to low CO_2 condition, the content of 18:3 increased and those of 18:1 and 18:2 decreased significantly. Thereafter the contents of these C_{18} acids gradually changed and at 48 h after the transfer, each level was very close to those observed in low-CO_2 cells. In contrast to the pronounced alterations among C_{18} acids, the variations in C_{16} acids were small and slow. The total contents of C_{16} and C_{18} acids were constant during the experimental period. Fig. 1 shows the amounts of major fatty acids per unit weight of cells and the cellular growth expressed by the incerease in the packed cell volume per unit volume of culture. When high-CO_2 cells were transferred to ordinary air, there was a lag of growth for the first 6 h after the transfer. The amounts of total fatty acids and 16:0 remained almost constant during 24 h-culture. It should be mentioned that the changes in the amounts of 18:1, 18:2 and 18:3 were observed during the first 6 h although the packed cell volume remained almost constant. The reciprocal relationships between the decreases in 18:1 and 18:2 and the incerase in 18:3 suggest that the pre-existing 18:1 and 18:2 might be desaturated to 18:3 without de novo synthesis of fatty acids.

Changes in fatty acid composition provoked by CO_2-concentration shift from 0.04 to 2% --- The transfer of low-CO_2 cells to high CO_2 condition caused complicated changes in fatty acid composition. The total content of C_{16} acids decreased and that of C_{18} acids increased. The most remarkable increase was observed in 18:1. The contents of 18:1 and 18:2 increased, while those of 16:0 and 18:3 decreased. Fig. 2 shows the changes in the amounts of fatty acids per unit weight of cells and the packed cell volume per unit volume of culture after the transfer of low-CO_2 cells to high CO_2 condition. Both the packed cell volume and the total amounts of fatty acids increased, but the increase in the latter was higher than the former. The increase of total fatty acids was due to the increases of 16:0, 18:1 and 18:2. The amounts of 18:3, on the other hand, remained almost constant during 48 h. A comparison of the amounts of individual fatty acids shows that the syntheses of 16:0, 18:1 and 18:2 were accelerated and that 18:3 was not decomposed during this period. These findings suggest that the changes in fatty acid composition are brought about by the accelerated

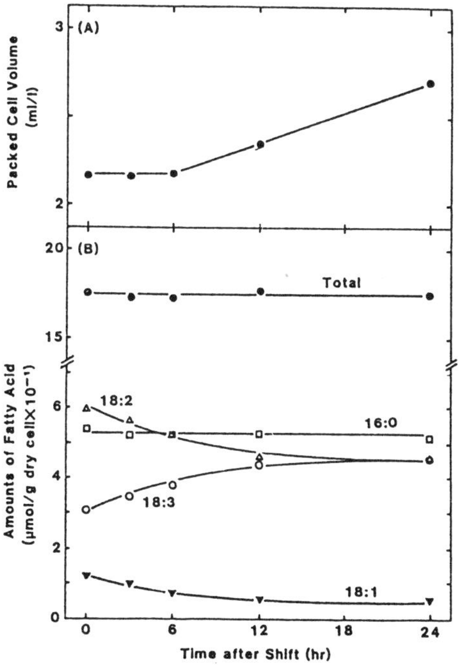

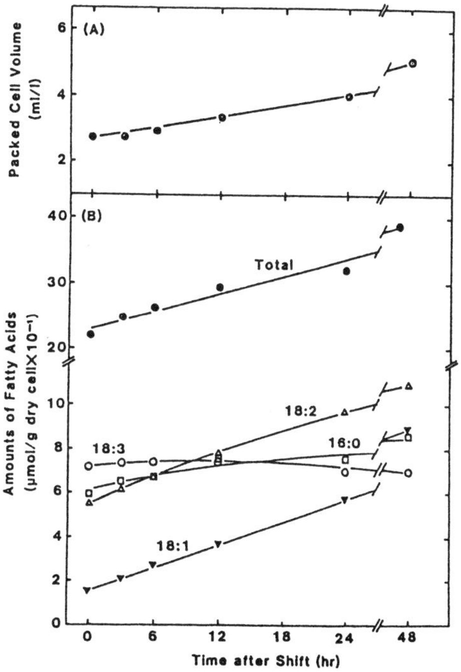

Fig. 1 (Left) Changes in packed cell volume (A) and amounts of fatty acids (B) after shift of CO_2-concentration from 2 to 0.04%.

Fig. 2 (Right) Changes in packed cell volume (A) and amounts of fatty acids (B) after shift of CO_2-concentration from 0.04 to 2%.

syntheses of 16:0, 18:1 and 18:2 and a dilution of the pre-existing 18:3 with newly synthesized more saturated fatty acids.

DISCUSSION

The fatty acid composition varied in response to CO_2 concentration given during growth of *C. vulgaris* 11h. The desaturation was lower when the cells were grown under high CO_2 condition. On the other hand, chain length of fatty acid was not influenced by CO_2 concentration. The most remarkable response to downward shift of CO_2 concentration was the desaturation of 18:2. Upon upward shift of CO_2 concentration, the response was represented by the decrease in unsaturation of fatty acids. With heterotrophically growing cells of *Chlorella fusca*, Dickson *et al*. [3] reported that the amount of 18:1 increased when the CO_2 concentration was raised from 1 to 30%.

The present study indicates that the environmetal CO_2 concentration affects on membrane lipids in addition to photosynthetic characteristics. Under low CO_2 concentration, *de novo* synthesis of carbonic anhydrase is induced and pyrenoid with starch sheath is developed in various eukaryotic microalgae. So far some intermediate(s) of photorespiratory pathway and/or CO_2/O_2 ratio has been proposed to be the initial reaction which induces the *de novo* synthesis of carbonic anhydrase [4,5]. Carbon dioxide is much more soluble in fats than water and the adsorption is quick [6]. Therefore, the changes in the degree of unsaturation of the fatty acyl chain in the membranes may be one of the initial reactions which occur in response to the decrease of CO_2 concentration.

REFERENCES

[1] Aizawa, K. and Miyachi, S. (1986) *FEMS Microbiol. Rev.* **39**, 215-233.
[2] Tsuzuki, M., Gantar, M., Aizawa, K. and Miyachi, S. (1986) *Plant Cell Physiol.* **27**, 737-739.
[3] Dickson, L.G., Galloway, R.A. and Patterson, G.W. (1969) *Plant Physiol.* **44**, 1413-1416.
[4] Ramazanov, Z.M., Pronina, N.A. and Semenenko, V.E. (1984) *Fiziol. Rastenii* **31**, 448-455.
[5] Ramazanov, Z.M. and Semenenko, V.E. (1986) *Fiziol Rastenii* **33**, 864-872.
[6] Mitz, M.A. (1979) *J. Theor. Biol.* **80**, 537-551.

BIOLOGICAL ROLE OF PLANT LIPIDS
P.A. BIACS, K. GRUIZ, T. KREMMER (eds)
Akadémiai Kiadó, Budapest and Plenum Publishing
Corporation, New York and London, 1989

EFFECTS OF CO_2 CONCENTRATION ON THE COMPOSITION OF LIPIDS IN CHLAMYDOMONAS REINHARDTII

Naoki Sato

Department of Botany, Faculty of Science, University of Tokyo, Hongo, Tokyo 113, Japan

INTRODUCTION

It is well known that growth temperature and O_2 concentration affect the composition of fatty acids in microorganisms, while the effects of CO_2 on the lipids have not been reported. During my study on the biosynthesis of the homoserine lipid in *Chlamydomonas*, I noticed that the composition of lipids changed with the concentration of CO_2. I thus analyzed the composition of lipids and fatty acids in the cells of *Chlamydomonas reinhardtii* which were grown under different concentrations of CO_2.

MATERIALS AND METHODS

Cells of *Chlamydomonas reinhardtii* strain C-238 (UTEX89) were grown in the modified Bristol medium with aeration by either air (0.04% CO_2), or air containing 1% or 3% CO_2. The growth temperature was 25 C. Light was provided by fluorescent lamps at a fluence rate of 10 W/m^2. Lipids were fractionated by two-dimensional TLC and the fatty acids were analyzed by GLC with a capillary column (Thermon 3000, 0.20 mm i.d. x 25 m).

RESULTS AND DISCUSSION

Lipid composition The contents of MGDG and PG per cell were 34 and 4.9 nmol/10^7cells, respectively in the cells grown in air, and 22 and 2.6 nmol/10^7cells, respectively in the cells grown in 1% CO_2. The contents of lipids were almost identical in the cells grown in either 1% or 3% CO_2. No marked differences were found in the content of other classes of lipids. This result was consistent with the fact that the CO_2-grown cells were slightly smaller in size than those grown in air. The detailed study on the ultrastructure of the cell is in progress.

Fatty acid composition Giroud et al. (1988) found that the 18:3(n-6) in *Chlamydomonas reinhardtii* strain 137, which had been believed to be 18:3 (6,9,12), was in fact 18:3(5,9,12). They also identified 18:4(5,9,12,15). I confirmed the occurrence of 18:3(5,9,12) and 18:4(5,9,12,15) in the strain C-238 by capillary column GLC and GC-MS.

The contents of 18:3(9,12,15) and 16:4(4,7,10,13) in MGDG was lower in the cells grown in 1% or 3% CO_2 than in the air-grown cells, while those of 18:1(9), 18:2(9,12) and 16:3(4,7,10) was higher in the former. The CO_2-grown cells contained: a lower level of 18:3(9,12,15) and higher levels of 18:1(9) and 18:2(9, 12) in DGDG; a lower level of 18:3(9,12,15) and a higher level of 16:0 in PG; lower levels of 18:4(5,9,12,15) and 18:3(9,12, 15) and higher levels of 18:3(5,9,12) and 18:2(9,12) in DGTS. On the other hand, the content of 18:3(5,9,12) in PE was higher in the CO_2-grown cells, and that of 18:1(11) was lower. The differences amounted to generally 5% or less of the total fatty acids in each class of lipid. These results show that CO_2 enrichment caused lower unsaturation in the fatty acids of MGDG, DGDG, PG and DGTS and higher unsaturation in the fatty acids of PE. No significant differences in the composition of fatty acids in SQDG was found. Although Kawaguchi et al. (this symposium) showed that CO_2 enrichment lowered the unsaturation of total fatty acids in several green algae, it should be noted that the direction of changes in unsaturation was not the same in different classes of lipids as shown here. For the moment, the biological meaning of these effects of CO_2 is not clear, but the present experimental system will provide an interesting material for the study of regulation of lipid biosynthesis.

REFERENCE

Giroud, C., Gerber, A. and Eichenberger, W. (1988) Lipids of *Chlamydomonas reinhardtii*. Analysis of molecular species and intracellular site(s) of biosynthesis. Plant Cell Physiol. 29: 587-595.

BIOLOGICAL ROLE OF PLANT LIPIDS
P.A. BIACS, K. GRUIZ, T. KREMMER (eds)
Akadémiai Kiadó, Budapest and Plenum Publishing
Corporation, New York and London, 1989

EFFECTS OF LOW TEMPERATURE ON LIPID AND FATTY ACID COMPOSITION OF SORGHUM ROOTS AND SHOOTS

A.T. Pham Thi, M.D. Sidibé-Andrieu, Y. Zuily-Fodil and J. Vieira da Silva

Laboratoire d'Ecologie Générale et Appliquée, Université Paris VII, 2 Place Jussieu
75251 Paris Cedex 05, France

When exposed to chilling, but non-freezing temperatures, plant cell membranes undergo a number of modifications in their lipid and fatty acid properties (see Lyons et al.1979). The most currently observed changes are an accumulation of phospholipids, particularly PC, and an increase in the level of fatty acid unsaturation. Sorghum are tropical plants, which show a chilling sensitivity differing from one species to another. Thus, Sorghum bicolor is sensitive to chilling, while Sorghum halepense could resist to temperatures lower than 5°C (Bagnall 1979).

Materials and methods

Plants of *Sorghum bicolor L., cv. Martin,* and *Sorghum halepense L.* are grown in controlled conditions (27°C/17°C, day-night temperature). When the plants have four leaves, they are transferred to a cold chamber (12°C/10°C) One week later, lipids from leaves and roots are extracted and analysed following classical methods of TLC on silica gel plates and GLC .

Results

Fig.1 and Tab.1 show the effects of low temperatures on the amount of polar lipids. *In the leaves,* galactolipid content decreases in the two sorghum species, particularly MGDG. In Sorghum bicolor, phospholipid content of leaves decreases, due essentially to a decrease in PC. On the contrary, in Sorghum halepense, the amount of phospholipids shows a slight increase. *In the roots,* no significant modifications are observed.

Tab.2 shows the percentage of the main fatty acids in the lipid classes of Sorghum leaves. For MGDG and DGDG, there was an increase in linolenic acid (18:3). Linoleic acid (18:2) percentage increases in PC. Concerning PG, the major change is a significant decrease in trans-hexadecenoic acid (16:1t). The fatty acid composition of PE does not show noticeable modifi-

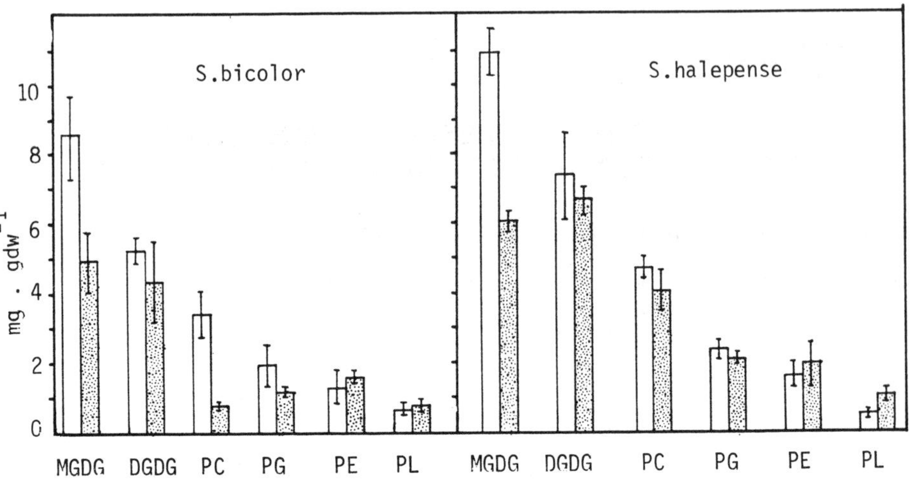

Fig. 1 - Effect of low temperature on polar lipid content of Sorghum leaves . MGDG : monogalactosyl diacylglycerol ; DGDG : digalactosyl diacylglycerol ; PC: phosphatidylcholine ; PE: phosphatidylethanolamine ; PG: phosphatidylglycerol ; PL: other phospholipids

control plants — plants subjected to low temperature

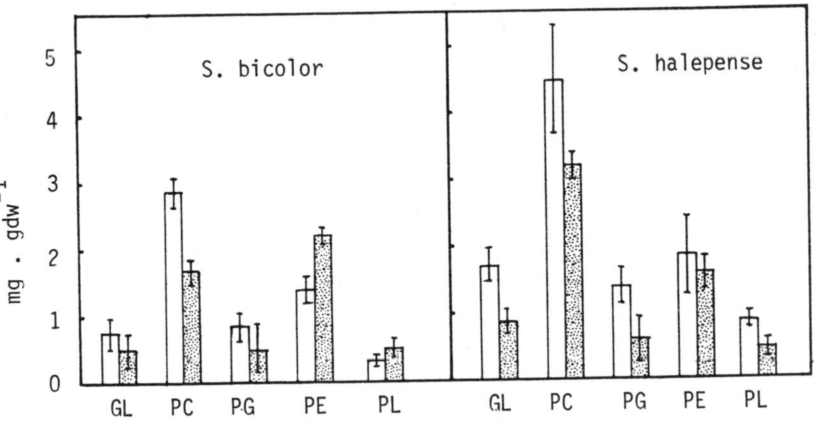

Fig.2 - Effect of low temperature on polar lipid content of Sorghum roots .

Tab.1 - Effect of low temperature on galactolipid and phospholipid content of Sorghum leaves and roots (in $mg.gdw^{-1}$). C= control ; S= plants subjected to low temperature .

		Leaves		Roots	
		GL	PL	GL	PL
S.bicolor	C	17.88 ± 1.45	10.25 ± 2.18	0.58 ± 0.13	4.53 ± 0.58
	S	12.03 ± 2.32	6.67 ± 0.46	0.51 ± 0.09	4.36 ± 0.70
S.halepense	C	23.53 ± 2.23	11.69 ± 1.64	1.70 ± 0.44	8.76 ± 2.99
	S	16.37 ± 0.88	13.39 ± 2.55	0.84 ± 0.21	6.17 ± 1.06

Tab. 2 - Effect of low temperature on the percentage of polar lipid main fatty acids from Sorghum leaves and roots. C= control ; S= stressed .

		LEAVES					
		C			S		
		16:0	18:2	18:3	16:0	18:2	18:3
bicolor	MGDG	1.6 ± 0.5	4.9 ± 0.4	90.8 ± 2.4	1.1 ± 0.1	3.1 ± 0.3	95.3 ± 0.2
	DGDG	14.7 ± 0.5	5.9 ± 1.0	74.8 ± 0.7	11.0 ± 2.1	5.0 ± 1.2	80.5 ± 2.1
	PC	30.5 ± 1.8	34.8 ± 1.1	25.9 ± 1.2	21.9 ± 4.0	47.9 ± 1.1	23.6 ± 0.2
	PE	22.7 ± 1.7	52.1 ± 1.3	18.6 ± 2.6	29.2 ± 2.6	49.8 ± 1.4	17.9 ± 1.5
	PG	36.8 ± 0.7	13.3 ± 2.1	31.1 ± 1.5	35.9 ± 3.6	19.7 ± 2.8	29.1 ± 7.5
		11.5 ± 2.9*			6.2 ± 1.3*		
halepense	MGDG	1.9 ± 0.3	10.3 ± 0.9	85.0 ± 2.1	1.4 ± 0.2	4.6 ± 0.1	93.1 ± 0.3
	DGDG	11.3 ± 1.1	14.4 ± 2.7	67.0 ± 4.7	10.6 ± 0.9	7.7 ± 0.9	78.0 ± 1.8
	PC	27.6 ± 4.1	38.6 ± 4.1	25.9 ± 1.8	22.2 ± 1.4	45.9 ± 1.9	25.7 ± 0.7
	PE	24.7 ± 3.1	53.2 ± 2.5	16.6 ± 1.3	23.2 ± 4.5	52.9 ± 7.7	20.7 ± 6.4
	PG	21.6 ± 3.6	18.5 ± 1.3	42.5 ± 3.7	34.1 ± 6.8	17.7 ± 2.3	38.8 ± 6.8
		9.2 ± 3.5*			5.7 ± 1.5*		
		ROOTS					
		C			S		
		16:0	18:2	18:3	16:0	18:2	18:3
bicolor	PC	24.1 ± 1.2	47.8 ± 1.6	19.8 ± 2.3	20.3 ± 1.1	53.4 ± 0.8	6.3 ± 2.3
	PE	33.2 ± 4.9	47.1 ± 3.0	10.7 ± 1.7	23.5 ± 5.6	53.5 ± 3.7	15.9 ± 4.0
	PG	37.5 ± 4.5	37.7 ± 4.6	12.9 ± 4.7	33.4 ± 1.3	40.2 ± 2.9	19.9 ± 2.3
halep.	PC	21.2 ± 6.3	47.9 ± 1.9	19.4 ± 2.9	19.9 ± 0.3	55.3 ± 0.8	16.8 ± 0.8
	PE	26.6 ± 4.9	49.7 ± 7.3	13.2 ± 3.7	27.4 ± 5.5	45.5 ± 7.0	14.5 ± 2.8
	PG	33.9 ± 5.1	40.4 ± 6.4	16.6 ± 4.7	32.7 ± 9.9	45.5 ± 7.5	16.4 ± 2.0

*16:1t

cations. In the roots (Tab.3), the proportion of 18:2 in PC increases in the two species. The percentage of 18:3 from PC decreases in Sorghum bicolor while it increases in Sorghum halepense .

Conclusion A temperature of 10-12°C,non-lethal for Sorghum species, provokes a decrease in polar lipid content of shoots and roots. In leaves, galactolipids, particularly MGDG, are the most sensitive. This degradation of membrane lipids by low temperatures is accompanied by an increase in the level of fatty acid unsaturation, which is due to an accumulation of 18:3 in galctolipids and of 18:2 in PC.

The differences between the chilling-resistant and the chilling-sensitive Sorghum are observed for PC: its content decreases in Sorghum bicolor leaves, and remains stable in S.halepense. PC unsaturation increases in roots from the resistant species and decreases in the susceptible species.

References

Lyons J.M.,Graham D., Raison J.K. (1979)- Low temperature stress in crop plants ; The role of the membrane . Academic Press , New York .

Bagnall D. (1979)- Low temperature responses of three Sorghum species . in " Low temperature stress in crop plants ; the role of the membrane" (Lyons J.M. et al.eds.) , pp. 67-80 , Academic Press, New York

BIOLOGICAL ROLE OF PLANT LIPIDS
P.A. BIACS, K. GRUIZ, T. KREMMER (eds)
Akadémiai Kiadó, Budapest and Plenum Publishing
Corporation, New York and London, 1989

TEMPERATURE-INDUCED DESATURATION OF FATTY ACIDS IN THE CYANOBACTERIUM, SYNECHOCYSTIS PCC 6803

H. Wada and N. Murata

National Institute for Basic Biology, Myodaiji, Okazaki 444, Japan

INTRODUCTION

Fatty acid composition of lipids in cyanobacteria is highly affected by environmental factors (1). The changes in fatty acid composition with growth temperature are regarded as an adaptive response to changes in the ambient temperature (2, 3).

In this study, we investigated the temperature-induced changes in glycerolipid and fatty acid composition under isothermal conditions and the desaturation of fatty acids after a downward temperature shift in the cyanobacterium, *Synechocystis* PCC 6803.

MATERIALS AND METHODS

Synechocystis PCC 6803 was grown photoautotrophically in the BG 11 medium with aeration of 1% CO_2 in air, as described previously (2). Total lipids were extracted from cells according to the method of Bligh and Dyer (4). Lipid classes were separated by TLC (6). The total lipids and separated lipid classes were subjected to methanolysis and the resultant methyl esters were analyzed by gas chromatography and gas chromatography-mass spectrometry. The positional distribution of fatty acids was analyzed by selective hydrolysis of the ester linkage at the *sn*-1 position with *Rhizopus delemer* lipase (5).

RESULTS AND DISCUSSION

The composition of major glycerolipids, monogalactosyl diaclyglycerol (MGDG), digalactosyl diacylglycerol (DGDG), sulfoquinovosyl diacylglycerol (SQDG) and phosphatidylglycerol (PG) was not affected by growth temperature. The C_{18} and C_{16} fatty acids were esterified to *sn*-1 and 2 positions, respectively, in all lipid classes, and the composition of C_{18} acids, but not C_{16} acids, depended on growth temperature. Under isothermal growth conditions the relative contents of γ-linolenic (γ-18:3) and octadecatetraenoic (18:4) acids at 22°C were higher than those at 38°C in MGDG and DGDG, and the relative content of α-linolenic acid (α-18:3) at 22°C was higher than that at 38°C in SQDG and PG. On the other hand, the relative contents of oleic (18:1) and linoleic (18:2) acids were lower at 22°C than those at 38°C in all lipid classes.

When the growth temperature was shifted from 38°C to 22°C, the C_{18} acids at the position 1 of *sn*-glycerol, but not the C_{16} acids at the position 2 of *sn*-glycerol, changed as shown in Fig. 1. The temperature-induced desaturations were stimulated as follows: 18:1 → 18:2 → γ-18:3 → 18:4 in MGDG, 18:1 → 18:2 → α-18:3 in SQDG and PG. These findings indicate that desaturase/s/ are specific with respect to the head group as well as the *sn*-glycerol positions.

The temperature shift-induced desaturation of fatty acids was inhibited by chloramphenicol, rifampicin and 3-(3',4'-dichlorophenyl)-1,1-dimethylurea and but not cerulenin. These findings indicate that the desaturase/s/ are inducibly synthesized after the temperature shift and that the desaturation is connected with the photosynthetic electron transport in the thylakoid membranes.

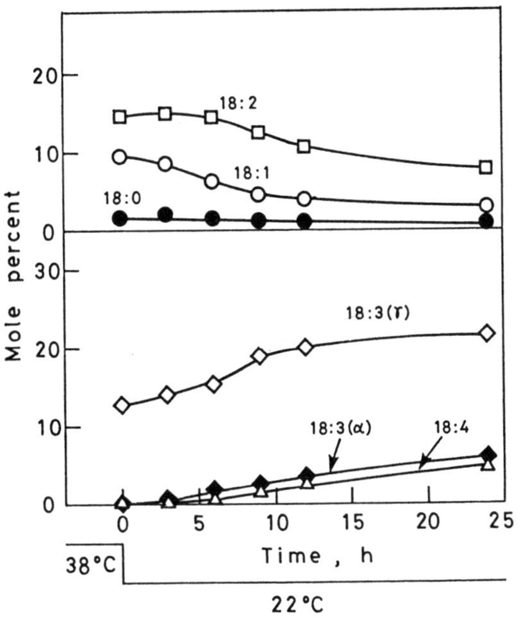

Fig. 1 Temperature-induced desaturation of fatty acids in *Synechocystis* PCC 6803.

REFERENCES

1. Murata, N. and Nishida, I. (1988) In "The Biochemistry of Plants" (P. K. Stumpf, ed.), vol. 9, pp. 315-347, Academic Press, London.
2. Sato, N. and Murata N. (1979) Biochim. Biophys. Acta 752: 19-28.
3. Sato, N. and Murata N. (1980) Biochim. Biophys. Acta 619: 353-366.
4. Bligh, E. G. and Dyer, W. J. (1959) Can. J. Biochem. Physiol. 37: 911-917.
5. Fischer, W., Heinz, E. and Zeus, M. (1973) Hoppe-seyler's Z. Physiol. Chem. 354: 1115-1123.
6. Sato, N. (1988) Method in Enzymol. in press.

BIOLOGICAL ROLE OF PLANT LIPIDS
P.A. BIACS, K. GRUIZ, T. KREMMER (eds)
Akadémiai Kiadó, Budapest and Plenum Publishing
Corporation, New York and London, 1989

CHOLINE TOLERANT NICOTIANA TABACUM CELL LINES ARE MORE RESISTANT TO COOL TEMPERATURES THAN THE COLINE SENSITIVE WILD TYPE

Myriam Gawer, Noémi Guern, Daisy Chevrin and Paul Mazliak

Université P. et M. Curie, UA1180, Laboratoire de Physiologie Cellulaire, Tour 53/3
4 Place Jussieu, 75252 Paris Cedex 05, France

INTRODUCTION

A strain of Tobacco cells resistant to choline (ChR) has been isolated from wild type cells sensible to choline (ChS). When choline chloride (CC) was added to the culture medium, ChS cells died after a growth cycle, whereas ChR constituted an homogene population the growth and viability of which were similar to those of ChS maintained in a medium deprived of choline. ChR cells differed from ChS cells by their low ability to accumulate choline during the stationary growth phase and a less succeptibility to cool culture temperature.

The regulation of intracellular choline accumulation may contribute to steady the survival of ChR and induce changes in lipid composition which favour suspension growth at cool temperature.

RESULTS AND DISCUSSION

Tobacco cell cultures and lipid analysis were as previously described (Gawer et al., 1987). The culture medium (MS) was according to Murashige and Skoog (1962) supplemented or not with choline chloride (CC).

1. Choline accumulation and cell size. As shown in Fig. 1 ChR cells, subcultured on their standard medium MS + 10 mM CC, slowly absorbed choline during exponential growth phase and limited intracellular accumulation during stationary growth phase. ChS cells, cultivated on MS + 10 mM CC absorbed, accumulated and metabolized more actively choline through their whole growth. During cell division phase, the ChS cell size gradually decreased and the recovery of the mean initial cell volume was delayed by the presence of exogenous choline during the stationary growth phase. This

growth inhibition was correlated with a high increase in the osmolarity of intracellular liquids, not noticeable in ChR cells. The lack of cell enlargement may be involved in the death of ChS cells.

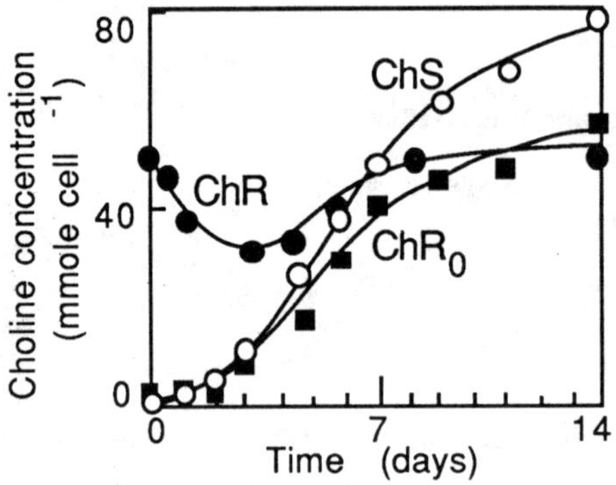

Fig. 1 Choline accumulation in Tobacco cells grown on 10 mM choline chloride.
ChRO are ChR cells exceptionally cultivated on MS during two growth cycles before transfer to MS + 10 mM CC.

2. Lipid composition of ChS and ChR cells. ChS and ChR cells, respectively cultivated at 26°C on their standard medium exhibited slight differences in their lipid composition and content (table 1). Even the molecular phosphatidylcholine species showed the same composition and distribution in both cell lines.

By contrast, the transfer of ChS cells from MS medium 10 MS + 10 mM CC medium resulted in an increase in fatty acid content and a decrease in the relative proportion of linolenic acid. Moreover transferring ChR cells to MS medium did not carry any difference in cell lipid composition.

Table 1 - Changes in lipid content of ChS and ChR tobacco cells cultivated with or without choline chloride, at 26°C.

Cell line	CC (mM)	Time (days) 0	3	5	7	14
		Fatty acid content (pg cell^{-1})				
ChS	0[(a)]	185	291	404	518	176
	10[(b)]	185	475	772	1136	278
ChR	10[(a)]	206	304	420	515	217
	0[(b)]	206	270	409	486	186

ChS cells routinely maintained on MS (a) were once transferred on MS + 10 mM CC (b). ChR cells routinely maintained on MS + 10 mM CC (a) were once transferred on MS (b).

Culture of both cell lines at 17°C was followed by bleaching of cells, increasing of lipid content (2-3 fold relative to 26°C), decreasing of galactolipids. The decrease of linolenic acid was attribuable to the deterioration of the photosynthetic apparatus. However the compositions of PG, MGDG and DGDG, lipids associated with the plastid membranes were richer in unsaturated fatty acids in choline tolerant cells (ChR) which better survived at cool temperature (table 2). So according to Murata's hypothesis (1983), it is likely that an increase in unsaturated molecular species of PG may be correlated to cold resistance. It may be assumed that stored choline in ChR cells promoted this cold resistance. Such choline effect has also been reported by Hórvath et al., (1981) on wheat seedlings.

Table 2 - Total fatty acid cell content and unsaturated fatty acid distribution in plastidial lipids of ChS and ChR tobacco cells cultivated at 17°C.

	Days	Fatty acid content ($pg\ cell^{-1}$)	% (18:2 + 18:3) Total fatty acids PG	MGDG	DGDG
ChS	0	200 + 24			
	8	675 + 17	40,6	71	64
	14	854 + 60			
ChR	0	265 + 30			
	8	734 + 23	54,3	82	75
	14	931 + 67			

REFERENCES

M. Gawer, N. Guern, D. Chervin and P. Mazliak - Comparison between a choline-tolerant Nicotiana tabacum cell line and the corresponding wild type : 1. Growth, thermosensitivity and lipid composition. Plant Physiol. and Biochem., 27, 323-331 (1987).

I. Hórvath, L. Vigh and T. Farkas - The manipulation of polar head group composition in the wheat Miranovskaya 808 affects frost tolerance. Planta, 151, 103-108 (1981).

T. Murashige and F. Skoog - A revised medium for rapid growth and bioassay with tobacco tissue cultures. Physiol. Plantarum, 15, 473-497 (1962).

N. Murata - Molecular species composition of phosphatidylglycerols from chilling sensitive and chilling resistant plants. Plant Cell Physiol., 24, 81-86 (1983).

BIOLOGICAL ROLE OF PLANT LIPIDS
P.A. BIACS, K. GRUIZ, T. KREMMER (eds)
Akadémiai Kiadó, Budapest and Plenum Publishing
Corporation, New York and London, 1989

EFFECT OF HIGH TEMPERATURES ON OXYGEN EVOLVING ACTIVITY AND LIPID COMPOSITION OF ISOLATED CHLOROPLASTS FROM SPINACH AND JOJOBA LEAVES

Thérése Guillot-Salomon[1], Jacqueline Bahl[1], L. Ben-Rais[1], M.-J. Alpha[1], Catherine Cantrel[1] and J.-P. Dubacq[2]

[1]Laboratoire de Biologie Végétale IV. CNRS (UA 1180) Univesité P. et M. Curie
12 Rue de Cuvier, 75005 Paris, France
[2]Lab. Biomembranes et Surfaces Cellulaires Végétales ENS, 24 Rue Lhomond, 75231 Paris, France
75231 Paris, France

The effects of temperature stress on chloroplasts were investigated by comparison of spinach (*Spinacia oleracea* L.), a temperate plant, and jojoba (*Simmondsia chinensis* (Link) Schneider), a desert-adapted plant.

Chloroplasts were isolated from leaves of spinach from the local market and from 10-week-old jojoba plants. Jojoba plants were grown at 30/25°C day/night temperature under light (60 μEm^{-2} s^{-1}) with a 16 h photoperiod.

To obtain a good yield of intact jojoba plastids, a high osmolarity of the isolation medium was maintained using 0.5 M sorbitol. Then purification was performed according to Guillot-salomon *et al.* (1987).

Isolated chloroplasts were incubated for 5 min at temperatures ranging from 25 to 55°C, and then immediately submitted to measurement of steady-state O_2-evolution ($H_2O \rightarrow$ ferricyanide + benzoquinone) at 25°C and to lipid analysis according to Guillot-Salomon *et al.* (1987).

In jojoba and spinach chloroplasts oxygen evolution was maximum (120 μmol O_2 h^{-1} mg^{-1} Chl) at 25°C. From 25°C to 40°C the oxygen evolution slightly decreased up to 66-68% of the initial activity. At higher temperatures a rapid drop of the O_2 evolved occurred and led to a total inhibition at 47.5^{o}C for spinach and at 52.5^{o}C for jojoba.

Jojoba chloroplasts contained polar lipids more saturated than those of spinach chloroplasts and digalactosyldiacylglycerol (DGDG) was the major lipid at 25°C (table). A 5 min exposure at higher temperatures did not significantly change

Table: Effect of temperature on galactolipids of isolated plastids.

	SPINACH			JOJOBA			
	25°C	40°C	47.5°C	25°C	40°C	47.5°C	52.5°C
MGDG *% of total*	*41*	*42*	*42*	*33*	*37*	*41*	*45*
Fatty Acids (%)							
16:0	2	2	3	9	8	7	6
16:3	20	21	21	0	0	0	0
18:0	1	0	0	2	1	1	2
18:1	0	1	2	8	6	6	4
18:2	2	1	1	9	9	8	7
18:3	75	75	73	72	76	78	81
DGDG *% of total*	*34*	*35*	*35*	*48*	*44*	*39*	*33*
Fatty acids (%)							
16:0	9	11	13	28	29	30	29
16:3	3	3	4	0	0	0	0
18:0	1	2	2	5	5	6	6
18:1	4	6	3	12	10	11	11
18:2	4	3	3	7	7	7	7
18:3	79	75	75	48	49	46	47
MGDG / DGDG	*1.2*	*1.2*	*1.2*	*0.7*	*0.8*	*1.1*	*1.4*

DGDG, digalactosyldiacylglycerol; MGDG, monogalactosyldiacylglycerol; 16:0, palmitic acid; 16:3, hexadecatrienoic acid; 18:0, stearic acid; 18:1, oleic acid; 18:2, linoleic acid; 18:3, linolenic acid.

the lipid composition of spinach chloroplasts. On the contrary, in jojoba chloroplasts, beyond 25°C the amount of DGDG decreased and monogalactosyldiacylglycerol (MGDG) became predominant. A parallel increase in linolenic acid content was observed in MGDG.

In conclusion, jojoba chloroplasts appear photosynthetically more resistant at high temperatures than spinach chloroplasts. A 5-min heat shock increases the ratio MGDG/DGDG and the MGDG unsaturation (18:3 especially) only in jojoba chloroplasts. Obviously these changes correspond to an enhancement of the membrane fluidity and to molecular rearrangements in thylakoid membranes. Following a heat shock the increase in MGDG (18:3/18:3), a lipid more tightly involved in photosystem activities (Sierfermann-Harms *et al.* 1982, Laskay and Lehoczki 1986, Horvath *et al.* 1987), could appear as a trigger mechanism for maintaining the efficiency of photosystems in chloroplasts from desert plants.

REFERENCES

Guillot-Salomon T., Farineau N., Cantrel C., Oursel A. and Tuquet C. (1987) Isolation and characterisation of developing chloroplasts from light-grown barley leaves. Physiol. Plantarum *69* 113-122.

Horvath G., Melis A., Hideg E., Droppa M. and Vigh L. (1987) Role of lipids in the organization and function of photosystem II studied by homogeneous catalytic hydrogenation of thylakoid membranes *in situ*. Biochim. Biophys. Acta *891* 68-74.

Laskay G. and Lehoczki E. (1986) Correlation between linolenic-acid deficiency in chloroplast membrane lipids and decreasing photosynthetic activity in barley. Biochim. Biophys. Acta *849* 77-84.

Siefermann-Harms D., Ross J.W., Kaneshiro K.H. and Yamamoto H.Y. (1982) Reconstitution by monogalactosyldiacylglycerol of energy transfer from light-harvesting chlorophyll *a/b*-protein complex to the photosystems in Triton X-100-solubilized thylakoids. FEBS Lett. *149* 191-196.

BIOLOGICAL ROLE OF PLANT LIPIDS
P.A. BIACS, K. GRUIZ, T. KREMMER (eds)
Akadémiai Kiadó, Budapest and Plenum Publishing
Corporation, New York and London, 1989

TEMPERATURE-DEPENDENT ALTERATION IN CIS AND TRANS UNSATURATION OF FATTY ACIDS IN VIBRIO SP. STRAIN ABE-1

H. Okuyama[1], S. Sasaki[1], S. Higashi[2] and N. Murata[2]

[1]Department of Botany, Faculty of Science, Hokkaido University, Sapporo 060, Japan
[2]National Institute for Basic Biology, Okazaki 444, Japan

INTRODUCTION

The trans configuration has been regarded as a non-physiological double-bond of fatty acids. However, a considerably high level of Δ9-trans-hexadecenoic acid (9-trans-16:1) was found in a marine psychrophilic bacterium, Vibrio sp. strain ABE-1 (Vibrio ABE-1), which has an optimum and maximum growth temperature at 15°C and 20°C, respectively (1). The phase transition temperature of the membrane phospholipids from Vibrio ABE-1 depends on growth temperature (2). Since the phospholipid composition of Vibrio ABE-1 is unaffected by growth temperature (2), the changes in phase transition temperature of the phospholipids are likely to be associated with the temperature-dependent alteration in their fatty acid composition.

In this study, the occurrence and the growth temperature-dependent alteration of trans-unsaturated fatty acid in Vibrio ABE-1 are described. It is suggested that the isomerization of cis/trans unsaturation of fatty acids may result in the growth temperature-dependent shift in the phase transition temperature of the membrane phospholipids.

MATERIALS AND METHODS

Vibrio ABE-1 was grown at 5, 15, and 20°C in a rotary shaker as described previously (3). Total lipids were extracted from the cells by the method of Bligh and Dyer (4). Phospholipids were separated from total lipids by one-dimensional thin-layer chromatography with a solvent of chloroform/methanol/water (65:25:4, by vol.). Total phospholipids were separated into phospholipid classes by two-dimensional thin-layer chromatography (5).

Aliquots of the total lipids, total phospholipids, and separated phospholipid classes were subjected to methanolysis at 80°C for 3 hours in 5% methanolic HCl. Resultant fatty acid methyl esters were extracted with hexane and analyzed by gas chromatography using a bonded fused silica column (cyanopropylmethylsilicone, 0.25 mm i.d. × 50 m, 0.25 µm film, Tokyo Kasei Kogyo). The fatty acid methyl esters were identified by comparing their retention times with those of standards. The double-bond positions and the *cis* and *trans* configurations of fatty acid methyl esters were determined by the pyrrolidide method (6) and IR analysis (7), respectively.

When the effect of temperature shift on the fatty acid composition was studied, cells grown at 5°C were washed with 0.5 M NaCl and suspended in a non-growing medium (5 mM Tris, 0.4 M NaCl, 10 mM KCl, 5 mM $MgCl_2$, and 1 M glycerol, pH 7.5) at 5°C at a concentration of 1.5×10^8 cells/ml. After cerulenin was added at 10 µg/ml to inhibit de novo synthesis of fatty acids, the temperature of the cell suspension was shifted from 5°C to 20°C. A portion of the cell suspension was withdrawn at appropriate intervals and subjected to lipid extraction.

RESULTS AND DISCUSSION

The presence of Δ9-*trans*-hexadecenoic acid in phospholipids from the 20°C-grown cells was detected by capillary

Table 1. Effect of growth temperature on the fatty acid composition of the total phospholipids from *Vibrio* ABE-1

Growth temp.	10:0	12:0	14:0	14:1 7c	16:0	16:1 9t	16:1 7c	16:1 9c	18:0	18:1 9c	18:1 11c
						(%)					
5°C	2	1	1	3	19	T	6	64	T[1]	T	2
15°C	2	1	1	1	23	2	7	58	T	T	3
20°C	1	T	1	1	26	12	7	47	T	T	4

[1], Less than 0.5%.

gas chromatography of fatty acid methyl esters (Fig. 1) and confirmed by the IR analysis and pyrrolidide method. The fatty acid composition of the total phospholipids from *Vibrio* ABE-1 cells grown at 5, 15, and 20°C is shown in Table 1. The content of 9-*trans*-16:1 was negligible in 5°C-grown cells, but amounted to 12% in 20°C-grown cells. Table 2 shows that a temperature shift from 5°C to 20°C in

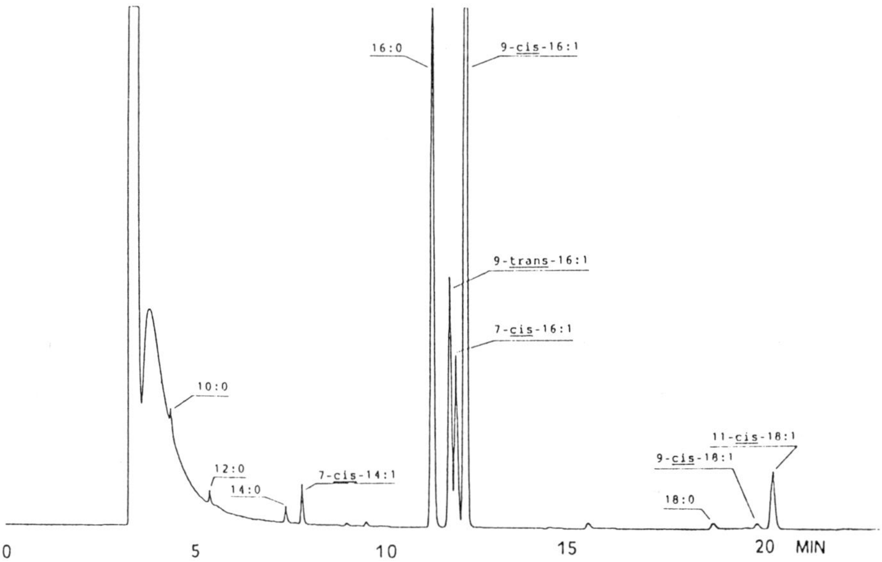

Fig. 1. Capillary gas-chromatogram of fatty acid methyl esters from total phospholipids of *Vibrio* ABE-1 grown at 20°C.

Table 2. Changes in fatty acid composition of the total lipids in *Vibrio* ABE-1 after temperature-shift up from 5°C to 20°C in the presence of cerulenin (10 µg/ml)

Time (hr)	10:0	12:0	14:0	14:1 7c	16:0	16:1 9t	16:1 7c	16:1 9c	18:0	18:1 9c	18:1 11c
						(%)					
0	4	2	1	4	21	5	6	51	T[1]	T	3
2	4	1	1	3	21	11	6	48	T	0	2
5	4	1	1	3	21	16	6	43	T	T	2
11	5	2	1	3	21	19	6	40	T	T	2
21.5	4	2	1	3	21	21	5	39	T	T	2

[1], Less than 0.5%.

the presence of cerulenin increased 9-*trans*-16:1 at the expense of 9-*cis*-16:1.

Vibrio ABE-1 is a bacterium which can grow over a temperature range such as below 0°C to 20°C. It is suggested that *trans*-unsaturated fatty acid besides the saturated and *cis*-unsaturated fatty acids might be needed for maintaining membrane fluidity at or near its maximum growth temperature.

The present study is the first report for *cis*/*trans* alteration in fatty acid unsaturation as a new strategy for microorganisms to adapt to changes in ambient temperatures.

This study was carried out under the NIBB Cooperative Research Program (87-161 and 88-136).

REFERENCES

1. Takada, Y., Ochiai, T., Okuyama, H., Nishi, K. and Sasaki, S., 1979, An obligately psychrophilic bacterium isolated on the Hokkaido coast. J. Gen. Appl. Microbiol., 25:11-19.
2. Okuyama, H., Fukunaga, N. and Sasaki, S., 1986, Homeoviscous adaptation in a psychrophilic bacterium, *Vibrio* sp. strain ABE-1. J. Gen. Appl. Microbiol., 32:473-482.
3. Hakeda, Y. and Fukunaga N., 1983, Effect of temperature stress on adenylate pools and energy charge in a psychrophilic bacterium, *Vibrio* ABE-1. Plant Cell Physiol., 24:849-856.
4. Bligh, E. G. and Dyer, W. J., 1959, A rapid method of total lipid extraction and purification. Can. J. Biochem. Physiol., 37:911-917.
5. Okuyama, H., Itoh, H., Fukunaga, N. and Sasaki, S., 1984, Separation and properties of the inner and outer membranes of a psychrophilic bacterium, *Vibrio* sp. strain ABE-1. Plant Cell Physiol., 25:1255-1264.
6. Andersson, B. A. and Holman, R. T., 1974, Pyrrolidides for mass spectrometric determination of the position of the double bond in monounsaturated fatty acids. Lipids, 9:185-190.
7. Chapman, D., 1965, Infrared spectroscopy of lipids. J. Am. Oil Chem. Soc., 42:353-371.

BIOLOGICAL ROLE OF PLANT LIPIDS
P.A. BIACS, K. GRUIZ, T. KREMMER (eds)
Akadémiai Kiadó, Budapest and Plenum Publishing
Corporation, New York and London, 1989

TEMPERATURE DEPENDENCE OF THE INHIBITORY EFFECT OF ANTIBODIES TO LIPIDS ON PHOTOSYNTHETIC ELECTRON TRANSPORT

A. Radunz and G.H. Schmid

Lehrstuhl Zellphysiologie, Universität Bielefeld, D-4800 Bielefeld 1, FRG

Summary

Antisera to lipids (monogalactolipid, digalactolipid, sulfolipid, phosphatidylcholine and phosphatidylglycerol) and antisera to carotenoids (lutein, zeaxanthin, violaxanthin and neoxanthin) inhibit photosynthetic electron transport on the donor side of photosystem I as well as of photosystem II. These inhibitory effects are dependent on the number of antibodies bound and on the temperature of the reaction medium. We have investigated the temperature influence between 10° and 45°C on the inhibition of photosystem I reactions. We observed that the maximum of the inhibitory effect for antisera to monogalactolipid, sulfolipid and phospholipids was located between 10° to 20°C and for antisera to lutein and zeaxanthin between 10° and 30°C. The antisera to the digalactolipid as well as those to violaxanthin and neoxanthin developed their highest inhibitory effect at 10°C. At higher temperatures, that is, between 40° and 45°C the degree of inhibition decreases strongly. A temperature dependence exists also for protein antisera (glycopeptide, 66 kDa polypeptide and cytochrome f).

Introduction

In our earlier publications we have used antibodies not only as reagents for the localisation of lipids and proteins in the thylakoid membrane (summarized by Radunz, 1981) but also for the functional characterization of compounds. Thus, by means of monospecific antisera to electron transport carriers of the photosynthetic electron transport chain, compounds such as plastoquinone (Radunz and Schmid, 1973), plastocyanin (Schmid et al., 1975), cytochrome f (Schmid et al., 1977), ferredoxin and ferredoxin-NADP-reductase (Schmid and Radunz, 1974) were localized at the sites in the electron transport chain which corresponded to their redox potential. Furthermore, the use of antisera demonstrated that polypeptides of the thylakoid membrane with molecular weights of 11 000 to 66 000 had the properties of electron transport components in the region of photosystem I and photosystem II (summarized Schmid et al., 1978, Bednarz et al., 1988, compare also the isolation of peptides, Menke and Koenig, 1980). Surprisingly, also the use of antisera to lipids (Radunz et al., 1984a,b ; Radunz, 1984) and carotenoids (Radunz and Schmid, 1979; Lehmann-Kirk et al., 1979) led to inhibitory effects on the donor sides of photosystem I and of photosystem II. These inhibitory effects of the lipid and pigment antisera are dependent on the physiological condition of the thylakoid membrane for instance induced by pH changes of the reaction medium. In this context we will report in the present publication that the inhibitory effect depends on the number of specific antibodies bound and on the temperature of the reaction medium.

Results

Antisera to constituents of the thylakoid membrane such as to glycolipids, phospholipids as well as to carotenoids inhibit photosynthetic electron transport in the region of photosystem I as well as of photosystem II. The degree of inhibition differs with the individual antisera used. If the degree of inhibition in the region of photosystem I and photosystem II is

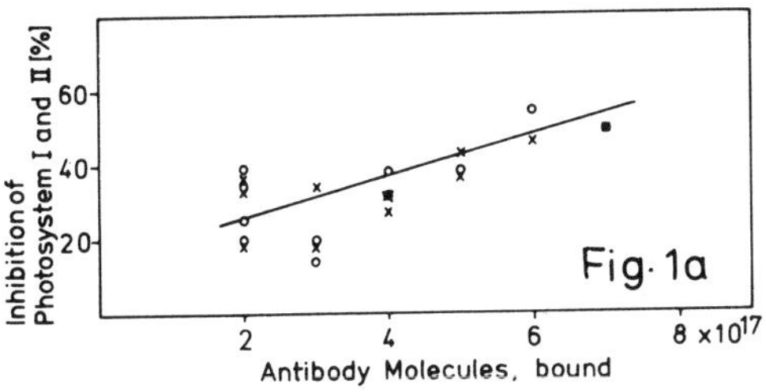

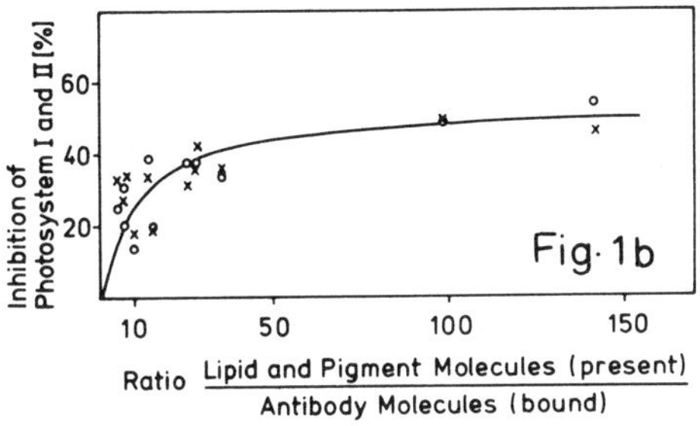

Fig. 1: Degree of inhibition of photosystem I (x) and photosystem II (o) in dependence on
a. the number of antibody molecules bound onto the thylakoid membrane of 1 g of chloroplasts
b. the ratio of the number of lipid and carotenoid molecules present and the number of antibody molecules bound.
Photosystem I reaction was measured as DCPiP/ascorbate → anthraquinone-2-sulfonate; photosystem II reaction: DPC → DCPiP

referred to the number of bound antibodies, it is seen, as shown in figure 1a, that a dependency of the inhibitory effect on the number of bound antibodies seems to exist. On the other side it should be borne in mind that the analysis of maximal binding of antibodies to lipids and proteins onto different chloroplast preparations has led to the observation that although all lipids are accessible to antibodies on the thylakoid surface directed to the stroma, the major portion of the lipids is located inside the thylakoid membrane or at the surface directed towards the inside of the thylakoid membrane. Moreover, lipids are located in the membrane in transversal as well as in lateral asymmetry (Radunz, 1980; Radunz, 1981). These facts taken together with the size difference between lipid molecules and antibody molecules anyway do not permit to expect a direct proportionality between number of antibody molecules bound and the caused inhibitory effect on electron transport. For these reasons, in figure 1b the degree of inhibition in the region of photosystem I and photosystem II was plotted as the function of the ratio of the amount of lipid and pigment molecules present and the number of antibody molecules bound. In this case, as expected a saturation curve is observed. The result shows that we have specific antisera to glycolipids, phospholipids and carotenoids, as the observed inhibitions of electron transport reactions are due to the specific binding of antibodies onto the respective antigenes.

The effect on photosynthetic electron transport is, as shown in earlier studies, not only dependent on the condition of the thylakoid membrane and on the pH of the reaction medium, but also on the temperature. We are able to show, that the inhibition of photosynthetic electron transport in the region of photosystem I is highest at temperatures from 10 to 20°C and 30°C respectively and that this inhibitory effect strongly decreases at temperatures of 40 and 45°C (Figures 2-4). The maxima of the inhibitory effects are very different for the individual lipid and pigment antisera. Thus, the maximum of inhibition is located for monogalactolipid, sulfolipid, phosphatidylcholine and phosphatidylglycerol antibodies between 10 to 20°C and for antibodies to lutein and zeaxanthin between 10 to 30°C. In contrast to this antibodies to the digalactolipid, as well as to violaxanthin and neoxanthin do not develop their highest effect of inhibition over such a broad temperature region. Instead, maximal inhibition lies here at 10°C and decreases from there more or less linearly until 45°C.

It should be noted that rates of photosynthetic electron transport are stimulated 6 to 10 fold when the temperature is increased from 10 to 45°C.

This temperature dependency of inhibition does not only occur with lipid and carotenoid antisera but is also observed with protein antisera. Figure 4 shows the temperature dependence of the degree of inhibition for a glycoproteid antiserum,

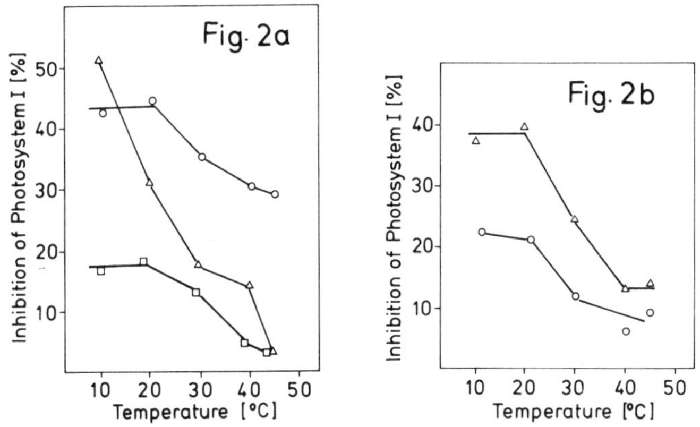

Fig. 2: Influence of temperature on the inhibition of the anthraquinone-2-sulfonate-Mehler reaction with DCPiP/ascorbate as electron donors with

a. o-o monogalactosyldigylceride antiserum
   △-△ digalactosyldiglyceride antiserum
   □-□ sulfoquinovosyldiglyceride antiserum

b. △-△ phosphatidylglycerol antiserum
   o-o phosphatidylcholine antiserum.

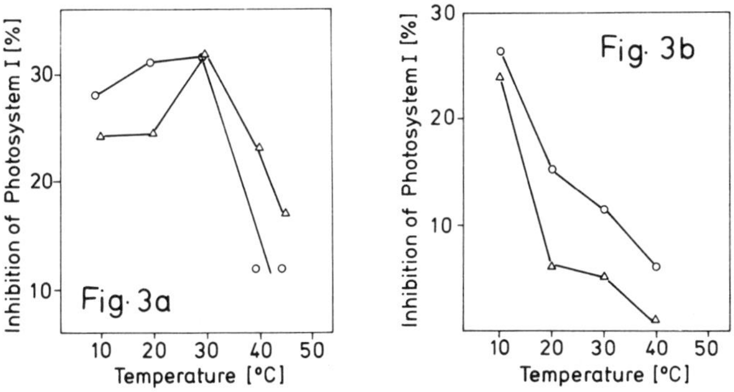

Fig. 3: Influence of temperature on the inhibition of the anthraquinone-2-sulfonate-Mehler reaction with DCPiP/ascorbate as electron donors with

a. o-o antiserum to lutein
   △-△ antiserum to zeaxanthin

b. o-o antiserum to violaxanthin
   △-△ antiserum to neoxanthin

(Radunz, unpublished) an antiserum to a 66 000 polypeptide (Schmid et al., 1978) and a antiserum to cytochrome f (Schmid et al, 1977). With all three protein antisera a pronounced inhibition maximum is observed. This maximum lies at different temperatures (Fig. 4)

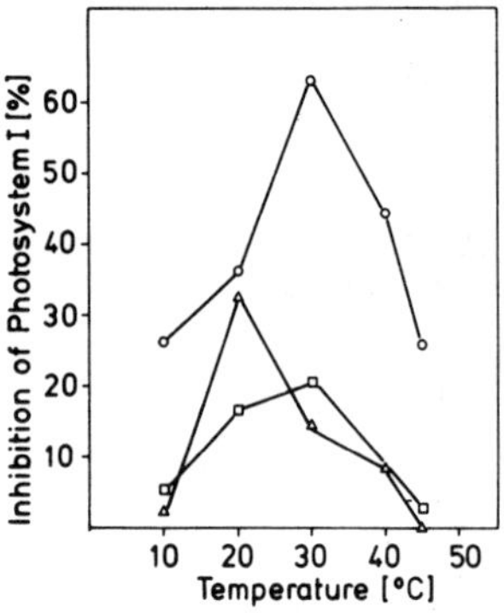

Fig. 4: Influence of temperature on the inhibition of the anthraquinone-2-sulfonate-Mehler reaction with DCPiP/ascorbate as electron donors with
o-o an antiserum to a glycoproteid
□-□ an antiserum to a 66 kDa polypeptide
△-△ the influence of the antiserum to cytochrome f was measured in the reaction $H_2O \rightarrow$ anthraquinone-2-sulfonate.

The decrease of inhibition of photosynthetic electron transport at higher temperatures is due to a decrease in binding affinity of the antibodies, which is verified by the fact that antibodies dissociate from their antigenic determinants at higher temperatures. Increased temperatures of 40 to 45°C do not denature immunoglobulins as far as their structure and binding capacity is concerned. Thus, we were able to demonstrate that antibodies to ribulose 1,5-bisphosphate carboxylase/ oxygenase heated for 20 minutes to 50°, 60° and 75°C were still able to form specific precipitates in agarose gel.

The inhibition of photosynthetic electron transport by antibodies to proteins and lipids is due to conformational changes of the proteins onto which antibodies specifically bind or onto which the corresponding lipids and pigments are adsorbed. The fact that the investigated antisera to lipids, pigments and proteins exhibit maximal inhibition at different temperatures points to the fact that these reactions are not only specific antibody-antigen binding reactions, but that the different lipids and pigments must be adsorbed onto quite different proteins and polypeptides since every protein undergoes at certain temperatures different conformational changes.

References

Bednarz, J., Radunz, A. and Schmid, G.H., 1988, Z. Naturforsch. 43c

Lehmann-Kirk, U., Bader, K.P., Schmid, G.H. and Radunz, A., 1979, Z. Naturforsch. 34c, 1218-1221

Menke, W. and Koenig, F., 1980, in: Methods in Enzymology, Vol. 69, 446-450, Photosynthesis and Nitrogen Fixation (A. San Pietro, ed.) Academic Press, New York

Radunz, A., 1980, Z. Naturforsch. 35c, 1024-1031

Radunz, A., 1981, Ber. Deutsch Bot. Ges. 94, 477-489

Radunz, A., 1984, in: Advances in Photosynthesis Research, Vol. III.2, 151-154 (C. Sybesma, ed.) Nijhoff/Junk, The Hague

Radunz, A. and Schmid, G.H., 1973, Z. Naturforsch. 28c, 36-44

Radunz, A. and Schmid, G.H., 1979, Ber. Deutsch Bot. Ges. 92, 437-443

Radunz, A., Bader, K.P. and Schmid, G.H., 1984a, Z. Pflanzenphysiol. 114, 227-231

Radunz, A., Bader, K.P. and Schmid, G.H., 1984b, in: Structure, Function und Metabolism of Plant Lipids, Vol. 9, 479-484 (P.A. Siegenthaler and W. Eichenberger, eds.) Elsevier Science Publishers, Amsterdam

Schmid, G.H. and Radunz, A., 1974, Z. Naturforsch. 29c, 384-391

Schmid, G.H., Radunz, A. and Menke, W., 1975, Z. Naturforsch. 30c, 201-212

Schmid, G.H., Radunz, A. and Menke, W., 1977, Z. Naturforsch. 32c, 271-280

Schmid, G.H. Menke, W., Radunz, A. and Koenig, F., 1978, Z. Naturforsch. 33c, 723-730

BIOLOGICAL ROLE OF PLANT LIPIDS
P.A. BIACS, K. GRUIZ, T. KREMMER (eds)
Akadémiai Kiadó, Budapest and Plenum Publishing
Corporation, New York and London, 1989

TEMPERATURE AND LIGHT EFFECTS ON THE EXPRESSION OF THE ARABIDOPSIS LINOLENATE MUTANT PHENOTYPE IN CALLUS CULTURES

J.A. Brockman and D.F. Hildebrand

Department of Agronomy, Univesity of Kentucky, Lexington, KY 40546, USA

Temperature and light have a large influence on linolenic acid levels. Low temperatures have been correlated with an increased ratio of polyunsaturated to saturated lipids in plant cell membranes, while plant tissues maintained in the dark show a significant increase in the desaturation of linoleate to linolenate upon illumination [1,2].

The *fadD* mutant of *Arabidopsis thaliana* L. which has a reduced amount of trienoic fatty acids provides a useful tool for the study of linolenic acid biosynthesis[3]. The maximum reduction of trienoic fatty acids in the *fadD* mutant is obtained at a growth temperature of 28°C. As the growth temperature decreases the effect of the mutation decreases to the point where the amount of trienoic fatty acid content of the *fadD* mutant resembles that of the wildtype[3]. In this study the differential temperature and light effects between the two genotypes are reported.

MATERIALS AND METHODS

Seeds of *Arabidopsis thaliana*, L. wildtype and the *fadD* mutant line were kindly provided by Dr. Chris Somerville (MSU-DOE Plant Research Laboratory, USA). Callus tissue was induced from leaf disks on Callus Inducing Media., transferred to Callus Growth Media after 10 days of induction[4] and grown at either 18°C or 28°C under continuous light at 150 uE m^{-1} s^{-1} or in complete darkness for three weeks.
Lipids were extracted by standard methods, methylated and analyzed by gas chromatography[5].

RESULTS AND DISCUSSION

Analysis of the linoleic and linolenic acid content of callus from the *fadD* mutant and wildtype *Arabidopsis* (Fig. 1) show that the reduction due to the mutation is seen only in the light at 28°C while in the dark at 28°C as well as in the light and dark at 18°C no differences between the genotypes were observed. This indicates that the desaturation process affected by the *fadD* mutation is sensitive to light.

Tissues grown at 18°C have a significantly higher ratio of linolenic to linoleic acid than tissues grown at 28°C (Fig. 1).

Furthermore, in contrast to the analogous experiment at 28°C, tissues derived from wildtype *Arabidopsis* and the *fadD* mutant do not show any significant differences with respect to the ratio of linolenic to linoleic acid when grown in the light at 18°C. The effect light has on the production of linolenic acid in callus tissue from the *fadD* mutant grown at 28°C seems to be overridden by the lowering of the growth temperature from 28°C to 18°C. This may be due to multiple pathways of linolenic acid production in *Arabidopsis* with the pathway predominant at 18°C being unaffected by the mutation.

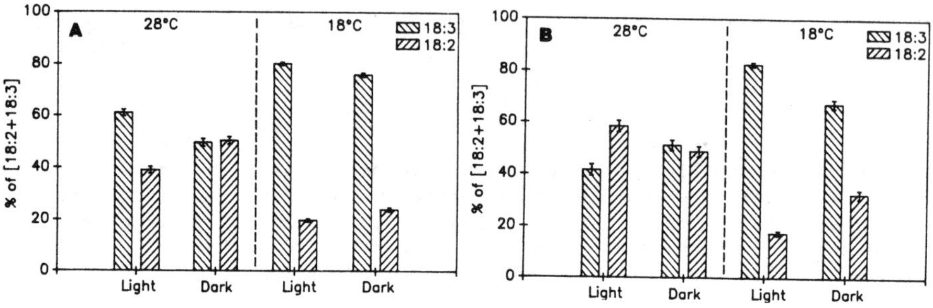

Fig. 1. Percentages of linoleic (18:2) and linolenic (18:3) acid relative to total amount of 18:2 and 18:3 present in *Arabidopsis* callus tissues of the wildtype (A) and *fadD* mutant (B). For growth conditions see Materials and Methods.

REFERENCES

[1.]Harwood, J.L. 1984. Effects of the environment on the acyl lipids of algae and higher plants. In Structure, function and Metabolism of Plant Lipids. Siegenthaler P.A. and Eigchenberger W., ed., Elsevier Science Publishers, Amsterdam, pp. 543-550.

[2.]Tremolieres, A. and M. Lepage. 1971. Changes in lipid composition during greening of etiolated pea seedlings. Plant Physiol. 47:329-379.

[3.]Browse, J., P. McCourt and C. Somerville. 1986. A mutant of *Arabidopsis* deficient in C18:3 and C16:3 leaf lipids. Plant Physiol. 81:859-864.

[4.]Feldmann, K.A. and M.D. Marks. 1986. Rapid and efficient regeneration of plants from explants of *Arabidopsis thaliana* Plant Science 47:63-69.

[5.]Wang, X.M., D.F. Hildebrand, H.A. Norman, M.L. Dahmer, J.B. St John and G.B. Collins. 1987. Reduction of linolenate content in soybean cotyledons by a substituted pyridazinone. Phytochemistry 26:955-960.

BIOLOGICAL ROLE OF PLANT LIPIDS
P.A. BIACS, K. GRUIZ, T. KREMMER (eds)
Akadémiai Kiadó, Budapest and Plenum Publishing
Corporation, New York and London, 1989

INFRASPECIFIC VARIABILITY IN COLD HARDINESS OF MARITIME PINE (PINUS PINASTER AIT.) AND FROST INDUCED CHANGES IN TERPENE HYDROCARBON COMPOSITION OF OLEORESIN

A. Marpeau[1], Ph. Baradat[2], P. Pastuszka[2], M. Gleizes[1], Th. Boisseaux[2], J. Walter[1] and J.P. Carde[1]

[1]Laboratoire de Physiologie Cellulaire Végétale, URA au CNRS 45, Université Bordeaux 1, 33405 Talence Cedex, France
[2]Laboratoire d'Amélioration des Arbres Forestiers, INRA-Bordeaux, 33610 Cestas, France

INTRODUCTION

In southwestern France (Aquitaine), maritime pine (*Pinus pinaster* Ait.) constitutes a monocultural forest of 1 million ha. After the unusually low temperatures of January 1985, 30 000 ha of stands were destroyed. The present paper reports some observations concerning the infraspecific variability of the response to low temperature conditions as well as the use of histological evaluation of damages as a good predictor of the surviving ability of pines. In addition, an unexpected consequence was a frost induced change in terpene metabolism in iberian provenances.

MATERIAL AND METHODS

In January 1985, minimal temperatures were about 17°C lower than normal ones, ten days long. Frost damaged maritime pines and healthy reference trees were sampled in 71 stands scattered throughout the whole Aquitaine forest area and also in comparative trials of various geographic origins. Naked eye observations of cortical tissues were made after notching the trunk. Histological studies were conducted on sections of cortical tissues fixed with glutaraldehyde and osmium and embedded in epoxy resin. Terpene analyses were carried out on cortical tissues of young shoots, according to Marpeau et al. (1975). Capillary columns were used for GLC analysis (Baradat and Marpeau, 1988). Data were computed from programs developed by Baradat (1985) and Baradat and Marpeau (1988).

RESULTS AND DISCUSSION

After the heavy cold conditions met in January 1985, damages of the pine forest occurred a few weeks later with different intensities according to the stands

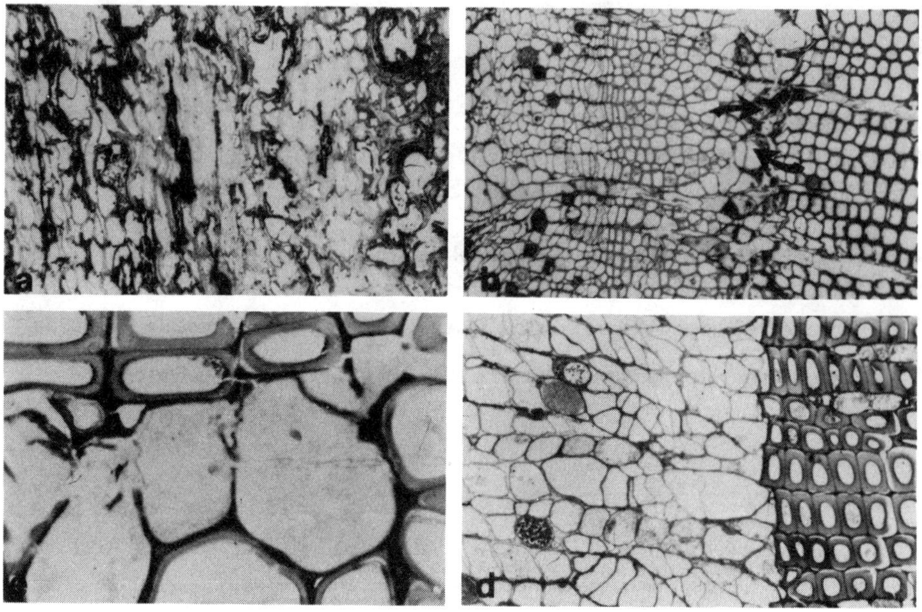

Figure 1a : Collapsed phloem tissue in the bark of a severely frozen tree (noted 5) x 370.
1b : Wood ring of the year in november 1985 (tree noted 4) : the ring is very thin and the first wood elements (arrow) are abnormal x 370.
1c : Incompletely lignified tracheids produced in early 1985, badly adhering to the 1984 ring (tree noted 4) x 920.
1d : Poorly developed phloem tissue with abnormal elements (november 1985, tree noted 4) x 370.

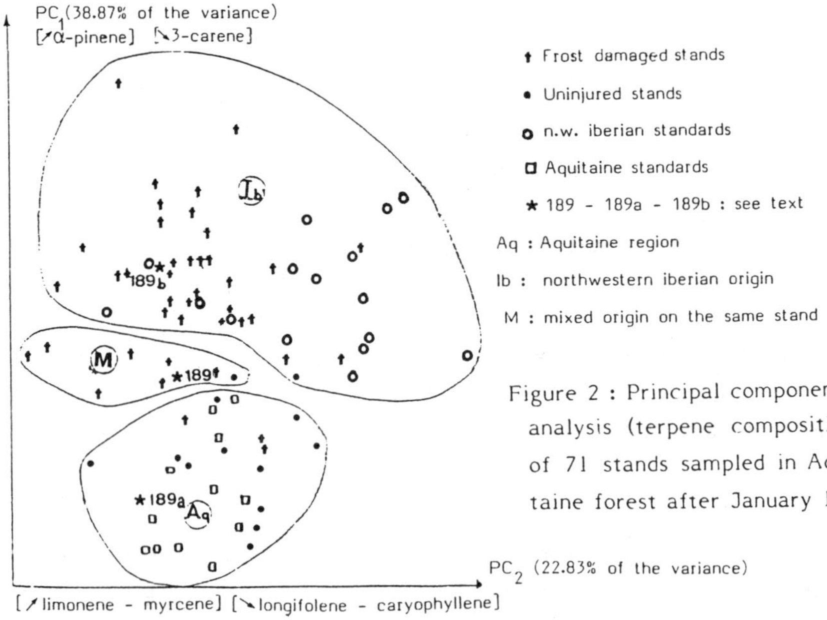

Figure 2 : Principal components analysis (terpene composition) of 71 stands sampled in Aquitaine forest after January 1985

and the individuals. The look of foliage and that of cambial-phloem tissues were used as external symptoms and noted from 1 to 5 according to severity of frost wounds. The correlation between both criteria was very high (0.98). The decay of trees noted 5 was very fast, whereas most trees noted 4 or 3 decayed later.

Frost injuries were interpreted at the histological level as more or less deep freezing of the trunk peripheral tissues resulting in destroying of the cell structure (Fig. 1a). Freezing of the cambial zone (stage 5) resulted in immediate death of the trees. In stages 4 and 3, frost induced alterations in the cambial zone and conducting tissue whereas in stage 2, only the peripheral layers of the bark were disrupted. Finally, in stage 1, there was no conspicuous modification of cell structure throughout the cortical tissues. Delayed modifications were observed with stages 3 and 4, namely disfunctioning of the cambial zone leading to the production of a limited number of conducting elements on both faces and therefore to very thin annual rings (Fig. 1b). In addition, many wood elements were not lignified and poorly adherent to the previous ring (Fig. 1c), whereas the presence of too few sieve cells (Fig. 1d) was a severe limitation to the conduction of photosynthates.

We tried to correlate the variations in the intensity of frost damages between stands and individuals to the geographic provenances of maritime pines. After the great fires of 1949, three geographical races of maritime pine seeds were used for regeneration of stands : mainly Aquitaine seeds (Aq) but also portuguese and northwestern spanish ones (Ib) which have been sown pure or mixed (M) in many stands.

Maritime pines which belong to different geographical races can be easily discriminated by the mean terpenic composition of oleoresin extracted from cortical tissues of their young shoots (Baradat et al., 1979). The origin of any pine stand in the natural area of *Pinus pinaster* can be identified by comparison with analyses made on stands of known origin.

The notation of the cambial-zone of 298 Aquitaine and 104 portuguese pines, on mature comparative trials submitted to the same low temperatures, was respectively 2.3 and 4.2, indicating a particular sensitivity of the "Ib" pines to cold conditions.

During 1985-1986 winter, 71 natural stands scattered in the Aquitaine forest area (Fig. 2) were analysed. Many of them were severely frost damaged. For each stand, the analysis of a mean sample was compared to standards by the way of a principal components analysis. The three principal components $PC_1 + PC_2 +$

PC_3 accounted for 77.46% of the total variance. It is clear that most of damaged populations are genetically identified to the northwestern iberian origin.

The case of mixed populations was clarified by the "189" analysis which was first located in the "M" cloud of points. Nevertheless, we observed a large heterogeneity in frost hardiness of the trees growing on this stand. Therefore, taking in account this dissimilarity, two sub-populations "189a" and "189b" were sampled and analysed for terpenes. The differential behaviour in respect to cold hardiness was explained by the analyses which indicated that the damaged population "189b" was a northwestern iberian provenance whereas the nearly uninjured "189a" was an aquitaine one.

All these results indicate that about 90% of the damages to maritime pines after the low temperatures of January 1985 can be attributed to exogenous pines with lower frost hardiness. This infraspecific response is consistent with the previous data of Bonneau et al. (1967) emphasizing the greater susceptibility of northwestern iberian pines to low temperatures. Similar results have been obtained with other Conifers, e.g. *Pinus sylvestris* (Eiche and Anderson, 1974), *Pinus contorta* (Rehfeldt, 1980), *Pinus strobus* (Maronek and Flint, 1974), *Pinus echinata* compared to *Pinus taeda* (Minckler, 1951). The heritability of frost hardiness was demonstrated for *Pinus sylvestris* (Norell et al., 1986), for *Pinus pinaster* (Tessier, 1986), and for other coniferous species.

*

* *

In the course of this study, we observed that many mean samples were abnormally rich in germacrene-D, a sesquiterpene hydrocarbon (Table 1).

Table 1 - Germacrene-D content in various maritime pine populations.

	Frost uninjured stands		Frost damaged stands		
Number of tested populations	12	16	3	32	8
Germacrene-D content	1-5%	1-8%	1-5%	13-35%	10-25%
Origin	"Aq"	n.w."Ib"	"Aq"	n.w."Ib"	"M"

The germacrene-D content, which is $\leqslant$ 5 to 8% in normal conditions, reaches 13 to 35% in "M" and "Ib" frost damaged populations.

In the case "M", a high germacrene content is observed only in "Ib" maritime pines (Table 2 - A : 189-189a-189b*). Interestingly, similar analytical were obtained two years later (winter 1987) on a heterogeneous stand (16) which survived to the same unusually low temperatures without great damages (Table 2 - B : 16, 16a, 16b*).

Table 2 - Germacrene-D content in the case of 2 heterogeneous populations "M" (189 and 16) subdivided in 2 sub-populations (a and b*)

	A - 1985		B - 1987	
Population Origin Germacrene-D content	189 "M" 15%		16 "M" 15%	
Sub-populations Origin Germacrene-D content	189a "Aq" 5%	189b* "Ib" 22%	16a "Aq" 0.8%	16b* "Ib" 33%

Therefore, the increase in germacrene-D content, observed only in "Ib" maritime pines submitted to low temperature conditions, is an infraspecific and steady deviation of the terpene metabolism.

It has been previously shown, using radioactive precursors, that E-farnesene is the first sesquiterpene hydrocarbon directly formed from farnesylpyrophosphate (FPP) (Gleizes et al., 1984). This acyclic precursor acts as a metabolic intermediate in the course of the formation of cyclic sesquiterpenes, particularly germacrene-D. In "Ib" frost damaged pines, a steady deregulation of the biosynthetic pathway should occur, leading to the production of an excess of farnesylpyrophosphate and its derivation towards the germacrene-D pathway, this compound acting as a "sink" for this hypothetical excess of FPP.

This modification seems to be the first report of changes in terpene biosynthesis induced by low temperature conditions.

REFERENCES

Baradat Ph., 1985 - OPEP. A conversational library of programs for tree breeding. Doc. Swedish Univ. of agricultural Sciences, Dept of Forest Genetic and Plant Physiology, Umeä, Sweden, 13 p.

Baradat Ph., Bernard-Dagan C. and Marpeau A., 1979 - Variation of terpenes within and between populations of maritime pine. Conference on biochemical genetics of forest tree, Umeä, pp. 151-169.

Baradat Ph. and Marpeau-Bezard A., 1988 - Le Pin maritime *Pinus pinaster* Ait. Biologie et génétique des terpènes pour la connaissance et l'amélioration de l'espèce. Thèse Doct. d'Etat es Sciences, Université Bordeaux I, 444 p. + annexes.

Boisseaux Th., 1986 - Influence de l'origine génétique (landaise ou ibérique) des peuplements de Pin maritime sur les dégâts causés par le froid de janvier 1985 au massif forestier aquitain. Mise au point d'un test variétal précoce utilisable pour le contrôle de lots de graines. Mémoire 3ème année ENITEF, Bordeaux, 70 p.

Bonneau M., Gelpe J. and Le Tacon F., 1969 - A propos du dépérissement du Pin portugais dans les Landes. Rev. forest. fr., 5 : 343-350.

Eiche V. and Andersson E., 1974 - Survival and growth in Scots pine (*Pinus silvestris* L.). Provenance experiments in Northern Sweden. Theor. Appl. Genet., 44 : 49-57.

Gleizes M., Marpeau A., Pauly G. and Bernard-Dagan C., 1984 - Sesquiterpene biosynthesis in maritime pine needles. Phytochemistry 23 : 1257-1259.

Maronek D.M. and Flint M.L., 1974 - Cold hardiness of needles of *Pinus strobus* L. as a function of geographic source. For. Sci., 20 : 135-141.

Marpeau A., Baradat Ph. and Bernard-Dagan C., 1975 - Les terpènes du Pin maritime : aspects biologiques et génétiques. IV. Hérédité de la teneur en deux sesquiterpènes : le longifolène et le caryophyllène. Ann. Sci. forest., 32 : 185-203.

Minckler L.S., 1951 - Southern pines from different geographic sources show different responses to low temperatures. J. For., 49 : 915-916.

Norell L., Eriksson G., Ekberg I. and Dormling I., 1986 - Inheritance of autumn frost hardiness in *Pinus sylvestris* L. seedlings. Theor. Appl. Genet., 72 : 440-448.

Teissier O., 1986 - Dégâts de gel sur les principales essences forestières dans le massif landais. Rapport de stage BTS, Option forêt. Ecole forestière de Mirecourt, 56 p. + annexes.

BIOLOGICAL ROLE OF PLANT LIPIDS
P.A. BIACS, K. GRUIZ, T. KREMMER (eds)
Akadémiai Kiadó, Budapest and Plenum Publishing
Corporation, New York and London, 1989

THE EFFECTS OF LIGHT AND OF OSMOTIC STRESS ON ENZYME ACTIVITIES ASSOCIATED WITH GALACTOLIPID SYNTHESIS

R.O. Mackender[1], C. Liljenberg[2] and C. Sundquist[2]

[1]Biology Department, Queen's University of Belfast, Belfast BT7 1NN, N. Ireland, UK
[2]Department of Plant Physiology, Botanical Institute, University of Göteborg, Carl Skottsbergs Gatta 22
S-413 19 Göteborg, Sweden

INTRODUCTION

The effects of light and of water stress in plant growth are well documented but less is known about their effects on plant enzymes. Chloroplast thylakoid membrane organization is particularly sensitive to light. The biosynthesis of galactolipids, the major acyl lipids of these membranes, is much investigated. However it is not known whether the enzymes involved are constitutive or inducible or are capable of being modulated directly by environmental factors. We wish to report some very preliminary data on UDPgalactosyl transferase (UDPgalT) activity (and thus indirectly on intergalactolipid transferase (IGT) activity) in plastids in relation to (1) the osmotic concentration (OC) of incubation media and (ii) the light regime in which plants were grown.

MATERIALS AND METHODS

See Mackender - this volume. High light (HL) and low light (LL) plants were raised on a 16 h photoperiod; red light (RL) and blue light (BL) plants were raised in continuous light at the same quantum flux density as LL plants.

RESULTS AND DISCUSSION

The effects of osmotic concentration. A typical result is shown in Table 1. Pretreatment at pH6.5/Ca^{2+} is unaffected by OC as is UDPgalT activity. The addition of phospholipase A_2 (PlA_2) either before or after the pH6.5/Ca^{2+} treatment has little effect on UDPgalT (and by implication IGT activity) in the absence of BSA, but in its presence PlA_2 would appear to inhibit IGT activity. The higher osmotic concentration during the PlC treatment seems to enhance subsequent UDPgalT activity.

Table 1. UDPgalT activity in isolated pea chloroplasts incubated at different osmotic concentrations (OC), following pretreatment for 20 minutes with either PlC or incubation at pH6.5/Ca^{2+} $\pm$ PlA_2 (+PlA_2 also $\pm$ BSA) under different osmotic conditions.

Pretreatment and conditions: Treatment	OC*	UDPgalT activity (dpm/μg chl/3 mins), OC* in assay medium: 0.4	1.0	0.5 -BSA	0.5 +BSA
PlC	0.4	1.68	2.15	-	-
0.1 μg/ml	1.0	3.97	1.93	-	-
pH6.5/10 mM Ca^{2+}	0.4	8.58	8.09	-	-
(IGT)	1.0	9.22	8.56	-	-
IGT PlA_2	0.5**	-	-	17.75	15.88
	1.0**	-	-	16.78	16.16
PlA_2 IGT	0.5**	-	-	17.96	13.62
	1.0**	-	-	17.48	14.25

* Molarity with respect to sorbitol. ** Only during PlA_2 treatment.

Table 2. The UDPgalT activity in plastids isolated from plants grown under HL, LL, RL or BL.

Parameter measured	Light regime: HL	LL**	RL	BL
Chl a/b ratio	3.89	3.76	3.58	3.49
Chl/plastid μg/10^6 plastids	0.53	0.72	0.23	0.27
UDPgalT activity*				
t_0	0.59	1.28	0.53	0.57
+ plC	0.74	2.36	0.99	1.06
+ pH6.5/10 mM Ca^{2+}	2.65	15.50	5.79	8.00

* As in Table 1. ** Discontinuous gradient of 30, 40 and 60% v/v Percoll used to purify these plastids.

The effects of light quality and quantity. A typical result is shown in Table 2. In untreated plastids UDPgalT activity is low but is highest in LL plastids. Following pretreatment with PlC, UDPgalT activity increases 100% in plastids from LL, RL and BL plants but only 50% in the HL plastids. The largest effect on UDPgalT activity is found following pH6.5/Ca^{2+} incubation (ie stimulation of IGT). Plastids from LL plants show the greatest effect followed by those from RL and BL plants. The activity in HL plastids is at least half that of plastids from other plants.

BIOLOGICAL ROLE OF PLANT LIPIDS
P.A. BIACS, K. GRUIZ, T. KREMMER (eds)
Akadémiai Kiadó, Budapest and Plenum Publishing
Corporation, New York and London, 1989

IMBIBITIONAL STRESS IN DRY POLLEN: INJURY IS CAUSED BY THE PRESENCE OF GEL PHASE PHOSPHOLIPID DURING IMBIBITION

F.A. Hoekstra[1], J.H. Crowe[2] and L.M. Crowe[2]

[1]Department of Plant Physiology, Agricultural University Wageningen, Holland
[2]Department of Zoology, University of California, Davis, USA

Dry anhydrobiotic organisms are often injured when they are plunged into water, and concomitantly they leak their soluble cellular constituents. Vital dry pollen also exhibits such behaviour. Simon (1974, New Phytol. 73:377-420) hypothesized that formation of $hexagonal_{II}$ phase lipid in the dry state and its reversal to the lamellar phase upon rehydration is the cause of the leakage. Temperature of imbibition and initial moisture content of the pollen determine the extent of the damage (Hoekstra, 1984, Pl. Physiol. 74:815-821). Improved survival of dry pollen after imbibition at elevated temperatures, however, is inconsistent with the $hexagonal_{II}$ phase being the cause of the leakage and death: its presence is favoured at elevated temperatures through an increase in acyl chain motion of PE.

Employing Fourier transform IR spectroscopy (Perkin Elmer 1750 FTIR + Perkin Elmer 7500 data station) we have recorded IR spectra of CH_2 stretching vibrations in intact pollen of Typha latifolia L. (Fig. 1). When pure lipids pass from the gel into the liquid crystalline (LC) phase the vibrational frequency of the symmetric CH_2-stretch increases. When pollen was slowly heated up a similar abrupt increase in wave number was observed. By plotting the frequency of the absorption peak versus the temperature, one can estimate the temperature at which half of the lipid has melted (T_m in Fig. 2). Germination capacity of and leakage from the pollen closely matched the shift in wave number.

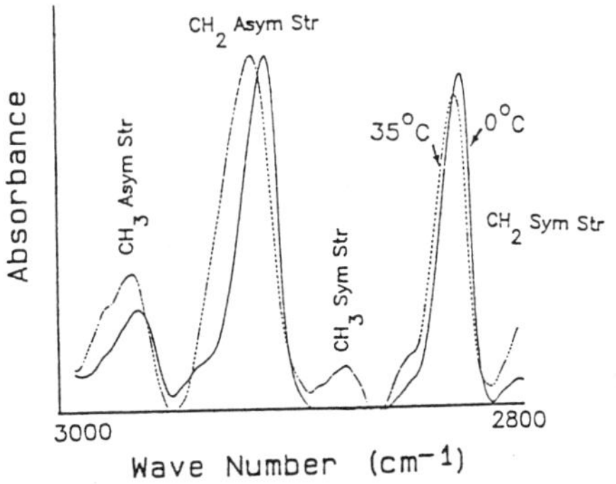

Fig. 1: IR-spectra showing the CH_2 and CH_3 symmetric and asymmetric stretch vibrations of intact pollen at 35°C (LC phase) and at 0°C (gel phase).

The transition temperature, however, was dependent on the moisture content of the pollen. Two decades ago Chapman et al. (Chem. Phys. Lipids 1:445-475, 1967) discovered hydration dependent phase behaviour in pure P-lipids. We found that purified pollen PC, the primary P-lipid present (50%), also showed hydration dependent phase behaviour, whereas purified pollen triglycerides did not. Surprisingly, the abundant NL in the pollen did not seem to contribute much to the frequency shift observed in intact pollen. Foregoing strongly suggests that the frequency shift in pollen can be assigned to a phase change in its P-lipids, which renders FTIR spectroscopy an excellent method to examine phase behaviour of membranes in intact cells.

We were able to construct a hydration-dependent phase diagram for intact pollen (Fig. 3). Below the apparent transition (P-lipids in gel phase) pollen is unable to imbibe without loss of vitality. Above the transition (P-lipids in LC-state) vitality is preserved and leakage minimized. In this light current techniques for optimizing germination <u>in vitro</u> of dry pollen such as imbibition at elevated temperatures and/or previous rehydration from the vapour phase become understandable: they promote LC-phase prior to the entry of bulk water. We suggest that a gel to LC phase transition during imbibition of dry pollen is the cause of the leakage and death.

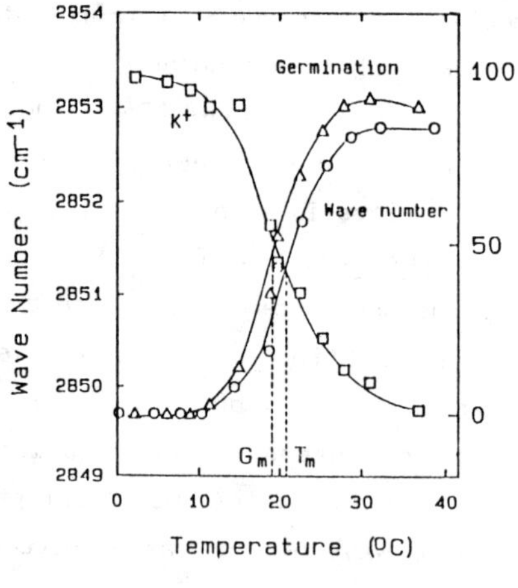

Fig. 2: Plot of the frequency of the absorbance maxima of the CH_2 symmetric stretch versus temperature. The germination and K^+-leakage data are also indicated. T_m and G_m are "mid" values.

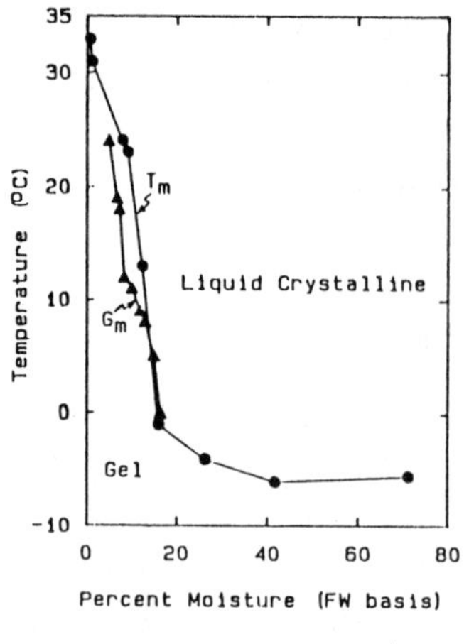

Fig. 3: Phase diagram of <u>Typha</u> pollen versus its moisture content. G_m and T_m are as in Fig. 2.

BIOLOGICAL ROLE OF PLANT LIPIDS
P.A. BIACS, K. GRUIZ, T. KREMMER (eds)
Akadémiai Kiadó, Budapest and Plenum Publishing
Corporation, New York and London, 1989

INCREASED LEAKAGE FROM AGEING POLLEN COINCIDES WITH INCREASED LEAKAGE FROM LIPOSOMES PRODUCED FROM ITS PURIFIED LIPIDS

Danielle G.J.L. van Bilsen and F.A. Hoekstra

Department of Plant Physiology, Agricultural University of Wageningen, The Netherlands

Dry pollen of cat-tail (Typha latifolia L.), containing 6-7% water on a fresh weight basis, has the considerable life span of approximately 150 days at 22°C. Decline of germination capacity coincides with increased leakage of K^+ from the re-imbibed grains, which is indicative of reduced plasma membrane integrity. Whereas polar lipid (PL) content hardly decreased during the loss of vitality, the content of free fatty acids (FFA's) increased three-fold, up to 0.22% of the dry weight (DW).

Incubation of pollen at higher relative humidity, giving a moisture content of approximately 15%, reduced longevity to less than 20 days (Fig. 1). Under these conditions the increase of K^+ leakage and FFA content seemed to preceed the decline of vitality. After 20 days, PL content had decreased to 65% of its original value (Fig. 1).

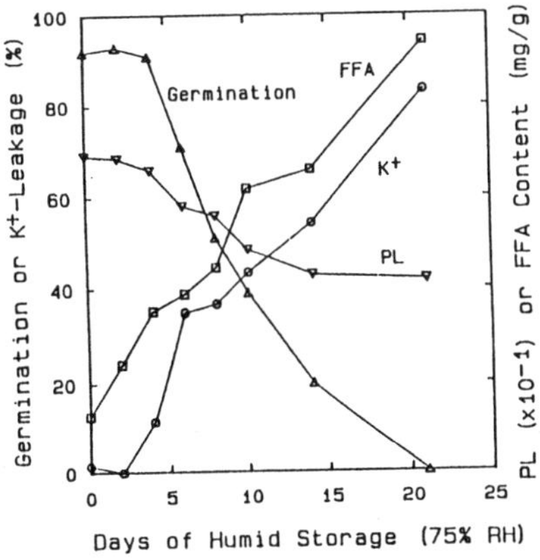

Fig.1. Effect of humid storage on viability, K^+-leakage, and contents of free fatty acids (FFA) and polar lipids (PL).

In order to analyse which lipidic compounds are involved in the altered membrane permeability, liposome studies were undertaken. Large unilamellar vesicles (80-90 nm; LUV) were prepared from purified pollen lipids, employing the technique of repeated extrusion at 500 psi through polycarbonate filters (0.1 µm pore diameter). The fluorescent dye, 5(6)carboxyfluorescein (CF), was entrapped during the production of the vesicles at 100 mMol, which is a self-quenching concentration. LUV were freed from exter-

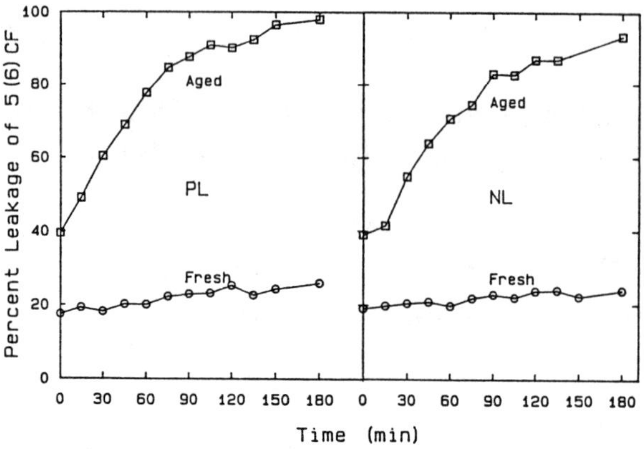

Fig.2. Leakage of CF from LUV during incubation at 20°C. LUV were composed of polar lipids (PL) or neutral lipids (NL) from fresh or aged (20 days, 75% RH at 22°C) pollen plus PC from fresh pollen (ratio 1:1).

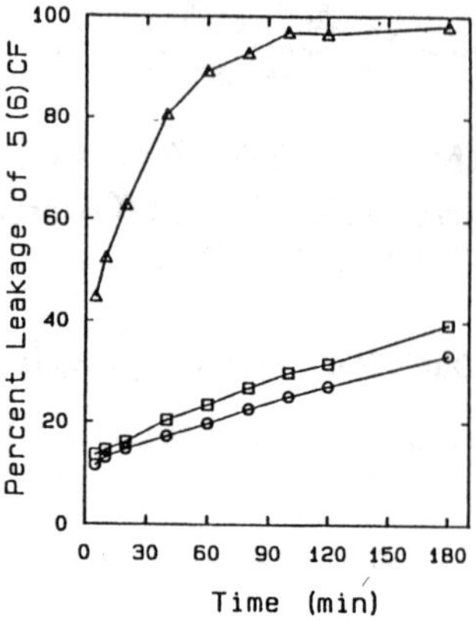

Fig.3. Leakage of CF as in Fig. 2. LUV were composed of fresh pollen PC and the indicated fractions (1:1), as follows: (○) total NL from fresh pollen; NL from fresh (△) and aged (□) pollen from which the FFA's were interchanged.

nal CF by passing them over a Sephadex G50 column. Leakage of CF from the vesicles was monitored during a 3 h storage period at 20°C.

Vesicles produced from mixtures (in the ratio 1:1) of fresh pollen PC and total PL, or PC and total neutral lipids (NL), slightly leaked CF into the surrounding medium when these fractions were purified from fresh pollen (Fig. 2). However, when the PL and NL fractions were purified from aged pollen that had lost vitality and excessively leaked K^+, leakage of CF from the vesicles rapidly reached maximal values. We conclude that during ageing changes occur in both the PL and NL fractions that may be associated with the reduced membrane integrity *in situ*.

To characterize the nature of the disturbing components the NL fraction was separated by TLC. As shown in Fig. 3, the increased FFA in the aged specimen turned out to be responsible for the increased leakage of CF from the vesicles. TLC separation of the PL fraction revealed the formation of lyso-PC and lyso-PE during the humid ageing up to 20% of the total amount of PL. The addition of these lysolipids to PL from fresh pollen somewhat increased the leakage of CF from LUV composed of this mixture plus fresh pollen PC (ratio 1:1). The proportion of linoleic acid, the major unsaturated fatty acid did not decrease significantly during ageing.

BIOLOGICAL ROLE OF PLANT LIPIDS
P.A. BIACS, K. GRUIZ, T. KREMMER (eds)
Akadémiai Kiadó, Budapest and Plenum Publishing
Corporation, New York and London, 1989

EFFECTS OF POWDERY MILDEW INFECTION ON THE LIPID METABOLISM OF CUCUMBER

J.K. Abood and Dorothy M. Lösel

Department of Plant Sciences, University of Sheffield, UK

The metabolic and physiological changes accompanying powdery mildew infection are less well understood than those induced by other biotrophic fungi, such as rusts. Among the powdery mildew diseases investigated, most attention has been paid to those affecting cereals rather than dicotyledon hosts and metabolic studies have tended to ignore lipid changes. *Sphaerotheca fuliginea*, which is responsible for substantial losses in cucumber production, is being investigated with a view to identifying metabolic features relating to the etiology and control of powdery mildew disease in this crop.

MATERIALS AND METHODS

Cucumber plants of variety Beit Alpha, grown in soil compost in growth room conditions (18-22°C, 16h day length) to the sixth leaf stage, were inoculated with spores from 7 to 10 days old infections of *Sphaerotheca fuliginea*. Tissue harvested from infected plants about 6 d after infection (early sporulation stage) and from healthy control plants, preliminary extraction in boiling isopropanol was extracted with chloroform: methanol solvents and washed (Folch *et al.*, 1957). Lipid contents were expressed on the basis of extracted dry weight (ext. d. wt.) after extraction of lipids, 80% ethanol-soluble fractions and, after treatment with amyloglucosidase, starch. Thin layer separation and estimations of lipid contents and investigation of the incorporation of ^{14}C-labelled photosynthate by healthy and infected leave was carried out as previously described (Lösel and Lewis, 1974). The radioactivity of total lipids and of lipid classes separated by thin layer chromatography was determined by scintillation counting.

RESULTS

Comparison of healthy and infected fifth (young mature) and sixth (newly expanded) leaves of cucumber plants, showed significantly higher lipid contents in diseased leaves than in healthy leaves (Fig. 1).

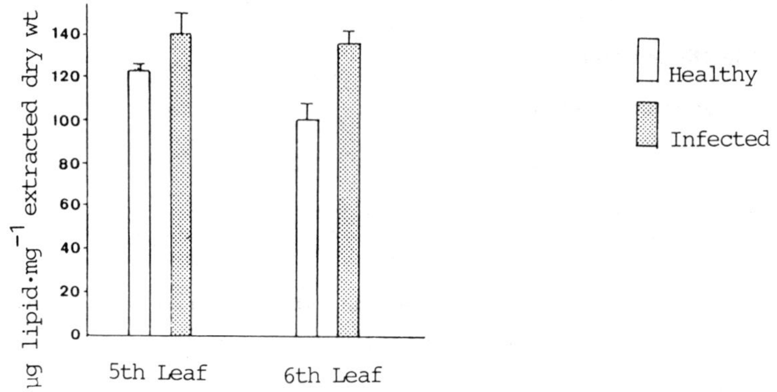

Fig. 1 Total lipid content of healthy and infected leaves of cucumber.

Table 1. Lipid composition of healthy and infected cucumber leaves. Means of samples from four replicate plants ± standard error; * = differences significant at 0.05 level.

	5th leaf		6th leaf	
	H	D	H	D
NEUTRAL LIPID ($\mu g.mg^{-1}$ ext. d. wt.)				
SE	4.0 ± 0.7	*14.5 ± 1.7	8.1 ± 2.0	12.9 ± 1.0
FAME	3.0 ± 0.3	* 8.6 ± 2.1	9.0 ± 2.1	5.1 ± 1.0
TG	5.6 ± 0.7	*12.0 ± 2.2	10.0 ± 0.7	7.1 ± 2.2
FA	5.4 ± 0.7	* 7.0 ± 1.4	3.8 ± 0.4	4.7 ± 0.6
DG	4.4 ± 1.3	*10.7 ± 0.1	5.7 ± 1.0	6.0 ± 0.8
S	4.7 ± 1.5	* 9.4 ± 0.9	2.3 ± 0.3	*6.3 ± 0.4
PHOSPHOLIPID (μg P. mg^{-1} ext. d. wt.)				
PC	0.7 ± 0.04	0.8 ± 0.08	0.3 ± 0.04	0.3 ± 0.01
PE	0.3 ± 0.01	*0.4 ± 0.03	0.3 ± 0.03	0.4 ± 0.04
PI	0.1 ± 0.01	*0.2 ± 0.01	0.1 ± 0.01	0.1 ± 0.01
PG	0.1 ± 0.02	*0.4 ± 0.05	0.3 ± 0.04	0.3 ± 0.01
Total	1.2 ± 0.07	*1.8 ± 0.19	1.0 ± 0.07	*1.1 ± 0.07

In the more mature leaves, infection resulted in substantial increases in sterol esters (SE), fatty acid methyl esters (MEFA), triacylglycerols (TG), diacylglycerols (DG), and sterols (S), with smaller increases in fatty acids (FA) (Table 1). In younger leaves, the sterol content was increased by infection but other neutral lipids did not change significantly. The total phospholipid content was significantly higher in infected than in healthy leaves of both ages. Phosphatidyl ethanolamine (PE), phosphatidyl inositol (PI) and phosphatidyl glycerol (PG) increased significantly in the older leaves following infection but in the younger leaves, where infection was less advanced, no significant differencesin individual phospholipids were detected at this stage. Infected leaves of both ages showed no significant change in the amounts of the major phospholipid, phosphatidyl choline (PC) or in phosphatidic acid which, in both healthy and infected leaves was present only in very small amounts.

The incorporation of ^{14}C-labelled photosynthate by lipid fractions of healthy and infected leaves of cucumber was investigated at early sporulation, immediately after photosynthesis and following chase periods of 4 h in light and 24h. Lipid of infected leaves contained less labelled carbon at each time of extraction than that from healthy leaves, particularly after the overnight chase (Fig. 2A). However, when related to total photosynthate (Fig. 2B), lipid ^{14}C accounted for increasing percentages of total carbon incorporation (lipids plus ethanol soluble fractions and starch) during the chase period in both healthy and diseased leaves, which differed only in a more rapid accumulation of labelled carbon by lipids of infected leaves during the first four hours.

In a study of the translocation of photosynthate within diseased plants, the distribution of ^{14}C-labelled photosynthate in lipids of leaves, shoot apex and root was compared. Although the radioactivity of lipids in leaves exposed to $^{14}CO_2$ was increased by infection, incorporation by lipids of the shoot apex and root tissues was greatly reduced in infected plants (Fig. 3A). The proportions of photosynthate present in lipids of both the shoot apex and leaves were, however, increased by infection while the proportion incorporated into root lipids remained unchanged (Fig. 3B).

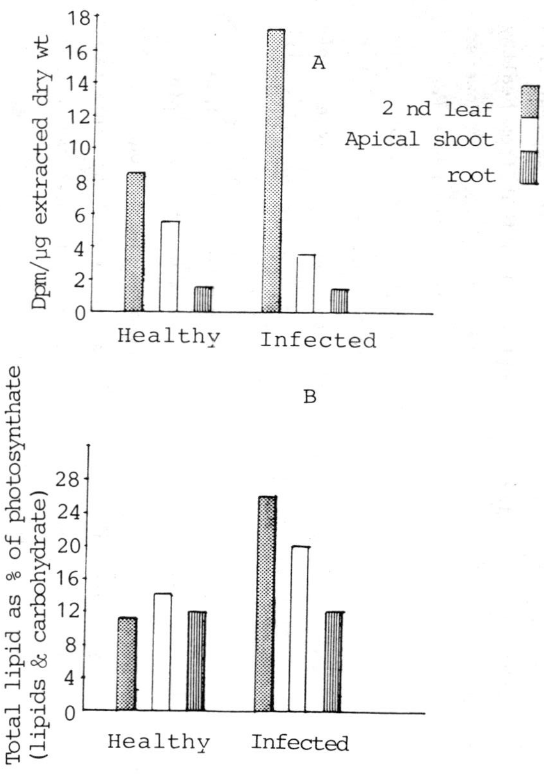
A
2 nd leaf
Apical shoot
root
Dpm/µg extracted dry wt
18
16
14
12
10
8
6
4
2
0
Healthy
Infected
B
Total lipid as % of photosynthate
(lipids & carbohydrate)
28
24
20
16
12
8
4
0
Healthy
Infected

Fig. 2

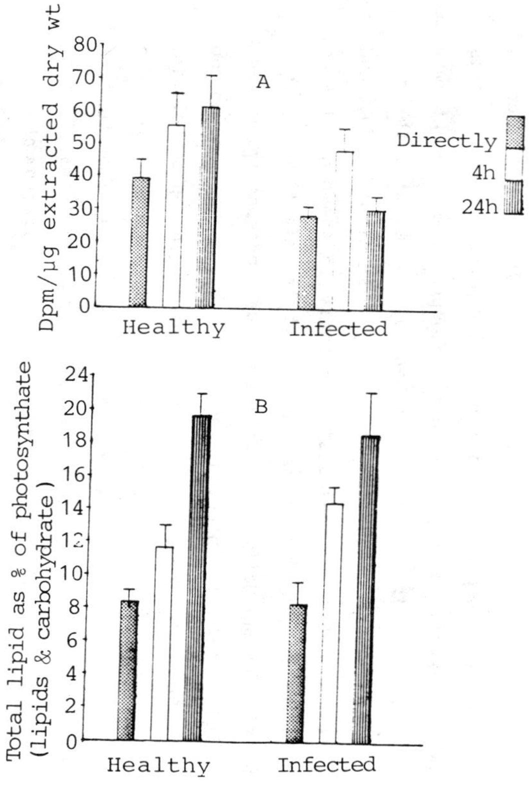
A
Directly
4h
24h
Dpm/µg extracted dry wt
80
70
60
50
40
30
20
10
0
Healthy
Infected
B
Total lipid as % of photosynthate
(lipids & carbohydrate)
24
20
18
16
14
12
10
8
6
4
2
0
Healthy
Infected

Fig. 3.

DISCUSSION

The alterations in lipid metabolism following powdery mildew infection of cucumber leaves and in the whole plant are less than those previously reported for infections by rusts and other biotrophic fungi (Brennan and Lösel, 1978, Lösel, 1980), as might be expected in this type of pathogen, which only penetrates epidermal cells, but greater than those recorded in a rigorous study by Sloss (1985) of lipid metabolism in *Erysiphe pisi* infections of pea cultivars. Both the small amount of fungal tissue present at the stage studied and microscopic observations of increased amounts of lipid in host mesophyll cells suggest that alterations in host in lipid metabolism are the major component detected here. More detailed studies of lipid and other metabolic changes accompanying powdery mildew infection are continuing.

Financial support for J.K.Abood from the Iraqi Ministry of Higher Education is gratefully acknowledged.

REFERENCES

Brennan, P.J. and Lösel, D.M. (1978) Physiology of fungal lipids. Adv. Microb. Physiol. 17, 47-179.

Folch, J., Lees, M. and Sloane-Stanley, G.H. (1957) J. biol. Chem. 226, 497-509.

Lösel, D.M. (1980) The effect of biotrophic fungal infection on the lipid metabolism of green plants. *In* Biogenesis and function of fungal lipids (Eds. P. Mazliac, P. Beneviste, C. Costes and R. Douce), pp. 263-268. Elsevier, Amsterdam, New York and Oxford.

Lösel, D.M. and Lewis, D.H. (1974) Lipid metabolism in leaves of *Tussilago farfara* during infection by *Puccinia poarum*. New Phytol. 73, 1157-1169.

Sloss, R.I. (1985) Lipid metabolism in *Pisum sativum* infected by Erysiphe pisi. Ph.D. Thesis, University of London.

BIOLOGICAL ROLE OF PLANT LIPIDS
P.A. BIACS, K. GRUIZ, T. KREMMER (eds)
Akadémiai Kiadó, Budapest and Plenum Publishing
Corporation, New York and London, 1989

THE EFFECT OF GROWTH CONDITIONS ON THE COMPOSITION OF THE SOLUBLE CUTICULAR LIPIDS AND THE WATER PERMEABILITY OF ISOLATED CUTICULAR MEMBRANES OF CITRUS AURANTIUM L.*

M. Riederer, Uta Geyer and J. Schönherr

Institut für Botanik und Mikrobiologie, Technische Universität München, Arcisstr. 21
D-8000 München 2, FRG

INTRODUCTION

The cuticular membrane is the interphase between all above-ground primary parts of higher plants and the atmospheric environment. Organic compounds with low volatility and inorganic ions leaving or entering primary parts of plants have to cross this barrier. The transport properties of cuticles have been found to be determined almost exclusively by wax-like cuticular lipids which are embedded within the cutin matrix or deposited on the outer surface of the cuticle. In contrast to the insoluble lipid polymer cutin made up of cross-linked hydroxy- and epoxyalkanoic acids, the cuticular waxes can be removed by organic solvents and thus are also called soluble cuticular lipids (SCL).

While the overall effect of the long-chain aliphatic SCL on the transport properties of plant cuticles is well documented, the influence of the chemical composition of cuticular waxes on the permeability of cuticles is less clear. Therefore, the authors have performed an extended study with the purpose of elucidating the effect of growth conditions on both the chemical composition of SCL and on the water permeability of isolated leaf cuticles. Selected results from this study are reported.

---

* This work was supported by the Deutsche Forschungsgemeinschaft. The help provided by Dr. R. Winkler, H. Krause and G. Schneider in identifying SCL components is gratefully acknowledged.

MATERIALS AND METHODS

Eight Citrus aurantium L. trees were grown in environmental chambers under controlled conditions. Air temperatures were controlled and varied from 15° to 35° C. Two levels of relative humidity (50 and 90 %)were used. Photon flux density was 0.5 to 1 mmol/(m^2 s) PAR at the level of the leaves. At the beginning of each treatment, the trees were pruned back. Five to ten fully expanded leaves were taken from the top of each shoot when the shoots had developed about 15 to 35 mature leaves under the conditions chosen for the particular experiment. Cuticular membranes were isolated enzymatically with pectinase solutions (Schönherr and Riederer 1986). The isolated cuticles were thoroughly washed with deionized water, flattened and dried.

For chemical analysis of SCL the isolated cuticular membranes were incubated in 0.02 M borax buffer in order to remove contaminants sorbed during incubation and were subsequently dried. From each combination of growth conditions ten cuticles (20 mm in diameter) were randomly sampled for analysis and extracted separately with chloroform. After removal of the solvent the extract was treated with N,N-bis-trimethylsilylfluoro-acetamid which forms the corresponding trimethylsilyl derivatives of hydroxyl and carboxyl compounds. The samples were analysed by temperature-programmed capillary gas chromatography without any further treatment. The coverages (amounts per unit cuticle area) of the 18 most prominent SCL components in the carbon-number range up to C_{40} were determined. The identity of the components had previously been established by thin layer chromatography and gas chromatography/mass spectrometry. The water permeability of cuticular membranes was determined gravimetrically according to Schönherr and Lendzian (1981).

RESULTS AND DISCUSSION

The direct analysis of SCL extracts from Citrus aurantium leaves yielded 18 prominent components in the carbon-number range up to C_{40} which were identified as primary n-alkanols (11 homologues), n-alkanes (5 homologues), tritriacontanal and friedel-3-on. The quantitative analysis of a series of n-alkyl esters (C_{38} to C_{52}) has not yet been fully completed. Primary alkanols which were present in a complete homologue series from C_{25} to C_{36} made up 80 to 90 % of the total amount of the compounds analysed, irrespective of growth conditions. The two most important homologues of this class were do- and tetratriacontanol while considerably lower coverages were observed for shorter and longer homologues. Though present in all samples, odd-chain homologues contributed only minor amounts to the total of alkanols. The cumulative amounts of alkanes were by factors of 4 to 13 smaller than those of the alkanols. Alkane homologues ranged from C_{29} to C_{33} with hen- and tritriacontane being the most important ones and the odd-numbered homologues clearly dominating their even-numbered counterparts. The amounts of the only alkanal homologue detectable in directly analysed extracts played minor roles. The triterpenoid ketone friedel-3-one was present in all samples while its relative and absolute contributions to the total of SCL greatly varied.

The absolute amounts and the relative composition of Citrus leaf SCL were significantly affected by growth conditions (Table 1). The totals of the alkanol and alkane fractions and the amounts of friedel-3-one varied by factors of 2.8, 3.2 and 17.5, respectively. However, no general dependence of SCL composition on a single environmental factor can be extracted from the present data even if analysed on a compound instead of a substance class basis. Exceptions are the distinct increases in the coverages of alkanols and friedel-3-one with increasing night temperatures, provided relative humidities were kept at the lower level during the day. A more detailed statistical analysis of the impact of growth conditions on SCL composition is currently under way.

The permeances for water of Citrus aurantium cuticles were determined for a total of 3174 isolated cuticular membranes individually. Permeances ranged from 0.5×10^{-10} to 3.5×10^{-10} m/s for all temperature and humidity combinations tested. The permeance frequency distributions of most treatments overlapped considerably while in some cases significant differences were found. These differences can be attributed exclusively to the effects of storage on water permeance of isolated cuticles since it has been found that water permeabilities drastically decreased as a function of time elapsed from isolation. No significant effect of growth conditions and consequently of SCL composition could be found when the influence of storage time on the water permeability isolated cuticles was eliminated.

Table 1: The effect of growth conditions on the composition of Citrus aurantium leaf SCL*

temperature (° C)	humidity (%)	alkanols (mg/m^2)	alkanes (mg/m^2)	alkanal (mg/m^2)	friedelone (mg/m^2)
25 / 15	50 / 50	10.6	1.8	0.3	0.4
25 / 15	90 / 50	21.2	2.1	0.3	0.5
25 / 15	90 / 50	19.3	1.9	0.0	0.7
25 / 15	90 / 50	16.0	1.8	0.0	0.4
25 / 15	50 / 90	18.6	2.4	0.0	0.7
25 / 15	90 / 90	13.8	1.1	0.3	0.7
25 / 20	50 / 90	22.9	1.9	0.0	1.9
25 / 20	90 / 90	22.9	2.3	0.0	0.2
30 / 20	50 / 50	26.8	3.0	0.8	1.7
30 / 20	90 / 90	19.8	2.7	0.4	0.7
30 / 25	50 / 50	29.6	2.8	0.0	23.5
30 / 25	90 / 50	25.2	2.4	0.0	3.5
30 / 25	90 / 90	15.6	3.5	0.3	0.5
35 / 25	90 / 90	17.6	2.1	0.5	1.4

* Growth conditions are given as day / night values.

REFERENCES

Schönherr J, Lendzian K (1981) A simple and inexpensive method of measuring water permeability of isolated plant cuticular membranes. Z. Pflanzenphysiol. 102:321-327

Schönherr J, Riederer M (1986) Plant cuticles sorb lipophilic compounds during enzymatic isolation. Plant Cell Environ. 9:459-466

BIOLOGICAL ROLE OF PLANT LIPIDS
P.A. BIACS, K. GRUIZ, T. KREMMER (eds)
Akadémiai Kiadó, Budapest and Plenum Publishing
Corporation, New York and London, 1989

EFFECTS OF SOME DETERGENTS ON THE STEROLS OF THE RED ALGA PORPHYRIDIUM

H. Nyberg and P. Saranpää

Department of Botany, University of Helsinki, Unioninkatu 44, 00170 Helsinki, Finland

INTRODUCTION

Detergents (surface-active components of washing agents) are known to inhibit the growth of unicellular algae (1,2). As detergents are in common household and industrial use, they contribute to water pollution (2,3,4). The harmful effects on the unicellular red alga *Porphyridium purpureum* have been reported in earlier detergent studies. Usually the presence of detergents in the culture medium induced the reduction of growth of *P. purpureum* (5) and changes in the amounts and fatty acid compositions of its glycolipids (6) and phospholipids (7,8). The most conspicuous lipid changes have been found to be an increase in the saturation grade of the membrane lipids, with a loss of especially arachidonic acid (6-8). This may result in the disruption of the vital membrane functions.

Sterols are important components in cell membranes and they are thought to function primarily in membrane structure stabilization (9). The sterols of *P. purpureum* are poorly known (10), partly because it seems to contain only small amounts of sterols; because of this *P. purpureum* was originally thought to lack sterols (11). In the red algae, the dominant sterol is usually cholesterol, but desmosterol and 22-dehydrocholesterol have been found to be prominent in some species (12). *Trans*-22-dehydrocholesterol has been reported to be the dominant sterol in *P. purpureum* (10).

In this study, the effects of the detergents Ufasan 65 (a linear alkylbenzene sulphonate (LAS), anionic), Triton X-100 (iso-octylphenoxypolyethoxyethanol, non-ionic) and TEGO-Betain L7 (an amphoteric betaine) on the sterol composition of *P. purpureum* were assessed using lipid fractionation techniques and GLC. The detergent concentrations used were 5-30 ppm. The main characteristics of the detergents used have been listed earlier (2).

MATERIALS AND METHODS

The CCAP 1380/1a *P. purpureum* strain was cultured on a solid 1 % agar synthetic nutrient medium (6). The detergents were added to the cooled autoclaved medium with sterile filters. Growth time was one month at 25°C under 100 $\mu E\ m^{-2}S^{-1}$ (400-700 nm) light with an 18:6h light:dark cycle. The algae were harvested and subjected fresh to silicic acid CC (13). The neutral lipid fraction was analysed by TLC (14) and the sterols were located by spraying with 0.001 % primuline (15). The sterols were then converted to TMS esters and analysed by GLC using a SE-30 glass capillary column (14). Sterol identification was done by co-chromatography with authentic sterols and by comparing chromatographic data with published Rt values (14,16).

RESULTS AND DISCUSSION

Sterols of Porphyridium purpureum

The dominant sterol in *P. purpureum* was found to be *trans*-22-dehydrocholesterol, which constituted about 3/4 of the total sterols (Tables 1-3). This is in accord with the results of the only earlier study on the sterol composition of *P. purpureum* (= *cruentum*) (10). Cholesterol was found to be the second most important sterol, but its amounts were only 4-5 % of the total in the controls. Cholesterol is known to be the dominant sterol in most red algae (12), but several species have been reported to contain desmosterol and 22-dehydrocholesterol. Red algae, with some exceptions, seem to contain C-27 sterols exclusively (12). We found desmosterol in small amounts also in *P. purpureum*. Cycloartenol is known to function in algae as a sterol precursor, and, accordingly, it was also found in *P. purpureum* (1.4 % in the controls). *P. purpureum* seemed to contain additionally small amounts of ergosterol, ß-sitosterol (comp. with (10)), and an unknown compound co-chromatographing with 5α -cholestan-3ß -ol. Other unknown sterols were certainly present, but only as minor components. Sterol identification in *P. purpureum* is difficult because of the small amount of sterols present and the large percentage of only one sterol, *trans*-22-dehydrocholesterol. For these reasons, no quantitation was attempted at this stage. In this respect, our results should be considered preliminary and the minor sterols of *P. purpureum* need further investigation. They might represent intermediate products in sterol biosynthesis (9). Chemotaxonomically the red algae can be clearly separated from other algal classes on the basis of their sterol composition, and, moreover, *P. purpureum* seems to be different from other red algae in having *trans*-22-dehydrocholesterol, and not cholesterol, as the dominant sterol. We could not find any steryl esters in *P. purpureum*.

Detergent influence on the sterol composition of P. purpureum

Detergents are known to cause increased saturation of *P. purpureum* glyco- and phospholipids (6-8), with the loss of especially arachidonic (20:4ω6), linoleic (18:2ω6) and eicosapentaenoic acids (20:5ω3). Usually an inhibition of growth has also been observed in the presence of detergents, especially the "hard" types (1,2,5). Of the detergents tested in this study, LAS (Ufasan 65) and TEGO-Betain L7 were "hard" and Triton X-100 "mild" in this respect.

Table 1. Relative amounts (%) of sterols in *Porphyridium purpureum* in the presence of LAS.

	conc. of detergent (ppm)			
sterol	0	10	20	30
trans-22-dehydrocholesterol	72.0	45.0	75.2	65.0
desmosterol	tr	1.7	1.2	2.7
unknown	tr	5.4	tr	3.4
cholesterol	4.3	13.8	20.0	17.5
ergosterol	tr	3.4	-	2.4
ß-sitosterol	2.1	5.3	tr	tr
cycloartenol	1.4	10.2	tr	tr

The effects of LAS on the sterol composition of *P. purpureum* are presented in Table 1. A large increase in the relative amounts of cholesterol, and in some cases also those of desmosterol and ergosterol, was observed, with a corresponding decrease in the amounts of 22-dehydrocholesterol at 10 and 30 ppm. At 10 ppm, the amounts of cholestanol (?), ß-sitosterol and cycloartenol were at their maximum. The results are not linear with increasing detergent concentration. This seems to be a feature often encountered in detergent effects on algae (5-8).

Table 2. Relative amounts (%) of sterols in *Porphyridium purpureum* in the presence of Triton X-100

	conc. of detergent (ppm)			
sterol	0	10	20	30
trans-22-dehydrocholesterol	72.0	49.4	59.1	55.2
desmosterol	tr	5.0	4.5	1.0
unknown	tr	2.3	3.5	2.4
cholesterol	4.3	5.8	5.4	3.7
ergosterol	tr	1.2	0.9	-
ß-sitosterol	2.1	2.8	1.5	1.8
cycloartenol	1.4	4.6	4.0	2.4

In spite of the known "mildness" of Triton X-100 (2, 5-8), it markedly decreased the amounts of 22-dehydrocholesterol in *P. purpureum* (Table 2). A corresponding increase in the amounts of cholestanol (?) and cycloartenol was observed. At 10 ppm, the amounts of desmosterol, cholesterol, ß-sitosterol and cycloartenol were at their maximum; this concentration of Triton X-100 often increases algal growth (1,2,5). A marked difference between the effects of LAS and Triton X-100 can be seen in the amounts of cholesterol.

Table 3. Relative amounts (%) of sterols in *Porphyridium purpureum* in the presence of TEGO-Betain L7

	conc. of detergent (ppm)		
sterol	0	5	10
trans-22-dehydrocholesterol	72.0	67.0	43.7
desmosterol	tr	2.4	5.1
unknown	tr	tr	1.1
cholesterol	4.3	15.2	3.4
ergosterol	tr	tr	3.3
ß-sitosterol	2.1	0.9	tr
cycloartenol	1.4	tr	4.8

TEGO-Betain L7 is an amphoteric betaine with a powerful capacity to decrease algal growth, in other respects the effects of betaines on algae are very poorly known (2,5). Its influence on the sterol composition of *P. purpureum* had the same observable features as the other detergents tested (Table 3). A decrease in the amounts of 22-dehydrocholesterol was again observed. A corresponding linear increase was at this time seen only with desmosterol. A large maximum of cholesterol was observed at 5 ppm and of cycloartenol at 10 ppm.

In conclusion it can be stated that all the detergents tested seemed to change the sterol composition of *P. purpureum* in a similar manner towards a lower percentage of the dominant sterol, *trans*-22-dehydrocholesterol and higher levels of other, minor sterols. A satisfactory explanation for these phenomena await further studies, but detergents probably have the ability to interfere with sterol biosynthesis, which may result in the accumulation of other compounds than the end product. The harmful effects of detergents on membranes and membrane enzymes have been observed earlier (5). As detergents are widely used by society, their possible role as water pollutants is a fact that cannot be ignored (2).

The English language was checked by Donald Smart.

REFERENCES

(1) Nyberg, H. 1976. The effects of some detergents on the growth of *Nitzschia holsatica* Hust. (Diatomeae). Ann. Bot. Fennici 13: 65-68.

(2) Nyberg, H. 1988. Growth of *Selenastrum capricornutum* in the presence of synthetic surfactants. Water Res. 22: 217-223.

(3) Azov, Y., Shelef, G. & Narkis, N. 1982. Effect of hard detergents on algae in a high-rate oxidation pond. Appl. Env. Microbiol. 43: 491-492.

(4) Zoller, U. 1985. The "hard" and "soft" surfactant profile of Israel municipal wastewaters. J. Am. Oil Chem. Soc. 62: 1006-1009.

(5) Nyberg, H. 1988. Growth and ATP levels of *Porphyridium purpureum* (Rhodophyceae, Bangiales) cultured in the presence of surfactants. Br. Phycol. J., in press.

(6) Nyberg, H. & Koskimies-Soininen, K. 1984. The glycolipid fatty acids of *Porphyridium purpureum* cultured in the presence of detergents. Phytochemistry 23: 751-757.

(7) Nyberg, H. & Koskimies-Soininen, K. 1984. The phospholipid fatty acids of *Porphyridium purpureum* cultured in the presence of Triton X-100 and sodium desoxycholate. Phytochemistry 23: 2489-2495.

(8) Nyberg, H. 1985. The influence of ionic detergents on the phospholipid fatty acid compositions of *Porphyridium purpureum*. Phytochemistry 24: 435-440.

(9) Nes, W.R. 1987. Multiple roles for plant sterols. In *The Metabolism, Structure and Function of Plant Lipids* (Stumpf, P.K., Mudd, J.B. & Nes, W.D., eds.), 3-9. Plenum Press, New York.

(10) Beastall, G.H., Rees, H.H. & Goodwin, T.W. 1971. Sterols in *Porphyridium cruentum*. Tetrahedron Lett. 52: 4935-4938.

(11) Patterson, G.W. 1971. The distribution of sterols in algae. Lipids 6: 120-127.

(12) Patterson, G.W. 1987. Sterol synthesis and distribution and algal phylogeny. In *The Metabolism, Structure and Function of Plant Lipids* (Stumpf, P.K., Mudd, J.B. & Nes, W.D., eds.), 631-636. Plenum Press, New York.

(13) Nyberg, H. 1986. GC-MS methods for lower plant glycolipid fatty acids. In *Modern Methods of Plant Analysis New Series Vol. 3: Gas Chromatography/Mass Spectrometry* (Linskens, H.F. & Jackson, J.F., eds.), 67-99. Springer, Berlin.

(14) Saranpää, P. & Nyberg, H. 1987. Lipids and sterols of *Pinus sylvestris L.* sapwood and heartwood. Trees 1: 82-87.

(15) Wright, R.S. 1971. A reagent for the non-destructive location of steroids and some other lipophilic materials on silica gel thin layer chromatography. J. Chromatogr. 59: 220-221.

(16) Patterson, G.W. 1971. Relation between structure and retention time of sterols in gas chromatography. Anal. Chem. 43: 1165-1170.

Author Index

Subject Index

D

M

R

S

T

U

Index of Taxa